Nanoencapsulation of Food Bioactive Ingredients

Principles and Applications

Nanoencapsulation of Food Bioactive Ingredients

Principles and Applications

Seid Mahdi Jafari
Gorgan University of Agricultural Sciences
and Natural Resources, Gorgan, Iran

ELSEVIER

ACADEMIC PRESS
An imprint of Elsevier

Academic Press is an imprint of Elsevier
125 London Wall, London EC2Y 5AS, United Kingdom
525 B Street, Suite 1800, San Diego, CA 92101-4495, United States
50 Hampshire Street, 5th Floor, Cambridge, MA 02139, United States
The Boulevard, Langford Lane, Kidlington, Oxford OX5 1GB, United Kingdom

Notices
Knowledge and best practice in this field are constantly changing. As new research and experience broaden our understanding, changes in research methods, professional practices, or medical treatment may become necessary.

Practitioners and researchers must always rely on their own experience and knowledge in evaluating and using any information, methods, compounds, or experiments described herein. In using such information or methods they should be mindful of their own safety and the safety of others, including parties for whom they have a professional responsibility.

To the fullest extent of the law, neither the Publisher nor the authors, contributors, or editors, assume any liability for any injury and/or damage to persons or property as a matter of products liability, negligence or otherwise, or from any use or operation of any methods, products, instructions, or ideas contained in the material herein.

British Library Cataloguing-in-Publication Data
A catalogue record for this book is available from the British Library

Library of Congress Cataloging-in-Publication Data
A catalog record for this book is available from the Library of Congress

ISBN: 978-0-12-809740-3

Publisher: Andre Gerhard Wolff
Acquisition Editor: Nina Rosa Bandeira
Editorial Project Manager: Mariana Kuhl
Production Project Manager: Anitha Sivaraj
Cover Designer: Vicky Pearson Esser

Typeset by MPS Limited, Chennai, India

Dedication

To my beloved mother and father.
Thanks for your kindness and devotion.

To my love, Elham and her family.
Thanks for your endless support.

To my fabulous son, AmirReza,
and my beautiful daughter, Elina.

"I am deathless, I am the eternal Lord
For I have spread the seed of the Word."

—Ferdowsi, Persian Poet (CE 935–1025)

Abu ʾl-Qasim Ferdowsi Tousi, or "Ferdowsi" was a Persian poet and the author of Shahnameh (the Persian "Book of Kings"), which is the world's longest epic poem created by a single poet, and the national epic of Greater Iran and the Persian-speaking world.

Tomb of Ferdowsi and his statue, located in Tous, close to Mashhad, Northeast of Iran

Contents

List of Contributors xvii
Preface xix

1 An Introduction to Nanoencapsulation Techniques for the Food Bioactive Ingredients
Seid Mahdi Jafari

1.1 Introduction 1
1.2 Nanoencapsulation Techniques 2
1.3 Lipid-Based Nanoencapsulation Techniques 5
1.3.1 Nanoemulsions 5
1.3.2 Nanoliposomes 8
1.3.3 Nanolipid Carriers 12
1.4 Nature-Inspired Nanoencapsulation Techniques 16
1.4.1 Caseins 16
1.4.2 Cyclodextrins 20
1.4.3 Amylose Nanocarriers 23
1.5 Special-Equipment-Based Nanoencapsulation Techniques 24
1.5.1 Electrospinning 25
1.5.2 Electrospraying 27
1.5.3 Nanospray Drying 29
1.6 Biopolymer-Based Nanoencapsulation Techniques 33
1.6.1 Single Biopolymer Nanoparticles 33
1.6.2 Complexation of Biopolymers 36
1.6.3 Nanogels 40
1.6.4 Nanotubes 42
1.7 Other Nanoencapsulation Techniques 43
1.7.1 Nanocrystals 43
1.7.2 Dendrimeters 46
1.8 Conclusions 47
References 47

2 Nanoencapsulation of Phenolic Compounds and Antioxidants
Afshin Faridi Esfanjani and Seid Mahdi Jafari

2.1 Introduction 63
2.2 Antioxidants 65

2.3 Phenolic Compounds 66
2.4 Nanoencapsulation Process of Phenolic Compounds and Antioxidants 69
2.5 Lipid-Based Formulations for Nanoencapsulation of Phenolics and Antioxidants 73
2.5.1 Nanoencapsulation of Phenolics and Antioxidants in Nanoemulsions 74
2.5.2 Nanoencapsulation of Phenolics and Antioxidants in Nanoliposomes 82
2.5.3 Nanoencapsulation of Phenolics and Antioxidants in Solid Lipid Nanoparticles 83
2.5.4 Nanoencapsulation of Phenolics and Antioxidants in Nanostructured Lipid Carriers 84
2.6 Biopolymer-Based Technologies for Nanoencapsulation of Phenolics and Antioxidants 85
2.7 Equipment-Based Technologies for Nanoencapsulation of Phenolics and Antioxidants 91
2.7.1 Electro-Spinning and Electro-Spraying 91
2.7.2 Nano-Spray Drying 93
2.8 Conclusion and Further Remarks 93
References 94

3 Nanoencapsulation of Fish Oil and Essential Fatty Acids
İsmail Tontul, Emrah Eroglu and Ayhan Topuz

3.1 Introduction to Essential Fatty Acids 103
3.1.1 Source of Essential Fatty Acids 104
3.1.2 Health Benefits of Essential Fatty Acids 104
3.1.3 Stability of the Fish Oil and Essential Fatty Acids 106
3.1.4 Nanoencapsulation of Fish Oil and Essential Fatty Acids 107
3.2 Lipid Formulation Nanoencapsulation of Fish Oil and Essential Fatty Acids 113
3.2.1 Encapsulation by Nanoemulsions 113
3.2.2 Encapsulation by Nanoliposomes 120
3.2.3 Encapsulation by Solid Lipid Nanoparticles 122
3.2.4 Encapsulation by Nanostructured Lipid Carriers 123
3.3 Natural Nanocarriers for Nanoencapsulation of Fish Oil and Essential Fatty Acids 125
3.3.1 Nanoencapsulation by Caseins 125
3.3.2 Nanoencapsulation by Cyclodextrins 125
3.4 Equipment Based-Nanoencapsulation of Fish Oil and Essential Fatty Acids 126
3.4.1 Nanoencapsulation by Electrospinning 126
3.4.2 Nanoencapsulation by Electrospraying 128
3.4.3 Nanoencapsulation by Nanospray Dryer 128
3.5 Biopolymer-Based Nanoencapsulation of Fish Oil and Essential Fatty Acids 129
3.6 Other Nanoencapsulation Technologies for Fish Oil and Essential Fatty Acids 131

3.7 Release, Characterization, and Safety of Nanoencapsulated Fish Oil and Essential Fatty Acid 131
3.7.1 Controlled Release of Fatty Acids From Nanocapsules 131
3.7.2 Instrumental Analysis and Characterization of Nanocapsules Containing Fish Oil 132
3.7.3 Safety and Regulatory Issues of Fish Oil-Loaded Nanocapsules 134
3.8 Conclusion and Further Remarks 135
References 135

4 Nanoencapsulation of Vitamins
Iman Katouzian and Seid Mahdi Jafari

4.1 Introduction 145
4.2 Nutritional Properties and Classification of Vitamins 146
4.2.1 Vitamin Deficiency Outcomes 147
4.2.2 Dosage of Vitamin Consumption 147
4.2.3 Vitamin Fortification in Food Products 148
4.2.4 Vitamin Retention under Food Processing Conditions 149
4.3 Microencapsulation of Vitamins 150
4.3.1 Spray Drying 151
4.3.2 Spray Chilling and Cooling 153
4.3.3 Emulsion Techniques 153
4.3.4 Fluidized Bed Coating 154
4.3.5 Liposome Carriers 154
4.3.6 Coacervation 154
4.4 Nanoencapsulation of Vitamins 155
4.4.1 Nanoemulsions 155
4.4.2 Nanohydrogels and Nano-Organogels 166
4.4.3 Solid Lipid Nanoparticles and Nanostructured Lipid Carriers (SLNs and NLCs) 167
4.4.4 Biopolymer Nanoparticles 168
4.4.5 Nanoliposomes 169
4.4.6 Cyclodextrins 171
4.4.7 Nanofibers and Electrosprayed Nanoparticles 171
4.5 Conclusions and Future Trends 173
References 173

5 Nanoencapsulation of Food Antimicrobial Agents and Essential Oils
Yue Zhang, Huaiqiong Chen and Kang Pan

5.1 Introduction 183
5.2 Traditional and Natural Food Antimicrobial Agents 184
5.2.1 Traditional Chemical Preservative Agents 184
5.2.2 Natural Antimicrobial Agents 187
5.2.3 Problems and Opportunities of Natural Antimicrobials 193

5.3 **Nanoencapsulation of Food Antimicrobial Agents** 194
5.3.1 Nanoencapsulation of Antimicrobials through Lipid-Formulation Techniques 194
5.3.2 Biopolymer-Based Nanoencapsulation of Food Antimicrobial Agents 200
5.3.3 Equipment-Based Nanoencapsulation of Food Antimicrobials 202
5.3.4 Nanocomposites of Food Antimicrobial Agents 203
5.3.5 Other Nanoencapsulation Technologies for Food Antimicrobial Agents 204
5.4 **Release, Characterization, and Safety of Nanoencapsulated Food Antimicrobial Agents** 204
5.4.1 Controlled Release of Antimicrobial Agents from Nanocapsules 204
5.4.2 Instrumental Analysis and Characterization of Antimicrobial-Loaded Nanocapsules 205
5.4.3 Safety of Nanoencapsulated Food Antimicrobial Agents 208
5.5 **Applications of Nanoencapsulated Food Antimicrobial Agents** 208
5.5.1 Aqueous Food Systems 208
5.5.2 Solid Food Matrix 209
5.5.3 Fresh Produces and Vegetables 210
5.6 **Conclusion** 211
References 211

6 Nanoencapsulation of Natural Food Colorants
Sahar Akhavan and Seid Mahdi Jafari

6.1 **Introduction** 223
6.2 **Classification of Food Colorants** 224
6.2.1 Heme Pigments 226
6.2.2 Chlorophylls 226
6.2.3 Carotenoids 226
6.2.4 Flavonoid Dyes 229
6.2.5 Betalains 230
6.2.6 Miscellaneous Colorants 230
6.3 **Stability and Bioavailability of Food Colorants** 231
6.4 **Encapsulation of Natural Colors in the Food Industry** 232
6.5 **Nanoencapsulation of Natural Food Colorants** 234
6.5.1 Lipid-Based Nanocarriers for Nanoencapsulation of Food Colors 239
6.5.2 Nature-Inspired Nanocarriers for Nanoencapsulation of Food Colors 245
6.5.3 Nanoencapsulation of Food Colorants by Special Equipment 247
6.5.4 Nanoencapsulation of Food Colorants by Biopolymer Nanoparticles 247
6.5.5 Other Methods of Nanoencapsulation for Food Colors 248

6.6 Safety Aspects of Nanoencapsulated Food Colorants 250
6.7 Application of Encapsulated Colorants in Food Formulations 250
6.8 Conclusion and Further Remarks 252
References 252
Further Reading 259

7 Nanoencapsulation of Flavors
Mohsen Asghari Ghajari, Iman Katouzian, Mohammad Ganjeh and Seid Mahdi Jafari

7.1 Introduction 261
7.2 Classification of Food Bioflavors 262
7.2.1 Primary Flavors 262
7.2.2 Secondary Flavors 263
7.3 Flavors from Different Origins 263
7.3.1 Flavors Derived from Saccharides 263
7.3.2 Flavors Derived from Amino Acids 264
7.3.3 Flavors Derived from Lipids 264
7.4 Flavors in Some Food Products 266
7.4.1 Flavors Derived from Cheeses 266
7.4.2 Flavors Derived from Red Meat 267
7.4.3 Flavors Derived from Bakery Products 267
7.4.4 Flavors Derived from Fruits and Vegetables 268
7.5 Microencapsulation of Food Flavors 268
7.5.1 Chemical Procedures for Encapsulation of Flavors 270
7.5.2 Mechanical Procedures for Encapsulation of Flavors 272
7.6 Nanoencapsulation Technologies for Food Flavors 278
7.6.1 Nanoemulsification of Food Flavors 278
7.6.2 Electrospraying and Electrospinning for Nanoencapsulation of Flavors 283
7.6.3 Nanoencapsulation of Flavors by Biopolymeric Nanocarriers 284
7.6.4 Nanoencapsulation of Flavors by Milk Proteins 287
7.7 Conclusion 287
References 288

8 Nanoencapsulation of Enzymes, Bioactive Peptides, and Biological Molecules
Muhammed Yusuf Çağlar, Mehmet Demirci, Kadir Bayrambaş, Bilal Çakır and İbrahim Gülseren

8.1 Introduction 297
8.2 Enzymes 299
8.2.1 Stability of Enzymes 299
8.3 Peptides 300
8.4 Phytosterols 300
8.5 Nucleic Acids 301

8.6 **Lipid Formulation Technologies for Nanoencapsulation of Biological Molecules** 302
 8.6.1 Nanoemulsions 302
 8.6.2 Nanoliposomes 303
 8.6.3 Nanostructured Lipid Carriers 303
8.7 **Natural Nanocarriers for Nanoencapsulation of Biological Molecules** 304
 8.7.1 Casein Nanocapsules 304
 8.7.2 Nanocrystal Nanocapsules 305
 8.7.3 Cyclodextrin Nanocapsules 306
8.8 **Equipment Based Technologies for Nanoencapsulation of Biological Molecules** 307
 8.8.1 Enzyme Immobilization Based on Electrohydrodynamic Processes 308
 8.8.2 Nucleic Acid Encapsulation Based on Electrohydrodynamic Processes 309
 8.8.3 Protein Encapsulation Based on Electrohydrodynamic Processes 310
8.9 **Biopolymer Based Technologies for Nanoencapsulation of Biological Molecules** 311
 8.9.1 Protein Nanogels 311
 8.9.2 Chitosan-Based Nanogels 312
 8.9.3 Alginate-Based Nanogels 312
 8.9.4 Nanotubes 312
 8.9.5 Starch Nanoparticles 313
8.10 **Release of Peptides From Nanoscale Delivery Systems** 314
8.11 **Safety of Nanoencapsulation Systems Loaded With Biological Molecules** 316
8.12 **Conclusion and Further Remarks** 316
Acknowledgments 317
References 318
Further Readings 329

9 Nanoencapsulation of Minerals

Seyed Mohammad Taghi Gharibzahedi and Seid Mahdi Jafari

9.1 **Introduction** 333
9.2 **Encapsulation Techniques of Minerals** 334
 9.2.1 Physicomechanical Methods 335
 9.2.2 Chemical Methods 343
9.3 **Nanoencapsulation of Minerals** 354
 9.3.1 Nanoliposomes 354
 9.3.2 Nanoemulsification 358
 9.3.3 Cyclodextrin Inclusion 359
 9.3.4 Solid Lipid Nanoparticles 360
 9.3.5 Biopolymer Nanoparticles 361
 9.3.6 Ionotropic Gelation 362
 9.3.7 Coacervation 364

9.4 **Application of Encapsulated Minerals in the Food Industry** 365
9.4.1 Fortification of Dairy Products with Encapsulated Minerals 365
9.4.2 Salt Fortification With Encapsulated Iron and Iodine 374
9.4.3 Use of Encapsulated Iron in Fortifying Cereals and Bakery Products 379
9.4.4 Encapsulated Minerals in Fortifying Other Foods 382
9.5 **Use of Mineral Nanoparticles** 384
9.6 **Conclusion and Further Remarks** 389
References 389

10 Release, Characterization, and Safety of Nanoencapsulated Food Ingredients

Seid Mahdi Jafari, Afshin Faridi Esfanjani, Iman Katouzian and Elham Assadpour

10.1 **Release of Nanoencapsulated Food Ingredients** 401
10.1.1 Controlled-Release Mechanisms 402
10.1.2 Analytical Methods for Measuring Release 404
10.2 **Biological Fate of Nanoencapsulated Food Ingredients** 404
10.3 **Mucoadhesive Delivery Systems** 407
10.3.1 Characterization of the Mucus Layer 407
10.3.2 Stages of Mucoadhesion 408
10.3.3 Mucoadhesion Theories 408
10.4 **Characterization of Nanoencapsulated Food Ingredients** 411
10.4.1 Morphology 411
10.4.2 Size 415
10.4.3 Electrical Characteristics 419
10.4.4 Surface Composition 420
10.4.5 Physicochemical Characterization 421
10.5 **Safety and Toxicity of Food NPs Employed in Food Products** 424
10.5.1 Genotoxicity and Epigenetic Regulations Modified by NPs 425
10.6 **Possible Health Risks and Carcinogenesis Induced by NPs in Some Organs and Body Systems** 433
10.6.1 Neurotoxicity 433
10.6.2 Hepatotoxicity 437
10.7 **Global Nanofood Regulations** 439
10.8 **Future Trends** 440
10.9 **Conclusion** 443
References 443

Index 455

List of Contributors

Sahar Akhavan Gorgan University of Agricultural Sciences and Natural Resources, Gorgan, Iran

Mohsen Asghari Ghajari Gorgan University of Agricultural Sciences and Natural Resources, Gorgan, Iran; Nano-encapsulation in the Food, Nutraceutical, and Pharmaceutical Industries Group (NFNPIG), Universal Scientific Education and Research Network (USERN), Tehran, Iran

Elham Assadpour Gorgan University of Agricultural Sciences and Natural Resources, Gorgan, Iran

Kadir Bayrambaş Tunceli University, Tunceli, Turkey

Muhammed Yusuf Çağlar İstanbul Sabahattin Zaim University (İZÜ), İstanbul, Turkey

Bilal Çakır İstanbul Sabahattin Zaim University (İZÜ), İstanbul, Turkey

Huaiqiong Chen Texas Tech University, Lubbock, TX, United States

Mehmet Demirci İstanbul Sabahattin Zaim University (İZÜ), İstanbul, Turkey

Emrah Eroglu Akdeniz University, Antalya, Turkey

Afshin Faridi Esfanjani Gorgan University of Agricultural Sciences and Natural Resources, Gorgan, Iran

Mohammad Ganjeh Gorgan University of Agricultural Sciences and Natural Resources, Gorgan, Iran

Seyed Mohammad Taghi Gharibzahedi Young Researchers and Elites Club, Science and Research Branch, Islamic Azad University, Tehran, Iran

İbrahim Gülseren İstanbul Sabahattin Zaim University (İZÜ), İstanbul, Turkey

Seid Mahdi Jafari Gorgan University of Agricultural Sciences and Natural Resources, Gorgan, Iran

Iman Katouzian Gorgan University of Agricultural Sciences and Natural Resources, Gorgan, Iran; Nano-encapsulation in the Food, Nutraceutical, and Pharmaceutical Industries Group (NFNPIG), Universal Scientific Education and Research Network (USERN), Tehran, Iran

Kang Pan Kellogg Company, Battle Creek, MI, United States

İsmail Tontul Akdeniz University, Antalya, Turkey; Necmettin Erbakan University, Konya, Turkey

Ayhan Topuz Akdeniz University, Antalya, Turkey

Yue Zhang University of Nebraska—Lincoln, Lincoln, NE, United States

Preface

Nutraceuticals are the link between nutrition and medicine. In other words, nutraceuticals are "food ingredients that have health benefits and inhibit the advancement of diseases." Recently, the application of different nutraceutical and bioactive compounds, such as essential fatty acids (omega 3), carotenoids (β-carotene and lycopene), vitamins (D, thiamin, riboflavin), antioxidants (tocopherols, flavonoids, polyphenolic compounds), phytosterols (stigmasterol and β-sitosterol), dietary fibers (inulin), minerals (Fe^{+2}, Mg^{+2}), and bioactive peptides (casein hydrolysates) has attracted the attention of many food scientists and industries for developing enriched healthy foods and functional products. Most of the nutraceuticals are sensible to decomposition during processing and storage after being incorporated into the food structures. Bioactive compounds are typically introduced into foods using different types of encapsulation (delivery) systems. Targeted delivery is one of the most important issues in the fields of encapsulation in pharmaceutical and food science. The goal of encapsulating nutraceuticals is to reduce the damage and undesirable changes through processing stages as well as digestion conditions. Hence, developing an appropriate carrier system is quite necessary.

Nanoencapsulation is a novel and practical branch of nanotechnology in the food industry. The term nanoencapsulation describes encapsulation on the nanometer scale using biopolymers, films, layers, or nanodispersions. The final capsule acts as a nanoscale shield for the food or nutraceutical molecules/ingredients. Often the bioactive ingredient is in the molecule or nanoscale state. The major benefit is the induced homogeneity, leading to better encapsulation efficiency in addition to the improved physical and chemical properties. Nanoencapsulation is flourishing with a rapid pace in the food sector. In fact, it is possible to fabricate valued-added food products by employing the nanoencapsulation technology in the field of food and nutritional sciences. Many comprehensive studies have focused on the application of nanoencapsulation in different aspects of the food industry, such as enhancing the public health, supporting the safety of food products, and designing the principles for the delivery of nutrients.

Hence, a practical means to develop the fabrication of functional foods is the process of nanoencapsulating sensitive food bioactive ingredients. Encapsulated ingredients can be formulated in a way to withstand the physical stresses in the digestive tract and deliver their payloads in a special site. Considering the semisolid and nonsolid food products, the reduction in the size of their network modifies their encapsulation potency without leaving any change in the sensory properties. Besides, by reaching the nanoscale, highly controllable biochemical vehicles are obtained. Meanwhile, the delivery rate scales directly with the particle size. Throughout the body, specific types of cells can absorb submicron nanoparticles more efficiently. Larger particles tend to release their bioactive molecules more slowly and over longer time periods. In addition, the size reduction in the particles elevates the bio-adhesive properties due to the increase in surface to volume ratio, which lastly heightens the bioavailability of the bioactive molecules by the prolonged transfer through the gastrointestinal tract.

In our previous book titled "*Nanoencapsulation Technologies for the Food and Nutraceutical Industries*" (Elsevier, 2017), we covered the nanoencapsulation techniques applicable to the food and nutraceutical industries plus their classification to make the foundation of next studies. In the mentioned book for the first time, we have classified nanoencapsulation technologies into five groups based on the main mechanism/ingredient, which is being used to make nanocapsules. They include lipid-based formulations (nanoemulsions, nanoliposomes, nanostructured lipid carriers), natural nanocarriers (caseins, cyclodextrins, nanocrystals), nanocarriers made with specialized equipment (electrospinning, electrospraying, nano spray dryer), bio-polymeric nanoparticles (individual protein and polysaccahride nanoparticles, their complexes), and miscellaneous techniques.

This book presents the cutting-edge research in the field of nanoencapsulation, which has been applied for different food bioactive components. The main goal of the present book has been providing recent research activities of nanoencapsulation in the food industry based on special and categorized food bioactive components. After giving an overview of nanoencapsulation techniques in the food sector (Chapter 1), we have discussed nanoencapsulation of phenolic compounds and antioxidants (Chapter 2), fish oil and essential fatty acids (Chapter 3), vitamins (Chapter 4), food antimicrobial agents and essential oils (Chapter 5), natural food colorants (Chapter 6), flavors (Chapter 7), enzymes, bioactive peptides, and biological molecules (Chapter 8), and minerals (Chapter 9). Finally, in Chapter 10, release, characterization, and safety of nanoencapsulated food ingredients have been presented.

This book would be useful for a diverse group of readers including food technologists, food engineers, nanotechnologists, nutritionists, food colloid experts, pharmacists, cosmetic experts, physicists, chemists, microbiologists, biotechnologists, and those who are interested in novel technologies in the area of food formulations, functional foods, and nutraceutical delivery systems.

Finally, I would like to appreciate all the contributors for sharing their vast knowledge in this book for both researchers and those who are interested in the area of nanoencapsulation. We also thank the staff of Elsevier for their fruitful cooperation and patience during the preparation and publication task.

Seid Mahdi Jafari
April 2017, Iran

Chapter 1

An Introduction to Nanoencapsulation Techniques for the Food Bioactive Ingredients

Seid Mahdi Jafari
Gorgan University of Agricultural Sciences and Natural Resources, Gorgan, Iran

1.1 INTRODUCTION

Nanocapsules are defined in the literature as mostly an oily or hydrophilic cavity surrounded by a thin wall material (Jafari, Fathi, & Mandala, 2015). A broad variety of wall materials, such as biopolymers (proteins, carbohydrates), lipids, chemical polymers, surfactants, etc., can be used for the preparation of nanocapsules (Fathi, Martín, & McClements, 2014; Fathi, Mozafari, & Mohebbi, 2012). Nanocapsules are promising applications, since they are ideal for the encapsulation of many different bioactive ingredients such as antioxidants, antimicrobial agents, phenolic compounds, natural pigments, peptides, essential fatty acids, minerals, etc. (Borel & Sabliov, 2014; Ezhilarasi, Karthik, Chhanwal, & Anandharamakrishnan, 2013).

The implementation of nanotechnology that basically focuses on the medical science and diagnostic field is well known as nanomedicine. The term has been defined as "monitoring, repair, construction, and control of human biological systems at the molecular level using engineered nanodevices and nanostructures" (Kumari, Singla, Guliani, & Yadav, 2014; Nitta & Numata, 2013). Some applications of nanotechnology in medicine are promising and many advantages have been offered in various medical areas such as targeted drug-delivery system and gene-targeted therapy. As many other medical areas, drug-delivery system experiences substantial growth concomitant with the advancement of nanotechnology over the past few years. Similarly, nutraceutical delivery system is an interdisciplinary field of studies applying knowledge from vast array of disciplines including biology, chemistry, pharmaceutical, and food sciences in combined fields of engineering and

Nanoencapsulation of Food Bioactive Ingredients. DOI: http://dx.doi.org/10.1016/B978-0-12-809740-3.00001-5

technology (Livney, 2015; McClements, 2015). In fact, it is one of the current nanotechnology advances employing nanosized particles for various manipulations either for technological or therapeutic purposes based on the nanotechnology concept.

Supposedly, an ideal nutraceutical-delivery system should transport bioactive molecules at particular sites without releasing its cargo at the previous points. Lately, many extensive efforts have been devoted to search for an appropriate technique to overcome some of the problems via research and development programs globally. It is then a practical role of delivery carrier being introduced in the process and the exploration for remarkable delivery vehicles has been interesting yet challenging area of research over the past decades. In this regard, the demanding exploration of delivery agents remains ongoing covering many aspects of research including on types of materials, physical and chemical properties as well as surface characterization of delivery carriers. In fact, sorts of materials have been profoundly studied since the past few years to efficiently construct an effective carrier such as inorganic nanomaterials, carbon nanotubes, gold, silver, and polymer-based nanoparticles (NPs) (Neves, Hashemi, & Prentice, 2015; Paredes, Asencio, Manuel, Allemandi, & Palma, 2016). Therefore, there is a broad range of nanosized encapsulation systems, most of them are still in the academic labs, only a very few made it to the market. A nanosized delivery system is scientifically defined in the food and pharmaceutical area as particles with a size of a few nanometers to just below 1000 nm (= 1 μm). It should not be mixed up with the legal definition of NPs/nanomaterial for labeling consumer products, e.g., labeling of cosmetic ingredients in the European Union with "nano" are necessary when more than 50% (by number distribution) of the particles have sizes below 100 nm (Quintanilla-Carvajal, Camacho-Díaz et al., 2010; Yada, Buck et al., 2014).

1.2 NANOENCAPSULATION TECHNIQUES

Various nanoencapsulation techniques for food bioactive components and nutraceuticals have been studied in the last couple of years including nanoemulsions; nanostructured lipid carriers (NLCs); nanosuspensions, solid–lipid nanoparticles (SLNs); nanosized liposomes; biopolymer NPs; and micelles made of proteins, polysaccharides, and their complexes or conjugates. These techniques yield nanoscale carriers (10–1000 nm).

From a classical viewpoint, nanoencapsulation technologies can be divided into two main approaches: "top-down" and "bottom-up." Regarding the "top-down" method, particle size is decreased during the encapsulation process, e.g., by utilizing various mechanical forces; on the contrary, in the "bottom-up" process, particle size is increased by methods, such as self-assembly. Furthermore, in some cases, a combination of both approaches has been used. This classification is not applicable nowadays, and it is better to categorize nanoencapsulation techniques in more groups based on some other indices.

In this book, we have classified nanoencapsulation technologies into five groups based on the main mechanism/ingredient, which is being used to make nanocapsules. They include lipid-based techniques, nature-inspired techniques, specialized-equipment techniques, biopolymer-based techniques,

TABLE 1.1 An Overview of Nanoencapsulation Techniques for the Food Industry

No.	Main Groups	Techniques	Different Strategies
1	Lipid-based techniques	Nano/ Microemulsions	*Single emulsions*: oil in water (O/W); water in oil (W/O)
2			*Double emulsions*: W/O/W; O/W/O
3			*Structural emulsions*: single interface layer; double interface layer
4		Nanostructured phospholipid carriers	*Liposomes*: monolayer; multilayer
5			*Phytosomes*: monolayer; multilayer
6			*Structural liposomes/ phytosomes*: with coatings
7		Nanolipid carriers	*Solid lipid nanoparticles* (SLNs)
8			*Nanostructured lipid carriers* (NLCs)
9			*Smart lipid carriers*
10	Nature-inspired techniques	Caseins	Alpha, beta, gamma-caseins
11		Cyclodextrins	Alpha, beta, gamma-cyclodextrins
12		Amylose	Single helix; double helix
13	Special equipment-based techniques	Electrospinning	Single injection nozzle; double injection
14		Electrospraying	
15		Nanospray dryer	
16		Microfluidics/ Nanofluidics	

(*Continued*)

TABLE 1.1 (Continued)

No.	Main Groups	Techniques	Different Strategies
17	Biopolymer-based techniques	Single biopolymer nanocarriers	*Protein nanoparticles* made by desolvation
18			*Polysaccharide nanoparticles* made by precipitation
19		Complexd biopolymer nanocarriers	Protein + protein
20			Polysaccharide + polysaccharide
21			Protein + polysaccharide
22		Nanogels	Hydrogels
23			Organogels/Oleogels
24			Mixed gels
25		Nanotubes	Protein nanotubes made with alpha-lactalbumin
26	Other techniques	Nanocrystals	Bioactives within crystals
27			Bioactive crystals within other nanocarriers
28		Nanostructured surfactants	Niosomes
29			Dendrimeters

TABLE 1.2 Different Morphologies and Structures for Nanocarriers of Food Ingredients and Nutraceuticals

No.	Main Groups	Structures	Different Forms
1	Spherical nanocarriers	Nanocapsules	Liquid form: Nanocarriers made with nanoemulsions; nanostructured lipid carriers; nanoliposomes/phytosomes; cyclodextrins; electrospraying; microfluidics; biopolymer nanoparticles; niosomes; dendrimeters
2			Dried form (containing a void): nanocarriers made with nanospray dryer, nanocrystals
3		Nanospheres	Nanogels; casein nanocarriers
4	Tubular nanocarriers	Filled fibers	Electrospun fibers
5		Hollow tubes	Protein nanotunes
6		Helix forms	Amylose nanocarriers

and disparate techniques as shown in Table 1.1. Also, different morphologies and structures of final nanocarriers has been reviewed in Table 1.2.

In the following sections, we will discuss briefly different techniques used for the nanoencapsulation of food bioactive ingredients. Much more details are provided in our previous book titled "*Nanoencapsulation technologies for the food and nutraceutical industries*," in which each chapter has been devoted to one nanoencapsulation technique (Jafari, 2017). It should be mentioned that selection of a nanoencapsulation technology depends on several parameters, such as physicochemical features, required particle size, release type, delivery method, process cost, etc.

1.3 LIPID-BASED NANOENCAPSULATION TECHNIQUES

Lipid-based nanoencapsulation systems are mostly used in the pharmaceutical and food industries and research programs. Despite the major advantages of biopolymer nanocapsules, they cannot be mass produced due to the demand for complicated chemical and thermal processes that should be monitored permanently. On the contrary, water-insoluble nanocarriers have the possibility to be scaled up plus the potential of more encapsulation efficiency and low toxicity (Fathi, Mozafari et al., 2012; Tamjidi, Shahedi, Varshosaz, & Nasirpour, 2013).

Most of the bioactive compounds, such as hydrophobic vitamins, fatty acids, flavonoids, aromas, preservatives, etc. have hydrophobic natures. For encapsulating these lipophilic compounds, the addition of emulsifiers (typically, oil in water) are often required. In addition, the presence of digestible lipids simplifies the absorption of nutrients since it raises the content of mixed micelles needed to dissolve and transport hydrophobic materials (Jafari & McClements, 2017). For this reason, numerous studies have been centered on the application of lipid-based nanocarriers. Different categories of lipid-based nanocarriers have been reported and reviewed in the last decade. The main types which have been studied and developed in recent years include nanoemulsions, nanoliposomes, NLCs, and SLNs.

1.3.1 Nanoemulsions

The droplet size of an emulsion, in the order of 10 to a few hundred nanometers, is indicative for its characterization as nanoemulsion. However, it should not be confused with microemulsions, which have smaller droplet sizes (i.e., 5–100 nm) and are thermodynamically stable but are created by low-energy emulsification systems (Gupta, Eral, Hatton, & Doyle, 2016; Mehrnia, Jafari, Makhmal-Zadeh, & Maghsoudlou, 2016). Oil-in-water (O/W) nanoemulsions serve as carriers of lipophilic bioactive agents, such as fish oil, essential fatty acids, plant sterols, carotenoids (like β-carotene), α-tocopherol, dietary fats, or a combination of these compounds via different methods. Generally, the lipophilic compounds are protected in these emulsions since the oil phase entails nutritional oils. On the other hand, water-in-oil (W/O) nanoemulsions can be used to encapsulate hydrophilic compounds including most polyphenols, water soluble vitamins, minerals, etc. (Donsì, Annunziata, Sessa, & Ferrari, 2011; Donsì, Sessa, Mediouni, Mgaidi, & Ferrari, 2011). The benefits of nanoemulsions concerning their high stability and bioavailability are extensively reported accordingly in the literature.

Recently, double nanoemulsions, a more complex type of liquid–liquid dispersion, have been receiving increased interest (Assadpour, Maghsoudlou, Jafari, Ghorbani, & Aalami, 2016a, 2016b; Faridi Esfanjani, Jafari, & Assadpour, 2017; Mehrnia, Jafari, Makhmal-Zadeh, & Maghsoudlou, 2017). Double nanoemulsions belong to a bigger group of multiple emulsions that themselves belong to structural emulsions. Double nanoemulsions due to their special structure could be used for nanoencapsulation of hydrophilic compounds in W/O/W emulsions (Assadpour, E., Maghsoudlou, Y., Jafari, S.-M., Ghorbani, M., & Aalami, M. 2016a, Assadpour, E., Maghsoudlou, Y., Jafari, S.-M., Ghorbani, M., & Aalami, M., 2016b; Esfanjani, Jafari, Assadpoor, & Mohammadi, 2015; Hemar, Cheng, Oliver, Sanguansri, & Augustin, 2010) or lipophilic compounds in O/W/O emulsions or both of them in a single double emulsion (Aditya, Aditya et al., 2015).

1.3.1.1 Ingredients Used for Preparing Nanoemulsions

O/W or W/O nanoemulsions have an oil or a water phase, dispersed in a continuous phase of water or oil accordingly. Double nanoemulsions (also duplex or multiple emulsions) are emulsions of emulsions such as an oil-in-water-in-oil ($O_1/W/O_2$) emulsion or a water-in-oil-in-water ($W_1/O/W_2$) emulsion. In the latter case, water droplets (W_1) are dispersed in a larger oil droplet (O) which in turn is dispersed in a continuous aqueous phase (W_2) (Mohammadi, Jafari, Esfanjani, & Akhavan, 2016; Mohammadi, Jafari, Assadpour, & Esfanjani, 2016). Such structure has been called emulsion in emulsion or primary emulsion in a secondary emulsion (Dickinson, 2011).

A brief review of the major components that can be used to formulate nanoemulsions is presented in the current section.

1.3.1.1.1 Oil

When an O/W or W/O nanoemulsion is to be prepared, the oil phase is quite important as its attributes influence the stability and qualitative properties of the final emulsion, including the emulsion sensory characteristics upon the final application (Aboalnaja, Yaghmoor, Kumosani, & McClements, 2016). Triacylglycerols, free fatty acids, essential oils, mineral oils, fat substitutes, waxes, or combination of them can be used. Some oils that are often found in the food industry include corn, olive, soybean, sunflower, coconut, canola, peanut, cottonseed, fish, and flaxseed or algae oils. Oils containing long-chain triglycerides are mostly used, nevertheless medium-chain triglycerides, and in certain cases short-chain triglycerides oils are also applied (Ozturk, Argin, Ozilgen, & McClements, 2014).

1.3.1.1.2 Emulsifiers

A successful design of nanoemulsions involves the use of the appropriate emulsifier(s) that will act on the surface of the dispersed phase, been adsorbed on it, leading to further droplet disruption and hindering droplet recoalescence or aggregation (Ziani, Fang, & McClements, 2012).

There are several emulsifiers been exploited in the food industry, categorized as small molecule surfactants, phospholipids, proteins [e.g., casein, β-lactoglobulin (β-lg)] or even polysaccharides (e.g., carboxymethyl cellulose) (Assadpour, Jafari, & Maghsoudlou, 2017). Food-grade emulsifiers such as polysaccharides are recently used in the food industry including hydrocolloids or modified starches. Emulsifiers have a specific interfacial behavior; they decrease the interfacial tension of the oil and water phases, but could balance electrostatic forces or result in steric repulsion, rheology change, and increased loading capability (Anton, Benoit, & Saulnier, 2008; Hategekimana, Masamba, Ma, & Zhong, 2015). When formulating nanoemulsions, the surfactant (emulsifier) concentration is quite important. As the particle surface increases, a greater amount of emulsifier is required to completely cover the oil droplets avoiding any destabilization.

Concerning the small molecule surfactants used in nanoemulsion formulation, there is a great variety of choices. Some examples of nonionic surfactants are sorbitan esters (e.g., span and tween series, mostly 40, 60, or 80), polyoxyethylene ethers (e.g., Brij), monoglycerides, sugar esters, and polyglycerol esters of monolaurate (Rao & McClements, 2012). Small molecule surfactants can form easily small droplets, compared to biopolymers. Due to their large molecules, protein- or polysaccharide-based emulsifiers, adsorbed at the interfacial layer, lead to a significant droplet size increase.

1.3.1.1.3 Stabilizers

Substances with thickening or gelling properties act as stabilizers. In emulsions, they are added into the aqueous phase and due to a viscosity increase, oil droplets movement is then retarded. However, modifying the aqueous phase rheology, changes also the mouth feel and texture (e.g., creaminess, gel strength) and new structures are created (Tang, Sivakumar, & Nashiru, 2013). Most stabilizers used in emulsions are hydrocolloids such as gum Arabic, pectins, xanthan, modified starch, alginates, galactomannans, and chitosan.

1.3.1.2 Preparation Methods

Nanoemulsions could be fabricated utilizing different or combined techniques. Generally, there are two main categories known as either high-energy or low-energy approaches depending on the energy input requirements (Jafari, He, & Bhandari, 2006; Jafari, He, & Bhandari, 2007a, 2007b).

High-energy techniques use a high mechanical energy. Their main advantage is the better control of size distribution and composition of the final emulsions (Jafari, He et al., 2007, 2007b; Shamsara, Muhidinov et al., 2015). Typically, the droplet size is controllable as it decreases at higher energy input and duration (Miastkowska, Banach et al., 2017), although there are some reports about droplet recoalescence during a high-energy emulsification (Jafari, Assadpoor, He, & Bhandari, 2008). Normally, high-energy methods alone do not result in oil droplets of very small size (<100 nm). In particular, application of natural biopolymer emulsifiers with large molecular weights hinders the formation of very small droplets. High-energy emulsification methods could be implemented by the following equipment:

- high-speed and high-shear homogenizers (rotor–stator devices);
- high-pressure valve homogenizers;
- microfluidizers; and
- ultrasound-based devices.

On the other hand, low-energy emulsification methods depend on the surfactant properties and the oily phase. Their main advantages are low-energy consumption and more straightforward scale-up of the process since the costs are minimized as well as employing simple production methods (Zhu, Zhang et al., 2015; Dasgupta, Ranjan, Mundra, Ramalingam, & Kumar, 2016). Low-energy emulsification drawbacks, regardless of the method, can be the requirement of large amounts of surfactants, typically needed in some cases. Natural surfactants have to be further investigated for food applications. Moreover, emulsifiers that are natural biopolymers, such as proteins or polysaccharides, can only be used to form nanoemulsions by high-pressure methods.

Briefly, the low-energy emulsification techniques can be distinguished in two main categories: phase inversion by simple phase mixing, and phase inversion by changing the hydrophilic–lipophilic balance through altering the system conditions. Accordingly, two commonly low-energy emulsification methods are the phase inversion and the spontaneous emulsification. For the first group, phase inversion temperature and phase inversion composition are applicable (Jafari, Paximada, Mandala, Assadpour, & Mehrnia, 2017).

It should be noted that another important and novel method of low-energy emulsification for producing nanoemulsions is membrane emulsification.

1.3.2 Nanoliposomes

Nanoliposomes are among the most commonly investigated colloidal delivery systems in the food and nutraceuticals research topics (Ghorbanzade, Jafari, Akhavan, & Hadavi, 2017). Since they have the capacity to encapsulate and deliver both hydrophilic and hydrophobic bioactives, nanoliposomes have been utilized in a number of studies and industrial products (Reza Mozafari, Johnson, Hatziantoniou, & Demetzos, 2008).

Liposomes are vesicles composed of polar lipid bilayers, primarily phospholipids, that surround an aqueous core. In order to form liposomal structure, the investigators can rely on the self-assembly of amphiphilic lipids in aqueous solutions. Concentric packing of amphipathic lipids forms an interior aqueous phase, which acts as a reservoir for encapsulation of hydrophilic compounds (Mozafari, Khosravi-Darani et al., 2008). In addition, their lipid bilayers provide a relatively lipophilic environment compared to the liposomal core. Consequently, they are quite versatile in the sense that they have the potential to encapsulate and stabilize aqueous compounds in their aqueous core and hydrophobic compounds in the lipid bilayers, respectively.

The nanoscale version of liposomes are called as nanoliposomes. Although they have many physicochemical similarities to conventional liposomes, nanoliposomes bring along all the benefits of the nanocarriers such as a high surface area and better penetration potential. The formation of a nanoliposome generates considerably more fresh surface areas which require an increased amount of energy input to generate nanoliposomes (Cadena, Pereira et al., 2013). In most of the recent applications, mean diameter range of nanoliposomes are referred to as approx. 50–150 nm. However, it must be noted that liposomes or nanoliposomes are not thermodynamically stable. Liposome manufacturing practices need to address the destabilization of liposomes; degradation of encapsulated materials; the influence of environmental variables such as composition, storage temperature, and pH; the nature of the encapsulated materials, ionic strength, and exposure to light and oxygen (Rasti, Jinap, Mozafari, & Yazid, 2012).

Liposomes can be classified based on their size, lamellarity and vesicularity characteristics (Mozafari, Khosravi-Darani et al., 2008). These include

- large unilamellar vesicles (LUVs) which mainly have a single lipid bilayer;
- multilamellar vesicles which have a larger number of concentric lipid bilayers;
- multivesicular vesicles which are composed of multiple vesicles coated with a shared bilayer;
- double bilayer vesicles as vesicles with a double bilayer; and
- small unilamellar vesicles (SUVs) and LUVs are collectively named as ULVs (unilamellar vesicles).

A relatively new group of nanoliposomes are called **nanophytosomes**, which are highly efficient delivery systems of phytochemicals (Katouzian & Jafari, 2016). The development of phytosomes has been a major enhancement in the nanoliposomal delivery of phytochemicals. Phytosomes are innovative delivery systems that are manufactured from solid dispersions of plant extracts in a phospholipid matrix. They are utilized in the delivery of compounds with low solubility and low bioavailability where the production methods such as solvent evaporation/antisolvent precipitation and supercritical fluids may be utilized. Phytosomes are encapsulation systems that are related to liposomes in terms of their structural attributes. However, nanophytosomes differ from nanoliposomes based on an appropriate stoichiometric ratio that bioactive compounds are chemically bound to the carrier structure, e.g., through H bonds which enhances the storage and digestive stability of the system (Demirci, Çağlar, Çakır, & Gulseren, 2017). Since in liposome technology, bioactive compounds are not bound to the particle, this renders liposomes leaky and causes the loss of encapsulated bioactives over time. The chemical bonding ensures the stability of phytosomes, enhances the encapsulation efficiency and stability of bioactives, generally at a stoichiometric molar ratio of 1:1 or 1:2 (phospholipids:phytochemicals. In this process, phosphatidylcholine is the most commonly utilized phospholipid. Phytosomes exhibit better pharmacokinetic and pharmacodynamic profiles than both liposomes and free plant extracts.

1.3.2.1 Ingredients Used for Preparing Nanoliposomes

1.3.2.1.1 Phospholipids

Soy lecithin is composed primarily of phosphatidylcholine which is probably the most commonly employed phospholipid system in nanoliposomes (Lu, Li, & Jiang, 2011). Over the recent years, a number of alternative raw materials have been proposed for the preparation of liposomes such as egg lecithin and milk fat globule membrane phospholipids. Relatively new or completely novel materials from alternative sources are also being promoted (sunflower phospholipids, marine lipids, rapeseed lecithin, etc.) in the literature. In various studies, the phospholipid composition was shown to determine the structural attributes of the liposomes. For example, egg-yolk lecithin liposomes demonstrated a highly complex structure compared to soy liposomes which were fairly uniform particles (Bryła, Lewandowicz, & Juzwa, 2015). Furthermore, the lipid composition of the liposomal bilayers influences the particle size of liposomes, which in turn could alter the surface charge (i.e., zeta-potential) and consequently the colloidal stability of the system. Due to the crowding effects, phospholipids with large head groups were hypothesized to pack within the outer layer (i.e., instead of the inner layer) which could in turn increase the absolute valued surface charge of nanoliposomes (Liu et al., 2011). Consequently, it is critical to select the appropriate phospholipid source for the design of liposomes.

1.3.2.1.2 Cholesterol

Cholesterol has been widely utilized in nanoliposomes as a primary component. The inclusion of cholesterol in liposomes increases the membrane rigidity and limits the conformational alterations in the liposomal bilayer and eventually, decreases the rate of bioactive compound release, especially for hydrophilic compounds. In order to limit cholesterol usage, the replacement of cholesterol with plant sterols have also been considered in nanoliposome formulations. Plant sterols increased both the encapsulation of ascorbic acid (i.e., a hydrophilic solute) and the mean diameter of liposomes, therefore fine tuning of production methods would need to be carried out (Alexander, Lopez, Fang, & Corredig, 2012).

1.3.2.1.3 Phenolic Compounds

Similar to cholesterol, certain other compounds, such as phenolic compounds, have a bearing on the encapsulation, physical properties, and delivery characteristics of nanoliposomes (Demirci, Çağlar et al., 2017). The variations in the structures of phenolic compounds can also be anticipated to alter encapsulation and stability characteristics of liposomal dispersions. In the case of tea catechins, e.g., the presence of gallic acid esters enhanced the affinity of EGCG molecules to the lipid/liposomal bilayers and induced the incorporation of significant amounts of polyphenols in the bilayers (Nakayama, Hashimoto, Kajiya, & Kumazawa, 2000). That also affected the membrane fluidity and morphology (Ikigai, Nakae, Hara, & Shimamura, 1993). The presence of catechins in the membranes enhanced the rigidity of cellular membranes as well. All these findings pointed out to liposomal encapsulation and stabilization of phenolic compounds.

1.3.2.1.4 Solvent

The solvents applied in nanoliposome preparation (i.e., thin film hydration, solvent injection methods, etc.) also deeply affect the basic characteristics and liposomal performance. For example, methanolic extracts of polyphenols from cyanobacteria and algae were more concentrated in terms of polyphenol content compared to ethanolic extracts, whereas the encapsulation efficiency of the ethanolic extracts was found to be higher (de Assis, Machado, da Motta, Costa, & de Souza-Soares, 2014). However, in the context of food products and nutraceuticals, the extent of solvent and surfactant usage has to be clearly limited (i.e., toxicity, irritancy, etc.). Also at high surfactant concentrations, vesicle size growth and bilayer solubilization might take place (Alonso, Villena, & Goñi, 1981) which could rupture liposomal structures and render the encapsulation process unsuccessful.

1.3.2.2 Preparation Methods

Conventionally, there are four major methods of nanoliposome manufacture (Demirci, Çağlar et al., 2017):

1.3.2.2.1 Thin Film Hydration

It is based on the aqueous rehydration of thin phospholipid films which have been initially dissolved in an appropriate organic solvent (e.g., chloroform) and then subjected to solvent evaporation. Afterward, the dehydrated film is being hydrated in an appropriate buffer solution that contains the hydrophilic solutes, which is also instrumental for the encapsulation of the solutes. Generally, extrusion membranes or other homogenizers are utilized in order to generate ULVs, and preferably SUVs in this technique.

1.3.2.2.2 Solvent Injection Method

During the application of this technique, lipids are initially solubilized in an appropriate solvent, possibly ethanol, and ethanolic solutions are directly administered into aqueous systems where liposomes are formed.

1.3.2.2.3 Detergent Removal Method

This method utilizes dialysis technique in the removal of surfactants from surfactant–phospholipid mixed micelles and leads to their transformation to liposomes.

1.3.2.2.4 Reverse Phase Method

In this technique, polar lipids are solubilized in an appropriate organic solvent and a water-in-oil emulsion is manufactured. Formation of liposomes will be preceded by the removal of the solvent, formation of a gel phase and aqueous dispersion.

It should be noted that after preparing liposomal structures by these methods, some types of high-energy devices such as a microfluidizer or an ultrasound probe should be applied to further decrease the size of liposomes into nanoscale capsules.

The techniques utilized in liposome manufacture have been recently reviewed by Huang et al. (2014). In addition to the above conventional techniques, some novel and specialized methods have been presented in recent years, such as

- dual asymmetric centrifugation,
- membrane contactor technology,
- cross-flow filtration detergent depletion method,
- freeze-drying double-emulsion method,
- supercritical CO_2 technologies,
- layer-by-layer deposition, and
- combinatorial (hybrid) methods.

1.3.3 Nanolipid Carriers

In comparison to other lipid-based nanoencapsulation systems, such as nanoliposomes and nanoemulsions, NLCs have lately attracted much attention from the food and pharmaceutical industries due to their advantages. They can be classified in two subgroups:

Solid–lipid nanoparticles: SLNs are novel nanoparticulate vehicles containing lipid droplets that are fully crystallized with the bioactive components being a part of the lipid matrix (Pandita, Kumar, Poonia, & Lather, 2014; Svetlichny, Külkamp-Guerreiro et al., 2015). SLNs were designed to combine the advantages of polymeric particles, liposomes, and emulsions and to avoid some of their disadvantages. Advantages of these lipid-based systems are the ease of production and the unproblematic regulatory status of the excipients used (excipients often natural, accepted status, well tolerated). A disadvantage of conventional emulsions and liposomes is the limited ability to stabilize chemically labile bioactives and the lack of controlled release. Due to the relatively low viscosity of the oils, the bioactives can easily diffuse into the surrounding water phase, where they can be degraded. These limitations can be overcome by using NPs having a solid matrix with a high viscosity (de Carvalho, Noronha et al., 2013). The lipids used are solid at body temperature (i.e., they keep their properties also after the administration to the body), but for production by high-pressure homogenization (HPH), they are processed in the melted state at elevated temperatures. Of course, the solubility of bioactives in solid lipids is lower than in liquid lipids; thus, the SLNs have a lower loading capacity than emulsions. The solubility of bioactives also depends on the polymorphic modification of the lipid matrix (Qian, Decker, Xiao, & McClements, 2013).

Nanostructured lipid carriers: NLCs entered into the nanoencapsulation field later to cover the deficiencies found in SLNs. The main aim was to enhance the loading capacity and to inhibit the expulsion of bioactive compounds. They are produced by dispersing a mixture of solid and liquid lipids, along with bioactive ingredients (as inner phase) in water containing emulsifiers (as outer phase). In contrast to SLNs, the presence of liquid lipid in inner phase of NLCs causes the possibility of better entrapping of bioactive ingredients which are more solubilized in liquid lipid (Beloqui, Solinís, Rodríguez-Gascón, Almeida, & Préat, 2016; Das, Ng, & Tan, 2012). In other words, the philosophy of NLC was to combine spatially different lipid molecules which are not or less able to form perfectly ordered lipid structures in the particle matrix. Preferentially, they should have as many imperfections as possible. For this, typically a solid lipid based on long-chain fatty acids (e.g., behenic acid, C22) is mixed with an oil based on short-chain fatty acids (e.g., Miglyol 812, containing C8 and C10). The differences in fatty acid chain length should distort perfect ordering (Chia-Lang, Saleh, & Jia-You, 2013). Typically, 1 solid lipid is mixed with 1 oil. Thus, NLCs are a blend of solid and liquid fat (oil).

1.3.3.1 Ingredients Used for Preparing SLNs and NLCs

Selection of ingredients need to be made first under regulatory aspects, and then subsequently under technical aspects (Pyo, Müller, & Keck, 2017). For commercial product development, only ingredients with regulatory accepted status should be used for the product development process, which reduces the screening parameters. Most important is the selection of the solid and liquid lipids to provide maximum solubility for the bioactive to be loaded. Often a lipid is just taken randomly, without assessing the solubility at all. A systematic solubility screening is recommended, determining the solubility in the melted state and the solid state of the solid lipid, and the solubility in oils at both production temperature and at room temperature.

1.3.3.1.1 Liquid Oil

When the production temperature is 60°C, the oils are heated to this temperature (e.g., in vials) and the bioactive is added successively in steps to generate increasing concentrations. After shaking, the amount of bioactive which can be completely dissolved is determined visually. In the next step, the oil solutions are cooled down to room temperature (Tamjidi, Shahedi et al., 2013; Weber, Zimmer, & Pardeike, 2014). It is necessary to determine what amount of bioactive remains completely in solution also after cooling down to room temperature and in addition after, e.g., 1 day of storage (in case of delayed crystallization).

1.3.3.1.2 Solid Lipid

The lipids are melted in glass vials and bioactive is added in increasing amounts to investigate its maximum solubility in the melted lipid. In the next step, the lipid is cooled down and the solubility is checked in the solid lipid (which is logically lower than in the melted and hot state). The evaluation can be done macroscopically by spinning the vial during the solidification process by hand, that the lipid solidifies in a thin film on the vial wall (Tamjidi, Shahedi, Varshosaz, & Nasirpour, 2014). This film can then be inspected for homogeneity or separation in two phases (crystallized bioactive and separate lipid phase with maximum dissolved bioactive). Alternatively, or in parallel, a drop of the hot melted lipid solution can be placed on a microscope slide, coated with a cover glass and pressed to a thin film. After solidification, microscopic inspection for phase separation can be performed. Result of this lipid solubility study is the knowledge of what concentration leads to solid solution within the particle matrix, and which sample creates a bioactive core with prolonged release (Zheng, Falkeborg, Zheng, Yang, & Xu, 2013).

1.3.3.1.3 Stabilizer

The next important point is the selection of stabilizer (Pyo, Müller et al., 2017). Two options are possible: electrostatic stabilizers (e.g., charged ionic surfactants) or steric stabilizers (e.g., poloxamer type). In case of oral or peroral delivery, the particles are generally exposed to an electrolyte-containing environment. The electrolytes reduce the zeta-potential; thus, the electrostatic stabilization leads to aggregation which impairs the functioning of the carriers (special nanoproperties are lost). Thus, it is generally recommended to go for steric stabilization via polymers, or a combined electrostatic and steric stabilization. As a rule of thumb, typical stabilizer concentrations are about 1.0%–1.5% for 10% lipid NP suspensions, and about 2% for 20% suspensions.

1.3.3.2 Preparation Methods

Many different production processes are described in the literature which are typically used on lab scale when expensive equipment is not available. The most relevant processes are the microemulsion technology and the HPH (da Silva, Nora et al., 2016; Rao & Khanum, 2016).

1.3.3.2.1 Microemulsion Method

In this process, a hot microemulsion is formed with the melted lipids and bioactive, then this mixture is poured into cold water. The dilution leads to breaking of the microemulsion, forming lipid droplets in the nanometer range (e.g., 100–200 nm), which immediately solidify due to the low temperature.

1.3.3.2.2 High-Pressure Homogenization

Much more feasible is the HPH, which allows the production of suspensions with particle contents up to about 40% (Pyo, Müller et al., 2017). Homogenizers are available from very small volumes to safe industrial scales. The production process is very simple, typically the so-called hot homogenization technique is used. The lipid blend is melted (typically 5–10°C above the melting point), the bioactive is dissolved in this melt, and the final sample is dispersed in a hot surfactant/stabilizer solution of equal temperature by high-speed stirring. The obtained preemulsion is subsequently passed typically 3 times through a temperature controlled HPH at 600 bar pressure (=8702 psi). Alternatively, often one passage at 800 bar (=11,603 psi) is sufficient. The mean particle size typically obtained with most surfactants or polymeric stabilizers is about 200–300 nm.

Medium-scale production can be performed in homogenizers with a capacity of about 50 kg (liter) per hour. A batch size of 1 kg can be run in the continuous mode, i.e., the product circulates from the source container into the homogenizer and from there back to the source container. After about 20 min, the production is finished; about at least 99.9% of the droplets have passed the homogenizer at least once.

Large-scale production is performed on larger homogenizers. Typically, one passage at 800 bar is sufficient, equivalent to three passages at 600 bar in a lab-scale homogenizer (Pyo, Müller et al., 2017). However, it needs about 1 h to process 150 kg, that means the respective bioactive is exposed to heat for 1 h. This might be no problem for many bioactives but not for labile compounds. In this case, the melted solid lipid should be admixed to cold oil containing the bioactive by a static blender before subsequently being admixed to the aqueous stabilizer phase in a second static blender.

The lipid melt should have a higher temperature to compensate the room temperature of the admixed bioactive-containing oil, yielding finally the required temperature for homogenization. To minimize heat exposure, the product should be cooled immediately by a heat exchanger.

1.4 NATURE-INSPIRED NANOENCAPSULATION TECHNIQUES

The idea of bioactives encapsulation using natural nanocarriers results from taking into account the nature-made functionalities of these NPs (Faridi Esfanjani & Jafari, 2016). As an example, casein micelles can naturally encapsulate and transport vital nutrients, amylose chains can bind a variety of flavor compounds developed during bread making, which can be released during subsequent heating, and starch derivatives [granules and cyclodextrins (CDs)] can form inclusion complexes with many nutraceuticals. Nature-inspired nanocarriers have received great attention for the delivery of nutrients and drugs due to their safety and low cost. The functional properties of naturally occurring nanovehicles are characterized by their biological source, isolation methods, or physical and biochemical alterations. These tiny natural NPs such as caseins, CDs, amylose nanostructures, and starch granules are employed in the nanoencapsulation of bioactive food ingredients and nutraceuticals.

1.4.1 Caseins

Milk caseins have unique structural and functional properties which enables them as suitable nanocarriers for bioactive components and nutraceuticals (Semo, Kesselman, Danino, & Livney, 2007). Some of these properties are binding with molecules and ions (e.g., ion binding), binding with hydrophobic molecules, interactions with other proteins/polymers (e.g., covalent conjugations and noncovalent interactions), reassembly and self-assembly properties, gelation properties, and surface activity (Abd El-Salam & El-Shibiny, 2012; Livney, 2010). The protein assembly of milk is composed of insoluble proteins named as caseins and soluble proteins named as whey proteins [β-lg, α-lactalbumin (α-La), and bovine-serum albumin (BSA)] which range from 80% and 20% in bovine milk, as to 60% and 40% in humans. The four casein subspecies account differences on the degree of phosphorylation, glycosylation (i.e., κ-casein), amino-acid profile, and their hydrophobicity. The dry matter in the casein micelle consists of 94% protein and the remaining 6% is composed of minerals, mainly calcium, phosphate magnesium, citrate, and few others in lower amounts. They interact with calcium phosphate nanoclusters, or so-called colloidal calcium phosphate and self-assemble in stable, and polydispersed supramolecular structures called casein micelles (Tavares, Croguennec, Carvalho, & Bouhallab, 2014).

Caseins appear of spherical shape with molecular mass of 106–109 Da and vary in diameter from 80 to 400 nm (Livney, Semo, Danino, & Kesselman, 2006).

The four individual types of casein molecules (α_{s1}, α_{s2}, β, κ-casein) found in a micelle are in relative proportions of 4:1:3.5:1.5, respectively. αs-Casein and β-casein are the main casein subspecies that interact with calcium-phosphate cluster due to serine-phosphate residues present within casein molecules, while κ-casein is positioned on the outside layer of the associated casein nanoclusters. It is believed that since there is an insufficient number of κ-casein molecules to stabilize the entire surface of the micelles, β-casein may play a role in stabilizing the protein rich regions in the inner core of the micelle (Elzoghby, Abo El-Fotoh, & Elgindy, 2011). The primary structure of caseins is characterized by a high amount of proline, which impedes the formation of significant ordered structures. For this reason, caseins are very flexible and are defined rheomorphic, as they can adapt their conformation to changing environmental conditions. Casein micelles are negatively charged, representing a ζ-potential of about −20 mV and an isoelectric point of 4.6; thus, precipitation occurs under acidic conditions (pH ~4.6). However, when pH is increased above the isoelectric point closer to the neutral conditions, caseins have the ability to resolubilize (Pan, Luo, Gan, Baek, & Zhong, 2014).

It has been proposed that the composition of casein micelles may be somehow correlated with their size. Therefore, smaller micelles may contain higher percentage of κ-casein, and relatively low β-casein, while such correlation has not been observed for α_s-caseins. Thereby, it is assumed that κ-caseins are mostly distributed on the surface of the micelles, β-caseins mainly in the inner core, and α_s-caseins can be found anywhere in the structure (Hasni, Bourassa et al., 2011).

Certain factors could influence the stability of casein micelles such as specific proteolitic enzymes, acids and addition of excessive amounts of calcium ions or ethanol. All these procedures are essential in many dairy productions (Ghasemi & Abbasi, 2014). Casein micelles are relatively stable structures that can withstand moderate temperature changes (heating or cooling) without significant changes in their basic structures (e.g., aggregation or disruption). At temperatures above 60°C, caseins still remain relatively stable to heating, while whey proteins are denatured via self-aggregation and thus aggregate with casein micelles ruled by disulphide bonding and hydrophobic interactions. Therefore, heat stability of caseins makes them ideal nanocarriers for heat-sensitive bioactive compounds.

The way how caseins self-assemble into micelles due to hydrophobic interactions and electrostatic interactions renders them very suitable for incorporation of hydrophobic active molecules. These protein assemblies are perfectly designed to transport all the essential nutrients from the mother to the offspring in the most precised way. Regarding the economical, nutritional, and safety issues, caseins have been proven to be one of the most reliable nanovehicles. Several reviews have been published on the functionality of the milk proteins more specifically, caseins (Abd El-Salam & El-Shibiny, 2012; Livney, 2010; Tavares, Croguennec et al., 2014).

Formerly, casein micelles were utilized as nanocarries within pharmaceutical research, afterward, they became established in the field of food and nutrition, especially the delivery systems prepared with natural biopolymers. The presence of acid-soluble calcium-phosphate bridging, makes caseins readily available to proteolysis during gastric digestion. The assembly of nutrients provided from casein micelles is a unique example of a successful delivery in nanoscale. Although the principal mechanisms are quite similar in both fields, the limitation to food grade materials requires considerable attention and can be often challenging in food processing and product development (Elzoghby, Abo El-Fotoh et al., 2011; Semo, Kesselman et al., 2007).

1.4.1.1 Ingredients Used for Preparing Casein Nanodelivery Vehicles

Caseins existing in their natural environment are extremely stable displaying good solubility, heat resistance, and highly surface active. However, caseins are also available commercially, in the form of sodium caseinate and calcium caseinate (Haratifar & Guri, 2017). They are prepared industrially by separating native caseins from skimmed milk fraction in the presence of acids, then resolubilized under alkaline conditions and finally processed through a spray-drying process in order to yield caseinates. Likewise, native structures, caseinate powders have high solubility in water, additionally they are able to rearrange and form aggregated casein clusters in ionic environments. Studies being carried out on caseins reassembly prepared by various techniques have proven that the organization of casein micelles in micellar substructures, when calcium, citrate, or phosphate ions were added to aqueous casein solutions, have been very similar to that of natural casein assemblies (Livney, Semo et al., 2006). The aggregation dynamics of caseins after solubilization can be achieved by simply controlling the pH of the solution during synthesis. Hence, studies have shown that during the aggregation process of caseinate solutions considering the pH changes, hydrophobic substances can be easily entrapped in the clusters (Sahu, Kasoju, & Bora, 2008). The mechanism of micelle formation and casein aggregation during the encapsulation process should not be neglected, as it is a crucial factor that directly influences the encapsulation efficiency and core materials stability.

Additionally, caseinate solutions have been introduced as stability and solubility amplifiers of lipophilic substances such as curcumin (Pan, Luo et al., 2014), β-carotene (Sáiz-Abajo, González-Ferrero, Moreno-Ruiz, Romo-Hualde, & González-Navarro, 2013), vitamin D (Haham et al., 2012), omega-3 fatty acids (Zimet & Livney, 2011), and essential oils (Arranz et al., 2015).

Other forms of purified casein proteins have also been used successfully in the encapsulation procedure to enhance the bioavailability of hydrophobic substances like, pure αs1-casein or β-casein (Sahu et al., 2008; Sneharani, Karakkat, Singh, & Appu Rao, 2010). To sum up, the aforementioned studies suggest that the entrapment of hydrophobic compounds in micellar fractions protects them against degradation induced by oxidation and processing.

1.4.1.2 Preparation Methods

The main nanoencapsulation techniques based on casein functional properties are as follows (Haratifar & Guri, 2017):

Binding with ions: As mentioned, all three types of caseins (α_{s1}, α_{s2}, and β-casein) have the ability to bind calcium, and calcium phosphate in their structure due to the presence of phosphate residues. Knowing this, researchers have practically looked at likelihood of utilizing caseins as carriers of other metal ions such as iron (Fe^{2+}) and magnesium (Mg^{2+}).

Hydrophobic interactions: Based on the chemistry of casein micellar organization, it is believed that the core of casein micelles provides an environment where nonpolar compounds could be accommodated for a safe delivery. It is well known that milk proteins mainly bind with hydrophobic components via hydrophobic and hydrogen bonding combined with van der Waals forces (Livney, 2010). The reversible binding of hydrophobic structures (e.g., vitamins, polyphenols, and fatty acids) can be through noncovalent interactions between the protein and the ligand. To ensure the efficient binding and delivery of hydrophobic molecules generally, irreversible mechanisms are used through strong covalent binding with amino and thiol groups which can either be situated in the inner core or the surface of caseins (Kimpel & Schmitt, 2015).

Self-assembly and reassembly of caseins: In many studies, casein micelles and their isolated individual forms (i.e., α- and β-caseins) have not required any specific modification to encapsulate and/or interact with hydrophobic compounds, which makes them even more appealing as bioactive nanocarriers. However, in some cases, modifications are applied to enhance the encapsulation properties of caseins. Different techniques have been applied to take advantage of caseins reassembling functional properties and self-assembly for β-casein (Tavares et al., 2014). The self-assembly of casein in micelles have been classified as a "bottom up" approach (Livney, 2015). In the case of β-casein, the presence of two domains, one negatively charged highly polar (N-terminal) and a highly nonpolar (C-terminal), justify its highly amphiphilic character displaying a self-association behavior. At low temperatures, this protein is monomeric and at concentrations of 0.5–2 mg/mL, and a neutral pH at 25 °C which is considered as critical micellization concentration, it assembles into nanosized micelles in water medium enclosing a hydrophobic domain which is surrounded by hydrophilic patches (Moeiniafshari, Zarrabi, & Abdol-Khalegh, 2015; Portnaya et al., 2006). Due to these structural and physicochemical characteristics and their ability for hydrophobic interactions, caseinates and β-caseins have been used to encapsulate different hydrophobic bioactive molecules, such as hydrophobic vitamins, curcumin, fatty acids, etc.

1.4.2 Cyclodextrins

CDs are interesting and applicable nanocarriers in the form of cyclic oligosaccharides produced from the hydrolytic process of starch degradation due to their cage-like supramolecular structure (Faridi Esfanjani & Jafari, 2016; Katouzian & Jafari, 2016). Also, they are considered as valuable solubilizing agents due to their functionalities and advantageous characteristics. The parent CDs containing 6, 7, and 8 glucopyranose units are α-, β-, and γ-CDs (Pinho, Grootveld, Soares, & Henriques, 2014).

Molecular inclusion of the host–guest complex can be considered as a unique "molecular nanoencapsulation" for many bioactive components. In other words, CDs are unfilled molecular nanocapsules with certain sizes to encapsulate many different ingredients through the insertion of appropriate "guest" structure into CDs cavity. This apolar cavity is less polar than water (Hădărugă et al., 2008), because the CD cyclic oligosaccharides with glucopyranose units in a cylindrical configuration have a water-insoluble inner and a water-soluble-outer linked to the hydrogen and glycosidic oxygen bands. The internal diameter of the water-insoluble cavity with a certain depth (0.78 nm) depends on the number of glucose units in α-, β-, and γ-CDs, which is 0.57, 0.78, and 0.95 nm, respectively (Pinho, Grootveld et al., 2014).

Generally, the inclusion is formed by complexing one CD molecule and one guest molecule in a stoichiometry ratio of 1:1. However, it is possible in the design of CD complex to include over one guest with low and/or high-molecular-weights into the cavity. Noncovalent interactions, such as van der Waals and hydrophobic forces compared with the steric effects and hydrogen bonding have a key role in forming the stable inclusion complexes or encapsulation processes (Hill, Gomes, & Taylor, 2013). This unique structure can potentially protect guest molecule(s) against negative effects of many reactions, such as photodegradation, enzymatic decomposition, steric rearrangement, racemization, oxidation, and even hydrolysis (Choi, Ruktanonchai, Min, Chun, & Soottitantawat, 2010).

Apolar nature of the CD cavity is due to the presence of three rings including C–H groups in C3 and C5 and glycosidic oxygen atoms are located in. Lewis base character and considerable electron mass confirm this intrinsic feature. This exclusive structure is responsible for their physicochemical characteristics, which can provide full or fractional inclusion of extensive numbers of functional ingredients (Haidong, Fang, Zhihong, & Changle, 2011). Change of guest's solubility and protection against effects of heat, light, oxidation and reduction, inhibition of its volatility, and guest's physiological responses can be formed with inclusion complexes of the host molecules (Fang, Comino, & Bhandari, 2013).

1.4.2.1 Ingredients Used for Preparing CD Nanocarriers

There are over 1500 derivatives for CDs which are generally categorized into two main groups of natural and modified (GharibZahedi & Jafari, 2017). Starch as the easily available carbohydrate polymer is the raw material for the industrial-scale production of CDs. These natural structures are formed from the starch transformation by certain bacteria, such as *Bacillus macerans* and *Bacillus circulans* through the CD-glycosyltransferase (CGTase) enzyme, which partially hydrolyzes starch into α-($C_{36}H_{60}O_{30}$), β-($C_{42}H_{70}O_{35}$), and γ($C_{45}H_{80}O_{40}$)-CDs (Pinho, Grootveld et al., 2014). Among these CDs, β- and γ-CDs significantly have the highest stability and water solubility, respectively (Del Valle, 2004). However, β-CD has the largest contribution (97%) in the international markets of CD sales due to its lower purification cost compared with that of α- and γ-CDs (Lee, Min et al., 2013).

Chemical modification of parent CDs results in numerous safe derivatives (e.g., 2-HP-, maltosyl-, and SBE-β-CDs) with improved physiochemical properties and complexing abilities (Mercader-Ros, Lucas-Abellán, Fortea, Gabaldón, & Núñez-Delicado, 2010). CD modification not only can provide its safety, but also can keep the formation ability of inclusion complexes with the different substrates. The modification usually converts natural CDs to their amorphous derivatives. Two primary chemical parameters including nucleophilicity of −OH groups and the production potential of complexes between CDs and applied reagents should be considered for their modification process.

Many chemical modifications on CDs were conducted through the grafting replacements to various sites namely primary and secondary rims or both faces in order to provide nonionic and cationic derivatives capable of producing nanoscale molecular cloud aggregates used in bio- and pharmaindustries. Two neutral CD derivatives of methylated (Me) and HP and also negatively-charged sulfobutylated CD derivative are frequently used (Duchêne, 2011). The most fundamental molecular units are −OH nucleophilic groups on the C6 site. The most acidic −OH groups are placed at the C2 site. Moreover, the −OH groups at the C3 position are the most unreachable.

In addition to the main modifications, fatty acids with various chain lengths can be chemically/enzymatically grafted on the −OH groups of the secondary face in order to form new amphiphilic CDs (GharibZahedi & Jafari, 2017). These CDs have been mostly derived from β- and γ-CDs to reduce their hydrophilic properties which may increase the contact of CDs with biological membranes. These CD derivatives offer an extensive diversity in particle shapes, such as reservoir, rather-lamellar, and matrix structures. Nanospheres and nanocapsules have matrix and reservoir arrangements, respectively.

Chemical derivatives produced from parent CDs (α-, β-, and γ-CDs) are called amphiphilic CDs. These CDs are obtained by modifying the primary and/or secondary structures through linking aliphatic linear or branched chains having different lengths from 2 to 18 carbons with a dissimilar chemical bond (e.g., ester, ether, thiol, or amide) types.

Surface charge (nonionic and ionic) is the most important factor to classify amphiphilic CDs. Based on this factor, three categories of amphiphilic CDs can be prepared (GharibZahedi & Jafari, 2017):

- nonionic amphiphilic CDs;
- cationic amphiphilic CDs; and
- anionic amphiphilic CDs

In conclusion, the aim of CD modification is to overcome some of the concerns of parent CDs which restrict industrial utilizations including (Bilensoy & Hincal, 2008)

1. interaction enhancement between CDs and biological membranes with the help of a relative external hydrophobicity,
2. permitting self-assembly of CDs as a result the spontaneous development of nanoscale materials (nanospheres/nanocapsules), and
3. a binding increase between CDs and water-insoluble guests due to having high quantities of various aliphatic long chains and hydrophobic sites to establish potential molecular links with water-insoluble components.

1.4.2.2 Preparation Methods

Generally, there is no single methodology for forming the complexes as a process must be developed for each guest to be nanoencapsulated by the CD. A critical classification on the host–guest complexes is based on the phase–solubility profiles in a solution. The most important methods involved in the complexes formation with CDs have been listed below (GharibZahedi & Jafari, 2017):

- dry mixing (physical blending) method,
- milling/cogrinding method,
- kneading method,
- coprecipitation method,
- slurry-complexation method,
- paste-complexation method,
- solvent evaporation (coevaporation/solid dispersion) method,
- damp mixing and heating method,
- neutralization precipitation method,

- HPH method,
- freeze-drying (lyophilization) method,
- spray-drying (atomization) method,
- microwave irradiation method,
- supercritical antisolvent method,
- extrusion method,
- gas–liquid method,
- melting method, and
- sealing method.

Also, formation techniques of the CD-based polymeric NPs are as following (GharibZahedi & Jafari, 2017):

1. Physicochemical processes
   a. Dialysis method,
   b. Emulsion solvent diffusion (spontaneous emulsification) method,
   c. Ionic gelation (coacervation) method, and
   d. Spray-drying method
2. Chemistry based processes
   a. Emulsion polymerization method,
   b. Microemulsion polymerization method, and
   c. Interfacial polycondensation method.

1.4.3 Amylose Nanocarriers

Different bioactive ingredients as ligands can be nanoencapsulated within amylose nanostructures. The presence of ligands induces a conformational change in amylose namely the conversion of double helices into a unit strand (Carbinatto, Ribeiro, Colnago, Evangelista, & Cury, 2016). The V-shaped formed amylose is dense and has a central hydrophobic cavity, capable of binding lipophilic molecules via hydrophobic interactions (Xu, Zhao et al., 2013). It seems that the ligands are present in specific sites inside the structure. As an example, the acyl chains and the carboxylic groups are included in crystalline and amorphous areas, respectively. Therefore, amylose molecular inclusion complexes are appropriate nanocarriers for fatty acids (Gokmen, Mogol et al., 2011) and the other guest molecules (Yang, Zhang et al., 2013) by accommodating hydrophobic domains within the central cavity. Different types of complexes (polymorphs) can be formed under the effects of temperature and time of annealing. Regarding the long-chain fatty acids, the high level of unsaturation hinders the formation of inclusion complexation structures (less organized structures) and results in the formation of larger and more dispersed particle populations (Xu, Zhao et al., 2013). Increasing the amylose chain length (degree of polymerization, DP) up to certain amounts, raises the complexation yields, relative crystallinities and thermal stabilities. Moreover, by the augmentation of the amylose DP, lipid chain length, and processing temperature the stability of the structures increased against degradation by enzymes and acids (Gelders, Duyck, Goesaert, & Delcour, 2005).

Self-assembly of conjugated linoleic acid (CLA) with either amylose or β-CD was reported by Yang, Gu, and Zhang (2009). Amylose–CLA complex revealed more protection against oxidative degradation of CLA, empowering the reactions between CLA and amylose. By imitating the conditions of the small intestine, the release of ligand from amylose–CLA complex was higher than β-CD–CLA complex (Yang et al., 2009). The combinations of high amylose corn starch and amylose with genistein showed high encapsulation efficiency of genistein in simulated stomach ambient along with slow release during digestion by amylase in the pancreatin solution (Cohen, Orlova, Kovalev, Ungar, & Shimoni, 2008). The complexation method affects the complex organization of V-amylose. Water/dimethyl sulfoxide complexation method resulted in amylose–CLA complexes of 150 nm spherical shape, while KOH/HCl complexation technique led to the elongated structure with the width of 43–160 nm. The spherical structure exhibited superior protection to CLA against oxidation (Lalush, Bar, Zakaria, Eichler, & Shimoni, 2005). Recently, Marinopoulou, Kalogianni, and Raphaelides (2016) studied the dynamic interfacial tension of the host–guest networks containing amylose–fatty acid. By analyzing the systems, the dynamic interfacial tension of the complex dispersion was dramatically higher compared to the fatty acid solution, revealing that a portion of fatty acids are not adsorbed at the interface anymore and this point refers to the formation of the inclusion complex system. In addition, the dynamic interfacial tension of the combination was lower than the net amylose dispersion. Hence, portion of the fatty acid chains was not included in the complexes even at fatty acid/amylose contents underneath the amylose saturation level (Marinopoulou et al., 2016). Ryno, Levine, and Iovine (2014) reported the application of microwave heating to reduce the annealing time required to prepare these amylose complexes from hours to minutes. Besides, Raphaelides, Dimitreli, Exarhopoulos, Ilia, and Koutsomihali (2015) have reported a continuous method for the production of starch guest-host systems on an industrial scale.

1.5 SPECIAL-EQUIPMENT-BASED NANOENCAPSULATION TECHNIQUES

Although for the nanoencapsulation of food ingredients using different technologies, it is necessary to apply some general equipment including homogenizers, mills, mixing devices, etc., there are some nanoencapsulation techniques which are feasible to implement only by specialized developed equipment such as electrospinning, electrospraying, nanospray dryer, and microfluidics devices.

1.5.1 Electrospinning

Electrostatic force is used in electrospinning to draw a fine nanofiber out of a polymer solution droplet, which flies toward the nearest lower electrical potential target (Frenot & Chronakis, 2003). Fundamentally, an electrospinning assembly comprises of three main components which are supply of high voltage, a spinneret (typically made by repurposing a blunt hypodermic syringe) and a target-collector plate connected to an electrical earth (a metal screen or film). A kilovolt DC supply is used to induce a surplus electric charge in the liquid polymer (either a melt, or more commonly a solution). A fiber strand is pulled out of the droplet and accelerated toward the grounded collector. The liquid polymer is either pumped from the spinneret at a constant volume flow using a syringe pump or more traditionally fed from a reservoir at constant head (Anu Bhushani & Anandharamakrishnan, 2014).

When a droplet of liquid polymer is charged with a high potential, the droplet deforms, being attracted to a site of low potential. This droplet distortion is a characteristic structure dubbed the "Taylor cone" (Reneker, Yarin, Fong, & Koombhongse, 2000; Yarin, Koombhongse & Reneker, 2001). When the electric charge rises to a certain value, internal electrostatic repulsion overcomes the intrinsic molecular tension forces present at the surface of the droplet at the tip of the Taylor cone, and a liquid polymer stream, still carrying a high electric charge is thrown out toward the nearest lower charged point (the collector) that is usually held at earth potential to encourage fiber capture (Doshi & Reneker, 1995). The polymer jet continues to be stretched as the solvent evaporates and a continuous fiber deposits onto the earthed target collector (Yu, Zhu, White, & Branford-White, 2009).

Progress in understanding the mechanisms of electrospinning has been made recently and electrospun nanofiber morphologies have been developed to deliver bioactive components along with the fiber forming polymer; these bioactives significantly influence the mechanical properties and functionality of the fiber product (Ghorani & Tucker, 2015). For instance, the bioactives may result in fiber products containing an array of soluble additives, insoluble NPs or nanotubes. In addition, fibers and particulates can be made with multilayer structures, typically with a resistant outer coating around a chemically active interior.

A large number of different shaped electrospun fiber cross sections and fiber diameters result from using various polymers and solvents. Electrospun micro- and nanofibers are monofilaments, and if the fiber is being solidified in flight, it results in a cross section of cylinder form (Spasova, Mincheva, Paneva, Manolova, & Rashkov, 2006).

1.5.1.1 Preparation of Electrospun Fibers

When applying electrospinning for preparation of nanofibers containing bioactive components, it is necessary to consider different parameters. The main factors influencing the nanofiber structure and bead formation are (Ghorani, Alehosseini, & Tucker, 2017) as follows:

1.5.1.1.1 Polymer Solution Properties

Adjusting the properties of the liquid polymer has a key role over fiber morphology and size (Teo & Ramakrishna, 2006). Fiber size and the evolution of beads are influenced by solution viscosity and surface tension (Fong, Chun, & Reneker, 1999; Spasova, Mincheva et al., 2006). The nature of the solution is defined by the molecular mass and concentration of the polymer, and hence, by solution conductivity and surface tension.

Polymer concentration, molecular weight, and fiber morphology: Generally, polymer solubility increases as inversely to molecular mass. The degree of molecular chain entanglement also influences solubility tending to increase with molecular weight. The ideal spinnable polymer solution has the maximum polymer content whilst still being of sufficiently low viscosity to allow jet formation. There is a critical minimum level of polymer concentration for fiber formation (Huang, Zhang, Kotaki, & Ramakrishna, 2003; Shenoy, Bates, Frisch, & Wnek, 2005).

Electrical conductivity: Raising electrical conductivity and selecting a higher molecular weight polymer whilst lowering polymer concentration will result in a decrease in spun fiber diameter. The solution electrical conductivity is related to the charge density limit, and so to the amount of elongation of the jet through an increase in electrostatic force. Therefore, for given processing conditions, a more electrically conductive spinning solution may stretch more, resulting in a smaller diameter fiber product (Tan, Inai, Kotaki, & Ramakrishna, 2005).

Effect of solvents: The electrical and physical properties of the stock solution, from which the fibers are spun, are significantly influenced by the majority component which is the solvent; this, in turn affects the morphology of the electrospun fiber products (Eda, Liu, & Shivkumar, 2007).

1.5.1.1.2 Processing Conditions

Other parameters with a significant effect on the electrospinning process are the applied voltage, the distance from spinneret to target (tip-to-collector distance), feed rate, and spinneret orifice diameter (Ramakrishna, 2005; Mazoochi & Jabbari, 2011).

Voltage: The most basic part of the electrospinning process is the use of a high applied voltage to charge the polymer solution. Alteration of the applied voltage sets the strength of the electrical field formed between the spinneret and the target, and hence, the magnitude of the force drawing the fiber (Leach, Feng, Tuck, & Corey, 2011).

Flow rate: If a spinning machine uses a pump, the delivery flow rate governs the quantity of solution at the spinneret. If the machine uses a header tank, then the replenishment rate is effectively matched to the removal rate. Thus, for pump powered machine, at a given applied voltage, there will be corresponding flow rate set point to maintain a stable Taylor cone (Rutledge, Li et al., 2001).

Diameter of the spinneret orifice: Some laboratory electrospinning systems use a blunt hypodermic syringe needle as a spinneret. A blunt needle is preferred, as a sharp point encourages loss of power by coronal discharge. This kind of loss can be minimized by using a nonconducting spinneret such as a micropipette tip. In any event, the internal diameter of the spinneret orifice does influence the subsequent spinning process. As the orifice diameter increases, spinneret clogging may happen (Mo, Xu, Kotaki, & Ramakrishna, 2004) which is said to be an effect of surface tension. A smaller droplet radius increases the effective surface tension, and this effect can be readily achieved by reducing the spinneret diameter.

Spinneret tip to collector distance: The elongation of the polymer jet into a solid fiber ideally takes place between the spinneret, as the site of formation of the Taylor cone and the earthed collector plate. The tip to collector distance is related to the electrostatic field strength and thence to the time of jet flight. The deposition of wet fibers can be attributed to an insufficient distance between the spinneret and the collector, so that the drying time is too short for the evaporation of the solvent before the fibers are deposited on the collector and consequently partially dried fibers with fused fiber–fiber intersections are observed as well as part of a densely packed structure (Barhate, Loong, & Ramakrishna, 2006).

1.5.1.1.3 Ambient Conditions

Ambient parameters particularly humidity (best measured as absolute humidity rather than relative humidity, as the absolute measure is more directly related to the conductivity of air) and temperature are said to influence fiber morphology and the productivity of the electrospinning process. This is because a direct relationship exists between both solvent evaporation rate, liquid solvent conductivity, and temperature. Both will affect the electrospinnability of a polymer (Chen & Yu, 2010). The viscosity and surface tension of a polymer solution are also affected by temperature (De Vrieze, Van Camp et al., 2009; Bhardwaj & Kundu, 2010).

1.5.2 Electrospraying

The electrospraying technique, also known as electrohydrodynamic atomization has found its way in food nanotechnology as a simple and effective encapsulation method (Tapia-Hernández, Rodríguez-Félix, & Katouzian, 2017). The spheres gained from the equipment may differ due to parameters, such as flow rate, electrostatic potential, collector distance plus the solution parameters like viscosity, density, and concentration. Environmental factors, such as temperature and pressure, will also affect the characteristics of the yielding nanocapsules (Anu Bhushani & Anandharamakrishnan, 2014).

Electrospraying and electrospinning are both based on the same physics governing the ejection of a continuous stream; however, if the degree of molecular cohesion is below a critical level, droplets are formed from the ejecta, not a continuous fiber (Bock, Woodruff, Hutmacher, & Dargaville, 2011). The difference between electrospinning and electrospraying is controlled by how much molecular cohesion occurs—this is difficult to measure directly but is readily inferred from the polymer solution viscosity and less directly from polymer concentration (Chakraborty, Liao, Adler, & Leong, 2009; Enayati, Chang, Bragman, Edirisinghe, & Stride, 2011).

Basically, electrospraying includes applying a high voltage, flow rate, and variation of particle collector distance, in addition to the properties of polymer solution, i.e., concentration, viscosity, and density. The theory of electrospraying technique is based on the ability of an electric field to crush a droplet and transform them to the micron or nanometer scale depending on the control parameters (Tapia-Hernández, Torres-Chávez et al., 2015). When an electric field is applied to a drop, an electric charge is generated within the droplet, named as Coulomb force; this force competes with the cohesion force of the particle and when it dominates over the cohesive force, a diminution in the surface tension occurs and finally NPs are obtained (Kao, Chang et al., 2012).

1.5.2.1 *Preparation of Electrosprayed Nanocarriers*

Electrospraying has been proposed as a method for nanoencapsulation, where the addition of a high level of electrostatic force reduces the size of the formation of aerosol droplets to the nanoscale. In other words, the electrostatic force resulting from the application of high voltages increases the degree of atomization, resulting in finer droplets than from the effect of surface tension alone. The removal of solvent from the strand occurs at an increased rate during the flight of the droplets because the surface to volume ratio is higher (Hao, Wang et al., 2013; Zhang & Kawakami, 2010).

There are three stages in electrohydrodynamic atomization process required to obtain particles with smaller diameter (Tapia-Hernández, Rodríguez-Félix et al., 2017):

1. First, the fluid must be dynamic with an initial acceleration so that the Taylor cone is formed. The Taylor cone is generated through a balance of forces as surface tension, gravity, liquid surface and electric tension, inertia, and viscous stress.
2. In the second phase, the fluid is disintegrated within the cone jet forming tiny droplets.
3. Lastly, the droplets are sprayed onto a collector surface.

Today, this technique has been modified in order to obtain better results in the formation of microparticles and NPs. Moreover, the use of various materials, mainly natural biopolymers has also been investigated. The variations of the technique are electrospraying in solution, coaxial electrospraying, and electrospraying by deposition on a substrate (Tapia-Hernández, Rodríguez-Félix et al., 2017).

Recently, the application of electrospraying has expanded beyond food and nutraceutical technology, transforming to a novel technique that is developing rapidly. In this area, the electrospraying can be used for three purposes (Tapia-Hernández, Torres-Chávez et al., 2015): primarily, as an application in the area of inedible food, being used to develop intelligent packaging that can resist the entry of pathogens. Second, it is employed to fabricate nanostructures based in food polymeric materials, i.e., to produce particles encapsulating bioactive substances or food ingredients; and finally, it is applied in the use of polymeric matrices for encapsulation of nutraceuticals (Tapia-Hernández, Rodríguez-Félix et al., 2017).

Nano and microparticles produced from food polymeric materials have the characteristic of being biodegradable and can be made by the electrospraying technique. These polymers are commonly used in the pharmaceuticals industry, where they serve as drug-delivery systems and to deliver other compounds to the targeted area. It has been shown that particles can also serve as nanocarriers of bioactive components. Therefore, the electrospraying technique is a feasible method for the entrapment of nutraceutical compounds (Gomez-Estaca, Balaguer, Gavara, & Hernandez-Munoz, 2012; López-Rubio & Lagaron, 2012).

Electrospraying of viscous polymer solutions tends to occur in three basic modes a dripping mode, an intermittent ejection mode, and a single cone stream characterized by the presence of a stable Taylor cone (Jain, Scott, Zustiak, & Sell, 2015). The diagnostic characteristic of electrospraying is the presence of fine charged droplets of about half the size of the Rayleigh limit where repulsive electrostatic force is said to overcome the attractive force of surface tension (Jaworek, 2007).

In order to obtain desired NPs, the parameters which influence the morphology of the particles should be carefully monitored. Three groups of parameters comprise the equipment, solution, and environmental factors. These factors were described in the previous section for electrospinning.

1.5.3 Nanospray Drying

Spray drying is defined as the transformation of a fluid from a liquid state into a dried particulate form by spraying the fluid into a hot drying medium. It is a suitable one-step process for the conversion of various liquid formulations (e.g., aqueous and organic solutions, emulsions, and suspensions) into dry powders (Mahdavi, Jafari, Ghorbani, & Assadpoor, 2014). Spray drying is a simple, fast, and scalable drying technology that is well established in the chemical, food, and pharmaceutical industries. Spray-drying equipment is commercially available and the production cost is lower compared to that of other drying technologies such as freeze drying (Jafari, Mahdavi-Khazaei, & Hemmati-Kakhki, 2016; Mahdavee Khazaei, Jafari, Ghorbani, & Hemmati Kakhki, 2014).

In spray-drying process, the nozzle atomizes the feed into droplets. The reduction in particle size leads to a large increase in the surface area. In the drying chamber, the solvent, which can be water, organic solvents, or mixtures thereof, is quickly removed by the continuous flow of the hot drying gas (usually air or inert gas). The dried solid particles are separated from the gas stream by a cyclone and collected in a collection vessel (Pourashouri, Shabanpour et al., 2014; Rajabi, Ghorbani, Jafari, Sadeghi Mahoonak, & Rajabzadeh, 2015).

The recent boost in nanotechnology has increased the need for spray dryer systems required to fabricate a high yield of nanoscale particles with narrow distribution of size. So far, several expert opinions have been published about the nanospray-drying technology (Arpagaus, 2011). In order to produce nanoscale particles via spray-drying technology, some modifications on the experimental set-up of traditional spray dryers are necessary. A major constraint is the limited efficiency of separation and collection of submicron particles by cyclone separators. Typical cyclones are unable to collect particles below 2 μm. Even with a high-performance glass cyclone, the median particle size cannot be reduced below 1.4 μm (Li, Anton, Arpagaus, Belleteix, & Vandamme, 2010). In other words, submicron-sized particles cannot be collected using traditional spray dryers. The only feasible option for collection of NPs is to use electrostatic particle collectors (Arpagaus, 2011). Another limitation is the turbulent gas flow in the drying chamber, which results in particle depositions on the chamber wall. Furthermore, traditional atomizers do not allow the fine droplets being generated to reach submicron particle sizes. With the recent advances in nanospray dryer technology, particles <1 μm can now be generated.

The Nano Spray Dryer B-90 was introduced in 2009 to extend spray drying to the submicron scale. This laboratory spray dryer is based on a fundamentally new concept of spray-drying technology, involving the fabrication of submicron particles from a solution, nanoemulsion, or nanosuspension. Its technological novelty lies in the gentle flow of laminar drying gas, the vibrating mesh spray technology to form fine droplets, and the highly efficient electrostatic precipitator to collect NPs.

In the nanospray dryer, a very small amount (e.g., a few milligrams of solid) are sufficient. This is especially relevant when expensive materials are used, such as peptides under clinical investigations in the early stages of drug development (Haggag & Faheem, 2015).

1.5.3.1 Preparation of Nanospray-Dried Particles

The process consists of the following basic steps—droplet generation, heating of the drying gas, drying of the droplets, and particle collection (Arpagaus, John, Collenberg, & Rütti, 2017).

1.5.3.1.1 Droplet Generation

The droplet generation in the nanospray dryer is based on vibrating mesh technology in a spray head. The underlying physical principle of droplet formation is utilized in ink-jet printers and vibrating mesh nebulizers for inhalation therapy (Oliveira, Guimarães, Cerize, Tunussi, & Poço, 2013). It includes a piezoelectric actuator driven by an electronic circuit. A small exchangeable spray cap is screwed onto the lower part of the spray head at a defined torque. The spray cap is made of a thin stainless steel mesh which is perforated with an array of precisely micron-sized holes. Spray meshes are available with orifice diameters of 4.0, 5.5, and 7.0 μm. Electricity drives the piezoelectric actuator at an ultrasonic frequency of about 60 kHz, which is close to the resonance frequency of the spray cap, thus minimizing electrical energy consumption.

Plastic tubes connect the spray head to the feed reservoir. A peristaltic pump with variable speed circulates the fluid uniformly from the reservoir through the spray head to the spray mesh and back to the reservoir. This circulation mode enables efficient and continuous atomization.

1.5.3.1.2 Drying of Droplets

The drying gas flows through a compact porous metal disc that enables efficient heat transfer from the hot metal surface to the drying gas. The porous structure is embedded with an electrical heating coil. This coil ensures optimal energy transfer to the porous metal and uniform heat distribution in the entire heater volume. The heating system takes only a few minutes to raise the inlet temperature from ambient to 120°C (Arpagaus, John et al., 2017). It also ensures a laminar flow of gas in the spray chamber to dry the fine aerosol. This is crucial because turbulence would lead to uncontrolled spray formation and particle loss on the drying chamber walls. The laminar flow enables gentle drying, which makes the system suitable for drying heat-sensitive products.

The aerosol droplets are injected from the top of the nanospray dryer in the same downward direction as the flow of drying gas (Arpagaus, John et al., 2017). The aspirator speed maintains the drying gas flow rate within a range of 80–160 L/min. The vertical configuration guides the submicron particles toward the electrostatic particle collector while minimizing particle adhesion to the sidewalls of the drying chamber, thus resulting in high collection efficiency. Thanks to these features, i.e., mild drying conditions, laminar gas flow, and cocurrent operation, the product is treated with care and drying takes place gently. Nanospray drying is thus a very suitable technology for heat-sensitive products as it has low risk of degradation or activity loss.

1.5.3.1.3 Particle Collection

The particle collection mechanism in the nanospray dryer is based on electrostatic charging, which, unlike cyclones, is independent of particle mass (Arpagaus, 2012; Lee, Heng, Ng, Chan, & Tan, 2011). The working principle consists of the following basic steps:

1. generation of an electrical field of about 15 kV voltage between the discharge and collecting electrodes,
2. generation of negatively charged ions in the gas,
3. charging of the spray dried particles, and
4. deflection of the particles to the collection electrode.

The electrostatic particle collector can capture submicron particles (<1 μm) at a separation efficiency greater than 99% for small solid batches of 30–500 mg (Bürki, Jeon, Arpagaus, & Betz, 2011; Lee et al., 2011). The electrostatic precipitator can even collect thin-walled particles without breaking (Feng et al., 2011; Sun, Song, Wang, & Yu, 2011). The particles are gently removed from the internal surface of the collecting electrode cylinder by utilizing the particle scraper and particle collecting paper that are included in the delivery of the laboratory instrument. The particles are filled in airtight glass vials and normally stored in a desiccator until further usage and analysis.

1.5.3.2 Optimizing the Nanospray-Drying Process

Before the spray-drying application work begins, it is advantageous to conduct feasibility tests at the laboratory bench using a small sample to determine the feasibility of the application. Nanospray drying is a technology in which the properties of the fabricated powder depend on the correlation and interdependency of the process parameters and formulation variables.

The most important input and output parameters are (Arpagaus, John et al., 2017) as follows:

- drying gas flow rate, humidity, and temperature;
- droplet size;
- particle size;
- solid concentration;
- feed rate;
- product yield;
- organic solvent instead of water;
- particle morphologies;

- encapsulation efficiency and bioactive compounds loading;
- controlled release of bioactive compounds;
- stability of bioactive compounds during nanospray drying; and
- storage stability.

Some fluids may be too viscous to pass through the mesh system and suspensions may block the mesh holes, which could result in intermittent droplet generation or even cessation of droplets generation. The processing time may gradually increase. Sometimes the product may even build-up deposits on the vibrating mesh and reduce the product yield (Schmid, 2011). However, as stated by Schmid, Arpagaus, and Friess (2011), virtually any substance can be atomized successfully with the nanospray dryer, as long as the right process parameters and formulation variables are used.

1.6 BIOPOLYMER-BASED NANOENCAPSULATION TECHNIQUES

Biopolymer-based NPs are defined as submicron biopolymeric particles, which can be used for nanoencapsulation of bioactive compounds. Based on the preparation technique, various forms such as NPs, nanospheres, or nanocapsules can be obtained. Nanospheres can be defined as matrix systems utilized for the uniform dispersal of active components, while nanocapsules are vesicular systems where active compounds are encapsulated within a cavity consisting of an inner liquid core which in turn is surrounded by a polymeric membrane (Livney, 2015; Lohcharoenkal, Wang, Chen, & Rojanasakul, 2014). Utilization of individual biopolymer NPs and also complexes of biopolymer particles along with nanogels and nanotubes made with biopolymers are described in the following sections.

1.6.1 Single Biopolymer Nanoparticles

In this section, formation of the nanocapsules by either single protein or carbohydrate biopolymers via desolvation and precipitation methods will be explained briefly.

1.6.1.1 Protein Nanoparticles (Desolvation Method)

Desolvation is defined as a self-sorting process with the minimum of Gibbs free energy for biopolymers like proteins to prepare NPs. Based on this theory, replacing some or all water molecules (solvent) around protein molecules with a desolvating agent (another solvent) will cause conformational changes in protein structure, and its solubility decreases significantly; therefore, phase separation and precipitation of proteins will occur consequently. This makes them ready to interact with each other and form NPs (Gao, Kokini, Fern, & Mejia, 2014; Sadeghi, Moosavi-Movahedi et al., 2014).

This method, which is also known as drawing-out precipitation, reaching to the desired nanoparticulation, basically depends on balancing the attractive and repulsive forces and decreasing intramolecular hydrophobic interactions. In general, a desolvating agent is a solvent of different polarity and hydrogen-bond-forming capacity, which is added dropwise to the protein solution. The common used agents are ethanol, methanol, acetone, propanol, etc., and also a mixture of them which have been investigated by many researchers (Sadeghi, Mehryar, Karimi, & Kokini, 2017). For example, gradual adding of ethanol to protein solution unfolds the protein structure at first, followed by proteins folding and the possible increase of peptide–peptide interaction, and finally, specific nanostructures will be assembled to form NPs (C. J. Coester, 2000; Jun, Nguyen et al., 2011). This new structure has more hydrophobic surface, which leads to lower solubility of the produced aggregates or NPs in comparison with the native proteins. Subsequently, the produced NPs should be cross linked in order to avoid aggregation. Crosslinking is proved to be an important step in nanoparticulation process mainly due to its role in biodegradability and release of the encapsulated material. It also prevents aggregation and agglomeration of the resulted NPs (Chen & Zhong, 2015). The most commonly used crosslinking agent is glutaraldehyde.

The effect of the operational parameters on nanoparticulation prepared by the desolvation method is very critical to fabricate small and uniform NPs. These parameters include (Sadeghi, Mehryar et al., 2017) the following:

- upscaling,
- amount and addition rate of the desolvating agent,
- type of the desolvating agent,
- concentration of the desolvating agent,
- stirring rate,
- cross-linking condition (type, concentration, and time),
- temperature,
- the effect of pH value on the desolvation process,
- protein concentration,
- buffer type, and
- ionic strength.

Food proteins are promising nanocarriers for the delivery purposes mainly due to their unique characteristics. On the other hand, therapeutically relevant peptides and proteins increasingly took part in curing the viral, malignant, and autoimmune ailments. The common nanocarriers suffer from their labile nature, high manufacturing expenses, provoking immunogenic and allergic responses, along with the low bioavailability and sensitivity toward peptidases. Application of protein NPs produced by desolvation method has been studied for the various food ingredients and nutraceuticals such as curcumin, folic acid, carvacrol, resveratrol, retinoic acid, vitamin D_3, date extract, and many drugs and other bioactive compounds (Abbasi, Emam-Djomeh, Mousavi, & Davoodi, 2014; Bagheri, Madadlou, Yarmand, & Mousavi, 2013; Gülseren, Fang, & Corredig, 2012b). For this goal, proteins of animal origin (gelatin, collagen, albumin), milk proteins (casein, whey proteins), and silk proteins along with plant proteins such as zein and gliadin have been investigated.

1.6.1.2 Polysaccharide Nanoparticles (Nanoprecipitation Method)

A similar technique of desolvation, named nanoprecipitation method, has been developed for the fabrication of NPs from other biopolymers including polysaccharides (Tan, Xu et al., 2009; Sailaja & Amareshwar, 2012). There are many parameters determining the particle size such as concentration of the polysaccharide solution, in which the higher the concentration, the larger the size of formed particles. Polysaccharide molecules are in better dispersed state in lower concentrations which enables the preparation of smaller NPs during nanoprecipitation (Zhao, Shi et al., 2011). Higher molecular weight accompanied by the lower degree of substitution would favor the preparation of smaller size NPs. Another effective parameter in fabrication of composite NPs using nanoprecipitation technique is pH (Gan & Wang, 2007; Taheri & Razavi, 2015).

1.6.1.2.1 Starch Nanoparticles

Starch, which is a storage of carbohydrate in plants, is structurally made up of amylose and amylopectin chains. Due to its biodegradable and biocompatible properties, starch is a promising, versatile, and inexpensive polysaccharide for bioactive delivery applications. In a variety of investigations, starch NPs were utilized and their size range was approx. between 10 and 1000 nm (Kim, Park, & Lim, 2015). Starch NPs have distinct advantages such as lower viscosity and small particle size, while considerable amounts of bioactive ingredients can be loaded in these systems. Starch NPs can be easily obtained by hydrolyzing starch granules using enzymes or acids. However, during the acid hydrolysis, powdered NPs display a strong tendency to aggregation (Jiranuntakul, Sugiyama et al., 2013). Acid hydrolysis is a simple and controllable method that has been widely performed to obtain starch NPs. In starch granules, crystalline part tends to be more resistant to acid hydrolysis than the amorphous one. This problem can be solved through mild acid hydrolysis of crystalline regions by using sulfuric acid or hydrochloric acid. Starch NPs can also be obtained by enzymatic processes, physical treatments such as ultrasonication, HPH, gamma irradiation, and reactive extrusion (Molina-García, Horridge, Sanz, & Martino, 2007). In order to reduce the treatment time, some researchers have utilized alternative means of physical hydrolysis (i.e., ultrasonics).

1.6.2 Complexation of Biopolymers

The physical and chemical features of individual biopolymers as a function of intrinsic and extrinsic molecular key parameters have been extensively explored throughout the past few decades such as molecular weight, steric conformation, concentration, charge amount and distribution, size homogeneity, ionic strength, ion type, pH, solvent quality, temperature, and various molecular interactions within or between different biopolymers. As compared to single biopolymers, formation of biopolymer complexes offers different functional and promising properties due to the synergistic combination between the functional features of various biopolymers (Hosseini, Emam-Djomeh, Sabatino, & Van der Meeren, 2015; Ron, Zimet, Bargarum, & Livney, 2010). Engineered functionalities such as encapsulation, emulsion and foam stabilization, formation of lightening agents by fat mimicry, creation of novel mixed electrostatic gels at concentrations well below those required for common gel forming agents, formation of new structures at various (nano, meso, or macro)scales and waste management may be developed using covalent linkages, Coulombic interactions, hydrogen bonding, and hydrophobic interactions in biopolymer blends. The new functionalities resulting from self-assembling structures are generally the focus of researches studying soft condensation in material science (Zimet & Livney, 2009). The ability to control and modify involved interactions can assist food technologists to engineer supramacromolecular structures in order to develop food products having more desirable structural properties.

The application of biopolymers and their mixtures for designing advanced engineered delivery systems to entrap bioactive components and to enable subsequent fortification of food products is a highly desirable and promising technique. In addition to the vast food applications, biopolymers can be used as promising replacements for synthetic polymers in nonfood applications (e.g., wound dressings, drug delivery, and scaffold engineering). Biopolymers are capable of encapsulating substances of different hydrophilicity ranging from completely water soluble to water insoluble (Hosseini, Ghiasi, & Jahromi, 2017). In addition, mild preparation process without involving harsh chemicals and poisonous solvents applied in the formation of structured vehicles is of paramount importance in green (organic solvent-free) industries. Therefore, nature-made individual biopolymers and their mixtures introduce new possibilities for developing multifunctional nanometric encapsulation systems for protecting and targeted release of sensitive health-promoting substances (Jones & McClements, 2011).

1.6.2.1 Molecular Forces Between Biopolymers

The net (i.e., repulsive or attractive) interaction between biopolymers is a consequence of various forces act between different segments of two adjacent macromolecular species (Hosseini, Ghiasi et al., 2017). In a specific system, several factors including those related to intrinsic characteristics (e.g., molecular mass and conformation, charge density and distribution, flexibility, and hydrophilicity) and those related to solvent (e.g., ion type, ionic strength, pH, dielectric constant, and solvent quality) and processing (e.g., pressure, temperature, mixing conditions, and method of acidification) as external characteristics dictate the predominant form of interactions between biopolymers. The above-mentioned parameters can be modulated for controlling the interactions and designing novel structures (Arroyo-Maya & McClements, 2015; Gulseren, Fang, & Corredig, 2012a, 2012b).

In aqueous dispersions, biopolymers experience two different types of demixing phenomena namely segregative and associative phase separations. These two phenomena are mainly dependent on the charges of polymers as influenced by external and internal factors. The resultant phase separation phenomenon is the consequence of long- or short-range (low or high amplitude) intermolecular interactions between contributors as well as different affinity of biopolymers to the solvent (Schmitt, Aberkane, & Sanchez, 2009).

1.6.2.1.1 Cosolubility

This phenomenon takes place in the presence of similarly charged species. This phenomenon can be observed in a biopolymer mixture containing very low concentrations or in mixed systems containing biopolymer subunits (i.e., monosaccharides and amino acids). When biopolymers carry electrical charges of similar sign (or when one partner is neutral), macromolecules are segregatively phase separated into two separate immiscible biopolymer-rich phases. High concentrations and ionic strengths are essential to observe this phenomenon (also known as thermodynamic incompatibility) (Ilyasoglu & El, 2014; Luo, Pan, & Zhong, 2015). Under these conditions, the net interactions between biopolymers are repulsive. These interactions, which are mainly resulted from excluded volume effects, are nonspecific and generally of temporary duration. Individual mixtures of structurally different polysaccharides or proteins may also undergo segregative phase separation (e.g., mixture of alcohol-soluble prolamins and acid-soluble glutelins; branched and linear polysaccharides; and intact and heat-denatured forms of same proteins) (Tolstoguzov, 2002).

1.6.2.1.2 Complex Coacervation

When both partners carry electrical charges of opposite sign, thermodynamic compatibility (also known as complex coacervation) occurs at relatively low concentrations (in most cases <4 wt%) and low ionic strengths (<400 mM). The associative phase separation results in the spontaneous formation of an upper macromolecule-depleted (water-rich) phase (also called equilibrium solution) together with a lower phase rich in complexed contributors (Hosseini, Emam-Djomeh et al., 2015). For attractive columbic interactions, the pH of solution should be between the isoelectric point (pI) of protein and the pK_a of polysaccharide. Electrostatic interactions are the major molecular forces present in the biopolymer-rich phase. In bottom phase, the resultant structure derived from the complexed biopolymers can appear as coacervates, interpolymeric (insoluble) complexes (or precipitates), and electrostatic gels (Turgeon & Laneuville, 2009) depending on the biopolymer type. Complex coacervation of various biopolymer pairs (especially those of protein and polysaccharide) under the influence of different factors have been extensively studied previously. When one partner (particularly polysaccharide) is in excess amount, soluble nanocomplexes can be obtained instead of precipitates (Hosseini, Ghiasi et al., 2017).

1.6.2.2 Preparation of Different Complexes

Generally, physical entrapment and chemical complexation of bioactive molecules (particularly hydrophobic ones) are the main strategies applied in the development of supramolecular biopolymer-based carrier systems (Matalanis, Jones, & McClements, 2011). Actually, different types of biopolymer mixtures are capable of binding nutraceuticals of interest and creating molecular complexes. The bioactive molecules can be encapsulated and protected using protein–polysaccharide, polysaccharide–polysaccharide, and protein–protein (nano)complexes (Hosseini, Ghiasi et al., 2017).

1.6.2.2.1 Protein–Polysaccharide Interactions

Complexes and coacervates derived from protein and polysaccharide interactions are ideal tools in nanoencapsulation technology. A large spectrum of different combinations of protein–polysaccharide applied for (nano)encapsulation of different components can be seen in the literature.

Among different structures resulted from protein–polysaccharide complexation, the coacervates are superior to others for encapsulation technology. Coacervates have great interfacial properties resulting in utilization for taste masking and controlled release purposes due to their ability to form a protective film (Ron, Zimet et al., 2010; Zimet & Livney, 2009). It should be noted that coacervates and complexes are structurally different from each other. However, in literature, the words "complexes" and "coacervates" are sometimes used interchangeably. Structurally, coacervates are in liquid form, which may then separate as dense and structured liquid spherical droplets (vesicles); interpolymeric complexes (insoluble complexes or amorphous coprecipitates) exhibit fractal aggregates and gels are 3D interconnected complexes (Ducel, Pouliquen, Richard, & Boury, 2008; Turgeon & Laneuville, 2009). Indeed, the structure determination of biopolymer complex coacervation at various hierarchical scales, including molecular (nano), meso, and large scales, is one of the most challenging and exciting aspects of these demixing phenomena. Because of dispersing the coacervate droplets in a continuous phase, systems capable of forming coacervates have sometimes been considered as (W/W) emulsions (Ducel et al., 2008; Turgeon & Laneuville, 2009).

For each binary mixed system composed of various proteins and polysaccharides, there are some unique factors dictating formation either coprecipitates or coacervates. An increase in the biopolymer charge density and a decrease in the backbone flexibility (indirectly influenced by the charge density) generally results in the formation of complexes instead of coacervates (Turgeon, Schmitt, & Sanchez, 2007). The weight ratio of protein to polysaccharide (or vice versa) is another significant factor determining the morphological characteristics of the resultant complex.

1.6.2.2.2 Polysaccharide–Polysaccharide Complexes

Oppositely charged polysaccharides can be used for developing mixed-type biopolymer complexes so as to encapsulate different functional components. Nonneutral polysaccharides are almost anionic with some exceptions. Therefore, cationic chitosan is mainly the ubiquitous biopolymer within complexes resulted from polysaccharide–polysaccharide electrostatic interactions. Among different anionic polysaccharides, sodium alginate (because of having a standard structure as well as gelation characteristics in the presence of calcium ion) has been widely used for complexation with chitosan (Hosseini, Ghiasi et al., 2017).

Similar to the protein–polysaccharide systems, various internal and external parameters are involved in polysaccharide–polysaccharide complexation.

1.6.2.2.3 Protein–Protein Complexes

Coassembly of proteins with each other through noncovalent bonding is a powerful method for generating ordered supramolecular structures. However, in contrast to the polysaccharide–protein and/or polysaccharide–polysaccharide systems, some limited studies describe the spontaneous complexation between two oppositely charged proteins and their coassembly into nano or microsized structures composed of at least two different proteins (Chen, Zhang, & Zhong, 2015). Under specific physicochemical conditions (the correct mixing ratios and defined pH conditions), proteins of acidic character (i.e., having low pI values) such as β-lg, BSA, and caseins are able to spontaneously self-assemble with proteins containing strong basic character such as lysozyme, lactoferrin, zein, and avidin (Hosseini, Ghiasi et al., 2017). In addition to the electrostatic interactions, protein–protein complexes can be obtained through heat treatment. As an example, heat-treatment of milk causes interactions mainly between κ-casein and β-lg in a pH-dependent manner. Therefore, the pH of milk before heat treatment has a significant influence on the various types of gelation including acid- or rennet-induced gelation, which is of considerable importance to the dairy industry (Vasbinder & de Kruif, 2003).

1.6.3 Nanogels

Nanogels, also called macromolecular micelles (Gonçalves, Pereira, Schellenberg, Coutinho, & Gama, 2012), are defined as the nanosized systems structured by chemically or physically linked polymer networks crosslinked hydrogel particles within a size range between 10^{-9} and 10^{-6} m (Abreu, Oliveira, Paula, & de Paula, 2012). They have good swelling ability in appropriate solvents. When the solvent utilized is water, the term "hydrogels" is used (Bourbon, Cerqueira, & Vicente, 2016).

Nanogels are primarily used to deliver or transport the drugs which partly or wholly destabilize due to metabolic activities in the human body or simply do not permeate thorough the intestinal mucosal barrier (Prego, Torres, & Alonso, 2005). A nanogel that will be utilized as a drug carrier should demonstrate certain properties, such as effective encapsulation of the bioactive compounds, providing protection against enzymatic attacks or aggregation caused by serum components, offer pH stability even during the possible deformation of the nanogel, and avoid any toxic effects or sudden burst release of the active compounds. Furthermore, during the manufacture of the nanogel, the technologies that are utilized such as sonication or polymerization may be detrimental to the bioactive compounds and care should be exercised upon their application (Ghayempour & Mortazavi, 2015; Shen & Yang, 2013).

Based on their stimuli responsiveness, nanogels are divided into two main subcategories. If the swelling process of a nanogel occurs simply by absorbing water, they are called nonresponsive nanogels. In some cases, environmental changes such as pH, enzymes, ionic stresses, temperature, or magnetic fields may affect the swelling ability of nanogels, and consequently their release behavior. Such nanogels are referred to as being "stimuli-responsive." Also the definition of multiresponsive nanogel represents the responsiveness of a nanogel to more than one environmental stimulus (Sultana, Imran-Ul-Haque, Arafat, & Sharmin, 2013). There are also further classifications on the production methods of nanogels and bioactive molecule loading strategies (Dextraan, 2007).

Since the removal of solvents and surfactants from the final products is difficult and expensive, and leads to the occurrence of potential toxic effects, their utilization in nanogel manufacture is among the major disadvantages of these delivery systems.

1.6.3.1 Ingredients Used to Prepare Nanogels

Nanogels can be prepared from biopolymers such as proteins and polysaccharides, and chemical polymers.

1.6.3.1.1 Protein Nanogels

Ovalbumin and lysozyme, which are hen egg proteins, along with whey proteins can be utilized in order to manufacture nanogels. Once these proteins are thermally denatured, they are bonded by hydrophobic interactions, hydrogen bonds, and disulfide linkages. To stabilize surface structure of nanogels, electrostatic repulsive forces can be effectively tailored. Alkali reagents have been used in nanogel formation process for rendering protein nanogels edible and nutritional (Yu, Yao, Jiang, & Zhang, 2006).

1.6.3.1.2 Polysaccharide Nanogels

Chitosan is a widely utilized biopolymer in the delivery of bioactive compounds, especially macromolecules, due to its both physicochemical and biological properties (Abreu, Oliveira et al., 2012). Because chitosan is water soluble (i.e., in dilute acetic acid solutions) and positively charged, it can easily interact with negatively charged polymers or polyanions in aqueous environments. Furthermore, chitosan has been utilized as an absorption enhancer through the mucosal barrier in a variety of studies. In addition, chitosan based nanogels display low toxicity and good biocompatibility (Rahaiee, Shojaosadati, Hashemi, Moini, & Razavi, 2015).

Alginate-based nanogels represent structures formed by NPs within a range of 250–850 nm with sodium alginate solution, calcium chloride, and polylydine. The interest for these nanogels is increasing in many fields such as antitubercular and antifungal drugs, therapeutic agents, and even gene delivery (Lertsutthiwong, Rojsitthisak, & Nimmannit, 2009; Shen & Yang, 2013).

1.6.3.1.3 Chemical Polymer Nanogels

This type of nanogels are produced by different chemical polymers such as polyvinyl alcohol (PVA), polyethylene oxide (PEO), polyethyleneimine (PEI), polyvinylpyrrolidone (PVP), and poly-*N*-isopropylacrylamide (PNIPAM).

PVA-based nanogels are produced through crosslinking of PVA polymeric chains by some methods such as using chemical (electron beam, crosslinking agents) and physical methods like freezing/thawing (Li, Wang, & Wu, 1998). For PEO- and PEI-based nanogels, when any oligonucleotides or anionic/amphiphilic compounds interact with the PEO-cl-PEI, hydrophobic chains from polyion complex are joined to the hydrophilic PEO chain. This formation causes to the collapse of gel particles (Bronich, Vinogradov, & Kabanov, 2001). PVP hydrogel NPs that have 100 nm final diameter are produced by using aqueous cores of reverse micellar droplets (Bharali, Sahoo, Mozumdar, & Maitra, 2003). By controlling the size of revere micellar droplets, size of PVP nanogels can be modified (Guowei, Adriane, Chen, Jie, & Yinfeng, 2007). PNIPAM-based nanogels can be produced by covalently crosslinking of PNIPAM–coallylamine networks and PNIPAM-*co*-acrylic acid networks (Gan & Lyon, 2001; Huang et al., 2004).

1.6.3.2 Preparation Methods

Production methods of nanogels can be divided into two major categories which are production from polymer precursors and polymerization of monomers. Using an appropriate solution, amphiphilic copolymers may form nanoscopic structures, and based on their properties, nanogels can be synthesized. Some crosslinking mechanisms are involved in these methods including:

Disulfide-based cross-linking: For these nanogels, which are formed in aqueous solutions, an appropriate hydrophilic unit is polyethylene glycol, and hydrophobic and cross-linkable unit could be pyridyl disulfide in order to induce disulfide bonding. Different sized nanogels can be formed by cross-linking polymer chains at varying polymer concentrations (Ryu et al., 2010a, 2010b).

Amine-based cross-linking: Amine groups are known to demonstrate reactivity toward isocyanates, carboxylic acids, iodides, which in turn can induce amine-based crosslinking during the formation of nanogels. For example, hydrophilic and cross-linkable blocks such as acrylic acid can be used to produce different amphiphilic block copolymers. Carboxylic acid groups are amdidated by diamine crosslinkers, consequently block copolymers form self-assembled and cross-linked micellar assemblies (Duong, Marquis, Whittaker, Davis, & Boyer, 2011; Li, Du, Sun, & Wooley, 2008).

Similarly, isocyanate reactions contribute to the formation of nanogels. Cross-linked micelles, which are pH responsive, can be produced by the addition of 1,8-diaminooctane to micellar aggregates of 3-isopropenyl-α,α-dimethylbenzyl isocyanate (Kim, Pourgholami, Morris, & Stenzel, 2011).

In addition to these techniques, a variety of cross-linking techniques that are based on click chemistry, photoinduction, and other physical techniques can be mentioned (Yesilyurt, Ramireddy, & Thayumanavan, 2011; Zhang, Zhou, Zhu, Ge, & Liu, 2008). Although heterogeneous polymerization is another widely utilized method, it is not applicable to the food systems (Oh, Tang, Gao, Tsarevsky, & Matyjaszewski, 2006).

1.6.4 Nanotubes

Nanotubes demonstrate a characteristic tubular shape similar to the structure of an empty carbon tube (Mousavi, Amjad-Iranagh, Nademi, & Modarress, 2013). They display thermal resistance at elevated temperatures and have a solid and flexible structure that might be used in many areas including food industries, nanomedicine and medical devices, etc. As an example relevant to food applications, it is noteworthy that when milk protein α-La was hydrolyzed partially, it transformed into self-assembled nanotubes under suitable conditions (Graveland-Bikker & de Kruif, 2006).

Recently researchers indicated that in addition to α-La, some globular whey proteins such as β-lg and BSA also self-assembled into fibrillar structures at elevated temperatures and low pH values. These nanofibrils had a characteristic diameter of approx. 5 nm and range up to 15 μm in length. Protease enzymes from *Bacillus licheniformis* were found to partially hydrolyze α-La molecules, and immediately afterward hydrolyzed protein was exposed to calcium ions. Consequently, the formation of linear nanotubes was observed (Graveland-Bikker & de Kruif, 2006; Sadeghi, Kalbasi et al., 2013).

All of the major physicochemical characteristics of α-La nanotubes render this component a promising encapsulation agent (Gouin, 2004). First, since α-La is a milk protein, its utilization in the food industry is already quite common (Rajagopal & Schneider, 2004). Much similar to the casein micelle, which is a promising natural nanovehicle for delivering hydrophobic bioactives including probiotics, self-assembled α-La nanotubes could play similar roles (Augustin & Hemar, 2009).

Halloysite ($Al_2Si_2O_5(OH)_4 \cdot 2H_2O$) is a naturally occurring material and has a hollow tubular structure, so it represents a two-layered aluminosilicate structure. Halloysite tubes sizes range between 0.5–5 μm in length and 15–100 nm in width. Among other nanotubes, halloysite nanotubes have some advantages such as natural occurrence and nonhazardous production processes. Furthermore, loading of bioactives is fairly high in this system (Vergaro, Lvov, & Leporatti, 2012). Levis and Deasy (2003) indicated that halloysite nanotubes displayed slow release of hydrophilic drugs such as diltiazemhydrochloride and propanol hydrochloride.

1.7 OTHER NANOENCAPSULATION TECHNIQUES

In this section, other techniques that can be utilized for the formation of bioactive-loaded nanocarriers or NPs are briefly reviewed.

1.7.1 Nanocrystals

Transfer of materials into the nanodimension changes many properties. The relevant properties of nanocrystals to increase oral bioavailability are (Keck & Müller, 2008) as follows:

1. increase in saturation solubility C_s;
2. increase in dissolution velocity d*c*/d*t* (*c*—concentration, *t*—time); and
3. adhesiveness to surfaces (e.g., biological membranes)

The increase in saturation solubility is due to an increased dissolution pressure of the nanocrystals. In the equilibrium of saturation solubility, the number of molecules dissolving from crystals is equal to the number that recrystallizes. In lower nanodimension, the increased dissolution pressure shifts the equilibrium to the dissolved molecules. The dissolution velocity is proportional to the surface area *A*, the surface area increases by an order of magnitude when moving, e.g., from a 50-μm crystal to a 5-μm crystal (micronized powder) and finally to a 500-nm nanocrystal (Romero, Brysch, Keck, & Muller, 2017).

The adhesiveness of particles to surfaces increases with decreasing particle size. This can be explained by the increase in contact area between the particles and the respective surface (Stieß, 2013). The adhesion of nanocrystals to membrane surfaces increases the drug permeation through these membranes. As nanocrystals move randomly around in the gastrointestinal content, there is a statistical probability that they will come in contact with the GIT wall and adhere to it. This effect increases with the decrease in particle size, mucoadhesion of smaller particles is higher (Ponchel, Montisci, Dembri, Durrer, & Duchêne, 1997).

To maintain their bioavailability enhancing effect, the nanocrystals need to be physically stable in the drinks, i.e., they should not aggregate to micrometer-sized aggregates. Nanocrystals in functional drinks are from the technical point a "suspension," and as such, they are subjected to destabilizing effects typical within a suspension. The nanocrystals are stabilized by high zeta-potential (mV) absolute values, whereby equally charged particles repel each other and stay separated. In addition, they are stabilized by the surrounding stabilizer layer (Romero, Brysch et al., 2017). Destabilization is mainly caused by electrolytes, preservatives (in case they are electrolytes), and osmotic/dehydrating effects of additives (Ravichandran, 2013).

1.7.1.1 Encapsulated (Coated) Nanocrystals

In case of nanocrystals of bioactives which are prone to degradation due to environmental/processing conditions, e.g., pH, oxygen, light, temperature, they can be coated by a protective polymer layer. This is also applicable for nanocrystals of bioactives which are sensible to the different conditions in the GI tract after ingestion, e.g., degradation in the acidic environment of the stomach (Kakran, Shegokar et al., 2012).

Nanocrystals start dissolving immediately when brought in contact with unsaturated fluids. This can happen after ingestion or even during the manufacturing of the final food/nutraceutic product. As an example, for nanocrystals for food/nutraceutics products, the respective site of absorption in the gastrointestinal tract has not yet been reached. Dissolved nanocrystals cannot adhere any more to the mucosa of the absorption site, creating a high concentration gradient. Too early dissolution will lead to a distribution of dissolved molecules across the content of the GIT, the reduction in the concentration gradient will lead subsequently to a reduced bioavailability (Romero, Brysch et al., 2017). Also, if the bioactive is absorbed in the intestine, degradation in the stomach will reduce bioavailability. In case of nanocrystals in food products, dissolving already before ingestion, e.g., functional drinks, should be avoided as much as possible and delayed until the nanocrystals have reached the GIT.

The solution for this is coating the nanocrystals with polymers. A full range of various coating polymers is available, e.g., well known from pharma polymethacrylate copolymers (Eudragit series, Evonik). Polylactides or poly (lactic-*co*-glycolic) acid copolymers are also an option. It appears also feasible to use water swellable polymers (hydrocolloids) which generate a viscous coat after getting in contact with water, slowing down the nanocrystal dissolution process. Normally these are all preformed, but it is also possible to generate coatings from the monomers (analogous to producing latex particles). The monomers are dissolved in the water phase of the nanosuspension, and coating is performed after initiation of the polymerization process (Damge, Vranckx, Balschmidt, & Couvreur, 1997). Coating with polymers can also change their surface properties, from hydrophobic to hydrophilic, when using amphiphilic coatings (Pellegrino, Manna et al., 2004).

The coating of single nanocrystal is often a more tedious and costly process. A smarter and more cost effective solution is the encapsulation in larger units, more or less "containers" for nanocrystals—the "container approach." Such containers are typically pellets. These are useful, e.g., to protect the nanocrystals until they are incorporated into the final composition of the product being produced or, in case of nutraceuticals, they can be loaded into, e.g., hard gelatin capsules. The nanocrystals are admixed to an aqueous extrusion mass and the pellets are coated with a polymer dissolving in specific conditions, e.g., acid pH of the stomach or more alkaline pH in the intestine (Romero, Brysch et al., 2017). Differently coated pellets can also be mixed to generate continuous release over a longer part of the GIT. Eudragits are again polymers of first choice in many cases. In case of nutraceutics, the largest "container" is a tablet, which shows the appropriate coating, e.g., enteric coating. A simpler solution is using multilayered tablets, with layers dissolving in different parts or velocities, e.g., fast dissolution (=burst release for, fast onset of action) and then followed by slow release.

1.7.1.2 Preparation Methods

Nanocrystals can be produced by bottom-up methods (starting from the molecule and growing it to nanocrystals through precipitation), and by top-down methods (by reducing the size of micrometer crystals to the nanodimension) (Müller, Gohla, & Keck, 2011). The classical precipitation process consists addition of a nonsolvent to the compound solution to obtain amorphous or crystalline NPs. Another precipitation method is the NanoMorph technology (Auweter, Bohn et al., 2002) to produce amorphous NPs. An O/W emulsion is produced and after lyophilization, amorphous NPs are obtained. It is used to produce carotenoid NPs for the food industry, e.g., locarotin, lucantin.

Top-down methods are the most relevant for the industrial production of nanocrystals, as through bead milling, HPH, and the combination of these processes. All of these are wet-milling processes (Romero, Brysch et al., 2017).

1.7.1.2.1 The Bead-Milling Process

The powder is dispersed in a stabilizer solution, and a macrosuspension produced. This suspension is passed several times through a bead mill.

The milling chamber is filled with fine milling pearls, typical sizes are between 0.2 and 0.5 mm of diameter. The crystals are ground between the moving beads, and a nanosuspension is obtained (=suspension of nanocrystals in a liquid phase).

1.7.1.2.2 High-Pressure Homogenization

The process is based on passing a coarse presuspension (μm-sized) at a high velocity through a very tiny gap (e.g., 5 μm) at high pressures of typically 200–600 bar (2×10^4–6×10^4 kPa). It may also be shown that a suspension can be effectively diminished to the nanosize if it is passed 5–20 times through a high-pressure homogenizer, typical pressure of 1500 bar (15×10^4 kPa) (i.e., clearly higher than typically used for emulsions).

1.7.1.2.3 Combination Technologies

It typically includes two subsequent steps (Salazar, Müler, & Möschwitzer, 2014). An example of this is the combination of precipitation and high-energy input (e.g., HPH). Also, it has been reported that bead milling and subsequent HPH leads to nanocrystals that have a higher physical stability (Petersen, 2006).

1.7.2 Dendrimeters

The dendrimer is a symmetric and spherical 3D polymeric assembly of molecules which is highly monodisperse and highly branched (Kavyani, Amjad-Iranagh, Dadvar, & Modarress, 2016). Structure of dendrimers consists of three major parts which are core, branches, and surface moieties. Because of monodisperse and multivalent character of dendrimers, they can be used for drug or bioactive compound delivery.

Synthesis of dendrimers require the selection of a suitable initiator that can be transformed into the reactive core of the assembly, where repetitive reactions take place in order to generate new highly branched dendrimers. Each reaction results in the generation of a larger dendrimeric assembly (Kavyani, Amjad-Iranagh, & Modarress, 2014). Dendrimer synthesis methods control all of the critical processing parameters such as size, shape, flexibility, and topology. Dendrimers can be synthesized by three different methods which are named as divergent, convergent, and double-stage convergent methods (Crespo et al., 2005).

In the divergent approach method, dendrimer widens outward from its core toward periphery. Reaction steps can be modified to produce desirable size. Another method for dendrimer synthesis is convergent approach which is the opposite of the divergent. In convergent method, dendrimer grows from outside to inside. Formation of denrimer starts in periphery and proceeds to the core. The third method, namely the double stage convergent approach, is the combination of the first two methods. In this method, building blocks are formed by divergent method followed by convergent assembly.

Encapsulation of various molecules into dendrimers can be based on either covalent attachment to the particle periphery or coordination of the functional groups with ionic interactions (Crampton & Simanek, 2007).

1.8 CONCLUSIONS

Nanocapsules are promising applications, since they are ideal for the encapsulation of many different food bioactive ingredients such as antioxidants, antimicrobial agents, phenolic compounds, natural pigments, peptides, essential fatty acids, minerals, etc. Nutraceutical delivery system is an interdisciplinary field of studies applying knowledge from vast array of disciplines including biology, chemistry, pharmaceutical, food sciences in combined fields of engineering and technology. There is a broad range of nanosized encapsulation systems, most of them are still in the academic labs, only a very few made it to the market. A nanosized delivery system is scientifically defined in the food and pharmaceutical area as particles with a size of a few nm to just below 1000 nm (= 1 μm). It should not be mixed up with the legal definition of NPs/nanomaterial for labeling consumer products, e.g., labeling of cosmetic ingredients in the European Union with "nano" are necessary when more than 50% (by number distribution) of the particles have sizes below 100 nm. It should be mentioned that selection of a nanoencapsulation technology depends on several parameters, such as physicochemical features, required particle size, release type, delivery method, process cost, etc. Although there are many different nanoencapsulation technologies applied in the literature, some of them can be utilized in the food industry due to regulatory and economic issues. One major gap on applying nanoencapsulation technologies for the food products is research activities and studies on technological improvement of food products rather than nutraceutical delivery. In fact, encapsulation experts in the food industry should focus on improving the shelf life and quality of food products by micro/nanoencapsulated bioactive ingredients. For example, retarding the oxidation of oils and fats, improving the organoleptic properties of foods, replacing synthetic preservatives with natural encapsulated antimicrobials, and similar subjects.

REFERENCES

Abbasi, A., Emam-Djomeh, Z., Mousavi, M. A. E., & Davoodi, D. (2014). Stability of vitamin D3 encapsulated in nanoparticles of whey protein isolate. *Food Chemistry*, *143*, 379–383.

Abd El-Salam, M. H., & El-Shibiny, S. (2012). Formation and potential uses of milk proteins as nano delivery vehicles for nutraceuticals: A review. *International Journal of Dairy Technology*, *65*(1), 13–21.

Aboalnaja, K. O., Yaghmoor, S., Kumosani, T. A., & McClements, D. J. (2016). Utilization of nanoemulsions to enhance bioactivity of pharmaceuticals, supplements, and nutraceuticals: Nanoemulsion delivery systems and nanoemulsion excipient systems. *Expert Opinion on Drug Delivery*, *13*(9), 1327–1336.

Abreu, F. O. M. S., Oliveira, E. F., Paula, H. C. B., & de Paula, R. C. M. (2012). Chitosan/cashew gum nanogels for essential oil encapsulation. *Carbohydrate Polymers*, *89*(4), 1277–1282.

Aditya, N. P., Aditya, S., Yang, H., Kim, H. W., Park, S. O., & Ko, S. (2015). Co-delivery of hydrophobic curcumin and hydrophilic catechin by a water-in-oil-in-water double emulsion. *Food Chemistry*, *173*, 7–13.

Alexander, M., Lopez, A. A., Fang, Y., & Corredig, M. (2012). Incorporation of phytosterols in soy phospholipids nanoliposomes: Encapsulation efficiency and stability. *LWT—Food Science and Technology*, *47*(2), 427–436.

Alonso, A., Villena, A., & Goñi, F. M. (1981). Lysis and reassembly of sonicated lecithin vesicles in the presence of Triton X-100. *FEBS Letters*, *123*(2), 200–204.

Anton, N., Benoit, J.-P., & Saulnier, P. (2008). Design and production of nanoparticles formulated from nano-emulsion templates—A review. *Journal of Controlled Release*, *128*(3), 185–199.

Anu Bhushani, J., & Anandharamakrishnan, C. (2014). Electrospinning and electrospraying techniques: Potential food based applications. *Trends in Food Science & Technology*, *38*(1), 21–33.

Arpagaus, C. (2011). Nano Spray Dryer B-90: Literature review and applications. *Buchi*, *63*, 8.

Arpagaus, C. (2012). A novel laboratory-scale spray dryer to produce nanoparticles. *Drying Technology*, *30*, 1113–1121.

Arpagaus, C., John, P., Collenberg, A., & Rütti, D. (2017). Nano-capsule formation by nano spray drying. In S. M. Jafari (Ed.), *Nanoencapsulation technologies for the food and nutraceutical industries* (pp. 346–402). Amsterdam: Elsevier.

Arranz, E., Santoyo, S., Jaime, L., Fornari, T., Reglero, G., Guri, A., & Corredig, M. (2015). Improved bioavailability of supercritical rosemary extract through encapsulation in different delivery systems after in vitro digestion. *Food Digestion: Research and Current Opinion*, *6* (1–3), 30–37.

Arroyo-Maya, I. J., & McClements, D. J. (2015). Biopolymer nanoparticles as potential delivery systems for anthocyanins: Fabrication and properties. *Food Research International*, *69*, 1–8.

Assadpour, E., Jafari, S.-M., & Maghsoudlou, Y. (2017). Evaluation of folic acid release from spray dried powder particles of pectin–whey protein nano-capsules. *International Journal of Biological Macromolecules*, *95*, 238–247.

Assadpour, E., Maghsoudlou, Y., Jafari, S.-M., Ghorbani, M., & Aalami, M. (2016a). Evaluation of Folic acid nano-encapsulation by double emulsions. *Food and Bioprocess Technology*, *9*(12), 2024–2032.

Assadpour, E., Maghsoudlou, Y., Jafari, S.-M., Ghorbani, M., & Aalami, M. (2016b). Optimization of folic acid nano-emulsification and encapsulation by maltodextrin–whey protein double emulsions. *International Journal of Biological Macromolecules*, *86*, 197–207.

Augustin, M. A., & Hemar, Y. (2009). Nano- and micro-structured assemblies for encapsulation of food ingredients. *Chemical Society Reviews*, *38*(4), 902–912.

Auweter, H., Bohn, H., Heger, R., Horn, D., Siegel, B., & Siemensmeyer, K. (2002). Precipitated water-insoluble colorants in colloid disperse form, US Patent 6494924.

Bagheri, L., Madadlou, A., Yarmand, M., & Mousavi, M. E. (2013). Nanoencapsulation of date palm pit extract in whey protein particles generated via desolvation method. *Food Research International*, *51*(2), 866–871.

Barhate, R. S., Loong, C. K., & Ramakrishna, S. (2006). Preparation and characterization of nanofibrous filtering media. *Journal of Membrane Science, 283*(1–2), 209–218.

Beloqui, A., Solinís, M. Á., Rodríguez-Gascón, A., Almeida, A. J., & Préat, V. (2016). Nanostructured lipid carriers: Promising drug delivery systems for future clinics. *Nanomedicine: Nanotechnology, Biology and Medicine, 12*(1), 143–161.

Bharali, D. J., Sahoo, S. K., Mozumdar, S., & Maitra, A. (2003). Cross-linked polyvinylpyrrolidone nanoparticles: A potential carrier for hydrophilic drugs. *Journal of Colloid and Interface Science, 258*(2), 415–423.

Bhardwaj, N., & Kundu, S. C. (2010). Electrospinning: A fascinating fiber fabrication technique. *Biotechnology Advances, 28*(3), 325–347.

Bilensoy, E., & Hincal, A. A. (2008). Cyclodextrin-based nanomaterials in pharmaceutical field. In S. C. Gad (Ed.), Pharmaceutical manufacturing handbook: Production and processes (pp. 1225–1247). NJ, USA: John Wiley & Sons.

Bock, N., Woodruff, M. A., Hutmacher, D. W., & Dargaville, T. R. (2011). Electrospraying, a reproducible method for production of polymeric microspheres for biomedical applications. *Polymers, 3*(1), 131–149.

Borel, T., & Sabliov, C. M. (2014). Nanodelivery of bioactive components for food applications: Types of delivery systems, properties, and their effect on ADME profiles and toxicity of nanoparticles. *Annual Review of Food Science and Technology, 5*, 197–213.

Bourbon, A. I., Cerqueira, M. A., & Vicente, A. A. (2016). Encapsulation and controlled release of bioactive compounds in lactoferrin-glycomacropeptide nanohydrogels: Curcumin and caffeine as model compounds. *Journal of Food Engineering, 180*, 110–119.

Bronich, T. K., Vinogradov, S. V., & Kabanov, A. V. (2001). Interaction of nanosized copolymer networks with oppositely charged amphiphilic molecules. *Nano Letters, 1*(10), 535–540.

Bryła, A., Lewandowicz, G., & Juzwa, W. (2015). Encapsulation of elderberry extract into phospholipid nanoparticles. *Journal of Food Engineering, 167*, 189–195.

Bürki, K., Jeon, I., Arpagaus, C., & Betz, G. (2011). New insights into respirable protein powder preparation using a nano spray dryer. *International Journal of Pharmaceutics, 408*, 248–256.

Cadena, P. G., Pereira, M. A., Cordeiro, R. B. S., Cavalcanti, I. M. F., Barros Neto, B., Pimentel, Md. C. C. B., … Santos-Magalhães, N. S. (2013). Nanoencapsulation of quercetin and resveratrol into elastic liposomes. *Biochimica et Biophysica Acta (BBA)—Biomembranes, 1828*(2), 309–316.

Carbinatto, F. M., Ribeiro, T. S., Colnago, L. A., Evangelista, R. C., & Cury, B. S. F. (2016). Preparation and characterization of amylose inclusion complexes for drug delivery applications. *Journal of Pharmaceutical Sciences, 105*(1), 231–241.

Chakraborty, S., Liao, I. C., Adler, A., & Leong, K. W. (2009). Electrohydrodynamics: A facile technique to fabricate drug delivery systems. *Advanced Drug Delivery Reviews, 61*(12), 1043–1054.

Chen, H.-M., & Yu, D.-G. (2010). An elevated temperature electrospinning process for preparing acyclovir-loaded PAN ultrafine fibers. *Journal of Materials Processing Technology, 210*(12), 1551–1555.

Chen, H., Zhang, Y., & Zhong, Q. (2015). Physical and antimicrobial properties of spray-dried zein–casein nanocapsules with co-encapsulated eugenol and thymol. *Journal of Food Engineering, 144*, 93–102.

Chen, H., & Zhong, Q. (2015). A novel method of preparing stable zein nanoparticle dispersions for encapsulation of peppermint oil. *Food Hydrocolloids, 43*, 593–602.

Chia-Lang, F., Saleh, A. A.-S., & Jia-You, F. (2013). Nanostructured lipid carriers (NLCs) for drug delivery and targeting. *Recent Patents on Nanotechnology*, *7*(1), 41–55.

Choi, M.-J., Ruktanonchai, U., Min, S.-G., Chun, J.-Y., & Soottitantawat, A. (2010). Physical characteristics of fish oil encapsulated by β-cyclodextrin using an aggregation method or polycaprolactone using an emulsion–diffusion method. *Food Chemistry*, *119*(4), 1694–1703.

Coester, C. J., Langer, K., van Briesen, H., & Kreuter, J. (2000). Gelatin nanoparticles by two step desolvation a new preparation method, surface modifications and cell uptake. *Journal of Microencapsulation*, *17*(2), 187–193.

Cohen, R., Orlova, Y., Kovalev, M., Ungar, Y., & Shimoni, E. (2008). Structural and functional properties of amylose complexes with genistein. *Journal of Agricultural and Food Chemistry*, *56*(11), 4212–4218.

Crampton, H. L., & Simanek, E. E. (2007). Dendrimers as drug delivery vehicles: Non-covalent interactions of bioactive compounds with dendrimers. *Polymer International*, *56*, 489–496.

Crespo, L., Sanclimens, G., Pons, M., Giralt, E., Royo, M., & Albericio, F. (2005). Peptide and amide bond-containing dendrimers. *Chemical Reviews*, *105*, 1663–1681.

Damge, C., Vranckx, H., Balschmidt, P., & Couvreur, P. (1997). Poly(alkyl cyanoacrylate) nanospheres for oral administration of insulin. *Journal of Pharmaceutical Sciences*, *86*(12), 1403–1409.

Das, S., Ng, W. K., & Tan, R. B. H. (2012). Are nanostructured lipid carriers (NLCs) better than solid lipid nanoparticles (SLNs): Development, characterizations and comparative evaluations of clotrimazole-loaded SLNs and NLCs. *European Journal of Pharmaceutical Sciences*, *47*(1), 139–151.

Dasgupta, N., Ranjan, S., Mundra, S., Ramalingam, C., & Kumar, A. (2016). Fabrication of food grade vitamin E nanoemulsion by low energy approach, characterization and its application. *International Journal of Food Properties*, *19*(3), 700–708.

da Silva, M. M., Nora, L., Cantillano, R. F. F., Paese, K., Guterres, S. S., Pohlmann, A. R., ... Rios, Ad. O. (2016). The production, characterization, and the stability of carotenoids loaded in lipid-core nanocapsules. *Food and Bioprocess Technology*, *9*(7), 1148–1158.

de Assis, L. M., Machado, A. R., da Motta, A. S., Costa, J. A. V., & de Souza-Soares, L. A. (2014). Development and characterization of nanovesicles containing phenolic compounds of microalgae *Spirulina* strain LEB-18 and *Chlorella pyrenoidosa*. *Advances in Materials Physics and Chemistry*, *4*, 6–12.

de Carvalho, S. M., Noronha, C. M., Floriani, C. L., Lino, R. C., Rocha, G., Bellettini, I. C., ... Barreto, P. L. M. (2013). Optimization of α-tocopherol loaded solid lipid nanoparticles by central composite design. *Industrial Crops and Products*, *49*, 278–285.

Del Valle, E. M. M. (2004). Cyclodextrins and their uses: A review. *Process Biochemistry*, *39*, 1033–1046.

Demirci, M., Çağlar, M. Y., Çakır, B., & Gulseren, I. (2017). Encapsulation by nano-liposomes. In S. M. Jafari (Ed.), *Nanoencapsulation technologies for the food and nutraceutical industries* (pp. 74–113). Amsterdam: Elsevier.

De Vrieze, S., Van Camp, T., Nelvig, A., Hagström, B., Westbroek, P., & De Clerck, K. (2009). The effect of temperature and humidity on electrospinning. *Journal of Materials Science*, *44*(5), 1357–1362.

Dextraan, B. (2007). Biodegradable dextran and PEG micro-and nanogels: Physicochemical properties and sustained drug delivery features. Doctoral dissertation, Ghent University.

Dickinson, E. (2011). Double emulsions stabilized by food biopolymers. *Food Biophysics*, *6*(1), 1–11.

Donsì, F., Annunziata, M., Sessa, M., & Ferrari, G. (2011). Nanoencapsulation of essential oils to enhance their antimicrobial activity in foods. *LWT Food Science and Technology*, *44*(9), 1908–1914.

Donsì, F., Sessa, M., Mediouni, H., Mgaidi, A., & Ferrari, G. (2011). Encapsulation of bioactive compounds in nanoemulsion- based delivery systems. *Procedia Food Science*, *1*, 1666–1671.

Doshi, J., & Reneker, D. H. (1995). Electrospinning process and applications of electrospun fibers. *Journal of Electrostatics*, *35*(2–3), 151–160.

Ducel, V., Pouliquen, D., Richard, J., & Boury, F. (2008). ^{1}H NMR relaxation studies of protein–polysaccharide mixtures. *International Journal of Biological Macromolecules*, *43*(4), 359–366.

Duchêne, D. (2011). Cyclodextrins and their inclusion complexes. In E. Bilensoy (Ed.), *Cyclodextrins in Pharmaceutics, Cosmetics, and Biomedicine: Current and Future Industrial Applications*. Hoboken, NJ: Wiley.

Duong, H. T., Marquis, C. P., Whittaker, M., Davis, T. P., & Boyer, C. (2011). Acid degradable and biocompatible polymeric nanoparticles for the potential codelivery of therapeutic agents. *Macromolecules*, *44*(20), 8008–8019.

Eda, G., Liu, J., & Shivkumar, S. (2007). Solvent effects on jet evolution during electrospinning of semi-dilute polystyrene solutions. *European Polymer Journal*, *43*(4), 1154–1167.

Elzoghby, A. O., Abo El-Fotoh, W. S., & Elgindy, N. A. (2011). Casein-based formulations as promising controlled release drug delivery systems. *Journal of Controlled Release*, *153*(3), 206–216.

Enayati, M., Chang, M.-W., Bragman, F., Edirisinghe, M., & Stride, E. (2011). Electrohydrodynamic preparation of particles, capsules and bubbles for biomedical engineering applications. *Colloids and Surfaces A: Physicochemical and Engineering Aspects*, *382*(1–3), 154–164.

Esfanjani, A. F., Jafari, S. M., Assadpoor, E., & Mohammadi, A. (2015). Nano-encapsulation of saffron extract through double-layered multiple emulsions of pectin and whey protein concentrate. *Journal of Food Engineering*, *165*, 149–155.

Ezhilarasi, P. N., Karthik, P., Chhanwal, N., & Anandharamakrishnan, C. (2013). Nanoencapsulation techniques for food bioactive components: A review. *Food and Bioprocess Technology*, *6*(3), 628–647.

Fang, Z., Comino, P. R., & Bhandari, B. (2013). Effect of encapsulation of d-limonene on the moisture adsorption property of β-cyclodextrin. *LWT—Food Science and Technology*, *51*(1), 164–169.

Faridi Esfanjani, A., & Jafari, S. M. (2016). Biopolymer nano-particles and natural nano-carriers for nano-encapsulation of phenolic compounds. *Colloids and Surfaces B: Biointerfaces*, *146*, 532–543.

Faridi Esfanjani, A., Jafari, S. M., & Assadpour, E. (2017). Preparation of a multiple emulsion based on pectin-whey protein complex for encapsulation of saffron extract nanodroplets. *Food Chemistry*, *221*, 1962–1969.

Fathi, M., Martín, Á., & McClements, D. J. (2014). Nanoencapsulation of food ingredients using carbohydrate based delivery systems. *Trends in Food Science & Technology*, *39*(1), 18–39.

Fathi, M., Mozafari, M. R., & Mohebbi, M. (2012). Nanoencapsulation of food ingredients using lipid based delivery systems. *Trends in Food Science & Technology*, *23*(1), 13–27.

Feng, A. L., Boraey, M. A., Gwin, M. A., Finlay, P. R., Kuehl, P. J., & Vehring, R. (2011). Mechanistic models facilitate efficient development of leucine containing microparticles for pulmonary drug delivery. *International Journal of Pharmaceutics*, *409*, 156–163.

Fong, H., Chun, I., & Reneker, D. H. (1999). Beaded nanofibers formed during electrospinning. *Polymer, 40*(16), 4585–4592.

Frenot, A., & Chronakis, I. S. (2003). Polymer nanofibers assembled by electrospinning. *Current Opinion in Colloid & Interface Science, 8*(1), 64–75.

Gan, D., & Lyon, L. A. (2001). Tunable swelling kinetics in core–shell hydrogel nanoparticles. *Journal of the American Chemical Society, 123*(31), 7511–7517.

Gan, Q., & Wang, T. (2007). Chitosan nanoparticle as protein delivery carrier—Systematic examination of fabrication conditions for efficient loading and release. *Colloids and Surfaces B: Biointerfaces, 59*(1), 24–34.

Gao, M., Kokini, J., Fern, L., & Mejia, O.M. (2014). Formation of lactalbumin nanoparticles by desolvation method. In *The Summer Undergraduate Research Fellowship (SURF) Symposium*, USA.

Gelders, G. G., Duyck, J. P., Goesaert, H., & Delcour, J. A. (2005). Enzyme and acid resistance of amylose-lipid complexes differing in amylose chain length, lipid and complexation temperature. *Carbohydrate Polymers, 60*(3), 379–389.

GharibZahedi, S. M. T., & Jafari, S. M. (2017). Nano-capsule formation by cyclodextrins. In S. M. Jafari (Ed.), *Nanoencapsulation technologies for the food and nutraceutical industries* (pp. 187–262). Amsterdam: Elsevier.

Ghasemi, S., & Abbasi, S. (2014). Formation of natural casein micelle nanocapsule by means of pH changes and ultrasound. *Food Hydrocolloids, 42*(Pt. 1), 42–47.

Ghayempour, S., & Mortazavi, S. M. (2015). Preparation and investigation of sodium alginate nanocapsules by different microemulsification devices. *Journal of Applied Polymer Science, 132*(17), n/a–n/a.

Ghorani, B., Alehosseini, A., & Tucker, N. (2017). Nano-capsule formation by electrospinning. In S. M. Jafari (Ed.), *Nanoencapsulation technologies for the food and nutraceutical industries* (pp. 263–319). Amsterdam: Elsevier.

Ghorani, B., & Tucker, N. (2015). Fundamentals of electrospinning as a novel delivery vehicle for bioactive compounds in food nanotechnology. *Food Hydrocolloids, 51*, 227–240.

Ghorbanzade, T., Jafari, S. M., Akhavan, S., & Hadavi, R. (2017). Nano-encapsulation of fish oil in nano-liposomes and its application in fortification of yogurt. *Food Chemistry, 216*, 146–152.

Gokmen, V., Mogol, B. e A., Lumaga, R. B., Fogliano, V., Kaplun, Z., & Shimoni, E. (2011). Development of functional bread containing nanoencapsulated omega-3 fatty acids. *Journal of Food Engineering, 105*(4), 585–591.

Gomez-Estaca, J., Balaguer, M. P., Gavara, R., & Hernandez-Munoz, P. (2012). Formation of zein nanoparticles by electrohydrodynamic atomization: Effect of the main processing variables and suitability for encapsulating the food coloring and active ingredient curcumin. *Food Hydrocolloids, 28*(1), 82–91.

Gouin, S. (2004). Microencapsulation: Industrial appraisal of existing Technologies andtrends. *Trends in Food Science & Technology, 15*(7), 330–347.

Gonçalves, C., Pereira, P., Schellenberg, P., Coutinho, P. J., & Gama, F. M. (2012). Self-assembled dextrin nanogel as curcumin delivery system. *Journal of Biomaterials and Nanobiotechnology, 3*, 178–184.

Graveland-Bikker, J. F., & de Kruif, C. G. (2006). Unique milk protein based nanotubes: Food and nanotechnology meet. *Trends in Food Science & Technology, 17*(5), 196–203.

Gulseren, I., Fang, Y., & Corredig, M. (2012a). Complexation of high methoxyl pectin with ethanol desolvated whey protein nanoparticles: Physico-chemical properties and encapsulation behaviour. *Food & Function, 3*(8), 859–866.

Gülseren, İ., Fang, Y., & Corredig, M. (2012b). Zinc incorporation capacity of whey protein nanoparticles prepared with desolvation with ethanol. *Food Chemistry, 135*(2), 770–774.

Guowei, D., Adriane, K., Chen, X., Jie, C., & Yinfeng, L. (2007). PVP magnetic nanospheres: Biocompatibility, in vitro and in vivo bleomycin release. *International Journal of Pharmaceutics, 328*(1), 78–85.

Gupta, A., Eral, H. B., Hatton, T. A., & Doyle, P. S. (2016). Nanoemulsions: Formation, properties and applications. *Soft Matter, 12*(11), 2826–2841.

Hădărugă, Nicoleta G., Hădărugă, Daniel I., Bandur, Geza N., Rivis, Adrian, Pârvu, Dorel, & Lupea, A. X. (2008). Protection and controlled release of fatty acids and essential oils by nanoencapsulation in cyclodextrins (a review). *Journal of Agroalimentary Processes and Technologies, 14*(2), 381–389.

Haggag, Y. A., & Faheem, A. M. (2015). Evaluation of nano spray drying as a method for drying and formulation of therapeutic peptides and proteins. *Frontiers in Pharmacology, 6*, 140.

Haham, M., Ish-Shalom, S., Nodelman, M., Duek, I., Segal, E., Kustanovich, M., & Livney, Y. D. (2012). Stability and bioavailability of vitamin D nanoencapsulated in casein micelles. *Food & Function, 3*(7), 737–744.

Haidong, L., Fang, Y., Zhihong, T., & Changle, R. (2011). Study on preparation of β-cyclodextrin encapsulation tea extract. *International Journal of Biological Macromolecules, 49*(4), 561–566.

Hao, S., Wang, Y., Wang, B., Deng, J., Liu, X., & Liu, J. (2013). Rapid preparation of pH-sensitive polymeric nanoparticle with high loading capacity using electrospray for oral drug delivery. *Materials Science and Engineering: C, 33*(8), 4562–4567.

Haratifar, S., & Guri, A. (2017). Nano-capsule formation by caseins. In S. M. Jafari (Ed.), *Nanoencapsulation technologies for the food and nutraceutical industries* (pp. 139–164). Amsterdam: Elsevier.

Hasni, I., Bourassa, P., Hamdani, S., Samson, G., Carpentier, R., & Tajmir-Riahi, H.-A. (2011). Interaction of milk α- and β-caseins with tea polyphenols. *Food Chemistry, 126*(2), 630–639.

Hategekimana, J., Masamba, K. G., Ma, J., & Zhong, F. (2015). Encapsulation of vitamin E: Effect of physicochemical properties of wall material on retention and stability. *Carbohydrate Polymers, 124*, 172–179.

Hemar, Y., Cheng, L., Oliver, C., Sanguansri, L., & Augustin, M. (2010). Encapsulation of resveratrol using water-in-oil-in-water double emulsions. *Food Biophysics, 5*(2), 120–127.

Hill, L. E., Gomes, C., & Taylor, T. M. (2013). Characterization of beta-cyclodextrin inclusion complexes containing essential oils (trans-cinnamaldehyde, eugenol, cinnamon bark, and clove bud extracts) for antimicrobial delivery applications. *LWT—Food Science and Technology, 51*(1), 86–93.

Hosseini, S. M. H., Emam-Djomeh, Z., Sabatino, P., & Van der Meeren, P. (2015). Nanocomplexes arising from protein–polysaccharide electrostatic interaction as a promising carrier for nutraceutical compounds. *Food Hydrocolloids, 50*, 16–26.

Hosseini, S. M. H., Ghiasi, F., & Jahromi, M. (2017). Nano-capsule formation by complexation of biopolymers. In S. M. Jafari (Ed.), *Nanoencapsulation technologies for the food and nutraceutical industries* (pp. 447–492). Amsterdam: Elsevier.

Huang, G., Gao, J., Hu, Z., John, J. V. S., Ponder, B. C., & Moro, D. (2004). Controlled drug release from hydrogel nanoparticle networks. *Journal of Controlled Release, 94*(2), 303–311.

Huang, Z., Li, X., Zhang, T., Song, Y., She, Z., Li, J., & Deng, Y. (2014). Progress involving new techniques for liposome preparation. *Asian Journal of Pharmaceutical Sciences, 9*(4), 176–182.

Huang, Z.-M., Zhang, Y. Z., Kotaki, M., & Ramakrishna, S. (2003). A review on polymer nanofibers by electrospinning and their applications in nanocomposites. *Composites Science and Technology*, *63*(15), 2223–2253.

Ikigai, H., Nakae, T., Hara, Y., & Shimamura, T. (1993). Bactericidal catechins damage the lipid bilayer. *Biochimica et Biophysica Acta (BBA)—Biomembranes*, *1147*(1), 132–136.

Ilyasoglu, H., & El, S. N. (2014). Nanoencapsulation of EPA/DHA with sodium caseinate–gum arabic complex and its usage in the enrichment of fruit juice. *LWT—Food Science and Technology*, *56*(2), 461–468.

Jafari, S. M. (2017). An overview of nano-encapsulation techniques and their classification. In S. M. Jafari (Ed.), *Nanoencapsulation technologies for the food and nutraceutical industries* (pp. 1–34). Amsterdam: Elsevier.

Jafari, S. M., Assadpoor, E., He, Y., & Bhandari, B. (2008). Re-coalescence of emulsion droplets during high-energy emulsification. *Food Hydrocolloids*, *22*(7), 1191–1202.

Jafari, S. M., Fathi, M., & Mandala, I. (2015). *Emerging product formation. Food waste recovery: Processing technologies and industrial techniques* (pp. 293–317). Amsterdam: Elsevier Inc.

Jafari, S. M., He, Y., & Bhandari, B. (2006). Nano-emulsion production by sonication and microfluidization—A comparison. *International Journal of Food Properties*, *9*(3), 475–485.

Jafari, S. M., He, Y., & Bhandari, B. (2007a). Optimization of nano-emulsions production by microfluidization. *European Food Research and Technology*, *225*(5–6), 733–741.

Jafari, S. M., He, Y., & Bhandari, B. (2007b). Production of sub-micron emulsions by ultrasound and microfluidization techniques. *Journal of Food Engineering*, *82*(4), 478–488.

Jafari, S. M., Mahdavi-Khazaei, K., & Hemmati-Kakhki, A. (2016). Microencapsulation of saffron petal anthocyanins with cress seed gum compared with Arabic gum through freeze drying. *Carbohydrate Polymers*, *140*, 20–25.

Jafari, S. M., & McClements, D. J. (2017). *Nanotechnology approaches for increasing nutrient bioavailability. Advances in food and nutrition research*. London: Academic Press.

Jafari, S. M., Paximada, P., Mandala, I., Assadpour, E., & Mehrnia, M.-A. (2017). Encapsulation by nano-emulsions. In S. M. Jafari (Ed.), *Nanoencapsulation technologies for the food and nutraceutical industries* (pp. 35–73). Amsterdam: Elsevier.

Jain, E., Scott, K. M., Zustiak, S. P., & Sell, S. A. (2015). Fabrication of polyethylene glycol-based hydrogel microspheres through electrospraying. *Macromolecular Materials and Engineering*, *300*(8), 823–835.

Jaworek, A. (2007). Micro- and nanoparticle production by electrospraying. *Powder Technology*, *176*(1), 18–35.

Jiranuntakul, W., Sugiyama, S., Tsukamoto, K., Puttanlek, C., Rungsardthong, V., Puncha-arnon, S., & Uttapap, D. (2013). Nano-structure of heat–moisture treated waxy and normal starches. *Carbohydrate Polymers*, *97*(1), 1–8.

Jones, O. G., & McClements, D. J. (2011). Recent progress in biopolymer nanoparticle and microparticle formation by heat-treating electrostatic protein–polysaccharide complexes. *Advances in Colloid and Interface Science*, *167*(1–2), 49–62.

Jun, J. Y., Nguyen, H. H., Paik, S.-Y.-R., Chun, H. S., Kang, B.-C., & Ko, S. (2011). Preparation of size-controlled bovine serum albumin (BSA) nanoparticles by a modified desolvation method. *Food Chemistry*, *127*(4), 1892–1898.

Kakran, M., Shegokar, R., Sahoo, N. G., Al Shaal, L., Li, L., & Müller, R. H. (2012). Fabrication of quercetin nanocrystals: Comparison of different methods. *European Journal of Pharmaceutics and Biopharmaceutics*, *80*(1), 113–121.

Kao, Y. h, Chang, S. j, Chuang, C. W., Kuo, S. m, Chen, S. y, & Lu, R. t (2012). Preparation of bio-polymeric nanoparticles by electrostatic field system. *IET Micro & Nano Letters*, *7*(10), 997–1000.

Katouzian, I., & Jafari, S. M. (2016). Nano-encapsulation as a promising approach for targeted delivery and controlled release of vitamins. *Trends in Food Science & Technology*, *53*, 34–48.

Kavyani, S., Amjad-Iranagh, S., Dadvar, M., & Modarress, H. (2016). Hybrid dendrimers of PPI (core)–PAMAM(shell), a molecular dynamics simulation study. *The Journal of Physical Chemistry B*, *120*(36), 9564–9575.

Kavyani, S., Amjad-Iranagh, S., & Modarress, H. (2014). Aqueous poly(amidoamine) dendrimer G3 and G4 generations with several interior cores at pHs 5 and 7: a molecular dynamics simulation study. *The Journal of Physical Chemistry B*, *118*(12), 3257–3266.

Keck, C. M., & Müller, R. H. (2008). smartCrystals—Review of the second generation of drug nanocrystals. In V. P. Torchilin (Ed.), Nanoparticlulates as drug carriers (pp. 555–579). London: Imperial College Press.

Kim, H.-Y., Park, S. S., & Lim, S.-T. (2015). Preparation, characterization and utilization of starch nanoparticles. *Colloids and Surfaces B: Biointerfaces*, *126*, 607–620.

Kim, Y., Pourgholami, M. H., Morris, D. L., & Stenzel, M. H. (2011). Triggering the fast release of drugs from crosslinked micelles in an acidic environment. *Journal of Materials Chemistry*, *21*(34), 12777–12783.

Kimpel, F., & Schmitt, J. J. (2015). Review: Milk proteins as nanocarrier systems for hydrophobic nutraceuticals. *Journal of Food Science*, *80*(11), 2361–2366.

Kumari, A., Singla, R., Guliani, A., & Yadav, S. K. (2014). Nanoencapsulation for drug delivery. *EXCLI Journal*, *13*, 265–286.

Lalush, I., Bar, H., Zakaria, I., Eichler, S., & Shimoni, E. (2005). Utilization of amylose–lipid complexes as molecular nanocapsules for conjugated linoleic acid. *Biomacromolecules*, *6*(1), 121–130.

Leach, M. K., Feng, Z.-Q., Tuck, S. J., & Corey, J. M. (2011). Electrospinning fundamentals: Optimizing solution and apparatus parameters. *Journal of Visualized Experiments: JoVE*, *January*(47), 2494.

Lee, M.-Y., Min, S.-G., You, S.-K., Choi, M.-J., Hong, G.-P., & Chun, J.-Y. (2013). Effect of β-cyclodextrin on physical properties of nanocapsules manufactured by emulsion–diffusion method. *Journal of Food Engineering*, *119*(3), 588–594.

Lee, S. H., Heng, D., Ng, W. K., Chan, H.-K., & Tan, R. B. H. (2011). Nano spray drying: A novel method for preparing protein nanoparticles for protein therapy. *International Journal of Pharmaceutics*, *403*, 192–200.

Lertsutthiwong, P., Rojsitthisak, P., & Nimmannit, U. (2009). Preparation of turmeric oil-loaded chitosan-alginate biopolymeric nanocapsules. *Materials Science and Engineering: C*, *29*(3), 856–860.

Levis, S. R., & Deasy, P. B. (2003). Use of coated microtubular halloysite for the sustained release of diltiazem hydrochloride and propranolol hydrochloride. *International Journal of Pharmaceutics*, *253*, 145–157.

Li, J. K., Wang, N., & Wu, X. S. (1998). Poly(vinyl alcohol) nanoparticles prepared by freezing–thawing process for protein/peptide drug delivery. *Journal of Controlled Release*, *56*(1), 117–126.

Li, X., Anton, N., Arpagaus, C., Belleteix, F., & Vandamme, T. F. (2010). Nanoparticles by spray drying using innovative new technology: The Büchi Nano Spray Dryer B-90. *Journal of Controlled Release*, *147*(2), 304–310.

Li, Y., Du, W., Sun, G., & Wooley, K. L. (2008). pH-responsive shell cross-linked nanoparticles with hydrolytically labile cross-links. *Macromolecules*, *41*(18), 6605–6607.

Liu, W., Liu, W. L., Liu, M. C., Liu, J. H., Yang, S. B., Zheng, H. J., ... Song, X. Y. (2011). Medium-chain fatty acid nanoliposomes for easy energy supply. *Nutrition*, *27*(6), 700–706.

Livney, Y., Semo, E., Danino, D., & Kesselman, E. (2006). Nanoencapsulation of hydrophobic nutraceutical substances within casein micelles. In *XIVth international workshop on bioencapsulation*, Lausanne CH.

Livney, Y. D. (2010). Milk proteins as vehicles for bioactives. *Current Opinion in Colloid & Interface Science*, *15*(1–2), 73–83.

Livney, Y. D. (2015). Nanostructured delivery systems in food: Latest developments and potential future directions. *Current Opinion in Food Science*, *3*, 125–135.

Lohcharoenkal, W., Wang, L., Chen, Y. C., & Rojanasakul, Y. (2014). Protein nanoparticles as drug delivery carriers for cancer therapy. *BioMed Research International*, *2014*, 12.

López-Rubio, A., & Lagaron, J. M. (2012). Whey protein capsules obtained through electrospraying for the encapsulation of bioactives. *Innovative Food Science & Emerging Technologies*, *13*, 200–206.

Lu, Q., Li, D.-C., & Jiang, J.-G. (2011). Preparation of a tea polyphenol nanoliposome system and its physicochemical properties. *Journal of Agricultural and Food Chemistry*, *59*(24), 13004–13011.

Luo, Y., Pan, K., & Zhong, Q. (2015). Casein/pectin nanocomplexes as potential oral delivery vehicles. *International Journal of Pharmaceutics*, *486*(1–2), 59–68.

Mahdavee Khazaei, K., Jafari, S. M., Ghorbani, M., & Hemmati Kakhki, A. (2014). Application of maltodextrin and gum Arabic in microencapsulation of saffron petal's anthocyanins and evaluating their storage stability and color. *Carbohydrate Polymers*, *105*(1), 57–62.

Mahdavi, S. A., Jafari, S. M., Ghorbani, M., & Assadpoor, E. (2014). Spray-drying microencapsulation of anthocyanins by natural biopolymers: a review. *Drying Technology*, *32*(5), 509–518.

Marinopoulou, A., Kalogianni, E. P., & Raphaelides, S. N. (2016). Amylose-fatty acid inclusion complexes as examined by interfacial tension measurements. *Colloids and Surfaces B: Biointerfaces*, *137*, 133–137.

Matalanis, A., Jones, O. G., & McClements, D. J. (2011). Structured biopolymer-based delivery systems for encapsulation, protection, and release of lipophilic compounds. *Food Hydrocolloids*, *25*(8), 1865–1880.

Mazoochi, T., & Jabbari, V. (2011). Chitosan nanofibrous scaffold fabricated via electrospinning: The effect of processing parameters on the nanofiber morphology. *International Journal of Polymer Analysis and Characterization*, *16*(5), 277–289.

McClements, D. J. (2015). Nanoscale nutrient delivery systems for food applications: Improving bioactive dispersibility, stability, and bioavailability. *Journal of Food Science*, *80*(7), N1602–N1611.

Mehrnia, M.-A., Jafari, S.-M., Makhmal-Zadeh, B. S., & Maghsoudlou, Y. (2017). Rheological and release properties of double nano-emulsions containing crocin prepared with Angum gum, Arabic gum and whey protein. *Food Hydrocolloids*, *66*, 259–267.

Mehrnia, M. A., Jafari, S. M., Makhmal-Zadeh, B. S., & Maghsoudlou, Y. (2016). Crocin loaded nano-emulsions: Factors affecting emulsion properties in spontaneous emulsification. *International Journal of Biological Macromolecules*, *84*, 261–267.

Mercader-Ros, M. T., Lucas-Abellán, C., Fortea, M. I., Gabaldón, J. A., & Núñez-Delicado, E. (2010). Effect of HP-β-cyclodextrins complexation on the antioxidant activity of flavonols. *Food Chemistry*, *118*(3), 769–773.

Miastkowska, M. A., Banach, M., Pulit-Prociak, J., Sikora, E. S., Głogowska, A., & Zielina, M. (2017). Statistical analysis of optimal ultrasound emulsification parameters in thistle oil nanoemulsions. *Journal of Surfactants and Detergents*, *20*(1), 233–246.

Mo, X. M., Xu, C. Y., Kotaki, M., & Ramakrishna, S. (2004). Electrospun P(LLA-CL) nanofiber: A biomimetic extracellular matrix for smooth muscle cell and endothelial cell proliferation. *Biomaterials*, *25*(10), 1883–1890.

Moeiniafshari, A. A., Zarrabi, A., & Abdol-Khalegh, B. (2015). Exploring the interaction of naringenin with bovine beta-casein nanoparticles using spectroscopy. *Food Hydrocolloids*, *51*, 1–6.

Mohammadi, A., Jafari, S. M., Assadpour, E., & Esfanjani, A. F. (2016). Nano-encapsulation of olive leaf phenolic compounds through WPC-pectin complexes and evaluating their release rate. *International Journal of Biological Macromolecules*, *82*, 816–822.

Mohammadi, A., Jafari, S. M., Esfanjani, A. F., & Akhavan, S. (2016). Application of nano-encapsulated olive leaf extract in controlling the oxidative stability of soybean oil. *Food Chemistry*, *190*, 513–519.

Molina-García, A. D., Horridge, E., Sanz, P. D., & Martino, M. N. (2007). Nanostructure of starch high-pressure treated granules discovered by low temperature scanning electron microscopy. In A. Méndez-Vilas, & J. Díaz (Eds.), *Modern research and educational topics in microscopy*. Badajoz: Formatex.

Mousavi, S. Z., Amjad-Iranagh, S., Nademi, Y., & Modarress, H. (2013). Carbon nanotube-encapsulated drug penetration through the cell membrane: An investigation based on steered molecular dynamics simulation. *The Journal of Membrane Biology*, *246*(9), 697–704.

Mozafari, M. R., Khosravi-Darani, K., Borazan, G. G., Cui, J., Pardakhty, A., & Yurdugul, S. (2008). Encapsulation of food ingredients using nanoliposome technology. *International Journal of Food Properties*, *11*(4), 833–844.

Müller, R. H., Gohla, S., & Keck, C. M. (2011). State of the art of nanocrystals—Special features, production, nanotoxicology aspects and intracellular delivery. *European Journal of Pharmaceutics and Biopharmaceutics*, *78*, 1–9.

Nakayama, T., Hashimoto, T., Kajiya, K., & Kumazawa, S. (2000). Affinity of polyphenols for lipid bilayers. *Biofactors*, *13*(1–4), 147–151.

Neves, M. A., Hashemi, J., & Prentice, C. (2015). Development of novel bioactives delivery systems by micro/nanotechnology. *Current Opinion in Food Science*, *1*, 7–12.

Nitta, S. K., & Numata, K. (2013). Biopolymer-based nanoparticles for drug/gene delivery and tissue engineering. *International Journal of Molecular Sciences*, *14*(1), 1629–1654.

Oh, J. K., Tang, C., Gao, H., Tsarevsky, N. V., & Matyjaszewski, K. (2006). Inverse miniemulsion ATRP: A new method for synthesis and functionalization of well-defined water-soluble/cross-linked polymeric particles. *Journal of the American Chemical Society*, *128*(16), 5578–5584.

Oliveira, A., Guimarães, K., Cerize, N., Tunussi, A., & Poço, J. (2013). Nano spray drying as an innovative technology for encapsulating hydrophilic active pharmaceutical ingredients (API). *Journal of Nanomedicine & Nanotechnology*, *4*(6), 186.

Ozturk, B., Argin, S., Ozilgen, M., & McClements, D. J. (2014). Formation and stabilization of nanoemulsion-based vitamin E delivery systems using natural surfactants: Quillaja saponin and lecithin. *Journal of Food Engineering*, *142*, 57–63.

Pan, K., Luo, Y., Gan, Y., Baek, S. J., & Zhong, Q. (2014). pH-driven encapsulation of curcumin in self-assembled casein nanoparticles for enhanced dispersibility and bioactivity. *Soft Matter*, *10*(35), 6820–6830.

Pandita, D., Kumar, S., Poonia, N., & Lather, V. (2014). Solid lipid nanoparticles enhance oral bioavailability of resveratrol, a natural polyphenol. *Food Research International*, *62*, 1165–1174.

Paredes, A. J., Asencio, C. M., Manuel, L. J., Allemandi, D. A., & Palma, S. D. (2016). Nanoencapsulation in the food industry: Manufacture, applications and characterization. *Journal of Food Bioengineering and Nanoprocessing*, *1*(1), 56–79.

Pellegrino, T., Manna, L., Kudera, S., Liedl, T., Koktysh, D., Rogach, A. L., ... Parak, W. J. (2004). Hydrophobic nanocrystals coated with an amphiphilic polymer shell: A general route to water soluble nanocrystals. *Nano Letters*, *4*(4), 703–707.

Petersen, R. (2006). Nanocrystals for use in topical cosmetic formulations and method of production thereof. Patent WO2008058755.

Pinho, E., Grootveld, M., Soares, G., & Henriques, M. (2014). Cyclodextrins as encapsulation agents for plant bioactive compounds. *Carbohydrate Polymers*, *101*, 121–135.

Ponchel, G., Montisci, M.-J., Dembri, A., Durrer, C., & Duchêne, D. (1997). Mucoadhesion of colloidal particulate systems in the gastro-intestinal tract. *European Journal of Pharmaceutics and Biopharmaceutics*, *44*(1), 25–31.

Portnaya, I., Cogan, U., Livney, Y. D., Ramon, O., Shimoni, K., Rosenberg, M., & Danino, D. (2006). Micellization of bovine β-casein studied by isothermal titration microcalorimetry and cryogenic transmission electron microscopy. *Journal of Agricultural and Food Chemistry*, *54*(15), 5555–5561.

Pourashouri, P., Shabanpour, B., Razavi, S. H., Jafari, S. M., Shabani, A., & Aubourg, S. P. (2014). Oxidative stability of spray-dried microencapsulated fish oils with different wall materials. *Journal of Aquatic Food Product Technology*, *23*(6), 567–578.

Prego, C., Torres, D., & Alonso, M. J. (2005). The potential of chitosan for the oral administration of peptides. *Expert Opinion on Drug Delivery*, *2*(5), 843–854.

Pyo, S. M., Müller, R. H., & Keck, C. M. (2017). Encapsulation by nano-structured lipid carriers. In S. M. Jafari (Ed.), *Nanoencapsulation technologies for the food and nutraceutical industries* (pp. 114–138). Amsterdam: Elsevier.

Qian, C., Decker, E. A., Xiao, H., & McClements, D. J. (2013). Impact of lipid nanoparticle physical state on particle aggregation and β-carotene degradation: Potential limitations of solid lipid nanoparticles. *Food Research International*, *52*(1), 342–349.

Quintanilla-Carvajal, M. X., Camacho-Díaz, B. H., Meraz-Torres, L. S., Chanona-Pérez, J. J., Alamilla-Beltrán, L., Jimenéz-Aparicio, A., & Gutiérrez-López, G. F. (2010). Nanoencapsulation: A new trend in food engineering processing. *Food Engineering Reviews*, *2*(1), 39–50.

Rahaiee, S., Shojaosadati, S. A., Hashemi, M., Moini, S., & Razavi, S. H. (2015). Improvement of crocin stability by biodegradeble nanoparticles of chitosan-alginate. *International Journal of Biological Macromolecules*, *79*, 423–432.

Rajabi, H., Ghorbani, M., Jafari, S. M., Sadeghi Mahoonak, A., & Rajabzadeh, G. (2015). Retention of saffron bioactive components by spray drying encapsulation using maltodextrin, gum Arabic and gelatin as wall materials. *Food Hydrocolloids*, *51*, 327–337.

Rajagopal, K., & Schneider, J. P. (2004). Selfassembling peptides and proteins for nanotechnological applications. *Current Opinion in Structural Biology*, *14*(4), 480–486.

Ramakrishna, S. (2005). *An introduction to electrospinning and nanofibers*. Hackensack, NJ: World Scientific.

Rao, J., & McClements, D. J. (2012). Lemon oil solubilization in mixed surfactant solutions: Rationalizing microemulsion & nanoemulsion formation. *Food Hydrocolloids*, *26*(1), 268–276.

Rao, P. J., & Khanum, H. (2016). A green chemistry approach for nanoencapsulation of bioactive compound—Curcumin. *LWT—Food Science and Technology*, *65*, 695–702.

Raphaelides, S. N., Dimitreli, G., Exarhopoulos, S., Ilia, E., & Koutsomihali, P. (2015). A process designed for the continuous production of starch inclusion complexes on an industrial scale. *Food and Bioproducts Processing*, *96*, 245–255.

Rasti, B., Jinap, S., Mozafari, M. R., & Yazid, A. M. (2012). Comparative study of the oxidative and physical stability of liposomal and nanoliposomal polyunsaturated fatty acids prepared with conventional and Mozafari methods. *Food Chemistry*, *135*(4), 2761–2770.

Ravichandran, R. (2013). Pharmacokinetic study of nanoparticulate curcumin: Oral formulation for enhanced bioavailability. *Journal of Biomaterials and Nanobiotechnology*, *4*(3), 291–299.

Reneker, D. H., Yarin, A. L., Fong, H., & Koombhongse, S. (2000). Bending instability of electrically charged liquid jets of polymer solutions in electrospinning. *Journal of Applied Physics*, *87*(9), 4531–4547.

Reza Mozafari, M., Johnson, C., Hatziantoniou, S., & Demetzos, C. (2008). Nanoliposomes and Their Applications in Food Nanotechnology. *Journal of Liposome Research*, *18*(4), 309–327.

Romero, G. B., Brysch, W., Keck, C. M., & Muller, R. H. (2017). Nano-capsule formation by nano crystals. In S. M. Jafari (Ed.), *Nanoencapsulation technologies for the food and nutraceutical industries* (pp. 165–186). Amsterdam: Elsevier.

Ron, N., Zimet, P., Bargarum, J., & Livney, Y. D. (2010). Beta-lactoglobulin–polysaccharide complexes as nanovehicles for hydrophobic nutraceuticals in non-fat foods and clear beverages. *International Dairy Journal*, *20*(10), 686–693.

Rutledge, G.C., Li, Y., Fridrikh, S., Warner, S.B., Kalayci, V.E., & Patra, P. (2001). Electrostatic spinning and properties of ultrafine fibers. *Annual report*, National Textile Center.

Ryno, L. M., Levine, Y., & Iovine, P. M. (2014). Synthesis, characterizationand comparative analysis of amylose–guest complexes prepared by microwave irradiation. *Carbohydrate Research*, *383*, 82–88.

Ryu, J. H., Chacko, R. T., Jiwpanich, S., Bickerton, S., Babu, R. P., & Thayumanavan, S. (2010a). Self-cross-linked polymer nanogels: A versatile nanoscopic drug delivery platform. *Journal of the American Chemical Society*, *132*(48), 17227–17235.

Ryu, J. H., Jiwpanich, S., Chacko, R., Bickerton, S., & Thayumanavan, S. (2010b). Surface-functionalizable polymer nanogels with facile hydrophobic guest encapsulation capabilities. *Journal of the American Chemical Society*, *132*(24), 8246–8247.

Sadeghi, R., Kalbasi, A., Emam-jomeh, Z., Razavi, S. H., Kokini, J., & Moosavi-Movahedi, A. A. (2013). Biocompatible nanotubes as potential carrier for curcumin as a model bioactive compound. *Journal of Nanoparticle Research*, *15*(11), 1931.

Sadeghi, R., Mehryar, L., Karimi, M., & Kokini, J. (2017). Nano-capsul formation by individual biopolymer nanoparticles. In S. M. Jafari (Ed.), *Nanoencapsulation technologies for the food and nutraceutical industries* (pp. 403–446). Amsterdam: Elsevier.

Sadeghi, R., Moosavi-Movahedi, A. A., Emam-jomeh, Z., Kalbasi, A., Razavi, S. H., Karimi, M., & Kokini, J. (2014). The effect of different desolvating agents on BSA nanoparticle properties and encapsulation of curcumin. *Journal of Nanoparticle Research*, *16*(9), 2565.

Sahu, A., Kasoju, N., & Bora, U. (2008). Fluorescence study of the curcumin – casein micelle complexation and its application as a drug nanocarrier to cancer cells. *Biomacromolecules*, *9*(10), 2905–2912.

Sailaja, A. K., & Amareshwar, P. (2012). Preparation of alginate nanoparticles by desovaltion technique using acetone as desolvating agent. *Asian Journal of Pharmaceutical and Clinical Research*, *5*(2), 132–134.

Sáiz-Abajo, M. J., González-Ferrero, C., Moreno-Ruiz, A., Romo-Hualde, A., & González-Navarro, C. J. (2013). Thermal protection of b-carotene in re-assembled casein micelles during different processing technologies applied in food industry. *Food Chemistry, 138*, 1581–1587.

Salazar, J., Müler, R. H., & Möschwitzer, J. P. (2014). Combinative particle size reduction technologies for the production of drug nanocrystals. *Journal of Pharmaceutics, 2014*, 14.

Schmid, K. (2011). Spray drying of protein precipitates and evaluation of the Nano Spray Dryer B-90. Ph.D. Thesis, Ludwig-Maximilians-University, Munich.

Schmid, K., Arpagaus, C., & Friess, W. (2011). Evaluation of the Nano Spray Dryer B-90 for pharmaceutical applications. *Pharmaceutical Development and Technology, 16*, 287–294.

Schmitt, C., Aberkane, L., & Sanchez, C. (2009). Protein–polysaccharide complexes and coacervates. In G. O. Phillips, & P. A. Williams (Eds.), *Handbook of hydrocolloids* (2nd ed., pp. 420–476). Boca Raton, FL: CRC Press.

Semo, E., Kesselman, E., Danino, D., & Livney, Y. D. (2007). Casein micelle as a natural nanocapsular vehicle for nutraceuticals. *Food Hydrocolloids, 21*(5–6), 936–942.

Shamsara, O., Muhidinov, Z. K., Jafari, S. M., Bobokalonov, J., Jonmurodov, A., Taghvaei, M., & Kumpugdee-Vollrath, M. (2015). Effect of ultrasonication, pH and heating on stability of apricot gum-lactoglobuline two layer nanoemulsions. *International Journal of Biological Macromolecules, 81*, 1019–1025.

Shen, B., & Yang, S. (2013). Composition and method for preparing alginate nanocapsules. USA, Google Patents.

Shenoy, S. L., Bates, W. D., Frisch, H. L., & Wnek, G. E. (2005). Role of chain entanglements on fiber formation during electrospinning of polymer solutions: Good solvent, non-specific polymer-polymer interaction limit. *Polymer, 46*(10), 3372–3384.

Sneharani, A. H., Karakkat, V. J., Singh, S. A., & Appu Rao, A. G. (2010). Interaction of curcumin with β-lactoglobulin—Stability, spectroscopic analysis, and molecular modeling of the complex. *Journal of Agricultural and Food Chemistry, 58*(20), 11130–11139.

Spasova, M., Mincheva, R., Paneva, D., Manolova, N., & Rashkov, I. (2006). Perspectives on: Criteria for complex evaluation of the morphology and alignment of electrospun polymer nanofibers. *Journal of Bioactive and Compatible Polymers, 21*(5), 465–479.

Stieß, M. (2013). *Mechanische Verfahrenstechnik 1*. Berlin Heidelberg: Springer-Verlag.

Sultana, F., Imran-Ul-Haque, M., Arafat, M., & Sharmin, S. (2013). An overview of nanogel drug delivery system. *Journal of Applied Pharmaceutical Science, 3*(8 Suppl. 1), S95–S105.

Sun, Y., Song, X., Wang, J., & Yu, J. (2011). Preparation of lithium carbonate hollow spheres by spray pyrolysis. *Crystal Research and Technology, 46*, 173–177.

Svetlichny, G., Külkamp-Guerreiro, I., Cunha, S., Silva, F., Bueno, K., Pohlmann, A., ... Guterres, S. (2015). Solid lipid nanoparticles containing copaiba oil and allantoin: Development and role of nanoencapsulation on the antifungal activity. *Die Pharmazie-An International Journal of Pharmaceutical Sciences, 70*(3), 155–164.

Taheri, A., & Razavi, S. M. A. (2015). Fabrication of cress seed gum nanoparticles, an anionic polysaccharide, using desolvation technique: An optimization study. *BioNanoScience, 5*(2), 104–116.

Tamjidi, F., Shahedi, M., Varshosaz, J., & Nasirpour, A. (2013). Nanostructured lipid carriers (NLC): A potential delivery system for bioactive food molecules. *Innovative Food Science & Emerging Technologies, 19*, 29–43.

Tamjidi, F., Shahedi, M., Varshosaz, J., & Nasirpour, A. (2014). Design and characterization of astaxanthin-loaded nanostructured lipid carriers. *Innovative Food Science & Emerging Technologies, 26*, 366–374.

Tan, S. H., Inai, R., Kotaki, M., & Ramakrishna, S. (2005). Systematic parameter study for ultra-fine fiber fabrication via electrospinning process. *Polymer*, *46*(16), 6128–6134.

Tan, Y., Xu, K., Li, L., Liu, C., Song, C., & Wang, P. (2009). Fabrication of size-controlled starch-based nanospheres by nanoprecipitation. *ACS Applied Materials & Interfaces*, *1*(4), 956–959.

Tang, S. Y., Sivakumar, M., & Nashiru, B. (2013). Impact of osmotic pressure and gelling in the generation of highly stable single core water-in-oil-in-water (W/O/W) nano multiple emulsions of aspirin assisted by two-stage ultrasonic cavitational emulsification. *Colloids and Surfaces B: Biointerfaces*, *102*, 653–658.

Tapia-Hernández, J. A., Rodríguez-Félix, F., & Katouzian, I. (2017). Nano-capsule formation by electrospraying. In S. M. Jafari (Ed.), *Nanoencapsulation technologies for the food and nutraceutical industries* (pp. 320–345). Amsterdam: Elsevier.

Tapia-Hernández, J. A., Torres-Chávez, P. I., Ramírez-Wong, B., Rascón-Chu, A., Plascencia-Jatomea, M., Barreras-Urbina, C. G., ... Rodríguez-Félix, F. (2015). Micro- and nanoparticles by Electrospray: Advances and applications in foods. *Journal of Agricultural and Food Chemistry*, *63*(19), 4699–4707.

Tavares, G. M., Croguennec, T., Carvalho, A. F., & Bouhallab, S. (2014). Milk proteins as encapsulation devices and delivery vehicles: Applications and trends. *Trends in Food Science & Technology*, *37*(1), 5–20.

Teo, W. E., & Ramakrishna, S. (2006). A review on electrospinning design and nanofibre assemblies. *Nanotechnology*, *17*(14), 89–106.

Tolstoguzov, V. (2002). Thermodynamic aspects of biopolymer functionality in biological systems, foods, and beverages. *Critical Reviews in Biotechnology*, *22*(2), 89–174.

Turgeon, S. L., & Laneuville, S. I. (2009). Protein + polysaccharide coacervates and complexes: From scientific background to their application as functional ingredients in food products. In S. Kasapis, I. T. Norton, & J. B. Ubbink (Eds.), *Modern biopolymer science* (pp. 327–363). New York: Elsevier.

Turgeon, S. L., Schmitt, C., & Sanchez, C. (2007). Protein–polysaccharide complexes and coacervates. *Current Opinion in Colloid & Interface Science*, *12*(4), 166–178.

Vasbinder, A. J., & de Kruif, C. G. (2003). Casein–whey protein interactions in heated milk: The influence of pH. *International Dairy Journal*, *13*, 69–677.

Vergaro, V., Lvov, Y. M., & Leporatti, S. (2012). Halloysite clay nanotubes for resveratrol delivery to cancer cells. *Macromolecular Bioscience*, *12*(9), 1265–1271.

Weber, S., Zimmer, A., & Pardeike, J. (2014). Solid lipid nanoparticles (SLN) and nanostructured lipid carriers (NLC) for pulmonary application: A review of the state of the art. *European Journal of Pharmaceutics and Biopharmaceutics*, *86*(1), 7–22.

Xu, J., Zhao, W., Ning, Y., Bashari, M., Wu, F., Chen, H., ... Xu, X. (2013). Improved stability and controlled release of ω3/ω6 polyunsaturated fatty acids by spring dextrin encapsulation. *Carbohydrate Polymers*, *92*(2), 1633–1640.

Yada, R. Y., Buck, N., Canady, R., DeMerlis, C., Duncan, T., Janer, G., ... Thurmond, S. (2014). Engineered nanoscale food ingredients: Evaluation of current knowledge on material characteristics relevant to uptake from the gastrointestinal tract. *Comprehensive Reviews in Food Science and Food Safety*, *13*(4), 730–744.

Yang, L., Zhang, B., Yi, J., Liang, J., Liu, Y., & Zhang, L.-M. (2013). Preparation, characterization, and properties of amylose–ibuprofen inclusion complexes. *Starch—Stärke*, *65*(7–8), 593–602.

Yang, Y., Gu, Z., & Zhang, G. (2009). Delivery of bioactive conjugated linoleic acid with self-assembled amylose – CLA complex. *Journal of Agricultural and Food Chemistry*, *57*(15), 7125–7130.

Yarin, A., Koombhongse, S., & Reneker, D. (2001). Bending instability in electrospinning of nanofibers. *Journal Applied Physics*, *89*(5), 3018–3026.

Yesilyurt, V., Ramireddy, R., & Thayumanavan, S. (2011). Photoregulated release of noncovalent guests from dendritic amphiphilic nanocontainers. *AngewandteChemie International Edition*, *50*(13), 3038–3042.

Yu, D.-G., Zhu, L.-M., White, K., & Branford-White, C. (2009). Electrospun nanofiber-based drug delivery systems. *Health*, *1*(02), 67–75.

Yu, Shaoyong, Yao, Ping, Jiang, Ming, & Zhang, Guangzhao (2006). Nanogels prepared by self-assembly of oppositely charged globular proteins. *Biopolymers*, *148–158*, 83.

Zhang, J., Zhou, Y., Zhu, Z., Ge, Z., & Liu, S. (2008). Polyion complex micelles possessing thermoresponsive coronas and their covalent core stabilization via "click" chemistry. *Macromolecules*, *41*(4), 1444–1454.

Zhang, S., & Kawakami, K. (2010). One-step preparation of chitosan solid nanoparticles by electrospray deposition. *International Journal of Pharmaceutics*, *397*(1), 211–217.

Zhao, L.-M., Shi, L.-E., Zhang, Z.-L., Chen, J.-M., Shi, D.-D., Yang, J., & Tang, Z.-X. (2011). Preparation and application of chitosan nanoparticles and nanofibers. *Brazilian Journal of Chemical Engineering*, *28*, 353–362.

Zheng, M., Falkeborg, M., Zheng, Y., Yang, T., & Xu, X. (2013). Formulation and characterization of nanostructured lipid carriers containing a mixed lipids core. *Colloids and Surfaces A: Physicochemical and Engineering Aspects*, *430*, 76–84.

Zhu, Y., Zhang, J., Zheng, Q., Wang, M., Deng, W., Li, Q., ... Yu, J. (2015). In vitro and in vivo evaluation of capsaicin-loaded microemulsion for enhanced oral bioavailability. *Journal of the Science of Food and Agriculture*, *95*(13), 2678–2685.

Ziani, K., Fang, Y., & McClements, D. J. (2012). Encapsulation of functional lipophilic components in surfactant-based colloidal delivery systems: Vitamin E, vitamin D, and lemon oil. *Food Chemistry*, *134*(2), 1106–1112.

Zimet, P., & Livney, Y. D. (2009). Beta-lactoglobulin and its nanocomplexes with pectin as vehicles for ω-3 polyunsaturated fatty acids. *Food Hydrocolloids*, *23*(4), 1120–1126.

Zimet, P. D. R., & Livney, Y. D. (2011). Re-assembled casein micelles and casein nanoparticles as nano-vehicles for omega-3 polyunsaturated fatty acids. *Food Hydrocolloids*, *25*(5), 1270–1276.

Chapter 2

Nanoencapsulation of Phenolic Compounds and Antioxidants

Afshin Faridi Esfanjani and Seid Mahdi Jafari
Gorgan University of Agricultural Sciences and Natural Resources, Gorgan, Iran

2.1 INTRODUCTION

Formulated *food* products via nanoencapsulated phenolic compounds and antioxidants are the more bioavailable in comparison with their nonencapsulated forms.

Through history, humans are looking for bioactive compounds originated from nature as healthy foods. The fruits, plants, and animals are the main sources of these compounds, which can be used directly as a food or by incorporation of them into food formulations (e.g., beverages). Antioxidants are a group of the most popular bioactive compounds in food and pharmaceutical industrials. Antioxidants have a key role in the promotion of health by inhibiting the oxidation of other molecules. These bioactive compounds have a large range of structures and functions, including repairing damaged molecules, blocking metal radical production, stimulating gene expression and endogenous antioxidant production, and preventing a variety of diseases such as cancer, atherosclerosis, cataracts, Alzheimer, and Parkinson.

Another group of the interesting bioactive compounds in food industry are phenolic compounds that are commonly found in plant foods (vegetables, fruits, cereals, olive, chocolate, legumes, etc.) and beverages (tea, coffee, etc.) (Bjelakovic, Nikolova, Simonetti, & Gluud, 2004; Scartezzini & Speroni, 2000). Phenolic compounds have higher antioxidant activity and also they can be used as an antiallergenic, antiinflammatory, antiatherogenic, antithrombotic, and antimicrobial agent in the food and other biological formulations (Bjelakovic et al., 2004). The in vivo effect of phenolic compounds and antioxidants is normally dependent on their bioavailability. As shown in Fig. 2.1, consumption of phenolic compounds and antioxidants as food formulation can be achieved in three methods including plant foods, extracts, and essences, and/or by encapsulated form of them.

Nanoencapsulation of Food Bioactive Ingredients. DOI: http://dx.doi.org/10.1016/B978-0-12-809740-3.00002-7

Food plants

Benefits:
- ✓ Dietary fiber intake provides many health benefits.

Restrictions:
- ✓ Negative interactions of phenolic compounds and antioxidants with the indigestible polysaccharides of cell walls
- ✓ Packaging and handling
- ✓ Low bioavailability
- ✓ Low shelf life
- ✓ The losses of them due to environmental stresses
- ✓ Untargeted release and instability in various conditions
- ✓ Undesirable flavor of phenolic compounds.
- ✓ Increasing the viscosity of the luminal contents

Extracts, beverages, etc.

Benefits:
- ✓ Easy access to phenolic compounds and antioxidants

Restrictions:
- ✓ Packaging and handling
- ✓ Low bioavailability
- ✓ Low shelf life
- ✓ The losses of them due to environmental stresses
- ✓ Untargeted release and instability in various conditions
- ✓ Undesirable flavor of phenolic compounds

Encapsulated products

Benefits:
- ✓ Protection of phenolic compounds and antioxidants against environmental and process stresses
- ✓ Targeted and controlled release
- ✓ Increasing bioavailability
- ✓ Facilitates the handling and packaging
- ✓ Masking undesirable flavors (e.g., phenolic compounds)
- ✓ High shelf life

FIGURE 2.1 Different sources of phenolic compounds and antioxidants.

Incorporation of fruits, stems, leaves, flowers, and roots of food plants into our daily diet is a simple and primal path for receiving phenolic compounds and antioxidants. For example, incorporation of the grape as a fruiting berry in the daily diet can provide some main needed antioxidants such as resveratrol (RSV). Consumption of phenolic compounds and antioxidants through eating fruits and plant foods has different restrictions including low bioavailability of them. Food plants are rich in dietary fibers and polysaccharides, proteins, and lipids so that phenolic compounds and antioxidants are dispersed in this matrix and are associated with some other compounds such as carbohydrates. Therefore, the bioavailability of phenolics in this form is low due to the physical and chemical interactions with the indigestible polysaccharides of cell walls (Saura-Calixto, Pérez-Jiménez, & Goñi, 2009). These problems can partly be solved by consumption of extracts, essences, beverages of food plants and fruits, etc., instead of whole fruits and vegetables. Indeed, in this way the accessibility of phenolic compounds increases and facilitates their absorption. However, there are some other drawbacks for using fruits and food plants such as difficultly in their packaging and handling, low solubility, low shelf life, losses due to environmental stresses (e.g., light, oxygen, high temperature, and pH) and food processes (e.g., baking, pasteurization, sterilization, mixing, drying, home preparation, storage, microwaving, steaming, and boiling), untargeted release and instability in various conditions (food formulations such as beverages), undesirable flavor of phenolic compounds, etc. (Fang & Bhandari, 2010; Larrauri, Ruperez, & Saura-Calixto, 1997; Manzocco, Calligaris, Mastrocola, Nicoli, & Lerici, 2000).

A promising approach, namely, encapsulation technology has been employed as an appropriate method for using phenolic compounds and antioxidants by researchers and producers. This method can overcome all the mentioned restrictions and can design the best delivery system in food and other biological formulations. In this process, bioactive compounds are loaded into capsules surrounded by a variety of food-grade materials as a wall. Therefore, incorporated ingredients into capsules are protected from different stresses by wall materials (as a shield) and their release is also controlled and targeted (Fang & Bhandari, 2010; Jafari, Assadpoor, He, & Bhandari, 2008b). There are many different techniques for encapsulation process such as spray drying (glassy solid fine powder), spray chilling (polymorph fine particles), melt extrusion (glassy solid particulates), melt injection (glassy solid threads), fluid bed coating (glassy solid fine powder), emulsification (vehicles), liposome (lipid bilayer vehicles), and inclusion (crystalline complex) (Jafari et al., 2008b; Mahdavi, Jafari, Ghorbani, & Assadpoor, 2014). Selection of the best technique depends on some main parameters such as the goal of encapsulation for the food formulation (as a colorant and flavoring agent or a nutrient and pharmaceutical agent), nature of the encapsulant (oil soluble or water soluble, sensitivity to environmental and process stresses), form of food formulation (liquid or solid), type of food processing (baking, pasteurization, sterilization, or mixing), and the cost of process (Benita, 2005; Jafari et al., 2008b).

Loading of the bioactive compounds into nano-size capsules has been possible with the advent of nanotechnology in the food industry. In this process, an increase in the surface area leads to the improvement in bioavailability of bioactive ingredients. This chapter provides recent information about nanoencapsulated phenolic compounds and antioxidants. First, structure and classification of antioxidants and phenolic compounds will be reviewed in a quick look. Then, we will cover the nanoencapsulation approaches for the phenolic compounds and antioxidants.

2.2 ANTIOXIDANTS

Daily dietary contains antioxidants which provide your healthy. This statement is emphasized by food specialists for many times. Indeed, the healthy role of antioxidants in our body is related to their free radical-scavenging activity. Free radicals such as hydroxyl radicals (·OH), peroxyl radicals (·OOR), superoxide anion (O_2^{-} ·), and peroxynitrite (·ONO_O) are our natural biological response to environmental toxins like cigarette smoke, sunlight, chemicals, cosmic and manmade radiation, and are even a key feature of pharmaceutical drugs. Overproduction of free radicals will cause aerophilic injury to biomolecules, (lipids, proteins, DNA), eventually leading to several chronic diseases like coronary artery disease, cancer, diabetes, autoimmune disorder, postischemic introduction injury, infarct, vas diseases, chronic inflammation, stroke and septic shock, aging, and other chronic diseases in humans. The antioxidants can prevent all of these by removing free radicals. On the other hand, antioxidants are a very popular additives in food formulations mainly because of their distinctive properties of extending the shelf life of food products without any adverse effect on their sensory or nutritional qualities (Heinonen, 2007; Misra, Dhillon, Brar, & Verma, 2014). Antioxidants are categorized as oil soluble and water soluble. Oil-soluble antioxidants are mostly located in cell membranes of body and commonly include carotenoids, vitamins A and E, and lipoic acid. On the contrary, water-soluble antioxidants are located in aqueous fluids of body such as blood and cytosol or cytoplasmic matrix. Some polyphenols, glutathione (GSH), and vitamin C are common water-soluble antioxidants (Misra et al., 2014; Shalaby & Shanab, 2013).

As shown in Table 2.1, antioxidants can also be classified in two groups: enzymatic and nonenzymatic. Enzymatic antioxidants can break down and remove free radicals; they include superoxide dismutase (SOD), catalase (CAT), glutathione peroxidase (GSHpx), and glutathione reductase. Nonenzymatic antioxidants include carotenoids, vitamin C, vitamin E, plant phenolic compounds, lipoic acid, and GSH (Misra et al., 2014; Shalaby & Shanab, 2013). Phenolic compounds as the main subgroup of nonenzymatic antioxidants are discussed in the following section.

2.3 PHENOLIC COMPOUNDS

Phenolic compounds are nonenzymatic antioxidants and can remove free radicals by reacting with them. Phenolics as a healthy agent are interesting in many food research areas because of their antioxidant activity, their abundance in nature, and their main role in the prevention of associated diseases from various oxidative stresses. Moreover, different biological effects can be provided by phenolic compounds, such as antimicrobial, antiviral, antiinflammatory, antiallergic, antithrombotic, and vasodilator actions. Phenolics can also contribute to organoleptic properties, flavor, and color of plant foods and beverages. For example, anthocyanins are responsible for the orange, red, blue, and purple colors of many fruits and vegetables such as pomegranates, berries, beets, and onions. Therefore, it is common that phenolic

TABLE 2.1 Antitoxins and Phenolic Compounds: Classification and Sources

Main Class	Sub-class	Examples	Common Sources
Antioxidants			
Enzymatic			
	Primary enzymes	Catalase, superoxide dismutase, glutathione peroxidase	Potatoes, beef Liver, cruciferous vegetables, melons
	Secondary enzymes	Glucose 6-phosphate, glutathione peroxidase, dehydrogenase	Whey proteins, cruciferous vegetables, nuts, milk thistle
Nonenzymatic			
	Minerals	Selenium, zinc	Nuts and legumes, seafood, whole-wheat bread, mushrooms, dairy foods
	Vitamins	Vitamin A, C, E	Carrots, dark leafy greens, dried apricots, seafood, fruits, nuts, dairy foods, whole grain products
	Carotenoids	β-Carotene, lycopene, lutein, zeaxanthin, astaxanthin	Carrots, dark green leafy vegetables, sweet red peppers, dried apricots, broccoli, seafood
	Organosulfurs	Allium, allyl sulfide, indoles	Garlic, onion, leek, chives, cruciferous vegetables
	Low molecular weight	Uric acid	Fruits, seafood, and fish, red-flag vegetables
	Cofactors	Coenzyme Q10	Beef, broccoli, pistachio nuts, chicken
Phenolic Compounds			
	Flavonols	Myricetin, quercetin, rutin, kaempferol and their glycosylated forms	Fruits and vegetables (onions, curly kale, leeks, broccoli blueberries), tea
	Flavones	Apigenin, luteolin, tangeretin, nobiletin, sinensetin	Parsley, celery, cereals (millet and wheat), skin of citrus

(Continued)

TABLE 2.1 (Continued)

	Flavanones	Hesperetin, hesperidin, homoeriodictyol, naringenin, naringin, eriodictyol	Citrus fruits (grapefruit, orange, lemon), tomatoes and some aromatic plants (mint)
	Isoflavones	Genistein, daidzein, glycitein	Leguminous plants soya and its processed products
	Anthocyanins	Cyanidin, delphinidin, malvidin, peonidin, pelargonidin, petunidin, and their glycosides	Red wine, some varieties of cereals, some leafy and root vegetables (aubergines, cabbage, beans, onions, radishes), flowers and fruits
	Hydroxybenzoic acids	Gallic acid, vanillic acid, protocatechuic acid, *p*-hydroxybenzoic acid	Tea, red fruits (raspberry, black currant, strawberry)
	Hydroxycinnamic acids	Caffeic acid, *p*-coumaric acid, sinapic acid, ferulic acid	Fruit (kiwis, blueberries, apples), cereal grains (wheat, rice, oat flours)
	Coumarins	Umbelliferone, aesculetin, scopoletin	Tonka bean, bark (chestnut), medicinal plants (*Melilotus officinalis, Angelica officinalis*)
	Stilbenes	Resveratrol M	Medicinal plants
	Lignans	Pinoresinol, podophyllotoxin, steganacin	Flax seed, sesame seed, cereals (rye, wheat, oat, barley), cruciferous vegetables (broccoli, cabbage) and fruits (apricots, strawberries)
	Tannins	Castalin, penta alloys glucose, procyanidins	Tea, berries, chocolate
	Stilbenes	Resveratrol	Red grape, cranberry, strawberry

compounds can be used as antioxidants, antidisease agents for oxidative stresses, colorants, and flavor agents in food formulations. As shown in Table. 2.1, there are many different phenolic compounds that can be obtained from our surrounding environment such as roots, stems, leaves, flowers and fruits of plants, beverages, etc. (Mohammadi, Jafari, Esfanjani, & Akhavan, 2016b; Rafiee, Jafari, Alami, & Khomeiri, 2012; Rahmanian, Jafari & Wani, 2015; Shahidi, Janitha & Wanasundara, 1992; Taghvaei & Jafari, 2015).

2.4 NANOENCAPSULATION PROCESS OF PHENOLIC COMPOUNDS AND ANTIOXIDANTS

In recent years, the increasing expansion of industrial life and appearance of different diseases such as diabetes, cancer, obesity, and Alzheimer have led people to diets rich in healthy plant foods, especially phenolic compounds and antioxidants. Currently, the emergence of nanotechnology in the food sector has also led to health promotion by fabricating functional food products.

Nanotechnology is defined as a process by which some main physicochemical properties of particles are changed by reducing their size to nanodimension. The various definitions of nanoparticle size range are presented in different industries such as food, pharmaceutical, and cosmetics. In the food industry, European Food Safety Authority (EFSA) has called the particles below 100 nm as nanoparticles. However, in general, particles with sizes less than 1000 nm can be considered as nanoparticles (Committee, 2011; Sanguansri & Augustin, 2006).

Nanoencapsulation is a main branch of nanotechnology which is mostly considered as a fundamental delivery system of phenolic compounds and antioxidants in functional food formulations. Nanoencapsulation can be applied in food sector by two common methods including "bottom-up" (self-assembly and self-organization) or "top-down" (physical processing). Also, phenolic compounds and antioxidants can be nanoencapsulated in various forms such as spherical or compact, tubular, complex nonspherical, compositionally heterogeneous (compositional variation core surface or distributed compositional variation), homogeneous aggregates/agglomerates (consisting of a single particle class or diverse particle types), micelle, and liposome (Bouwmeester et al., 2009; Sozer & Kokini, 2009).

Nanoencapsulation of phenolic compounds and antioxidants provides a large surface area and causes steerable release, protects them against different stresses during the processes and storage in comparison to their microencapsulation, and improves their bioavailability for removing free radicals, antidisease, and antimicrobial activity. Indeed, the bioavailability of phenolic compounds and antioxidants can be increased with higher solubility, absorption, and permeation of them in the body and food formulations through nanoencapsulation process. After consumption of nanoencapsulated phenolics and antioxidants in food, there are many physical and chemical changes during their passage through digestive system (mouth–stomach–intestine), which affect their ability to be digested and/or absorbed. Nanoencapsulated phenolics and antioxidants are more stable in stomach condition (low pH and enzymes) in comparison to their nonencapsulated form. Nanoencapsulation also allows phenolics and antioxidants to be absorbed passively from the lumen of the intestine into the lymphatic and blood circulatory system; therefore, their bioavailability is increased (Li, Jiang, Xu, & Gu, 2015b; McClements & Li, 2010; Palafox-Carlos, Ayala-Zavala, & González-Aguilar, 2011).

Generally, the nanoencapsulation of phenolic compounds and antioxidants can provide the following main benefits:

- Providing a larger surface area for interaction with the biological substrates.
- Providing a higher encapsulation efficiency in comparison to microencapsulation process.
- Enhancing the solubility of low water-soluble phenolics and antioxidants: this can occur in biopolymer-based nanoparticles. Indeed, low water-soluble compounds are trapped inside nanoparticles and are coated by polar groups on the surface of particles.
- Enhancing the absorption of phenolics and antioxidants by disrupting tight junctions and/or direct uptake by epithelial cells via endocytosis.
- Protecting the phenolics and antioxidants against the oxidation/degradation in the gastrointestinal tract.
- Protecting the phenolics and antioxidants from environmental and process stresses (e.g., temperature, pH, oxygen, light, mixing, enzymes, proteins, and metallic ions). This can promote their shelf life during storage and process (for example, their losses decrease during spray drying, pasteurization, sterilization, and similar processes).
- Targeted and controlled release, for example, a prolonged release of phenolic compounds in food formulation can be provided by multilayer nanoencapsulation systems.
- Alleviating unpleasant taste of the phenolics and antioxidants which limits their application in food formulations such as beverages and chewing gums.
- Production of the fortified transparent beverages by phenolics and antioxidants with high marketability; these products can be obtained by nanoemulsions (<100 nm) so that other flavor ingredients are disrupted in oil or water phase of the emulsion.
- Easy handling and packaging of nanoencapsulated phenolics and antioxidants, which can be achieved by nanoencapsulated powders produced via nano-spray drying, electro-spraying, and electro-spinning instruments.

Enthusiasm for using healthy and functional products by applying nanoencapsulation techniques has dramatically expanded in recent years. A report by the Global Industry Analysts (GIA) about nanoencapsulation for food products shows that global market for nanoencapsulation of food products is projected to reach US$8 billion by 2020, driven by strong demand for functional foods, and improvements in nanoencapsulation techniques. According to mentioned benefits of nanoencapsulated phenolics and antioxidants, this is an undeniable fact that these products play a main role in the global market of functional foods. As shown in Table 2.2, nanoencapsulated phenolics and antioxidants can be applied in the food industry through various sectors such as:

TABLE 2.2 Common Nanoencapsulation Methods of Phenolic Compounds and Antioxidants for Food Formulations

Capability	Technique	Morphology	Application
• Removal of free radicals • Antioxidant • Anticancer • Antidiabetes • Anti-Alzheimer • Antimicrobial • Antiallergenic • Antiinflammatory • Antiatherogenic • Antithrombotic • Antimicrobial • Flavor • Colorant	Nano-Spray drying	Spherical or compact	• A powder food product, namely, nanopowder contains phenolic compounds and antioxidants. • A liquid food product packaged in opaque glasses like the nanoemulsion and nanoliposome contain phenolic compounds and antioxidants. • Fortification of food powders such as bouillons, cake and jelly powders, juice powder, and ground cutlet (after production). • Fortification agents for functional food formulations including dairy products, beverages, bakery and confectionery products (after and before production). • Designing functional beverages, with considering the appearance and taste of formulated beverages. For example, O/W emulsions stabilized by gums is a good candidate for designing these beverages. • As a natural antioxidant agent in oil formulations (after production) instead of synthetic antioxidants. • Protection of some antioxidants by incorporating them into nanoencapsulated powders or fibers.
	Electro-spraying	Spherical or compact	
	Electro-spinning	Tubular	
	Biopolymer nanoparticle	Spherical or compact	
	Biopolymer micelle	Aggregated copolymers	
	Nano liposome	Bilayer lipid vesicles	
	Lipid nanostructure	Lipid vesicles	
	Nanoemulsion	Vesicles	
	Inclusion complexation	Crystalline complex	
	Lipid micelle	Single layer lipid vesicles	
	Nanotubes	Tubular	

a. *As a food product*: This can be provided in two forms including the solid (powder or fibers) and liquid. The nanoencapsulated powders and fibers of phenolics and antioxidants can be produced by nano-spray drying, electro-spraying, and electro-spinning. These powders or fibers can be used in industry (such as fortification of foods after and before production or formulation of food powders such as jelly powders, cake powders, ground cutlet, juice powders, and bouillons) and home cooking. Liquid nanoencapsulated forms can be obtained by nanoemulsions, nanoliposomes, etc. These products can be packed in opaque glasses; however, packaging and handling of solid nanoencapsulated products are easier than their liquid forms.
b. *Dairy sector*: The most popular dairy products are milk, yogurt, ice cream, and cheese. Nanoencapsulated phenolics and antioxidants can be incorporated into dairy formulation before, during, and after processing. In order to fortify these products with nanoencapsulated phenolics and antioxidants, it is necessary to design a nanoencapsulated system by using compatible compounds (such as whey proteins, caseins, or vegetable and mineral oils). These compounds should not change the taste, color, appearance, and rheology of dairy products. For example, Fathi, Varshosaz, Mohebbi, and Shahidi (2013) fortified the milk with loaded hesperetin in solid lipid nanoparticles (SLNs) and nanostructured lipid carriers (NLCs). They found the developed nanocarriers did not show any significant differences with blank milk sample and could well mask the bitter taste, after taste, and obviate poor solubility of hesperetin.
c. *Bakery sector*: Bread, muffin, biscuits, wafers, and cookies can be fortified with liquid or solid-based nanoencapsulated phenolics and antioxidants. For instance, loaded fish oil (a source of omega-3 fatty acids as an antioxidant) into nanocapsules was successfully incorporated in bread formulation and the fortified bread showed many advantages such as masking the unpleasant taste of the fish oil (Siegrist, Cousin, Kastenholz, & Wiek, 2007). Also, fortification of bread dough, muffin, etc. with nanoencapsulated phenolics and antioxidants can increase their resistance during cooking in comparison to their nonencapsulated form. The bakery products can also be fortified after cooking by injection of phenolics and antioxidants loaded in nanogels into them.
d. *Beverages and confectionery sector*: Recently, antioxidant- or phenolic-rich beverages have been introduced in food markets. These can be produced by two methods including fortification with nanoencapsulated phenolics and antioxidants or designing a nanoencapsulation system compatible with these products. Nanoemulsions, namely, "beverage emulsions" are a common method for designing beverages. Indeed, the beverage compounds (such as flavorings and colorants) are incorporated into stabilized nanoemulsions by food-grade materials such as gums, gelatins, and beverage oils. The nanoemulsions below 100 nm in size can

be applied in producing transparent beverages and nanoemulsions with droplets at a range of 100–1000 nm can be applied in opaque beverages. For example, the oil in water (O/W) nanoemulsions containing lipophilic antioxidants and phenolics are widely used for producing beverages (Mirhosseini, Tan, Hamid, & Yusof, 2008; Qian, Decker, Xiao, & Mcclements, 2012).

e. *Oil sector*: Some oils with antioxidant activity such as omega-3 fatty acids can be protected from environmental stresses of oxidation by nanoencapsulated phenolics and antioxidants. For these purpose, antioxidants are surrounded by wall materials into nanocapsules, nanofibers, nano-O/W emulsions, etc. (Jafari, Assadpoor, Bhandari, & He, 2008a; Ünlüsayin et al., 2016). Then, nanoencapsulated phenolics and antioxidants can be applied as natural antioxidant agents instead of synthetic antioxidants. For example, Mohammadi et al. (2016b) showed the high oxidative stability of fortified soy oil by nanoencapsulated phenolics of olive leaf in microemulsions (<100 nm) and multiple emulsions (400–1200 nm) in comparison to nonencapsulated phenolic compounds.

For the fabrication and development of nanoencapsulation systems, many different methods and compounds have been studied in the food industry. Therefore, it is necessary to be familiar with the classification, formulation, and principles of nanoencapsulation systems for phenolics and antioxidants which have been discussed in the following sections. For this chapter, we have classified applied nanoencapsulation technologies for phenolics and antioxidants into lipid formulations, biopolymer nanoparticles, and specialized equipment techniques based on the work of Jafari (2016).

2.5 LIPID-BASED FORMULATIONS FOR NANOENCAPSULATION OF PHENOLICS AND ANTIOXIDANTS

Lipids play a key role in nanoencapsulation systems of phenolic compounds and antioxidants. These compounds can be loaded into lipid matrix of O/W or W/O nanoemulsions, NLCs, and SLNs. The lipid phase can be applied as a continuous phase for distributing the nanodroplets of water containing phenolics and antioxidants in W/O nanoemulsions (McClements, 2015; Souto, Wissing, Barbosa, & Muller, 2004). Stabilizing agent is also another role of lipids for designing nanoencapsulation systems such as phospholipids in nanoliposomes (Fathi, Mozafari, & Mohebbi, 2012).

The applied food-grade lipids in nanoencapsulation systems are generally classified into two groups including liquid lipids (oils) and solid lipids. The liquid lipids are used in nanoemulsions and nanoliposome formulations, and include saturated oils such as medium-chain triglycerides (MCTs), paraffin oil, isopropyl palmitate, and capric/caprylic triglyceride, and unsaturated oils such as oleic acid, squalene, vegetable and seed oils. Solid lipids are used in SLN formulations and include glyceryl behenate (Compritol 888 ATO), glyceryl monostearate, glyceryl palmitostearate, and stearic acid. The mixture of solid and liquid lipids can also be utilized in nanoencapsulation systems such as NLCs (Souto et al., 2004). This section provides information about principles and applications of lipid-based nanoencapsulation systems for phenolic compounds and antioxidants (Table 2.3 and Fig. 2.2).

2.5.1 Nanoencapsulation of Phenolics and Antioxidants in Nanoemulsions

An emulsion system consists of two immiscible liquids, usually as oil and water, one of which is dispersed as droplets throughout the other by stabilizers. There are two basic methods for producing nanoemulsions including high energy methods (such as high-pressure homogenization and ultra-sonication) and low energy methods, e.g., using controllable mixture at low stirring named as spontaneous emulsification (Jafari, Assadpoor, He, & Bhandari, 2008c).

25.1.1 High-Energy Methods

The mass production, industrial potential, and fast-time process are main advantages of produced nanoemulsions by high energy instruments. With regard to the nature of phenolics and antioxidants, there are two groups of emulsion systems; oil in water (O/W) nanoemulsions and water in oil (W/O) nanoemulsions for nanoencapsulation of water-insoluble (lipophilic) and water-soluble (hydrophilic) compounds, respectively (Anton, Benoit, & Saulnier, 2008; Fang & Bhandari, 2010).

O/W nanoemulsions are basically composed of an oil phase (as a dispersed phase) and an aqueous phase containing emulsifiers (as continuous phase). The liposoluble phenolics and antioxidants (e.g., carotenoids, tannins, lignins and cell wall-bound hydroxycinnamic acids) can be nanoencapsulated in O/W emulsions by their dispersion in oil matrix surrounded with emulsifiers (as wall materials) (Mao et al., 2009; Wang et al., 2008). For this work, it is firstly needed to dissolve the target compounds into the oil phase to the extent that the sample becomes completely transparent with no evidence of crystals. Some compounds such as carotenoids easily dissolve in oils with mild heating and stirring (Qian et al., 2012). However, there are some compounds such as RSV which has to be diluted in an alcohol (e.g., ethanol) before dissolving in oils (Sessa et al., 2014). Then, the coarse emulsion is prepared by adding oil phase into aqueous phase containing emulsifiers while mixing by high-speed homogenizers, blenders, etc. Finally, nanoemulsions can be obtained by passing the produced coarse emulsion from high-pressure

TABLE 2.3 Lipid-based Systems for Nanoencapsulation of Phenolic Compounds and Antioxidants

Bioactive Ingredient	Technique	Wall Materials	Activity	Reference
Resveratrol	Nanoemulsions	Tween 80	Protected against degradation	Davidov-Pardo and McClements (2015)
β-carotene	Nanoemulsions	Octenyl succinate starch and whey protein isolate	Improved its bioavailability and broaden its applications	Mao et al. (2009)
β-carotene	Nanoemulsions	Sodium caseinate	Improved its physicochemical stability and in vitro bioaccessibility	Yi, Li, Zhong, and Yokoyama (2014)
Curcumin	Nanoemulsions	Modified starches	Optimizing delivery system	Abbas et al. (2014)
Grape marc polyphenols	Nanoemulsions	GMO and soy lecithin	Preventing degradation and improving the delivery through biological membranes	Sessa et al. (2014)
Flavonoids	Nanoliposomes	Egg PSC	Enhancing solubility in water	Goniotaki, Hatziantoniou, Dimas, Wagner, and Demetzos (2004)
Resveratrol	Nanoliposomes	Cholesterol and diacetyl phosphate	Enhancing antioxidant activity	Kristl et al. (2009)
Ellagic acid	Nanoliposomes	Soybean lecithin, chitosan, and dextran sulfate	Sustained release	Madrigal-Carballo et al. (2010)
Catechin	Nanoliposomes	Cholesterol and egg PSC	Enhancing permeation	Fang, Hwang, Huang, and Fang (2006)

(*Continued*)

TABLE 2.3 (Continued)

Bioactive Ingredient	Technique	Wall Materials	Activity	Reference
Quercetin	Nanoliposomes	Cholesterol and egg PSC	Enhancing bioavailability	Priprem, Watanatorn, Sutthiparinyanont, Phachonpai, and Muchimapura (2008)
Astaxanthin	SLN	Stearic acid, glycerin monostearate, and glycerol stearates	Enhancing bioavailability, water solubility, and stability against light, heat, and oxygen	Li, Zahi, Yuan, Tian, and Liang (2015a)
Curcuminoids	SLN	Stearic acid, glyceryl monostearate, and Poloxamer 188	A prolong release and protection against environmental stresses, e.g., light and oxygen	Tiyaboonchai, Tungpradit, and Plianbangchang (2007)
Quercetin	SLN	Glyceryl monostearate and soy lecithin	Enhancement of gastrointestinal absorption	Li et al. (2009a)
Resveratrol	SLN	Compritol 888ATO	Enhancing half-life of the molecule to achieve the therapeutic concentration at the site of action	Jose et al. (2014)
Quercetin	NLC	Soy lecithin, glyceryl tridecanoate, glyceryl tripalmitate, vitamin E acetate, and Kolliphor HS15	Enhancing anticancer activity	Sun et al. (2014)
Silymarin	NLC	Glycerol disintegrates (Precirol ATO-5), lecithin and oleic acid	Improving the oral bioavailability	Shangguan et al. (2014)
β-carotene	NLC	Sun flower oil and propylene glycol monostearate	Increased protection against degradation	Hentschel, Gramdorf, Muller, and Kurz (2008)
Resveratrol	NLC	Cetyl palmitate and miglyol-812	A controlled release after uptake	Neves, Lucio, Martins, Lima, and Reis (2013)

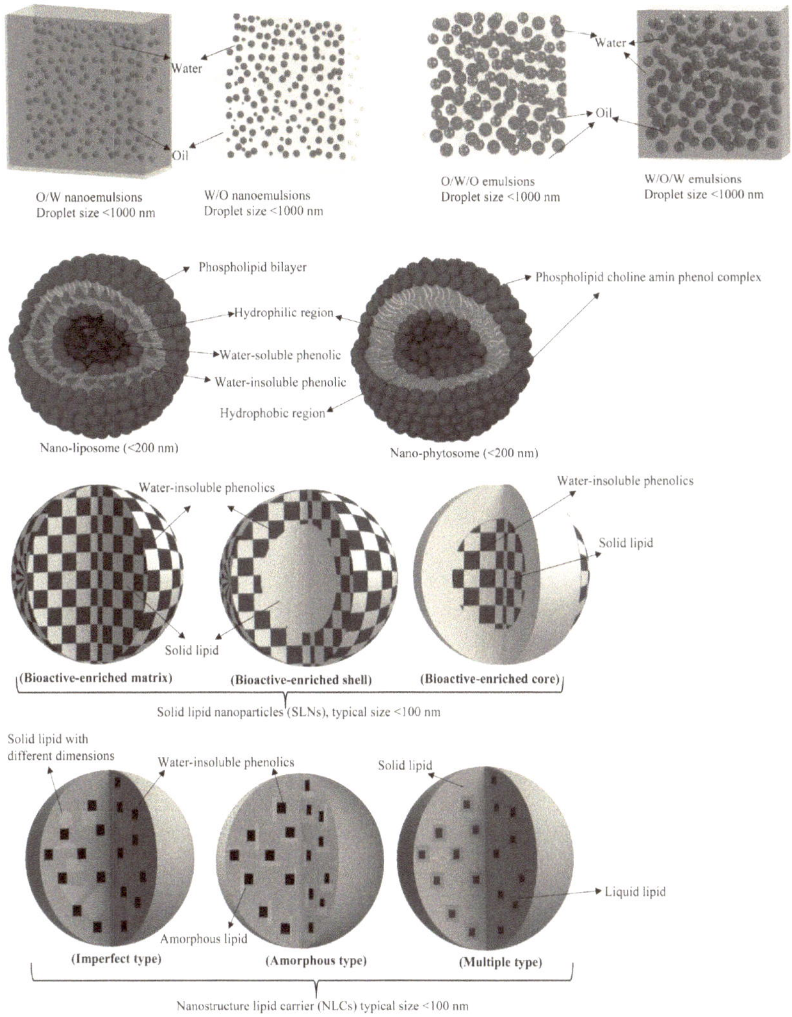

FIGURE 2.2 Schematic representation of nanoencapsulated phenolics by lipid-based formulation methods.

homogenization or ultra-sonication processes (Abbas, Bashari, Akhtar, & Zhang, 2014; Qian et al., 2012; Sessa et al., 2014).

O/W nanoemulsion systems loaded with phenolics and antioxidants can commonly be stabilized by three groups of food grade materials including proteins (e.g., whey proteins, sodium caseinate, gelatin, and soybean proteins), polysaccharides (e.g., gums, pectins, maltodextrin, and chitosan), and small-molecule surfactants (e.g., Tween series).

Nanoemulsions stabilized by proteins: Whey protein is a unique emulsifier that can be applied in O/W nanoemulsion formulations. For example,

Mao et al. (2009) showed that the loaded β-carotene in O/W nanoemulsions stabilized by whey protein isolate (WPI) had a good stability against environmental stresses. In another study, produced nanoemulsions with whey protein concentrate (WPC) and Tween 80 were applied for nanoencapsulation of curcumin in MCT oil droplets (Sari et al., 2015). The mean diameter, zeta potential and encapsulation efficiency of resulted droplets were 141.6 ± 15.4 nm, 6.9 ± 0.2 mV, and $90.56 \pm 0.47\%$, respectively. Eventually, they showed loaded curcumin in nanoemulsions was more stable than unencapsulated curcumin. The prepared nanoemulsions were stable against pasteurization, different ionic strengths (0.1e1 M) and pH ranging from 3 to 7. They found that the *in vitro* release kinetics of curcumin from nanoemulsion in simulated gastrointestinal conditions was slow which can increase the curcumin bioavailability (Sari et al., 2015).

Qian et al. (2012) loaded β-carotene in orange oil droplets through O/W nanoemulsions stabilized by β-lactoglobulin and Tween 20. They showed that β -carotene encapsulated within protein-coated lipid droplets was more stable to chemical degradation than that its counterpart encapsulated within non-ionic surfactant coated droplets. Also, they proposed these nanoemulsions can be used for the fortification of beverages or other food products.

Nanoemulsions stabilized by carbohydrates: Recently, carbohydrate based nanoemulsions have been introduced as the best nanoencapsulation materials for phenolic compounds and antioxidants. For example, application of gums such as Arabic gum is very popular for designing beverage nanoemulsions containing antioxidants. The complex of carbohydrate and protein as an emulsifier can also improve the stability of nanoemulsions.

In a study by Ru et al. (2012), epigallocatechin-3-gallate (EGCG) was successfully nanoencapsulated in stabilized nanoemulsions by ι-carrageenan and β-lactoglobulin, with a goal to develop biocompatible carriers for EGCG. The nanoencapsulated EGCG in nanoemulsions (~400 nm) showed an enhanced *in vitro* anticancer activity compared to the free EGCG (Ru, Yu, & Huang, 2010). Abbas et al. (2014) incorporated curcumin in MCT droplets by nanoemulsion stabilized with octenyl succinic anhydride modified starches, namely, Purity Gum 2000 (PG), Hi-Cap 100 (HC), and Purity Gum Ultra (PGU). The PGU-stabilized nanoemulsions provided the highest stability with small droplet sizes (~145.4 nm).

Nanoemulsions stabilized by small molecule surfactants: The stabilized nanoemulsions with small-molecule surfactants are smaller in size than those stabilized with proteins and carbohydrates. However, these nanoemulsions have a lower stability compared with macro-molecules (Jafari, He, & Bhandari, 2007). Tween and Span series are common small-molecule surfactants that have been applied for stabilization of nanoemulsions. For instance, quercetin was successfully loaded into limonene oil droplets in nanoemulsions stabilized with Tween 80 and Span 20. As a result, the stability of nanoemulsions was influenced by percentage of oil, surfactant, and homogenization conditions (Karadag, Yang, Ozcelik, & Huang, 2013). β-Carotene has also been loaded in nanoemulsions stabilized with Tween as surfactant by high-pressure homogenization. The β-carotene-loaded nanoemulsions had a good physical stability (Tubesha, Bakar, & Ismail, 2013).

There are some complex emulsion systems named *multiple emulsions* which can be used as an appropriated method for nanoencapsulation of hydrophilic or lipophilic phenolics and antioxidants. The most common form of multiple emulsions is water-in-oil-in-water (W/O/W), but oil-in-water-in-oil (O/W/O) emulsions can also be used in specific applications as shown in Fig. 2.2 (Esfanjani, Jafari, Assadpoor, & Mohammadi, 2015). In a study by Mohammadi, Jafari, Assadpour, and Esfanjani (2016a), they designed a stable and sustained-release nanoencapsulation system for olive leaf phenolic extracts through W/O/W multiple emulsions stabilized by WPC and pectin. They indicated that the antioxidant activity of this system as a natural antioxidant in soybean oil was more than nonencapsulated form of olive leaf extract (Mohammadi et al., 2016b).

25.1.2 Low-Energy Methods

Nanoemulsions can be produced by using the internal chemical energy of the system so that this method can provide smaller droplet sizes compared with high-energy methods. Longer stability, transparent appearance, higher energy efficiency, lower cost, and simplicity of implementation are the main advantages of low energy methods. However, these methods need a large amount of surfactants (Flanagan & Singh, 2006a). There are a variety of low-energy emulsification methods like microemulsion, emulsion inversion point (EIP), phase inversion temperature (PIT), and phase inversion composition (PIC) methods. This section provides some information about the application of produced nanoemulsions via low-energy methods for nanoencapsulation of phenolic compounds and antioxidants.

Microemulsions (5–100 nm) are one of the most popular low-energy methods that can be used as a nanoencapsulation system for phenolics and antioxidants (Flanagan & Singh, 2006a). The main characterizations of these systems include thermodynamic stability, low viscosity and isotropic state, and very small droplets. O/W and W/O microemulsions can be produced with regard to continuous and dispersed phases. Microemulsions are basically composed of oil, water, surfactant, and cosurfactant. The short- and medium-chain triglyceride oils are more appropriate than long-chain triglycerides in the formulation of microemulsions. Application of some cheap, food-grade, and commercial oils such as vegetable oils can also be considered in the formulation of microemulsions for the food industry. For food application, it is recommended to use low-toxicity surfactants including nonionic (e.g., Tween and Span series) and zwitterionic surfactants (e.g., phospholipids). Surfactants play a key role in producing highly stable microemulsions by lowering the interfacial tension. Cosurfactants such as short-chain alcohols are also commonly used for further reducing the interfacial tension (Flanagan & Singh, 2006a, b).

For the preparation of microemulsions, firstly, the area of microemulsion must be found by the construction of pseudo-ternary phase diagrams (Fan et al., 2014). Then, based on the type of microemulsion (O/W or W/O), the dispersed phase is slowly added into the continuous phase while stirring. After each addition, the system is given enough time to become transparent and isotropic. For example, Fan et al. (2014) constructed pseudo-ternary phase diagrams in order to find out the ratio of components for the existence area of microemulsions in order to load propolis flavone. The pseudo-ternary phase diagrams were constructed by using water titration method (Fig. 2.3). For this work, the mixture of oil with surfactant (S) and cosurfactant (CoS) was added into water and then, the microemulsion area was assessed when the final mixtures had a clear appearance and were isotropic. They showed that changing appearance of mixtures from clear to opaque and vice versa are the critical points between microemulsion region and other phase regions. The shadow area of the phase in prepared diagrams showed the microemulsion region and according to their results, this area was largest when S:CoS was 2:1 (Fan et al., 2014).

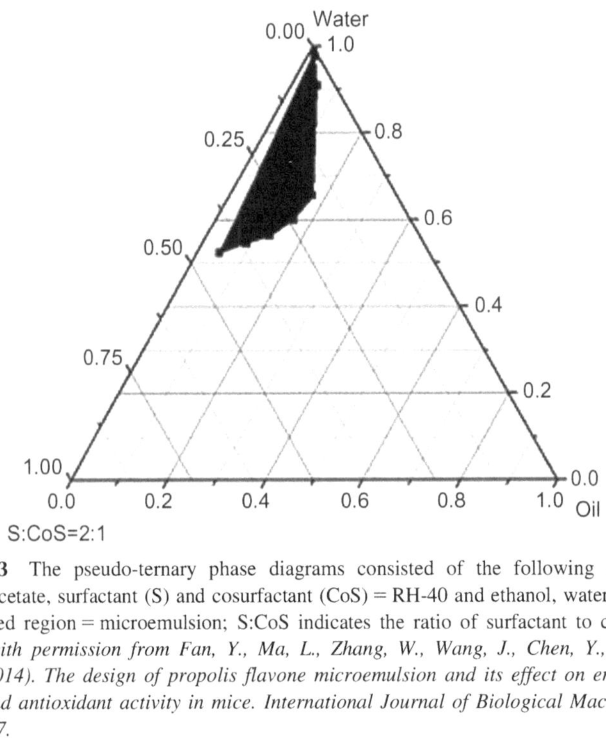

FIGURE 2.3 The pseudo-ternary phase diagrams consisted of the following components: oil = ethyl acetate, surfactant (S) and cosurfactant (CoS) = RH-40 and ethanol, water = deionized water. Shaded region = microemulsion; S:CoS indicates the ratio of surfactant to co-surfactant. *Reprinted with permission from Fan, Y., Ma, L., Zhang, W., Wang, J., Chen, Y., Gao, Y., & Song, X. (2014). The design of propolis flavone microemulsion and its effect on enhancing the immunity and antioxidant activity in mice. International Journal of Biological Macromolecules, 65, 200–207.*

Recently, microemulsions have been investigated in the food sector, especially in designing nanoencapsulation systems for phenolics and antioxidants. For instance, Chatzidaki, Mitsou, Yaghmur, Xenakis, and Papadimitriou (2015) fabricated a nanoencapsulation system for natural antioxidants (including gallic acid, p-hydroxybenzoic acid, protocatechuic acid, and tyrosol) by using food-grade W/O microemulsions based on lecithin, caprylic/capric triglycerides, isopropyl myristate, alcohols, and water. The droplet size of produced microemulsion was smaller than 10 nm.

EIP is based on a catastrophic phase inversion due to changes in a composition parameter (i.e., the water-to-oil ratio). Indeed, the EIP method simply involves titrating water into a mixture of oil and surfactant, which initially leads to the formation of a W/O emulsion and then, it inverts into an O/W emulsion (Ostertag, Weiss, & Mcclements, 2012). EIP as a low-energy method has also been applied in the formulation of nanoencapsulation systems for phenolics and antioxidants. For example, vitamin E acetate is one of the popular lipophilic materials (with antioxidant activity) that can be utilized in food beverage formulations. Therefore, it is important to design an optimum nanoencapsulation system for vitamin E due to its low water-solubility (Siro, Kapolna, Kapolna, & Lugasi, 2008). Mayer, Weiss, and Mcclements (2013) used EIP method for fabrication of a nanoencapsulation system loaded with vitamin E through a food-grade O/W nanoemulsion. They used variety of materials for stabilization of emulsions such as MCT, a series of nonionic surfactants (Tween 20, 40, 60, 80, and 85). They showed that the formulated emulsions with 2 wt% MCT, 8 wt% vitamin E acetate, and 20 wt% Tween 80 provide the smallest particle sizes (40 nm).

PIT is another low-energy technique based on thermal treatment. In this method, surfactant, lipids, and water mixture are heated around or slightly above the PIT and then quench cooling with continuous stirring is implemented. The mixture is normally treated with three temperature cycles from room temperature to PIT at the magnetic stirring. In the last cycle, when the mixture is cooled, ice cold deionized water (0°C) is added to the mixture. The fast cooling-dilution process results in nanoemulsions (Fernandez, Andre, Rieger, & Kuhnle, 2004).

PIT can be applied as a food-grade nanoencapsulation system for phenolic compounds and antioxidants. For instance, coenzyme Q10 as an antioxidant was loaded in PIT-nanoemulsions stabilized by lecithin and Tween 20 (Chen, Guan, & Zhong, 2015a). The droplet size of resulted nanoemulsions was <12 nm that did not change over 70 days' storage at 21°C. Also, the produced nanoemulsions provided a good dissolving environment with high stability for coenzyme Q10.

Some essential oils (EOs) such as orange essential oils have functional properties, e.g., antioxidant and antimicrobial activity (Frassinetti, Caltavuturo, Cini, Della Croce, & Maserti, 2011). Therefore, food products and beverages can be fortified by these oils. To achieve this, applying the best encapsulation and delivery system is needed. Souza, Almeida, Spinelli, and Mansur (2011) investigated the ability of prepared PIT nanoemulsions by nonionic surfactants for nanoencapsulation of orange oil. They showed that the oil phase concentration had an important effect on the stability of prepared nanoemulsions: the smallest droplets in the dispersed phase were provided by the emulsions formulated with an oil phase of 14 wt%.

2.5.2 Nanoencapsulation of Phenolics and Antioxidants in Nanoliposomes

Liposomes are colloidal spherical structures surrounded with polar lipids so that the orientation of hydrophilic heads is toward the water compartment and arrangement of lipophilic tails away from the water is toward the center of vesicle. Therefore, as shown in Fig. 2.2, this structure provides the encapsulation system for both hydrophilic and lipophilic phenolics and antioxidants in water core and lipid section, respectively. Liposomes are mainly produced by phospholipids as bilayer components, therefore, there are many food resources, e.g., egg, soy, and milk for preparing liposomes (Katouzian & Jafari, 2016; Reza Mozafari, Johnson, Hatziantoniou, & Demetzos, 2008). The supplying energy for hydrated phospholipids can provide bilayer structures of nanoscale liposomes. There are many methods for production of nanoliposomes which are provided in a number of books and reviews (Mozafari, 2010; Samad, Sultana & Aqil, 2007).

Some studies have shown the ability of nanoliposomes for loading phenolic compounds and antioxidants. For instance, Chen et al. (2015b) loaded curcumin in nanoliposomes prepared by combining thin film and dynamic high-pressure microfluidization (DHPM). They showed that the curcumin-loaded nanoliposomes exhibited good water solubility, physicochemical instability, and inadequate bioavailability in comparison to free curcumin. Coenzyme Q10 as a nonenzymatic antioxidant was also loaded in nanoliposomes fabricated by phospholipid/cholesterol/Tween 80 (Xia, Xu, & Zhang, 2006). The z-average diameter and encapsulation efficiency of resulted nanoliposomes were 68 nm and 95%, respectively.

Cadena et al. (2013) showed that antioxidant capabilities of typical flavonoids such as quercetin and RSV can be increased by incorporating them into nanoliposomes. The resulted nanoliposomes presented a small unilamellar form with a mean diameter of 149 nm, polydispersion index of 0.3, and a slightly negative zeta potential (−13.3 mV) which were stable and suitable for subcutaneous injection (Cadena et al., 2013). Kristl, Teskač, Caddeo, Abramović, and Šentjurc (2009) entrapped RSV into liposome by sonication. The high entrapment of RSV and rapid cellular internalization were provided by producing RSV nanoliposomes (with a mean diameter of 84 nm, polydispersity index of 0.2, and zeta potential of 40 mV). They showed that protecting cells from free radical damage was only possible when RSV was loaded into nanoliposomes.

Recently, the coating of liposome droplets with biopolymers (e.g., proteins) has been applied for improving the stability of liposomes. For example, ellagic acid was entrapped into layer by layer nanoliposomes (386.5 nm) fabricated by adding cationic (chitosan) and anionic (dextran sulfate) biopolymers to the suspension of anionic liposomes (Madrigal-Carballo et al., 2010). The results showed that these liposomes can be used as an appropriated delivery system for functional bioactive compounds, such as flavors, bioactive lipids, enzymes, peptides, antimicrobials, and antioxidants.

2.5.3 Nanoencapsulation of Phenolics and Antioxidants in Solid Lipid Nanoparticles

SLNs can be applied for designing nanoencapsulation systems of phenolic compounds and antioxidants. Indeed, these bioactive compounds are incorporated into a solid lipid matrix and are surrounded by surfactants. There are a number of review studies that provide main information about formulation, preparation, and characterization of SLNs (Muèller, Maèder, & Gohla, 2000; Müller & Lucks, 1996; Muller, Radtke & Wissing, 2002).

SLNs have more advantages such as controlled release, higher encapsulation efficiency, and the large-scale production without using organic solvents, in comparison to other lipid-based vehicles such as nanoliposomes and nanoemulsions (Müller & Lucks, 1996). Therefore, SLNs can be a good candidate for nanoencapsulation of phenolics and antioxidants. As shown in Fig. 2.2, bioactive compounds can be incorporated into SLNs in three forms: enriched matrix, enriched shell, and enriched core.

The loaded phenolic compounds and antioxidants into SLNs are more bioavailable compared with their free form. For example, curcumin as a valuable phenolic compound has many advantages such as antioxidant and anticancer properties. However, its application is restricted due to its poor solubility and stability in aqueous systems, as well as its rapid metabolism and systemic elimination (Anand, Kunnumakkara, Newman, & Aggarwal, 2007). Several studies have shown that these restrictions can be eliminated by loading curcumin into SLNs (Kakkar, Singh, Singla, & Kaur, 2011; Sun et al., 2013). In a study by Jose et al. (2014), RSV was loaded into SLNs formulated with Tween 80 or a combination of Tween 80 and polyvinyl alcohol (PVA) as surfactants. The diameter and zeta potential of produced particles were 248.30 nm and −25.49 mV, respectively. Their results showed that the RSV-loaded SLNs provided a sustained release and were promising therapeutic systems to treat neoplastic diseases located in the brain tissue (Jose et al., 2014).

Li et al. (2009a) incorporated quercetin into SLNs with an active encapsulation efficiency of 91.1% and 155.3 nm in mean particle diameter. The quercetin-SLN was composed of glyceryl monostearate, soya lecithin, Tween 80, and PEG 400 with emulsification and low-temperature solidification method. The results revealed that SLNs were the best delivery system to enhance the oral absorption of poorly water-soluble phenolics, like the quercetin.

Enzymatic antioxidants can also be protected from proteolysis by incorporating them into SLNs. Qi, Chen, Huang, Jin, and Wang (2012) loaded CAT as an enzymatic antioxidant in SLNs prepared with high-purity soybean phosphatidylcholine (SPC). They showed that these formulated SLNs can protect CAT against proteolysis, suggesting the potential application of SPC in delivery and protection of functional food enzymes.

2.5.4 Nanoencapsulation of Phenolics and Antioxidants in Nanostructured Lipid Carriers

NLCs are the optimal form of SLNs. Indeed, the formulation and production of NLCs is similar to SLNs as explained in the previous section, with the exception that the lipid phase in NLCs contains liquid lipid (oil) mixed with solid lipid (Müller & Lucks, 1996; Müller et al., 2002). Applying liquid lipids in NLCs provides main advantages such as high encapsulation efficiency, better morphological characterization, minimized encapsulated ingredient expulsion during storage, and controlled bioactive release properties, with smaller particle sizes. Therefore, NLCs can provide an optimum nanoencapsulation system for phenolic compounds and antioxidants. For example, the ability of SLNs and NLCs has been evaluated for nanoencapsulation of RSV by Gokce et al. (2012). SLNs were prepared using glyceryl behenate followed by homogenization, and NLCs were also produced using the same method, by replacing 5%, 15%, 30%, and 45% of the solid lipid with liquid migliol oil. The mean particle diameter of resulting SLNs and NLCs were ~287 nm and 110 nm, respectively. Also, the encapsulation efficiency was 18% higher in NLC systems. As a result, RSV-loaded NLCs with smaller particle sizes and higher loading appeared to be superior to SLNs for dermal applications (Gokce et al., 2012).

Phenolics and antioxidants with regard to their nature, goal of encapsulation, formulation, and production method can be incorporated into NLCs in three forms including imperfect crystal, amorphous, and multiple types as shown in Fig. 2.2 (Müller et al., 2002; Souto et al., 2004). Many water-insoluble phenolics and antioxidants have been reported to be successfully loaded into NLCs resulting in higher bioavailability (Table 2.3). For instance, lutein as a potential natural antioxidant, was loaded into NLCs prepared by high-pressure homogenization (Mitri, Shegokar, Gohla, Anselmi, & Muller, 2011). The results showed that these nanocarriers can protect lutein from environmental stresses. β-Carotene was also loaded into NLCs produced with Tween 20 and mixture of palmitic acid and corn oil by solvent diffusion method (Hejri, Khosravi, Gharanjig, & Hejazi, 2013). They found that the ultra-small NLCs with a high β-carotene retention can increase the bioavailability of β-carotene by its protection in lipid nanoparticles; therefore, NLCs could be an appropriate nanocarrier for β-carotene in food formulations (Hejri et al., 2013). In another study, Teeranachaideekul, Souto, Junyaprasert, and Muller (2007) successfully loaded coenzyme Q10 in a nanoparticle of acetyl palmitate and caprylic/capric triacylglycerols by using high-pressure homogenization in an NLC system. The entrapment efficiency was 100%. They showed that varying the lipid and oil ratio had no effect on the particle size of NLCs. However, increasing the amounts of oil loading led to a less ordered structure within the particles (Teeranachaideekul et al., 2007).

2.6 BIOPOLYMER-BASED TECHNOLOGIES FOR NANOENCAPSULATION OF PHENOLICS AND ANTIOXIDANTS

Some main properties of biopolymers such as their antimicrobial and antioxidant activities, and their role in improving textural and nutritional value of food products provide possibility of applying them as good nanoencapsulation materials for phenolic compounds and antioxidants in food formulations. Biopolymer nanoparticles are colloidal structures which are assembled from one or more types of biopolymer molecules, with a number of potential applications in the food and nutraceutical industries (Esfanjani & Jafari, 2016; Sundar, Kundu & Kundu, 2016).

There are different methods for producing biopolymer nanoparticles in encapsulating, protecting, and controlled release of bioactive compounds. Depending upon the preparation method, nanoparticles, nanospheres, or nanocapsules can be obtained (Fig. 2.4). The formulation, production, and characterization of biopolymer nanoparticles for nanoencapsulation of bioactive compounds, especially phenolic compounds, have been described in a number of review papers (Esfanjani & Jafari, 2016; Nitta & Numata, 2013; Sundar et al., 2016). For food applications, it is important to apply food-grade materials in the formulation of nanoparticles. For this purpose, there are two main groups of biopolymers including proteins (e.g., β-lactoglobulin, zein, gelatin, soy protein, collagen, and albumin) and carbohydrates (e.g., chitosan and other polysaccharides) (Jones, Lesmes, Dubin, & Mcclements, 2010; Nitta & Numata, 2013). Recent studies on application of biopolymer nanoparticles for encapsulation of phenolic compounds and antioxidants have been summarized in Table 2.4 and a schematic representation is provided in Fig. 2.4.

Zein nanoparticles: Zein, a prolamine protein found in maize, has long been applied in food and pharmaceutical industries because of its film-forming ability and superior biodegradability and biocompatibility (Woods & Selling,

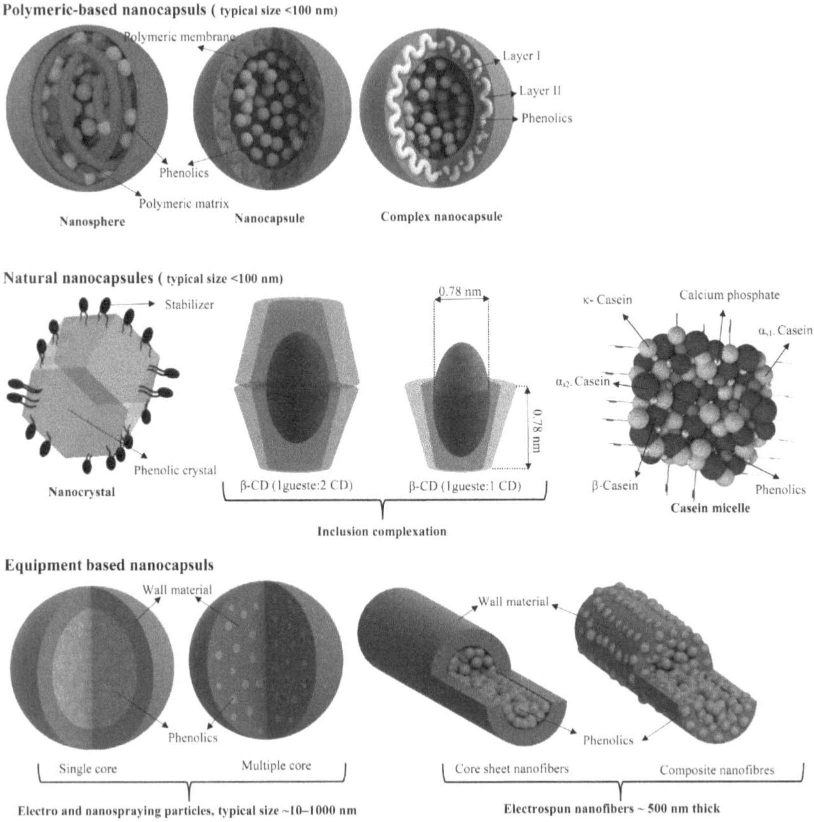

FIGURE 2.4 Schematic representation of nanoencapsulated phenolics by polymeric, natural, and equipment-based methods. *Reprinted with permission from Esfanjani and Jafari (2016).*

2007). For example, zein nanoparticles were applied for nanoencapsulation of essential oils (EOs) with antioxidant activity (Wu, Luo, & Wang, 2012). The results of this study showed that the water solubility of EOs can be increased by incorporation of them into zein nanoparticles without hindering their ability to scavenge free radicals or to control *E.coli* growth. Indeed, zein nanoparticles loaded with EOs facilitate application of EOs in food formulations (Wu et al., 2012). In another study, Chuacharoen and Sabliov (2016) successfully fortified milk with β-carotene loaded into zein nanoparticles. Improving stability and enhancing antioxidant activity of β-carotene under simulated gastrointestinal environments was obtained by using β-carotene entrapped in zein nanoparticles (Chuacharoen & Sabliov, 2016).

β-Lactoglobulin: β-Lactoglobulin (β-Lg) is the primary gelling agent and the main component of whey proteins. It was suggested to be a pH- and temperature-sensitive hydrogel former. β-Lg is a small (18.3 kDa) globular protein with two disulfide bonds and one free thiol group which is

TABLE 2.4 Biopolymeric and Instrument-based Systems for Nanoencapsulation of Phenolic Compounds and Antioxidants

Bioactive Ingredient	Technique	Wall Materials	Activity	Reference
Curcumin	Complex biopolymeric nanoparticles	β-Lactoglobulin and alginate	Delivery system for nutraceuticals in liquid foods	Hosseini et al. (2015)
Oleoresin	Inclusion complex	Cyclodextrins	Enhancing the antimicrobial and antioxidant activity	Teixeira, Ozdemir, Hill, and Gomes (2013)
Astaxanthin	Inclusion complex	Cyclodextrins	Increasing stability and water solubility	Yuan, Jin, Xu, Zhuang, and Shen (2008)
Naringenin	Casein nanoparticles	β-Casein	Enhancing solubility	Moeiniafshari, Zarrabi, and Bordbar (2015)
Catechin	Casein nanoparticles	β-Casein	An appropriate platform for delivery system	Haratifar and Corredig (2014)
Quercetin	Polymeric nanoparticles	Chitosan	Improving the bioavailability	Zhang, Yang, Tang, Hu, and Zou (2008)
Catechin	Polymeric nanoparticles	Chitosan/poly (γ-glutamic acid)	Enhancing the transport and antioxidant activity	Tang et al. (2013)
Curcumin	Polymeric nanoparticles	Zein	Controlling of release	Hu, Lin, Liu, Li, and Zhao (2012)
Procyanidins	Polymeric nanoparticles	Zein	Decreasing the cytotoxicity of procyanidins in HL-60 cells	Zou, Li, Percival, Bonard, and Gu (2012)

(*Continued*)

TABLE 2.4 (Continued)

Bioactive Ingredient	Technique	Wall Materials	Activity	Reference
Epigallocatechin Gallate	Electro-spinning	Zein	Resisted solubilization in water	Li, Lim, and Kakuda (2009b)
Curcumin	Electro-spinning	Chitosan/poly (lactic acid)	Improving the bioavailability	Dhurai et al. (2013)
Gallic acid	Electro-spinning	Zein	Preserving phenolic character	Neo et al. (2013)
Superoxide dismutase	Electro-spinning	Poly (lactic acid) (PLLA)	Controlled release system	Chen et al. (2010)
Epigallocatechin gallate (EGCG)	Electro-spraying	Gelatin	Application in functional foods	Gómez-Mascaraque, Lagarón, and López-Rubio (2015)
Folic acid	Nano-spray dryer	Whey protein concentrate matrix and a commercial resistant starch	Improving stability	Pérez-Masiá et al. (2015)

inaccessible to the solvent at or below neutral pH (Kontopidis, Holt, & Sawyer, 2004). Recently, designing of nanovehicles by β-Lg have been expanded due to its ability to bind many compounds, cost effectiveness, abundant availability, and acceptability (ABD EL-SALAM & EL-SHIBINY, 2012). For instance, preservation of the antioxidant activity of (−)-Epigallocatechin-3-gallate (EGCG) was obtained by loading into β-Lg nanoparticles (Li, Du, Jin, & Du, 2012). In another study, the EGCG as a natural phenolic compound was loaded into thermally modified β-Lg to form coassembled nanovehicles (Shpigelman, Israeli, & Livney, 2010). These nanovehicles provided the best protection to EGCG against oxidation and degradation. Also, the resulted ultra-small sized EGCG β-Lg nanovehicles (<50 nm) facilitated their application in clear beverage formulations.

Casein nanoparticles: Casein is the main milk protein (approximately 80% of total proteins in bovine milk) and is mainly composed of calcium, phosphate, and protein, for the neonate (Holt & Sawyer, 1993). Almost, 95% of milk caseins are established by micelles as colloid particles. The micelles are spherical colloids of calcium phosphate, 50−500 nm in diameter (150 nm in average). Casein micelles (CMs) can be reassembled into spherical CMs; the micelles are composed of αs1-, αs2-, and β-caseins bound to calcium phosphate nanoclusters by their phosphoserine domains. κ-Casein binds with αs-caseins so that the N-terminal section binds with hydrophobic parts and C-terminal sections are placed in hydrophilic parts as shown in Fig. 2.4 (Fox, 2003).

Caseins have open structures containing numerous proline residues evenly distributed throughout their amino acids and are very tendentious as binders of bioactive compounds such as phenolic compounds and antioxidants. For example, the hydrophobic interactions between curcumin and β-casein micelles can increase the solubility of curcumin and its bioavailability and antioxidant activity (Esmaili et al., 2011). Therefore, β-casein micelles can be a good candidate for designing nanocarriers applicable in food formulations.

In a study by Sáiz-Abajo, González-Ferrero, Moreno-Ruiz, Romo-Hualde, and González-Navarro (2013), β-carotene as a popular antioxidant and colorant agent in food formulations was encapsulated in β-casein micelles in the form of nanostructures. For this work, self-assembly method was applied to reassemble nanomicelles containing β-carotene. The encapsulated β-carotene in β-casein nanoparticles showed a high stability against common food processes like heat stabilization, high-pressure processing, and other processes (Sáiz-Abajo et al., 2013). These results revealed that β-casein nanoparticles can provide a good nanoencapsulation system for heat-sensitive ingredients (e.g., phenolic compounds) in bakery products.

Chitosan nanoparticles: Chitosan is a unique cationic polysaccharide, which has long been applied in food industries and encapsulation systems because of its film-forming ability and superior biodegradability and biocompatibility (Sabaghi, Maghsoudlou, & Habibi, 2015a; Sabaghi, Maghsoudlou, Khomeiri, & Ziaiifar, 2015b; Kumar, 2000). Chitosan nanoparticles can also be applied for nanoencapsulation of phenolic compounds and antioxidants because of its mucous adhesiveness, nontoxicity, biocompatibility, and biodegradability. In a study by Yadav, Lomash, Samim, and Flora (2012), curcumin was encapsulated in water-soluble chitosan nanoparticles (<50 nm). They indicated that encapsulated curcumin in chitosan nanoparticles provides high antioxidant and chelating potentials compared with free curcumin at a much lower dose.

Tang et al. (2013) successfully prepared a pH-responsive nanoparticle self-assembled by chitosan and poly γ-glutamic acid (γ-PGA) for oral delivery of tea catechins. The nanoparticles with a positive surface charge could transiently open the tight junctions between $CaCO_2$ cells and thus increase the paracellular transport of tea catechins. These results suggested that food formulations such as beverages can be fortified with chitosan/γ-PGA nanoparticle as a carrier for transmucosal delivery of tea catechins. In another study by Konecsni, Low, and Nickerson (2012), the chitosan (CH)–tripolyphosphate (TPP) submicron particles fabricated for encapsulation of rutin. Loaded rutin in CH-TPP nanoparticles had a good stability in simulated stomach fluids with high entrapment efficiencies. The results of this study suggested that the CH-TPP nanoparticles can be a good candidate for nanoencapsulation of phenolic compounds in food and natural health products.

Cyclodextrin nanoparticles: Cyclodextrins (CDs) are cyclic oligosaccharides consisting of 6, 7, or 8 glucopyranose units linked by a (1–4) glycosidic bond. Naturally, CDs have three cylindrical shapes of α-, β-, and γ-CD, but β form is commonly used for the encapsulation purpose (Fig. 2.4). Truncated cone shape of CDs, with a hydrophobic part inside and a hydrophilic external surface causes the formation of inclusion complexes (ICs) with water-insoluble ingredients (Challa, Ahuja, Ali, & Khar, 2005; Pinho, Grootveld, Soares, & Henriques, 2014). Indeed, phenolics and antioxidants can be surrounded by hydrophobic components in CDs, and could be protected from environment conditions (pH, light, oxygen, and temperature), with an improved water solubility. For instance, curcumin was incorporated into hydrophobic part of CDs for increasing its water solubility (Yallapu, Jaggi, & Chauhan, 2010). Curcumin encapsulation efficiency was improved by increasing the ratio of curcumin to CD. Results showed that encapsulation of curcumin in CD nanoparticles enhanced curcumin delivery and improved its therapeutic efficacy in prostate cancer cells compared with free curcumin (Yallapu et al., 2010). In another research, nanoencapsulation of catechins with β-cyclodextrin (BCD) at 1:1 ratio was achieved by using molecular inclusion method (Krishnaswamy, Orsat, & Thangavel, 2012). The size of resulted particles was 67–470 nm. In this study, the results of the interaction between catechins and BCD showed the supramolecular complex formation.

Complex biopolymer nanoparticles: Two or more different biopolymers such as proteins and polysaccharides can be applied in biopolymer nanoparticles. Encapsulated ingredients in these systems have higher stability against environmental conditions due to entrapment into strong complex walls as shown in Fig. 2.4 (Esfanjani & Jafari, 2016). Different nutraceuticals including β-carotene, curcumin, folic acid, and ergocalciferol have been encapsulated in biopolymer nanoparticles fabricated by β-lactoglobulin and sodium alginate complexes. The resulted soluble nanocomplexes had a good stability against aggregation (Hosseini, Emam-Djomeh, Sabatino, & Van Der Meeren, 2015). Anthocyanins have also been loaded in complex biopolymer nanoparticles prepared by using thermal processing and electrostatic complexation of WPI and beet pectin (BP). The mean particle diameter was relatively small (200 nm) around pH 4.0 but increased appreciably at lower and higher pH values due to particle flocculation. A higher loading efficiency (LE) was observed when anthocyanin was added before heating the WPI–BP solution (LE = 55%) rather than afterward (LE = 35%), which was attributed to increased protein–anthocyanin interactions (Arroyo-Maya & McClements, 2015). In a study by Ha, Kim, Lee, and Lee (2013), β-lactoglobulin and chitosan complex nanoparticles were applied as a nanoencapsulation system for quercetin. The formation and physical properties of produced nanoparticles were affected by charged amount of β-Lg and subambient temperature treatment. The result of this study provides a nonlipid delivery system of quercetin for food formulations, especially low-fat and nonfat food.

2.7 EQUIPMENT-BASED TECHNOLOGIES FOR NANOENCAPSULATION OF PHENOLICS AND ANTIOXIDANTS

Recently, specialized instruments such as electro-spinning, electro-spraying, and nano-spray dryer have been applied in nanoencapsulation of bioactive compounds. These instruments have some advantages including the fast and industrial production of nanoencapsulated ingredients, high encapsulation efficiency, and production of solid encapsulated forms of bioactive compounds like nanopowders or nanofibers, with a commercial and simple setup. These instruments are based on electricity and can be classified into thermal methods (e.g., nano-spray drying) and nonthermal methods (e.g., electro-spraying and electro-spinning) as described in Table 2.4 (Esfanjani & Jafari, 2016).

2.7.1 Electro-Spinning and Electro-Spraying

The phenolic compounds and antioxidants can be loaded into nanofibers by using electro-spinning (Esfanjani & Jafari, 2016). Biopolymers such as proteins (e.g., WPI or WPC, casein, and zein) and carbohydrates (e.g., starch, pectin, cellulose, gums, and chitosan) are good candidates for designing nanofibers as a nanoencapsulation system of bioactive compounds in food formulations through electro-spinning (Ghorani & Tucker, 2015). The electro-spinning system is composed of a high-voltage supplier, a feeding pump, a capillary tube with a small-diameter pipette or needle, syringe, polymer solution, Jet, Taylor cone, and a metal collecting screen. In the electro-spinning process, the injected polymer solution is charged while exiting from the capillary tip and starts to deform from the shape caused by surface tension alone. When the electric field attains a critical value, electrical force overcomes the surface tension and the charged jet of the fluid is ejected from the tip of the Taylor cone. The first jet has a straight-line motion but it is then elongated partly by means of the "whipping instability" during its transit from the tip to the collector, also during the transition is drown as an extended fiber by electrostatic filed and then quickly solidifies by solvent evaporation. After collision with the metal plate, nanostructured fibers are collected from the plate (Esfanjani & Jafari, 2016).

Quercetin and ferulic acid were incorporated within amaranth protein isolate and pullulan ultrathin fibers using the electro-spinning process (Aceituno-Medina, Mendoza, Rodríguez, Lagaron, & López-Rubio, 2015). The results suggested a sustained release of quercetin and ferulic acid through in vitro digestion from the produced fibers which can improve their antioxidant capacity in comparison with the free compounds. In another study by Neo et al. (2013), zein ultra-fine fibers enriched with gallic acid were produced by using electro-spinning technique. The diameter range of final fibers was 327–387 nm. The resulted information from XRD, DSC, and ATR-FTIR proved that interaction occurred between gallic acid and zein at the molecular level and also showed that the degradation temperature of the gallic acid–zein complex had shifted to a slightly lower temperature compared with the pure gallic acid powder due to the changes in its physical status. Totally, the results of this study demonstrated that the encapsulated gallic acid preserved its phenolic character and antioxidant activity after electro-spinning.

The phenolics and antioxidants can also be encapsulated in nanocapsules by using electro-spraying (electro-hydrodynamic spraying). The principle of this process is like the electro-spinning, except that the jet formed in electro-spraying breaks into droplets (Tapia-Hernández et al., 2015). López-Rubio and Lagaron (2012) successfully produced WPC nanocapsules containing β-carotene by using electro-spraying process. The results of this study showed that the size of capsules was affected by pH of the starting WPC aqueous solution so that the greater amounts of smaller capsules (submicron and nanocapsules) with narrower distribution were obtained by changing the pH of the solution from pH 6.4. The results also suggested that electro-spraying can provide the industrial production of nanoencapsulated bioactive compounds with applicability in food formulations.

2.7.2 Nano-Spray Drying

Spray dryer is one of the industrial, commercial, flexible, and economical instruments for microencapsulation of the bioactive compounds, which has been used frequently in the food industry. Indeed, this instrument is used for transforming liquids (solutions, emulsions, suspension, slurries, pastes, or even melts) into solid powders. The encapsulated powders obtained by traditional spray dryers are generally of micron size, and usually heterogeneous and amorphous. Two major morphologies of encapsulated capsules from spraying methods (like spray drying, nano-spray drying, and electro-spraying) can be single-core coated by a wall and multiple-core dispersed in a matrix as shown in Fig. 2.4 (Esfanjani et al., 2015; Jafari et al., 2008b; Mahdavi et al., 2014).

Recently, Büchi company has introduced a nano-spray dryer with the capability of producing nano-scale encapsulated powders with very narrow distributions and high formulation yields (Li, Anton, Arpagaus, Belleteix, & Vandamme, 2010). The technological novelty of this patented spray dryer lies in the gentle laminar drying flow, the vibrating mesh spray technology, and the highly efficient electrostatic particle collector (Arpagaus, 2011). There are some studies that have shown that nano-spray dryer can be used as an appropriate method for nanoencapsulation of phenolic compounds and antioxidants (Pérez-Masiá et al., 2015; O'Toole et al., 2012).

2.8 CONCLUSION AND FURTHER REMARKS

Phenolic compounds and antioxidants have some advantages such as disease prevention (e.g., anticancer), removing of free radicals, etc. These benefits have resulted in the production of functional and healthy foods by applying phenolics and antioxidants in food formulations. Therefore, it is important to implement the best nanoencapsulation system for delivery, protection, and controlled release of these bioactive compounds. In this chapter, we explained the nanoencapsulation systems as the best delivery systems of these compounds in food formulations compared with their microencapsulated forms. Selecting an appropriate nanoencapsulation system for food fortification is dependent on some main points including the goal of encapsulation and the type of food formulation that is supposed to be fortified by nanoencapsulation systems. Indeed, nanoencapsulation systems must not change the main properties of fortified food formulations such as appearance and rheology. For example, dairy products such as milk should be fortified by nanoencapsulated phenolics and antioxidants prepared by milk proteins (e.g., whey proteins and caseins). Other sections of this chapter also provided the main nanoencapsulation techniques applicable for phenolic compounds and antioxidants with the possibility to be incorporated into food formulations. There are a variety of food-grade nanoencapsulation systems for phenolics and antioxidants. These systems are classified into lipid-based nanoencapsulation methods (nanoemulsions, nanoliposome, SLNs, and NLCs), biopolymeric nanoparticles, and equipment-based nanoencapsulation technologies (electro-spinning, electro-spraying, and nano-spray dryer).

REFERENCES

Abbas, S., Bashari, M., Akhtar, W., Li, W. W., & Zhang, X. (2014). Process optimization of ultrasound-assisted curcumin nanoemulsions stabilized by OSA-modified starch. *Ultrasonics Sonochemistry, 21*, 1265–1274.

Abd El-Salam, M. H., & El-Shibiny, S. (2012). Formation and potential uses of milk proteins as nano delivery vehicles for nutraceuticals: a review. *International Journal of Dairy Technology, 65*, 13–21.

Aceituno-Medina, M., Mendoza, S., Rodríguez, B. A., Lagaron, J. M., & López-Rubio, A. (2015). Improved antioxidant capacity of quercetin and ferulic acid during in-vitro digestion through encapsulation within food-grade electrospun fibers. *Journal of Functional Foods, 12*, 332–341.

Anand, P., Kunnumakkara, A. B., Newman, R. A., & Aggarwal, B. B. (2007). Bioavailability of curcumin: problems and promises. *Molecular Pharmaceutics, 4*, 807–818.

Anton, N., Benoit, J.-P., & Saulnier, P. (2008). Design and production of nanoparticles formulated from nano-emulsion templates—A review. *Journal of Controlled Release, 128*, 185–199.

Arpagaus, C. (2011). Nano spray dryer B-90: Literature review and applications. *Büchi Information Bulletin, 63*, 8.

Arroyo-Maya, I. J., & Mcclements, D. J. (2015). Biopolymer nanoparticles as potential delivery systems for anthocyanins: Fabrication and properties. *Food Research International, 69*, 1–8.

Benita, S. (2005). Microencapsulation*: Methods and industrial* applications. New York: CRC Press.

Bjelakovic, G., Nikolova, D., Simonetti, R. G., & Gluud, C. (2004). Antioxidant supplements for prevention of gastrointestinal cancers: a systematic review and meta-analysis. *The Lancet, 364*, 1219–1228.

Bouwmeester, H., Dekkers, S., Noordam, M. Y., Hagens, W. I., Bulder, A. S., De Heer, C., ... Sips, A. J. (2009). Review of health safety aspects of nanotechnologies in food production. *Regulatory Toxicology and Pharmacology, 53*, 52–62.

Cadena, P. G., Pereira, M. A., Cordeiro, R. B., Cavalcanti, I. M., Neto, B. B., Maria Do Carmo, C., ... Santos-Magalhães, N. S. (2013). Nanoencapsulation of quercetin and resveratrol into elastic liposomes. *Biochimica et Biophysica Acta (BBA)-Biomembranes, 1828*, 309–316.

Challa, R., Ahuja, A., Ali, J., & Khar, R. (2005). Cyclodextrins in drug delivery: An updated review. *Aaps Pharmscitech, 6*, E329–E357.

Chatzidaki, M. D., Mitsou, E., Yaghmur, A., Xenakis, A., & Papadimitriou, V. (2015). Formulation and characterization of food-grade microemulsions as carriers of natural phenolic antioxidants. *Colloids and Surfaces A: Physicochemical and Engineering Aspects, 483*, 130–136.

Chen, H., Guan, Y., & Zhong, Q. (2015a). Microemulsions based on a sunflower lecithin–Tween 20 blend have high capacity for dissolving peppermint oil and stabilizing coenzyme Q10. *Journal of Agricultural and Food Chemistry, 63*, 983–989.

Chen, P., Sun, Y.-J., Zhu, Z.-C., Wang, R.-X., Shi, X.-D., Lin, C., & Ye, Y.-T. (2010). A controlled release system of superoxide dismutase by electrospun fiber and its antioxidant activity in vitro. *Journal of Materials Science: Materials in Medicine*, *21*, 609–614.

Chen, X., Zou, L.-Q., Niu, J., Liu, W., Peng, S.-F., & Liu, C.-M. (2015b). The stability, sustained release and cellular antioxidant activity of curcumin nanoliposomes. *Molecules*, *20*, 14293–14311.

Chuacharoen, T., & Sabliov, C. M. (2016). The potential of zein nanoparticles to protect entrapped β-carotene in the presence of milk under simulated gastrointestinal (GI) conditions. *LWT-Food Science and Technology*, *72*, 302–309.

Committee, E. S. (2011). Guidance on the risk assessment of the application of nanoscience and nanotechnologies in the food and feed chain. *EFSA Journal*, *9*, 2140.

Davidov-Pardo, G., & Mcclements, D. J. (2015). Nutraceutical delivery systems: resveratrol encapsulation in grape seed oil nanoemulsions formed by spontaneous emulsification. *Food Chemistry*, *167*, 205–212.

Dhurai, B., Saraswathy, N., Maheswaran, R., Sethupathi, P., Vanitha, P., Vigneshwaran, S., & Rameshbabu, V. (2013). Electrospinning of curcumin loaded chitosan/poly (lactic acid) nanofilm and evaluation of its medicinal characteristics. *Frontiers of Materials Science*, *7*, 350–361.

Esfanjani, A. F., & Jafari, S. M. (2016). Biopolymer nano-particles and natural nano-carriers for nano-encapsulation of phenolic compounds. *Colloids and Surfaces B: Biointerfaces*, *146*, 532–543.

Esfanjani, A. F., Jafari, S. M., Assadpoor, E., & Mohammadi, A. (2015). Nano-encapsulation of saffron extract through double-layered multiple emulsions of pectin and whey protein concentrate. *Journal of Food Engineering*, *165*, 149–155.

Esmaili, M., Ghaffari, S. M., Moosavi-Movahedi, Z., Atri, M. S., Sharifizadeh, A., Farhadi, M., ... Moosavi-Movahedi, A. A. (2011). Beta casein-micelle as a nano vehicle for solubility enhancement of curcumin; food industry application. *LWT-Food Science and Technology*, *44*, 2166–2172.

Fan, Y., Ma, L., Zhang, W., Wang, J., Chen, Y., Gao, Y., ... Song, X. (2014). The design of propolis flavone microemulsion and its effect on enhancing the immunity and antioxidant activity in mice. *International Journal of Biological Macromolecules*, *65*, 200–207.

Fang, J.-Y., Hwang, T.-L., Huang, Y.-L., & Fang, C.-L. (2006). Enhancement of the transdermal delivery of catechins by liposomes incorporating anionic surfactants and ethanol. *International Journal of Pharmaceutics*, *310*, 131–138.

Fang, Z., & Bhandari, B. (2010). Encapsulation of polyphenols—A review. *Trends in Food Science & Technology*, *21*, 510–523.

Fathi, M., Mozafari, M., & Mohebbi, M. (2012). Nanoencapsulation of food ingredients using lipid based delivery systems. *Trends in Food Science & Technology*, *23*, 13–27.

Fathi, M., Varshosaz, J., Mohebbi, M., & Shahidi, F. (2013). Hesperetin-loaded solid lipid nanoparticles and nanostructure lipid carriers for food fortification: Preparation, characterization, and modeling. *Food and Bioprocess Technology*, *6*, 1464–1475.

Fernandez, P., Andre, V., Rieger, J., & Kuhnle, A. (2004). Nano-emulsion formation by emulsion phase inversion. *Colloids and Surfaces A: Physicochemical and Engineering Aspects*, *251*, 53–58.

Flanagan, J., & Singh, H. (2006a). Microemulsions: A potential delivery system for bioactives in food. *Critical Reviews in Food Science and Nutrition*, *46*, 221–237.

Flanagan, J., & Singh, H. (2006b). *Recent advances in the delivery of food-derived bioactives and drugs using microemulsions. Nanocarrier Technologies* (pp. 95–111). Amesterdam: Springer.

Fox, P. (2003). *Milk proteins: General and historical aspects. Advanced Dairy Chemistry—1 Proteins* (pp. 1–48). New York: Springer.

Frassinetti, S., Caltavuturo, L., Cini, M., Della Croce, C., & Maserti, B. (2011). Antibacterial and antioxidant activity of essential oils from *Citrus* spp. *Journal of Essential Oil Research*, *23*, 27–31.

Ghorani, B., & Tucker, N. (2015). Fundamentals of electrospinning as a novel delivery vehicle for bioactive compounds in food nanotechnology. *Food Hydrocolloids*, *51*, 227–240.

Gokce, E. H., Korkmaz, E., Dellera, E., Sandri, G., Bonferoni, M. C., & Ozer, O. (2012). Resveratrol-loaded solid lipid nanoparticles versus nanostructured lipid carriers: Evaluation of antioxidant potential for dermal applications. *International Journal of Nanomedicine*, *7*, 1841.

Gómez-Mascaraque, L. G., Lagarón, J. M., & López-Rubio, A. (2015). Electrosprayed gelatin submicroparticles as edible carriers for the encapsulation of polyphenols of interest in functional foods. *Food Hydrocolloids*, *49*, 42–52.

Goniotaki, M., Hatziantoniou, S., Dimas, K., Wagner, M., & Demetzos, C. (2004). Encapsulation of naturally occurring flavonoids into liposomes: physicochemical properties and biological activity against human cancer cell lines. *Journal of Pharmacy and Pharmacology*, *56*, 1217–1224.

Ha, H.-K., Kim, J. W., Lee, M.-R., & Lee, W.-J. (2013). Formation and characterization of quercetin-loaded chitosan oligosaccharide/β-lactoglobulin nanoparticle. *Food Research International*, *52*, 82–90.

Haratifar, S., & Corredig, M. (2014). Interactions between tea catechins and casein micelles and their impact on renneting functionality. *Food Chemistry*, *143*, 27–32.

Heinonen, M. (2007). Antioxidant activity and antimicrobial effect of berry phenolics—A Finnish perspective. *Molecular Nutrition & Food Research*, *51*, 684–691.

Hejri, A., Khosravi, A., Gharanjig, K., & Hejazi, M. (2013). Optimisation of the formulation of β-carotene loaded nanostructured lipid carriers prepared by solvent diffusion method. *Food Chemistry*, *141*, 117–123.

Hentschel, A., Gramdorf, S., Muller, R., & Kurz, T. (2008). β-Carotene-loaded nanostructured lipid carriers. *Journal of Food Science*, *73*, N1–N6.

Holt, C., & Sawyer, L. (1993). Caseins as rheomorphic proteins: Interpretation of primary and secondary structures of the α S1-, β- and κ-caseins. *Journal of the Chemical Society, Faraday Transactions*, *89*, 2683–2692.

Hosseini, S. M. H., Emam-Djomeh, Z., Sabatino, P., & Van Der Meeren, P. (2015). Nanocomplexes arising from protein-polysaccharide electrostatic interaction as a promising carrier for nutraceutical compounds. *Food Hydrocolloids*, *50*, 16–26.

Hu, D., Lin, C., Liu, L., Li, S., & Zhao, Y. (2012). Preparation, characterization, and in vitro release investigation of lutein/zein nanoparticles via solution enhanced dispersion by supercritical fluids. *Journal of Food Engineering*, *109*, 545–552.

Jafari, S. M., Assadpoor, E., Bhandari, B., & He, Y. (2008a). Nano-particle encapsulation of fish oil by spray drying. *Food Research International*, *41*, 172–183.

Jafari, S. M., Assadpoor, E., He, Y., & Bhandari, B. (2008b). Encapsulation efficiency of food flavours and oils during spray drying. *Drying Technology*, *26*, 816–835.

Jafari, S. M., Assadpoor, E., He, Y., & Bhandari, B. (2008c). Re-coalescence of emulsion droplets during high-energy emulsification. *Food Hydrocolloids*, *22*, 1191–1202.

Jafari, S. M., He, Y., & Bhandari, B. (2007). Effectiveness of encapsulating biopolymers to produce sub-micron emulsions by high energy emulsification techniques. *Food Research International*, *40*, 862–873.

Jones, O. G., Lesmes, U., Dubin, P., & Mcclements, D. J. (2010). Effect of polysaccharide charge on formation and properties of biopolymer nanoparticles created by heat treatment of β-lactoglobulin–pectin complexes. *Food Hydrocolloids, 24*, 374–383.

Jose, S., Anju, S., Cinu, T., Aleykutty, N., Thomas, S., & Souto, E. (2014). In vivo pharmacokinetics and biodistribution of resveratrol-loaded solid lipid nanoparticles for brain delivery. *International Journal of Pharmaceutics, 474*, 6–13.

Kakkar, V., Singh, S., Singla, D., & Kaur, I. P. (2011). Exploring solid lipid nanoparticles to enhance the oral bioavailability of curcumin. *Molecular Nutrition & Food Research, 55*, 495–503.

Karadag, A., Yang, X., Ozcelik, B., & Huang, Q. (2013). Optimization of preparation conditions for quercetin nanoemulsions using response surface methodology. *Journal of Agricultural and Food Chemistry, 61*, 2130–2139.

Katouzian, I., & Jafari, S. M. (2016). Nano-encapsulation as a promising approach for targeted delivery and controlled release of vitamins. *Trends in Food Science & Technology, 53*, 34–48.

Konecsni, K., Low, N., & Nickerson, M. (2012). Chitosan–tripolyphosphate submicron particles as the carrier of entrapped rutin. *Food Chemistry, 134*, 1775–1779.

Kontopidis, G., Holt, C., & Sawyer, L. (2004). Invited review: β-lactoglobulin: binding properties, structure, and function. *Journal of Dairy Science, 87*, 785–796.

Krishnaswamy, K., Orsat, V., & Thangavel, K. (2012). Synthesis and characterization of nano-encapsulated catechin by molecular inclusion with beta-cyclodextrin. *Journal of Food Engineering, 111*, 255–264.

Kristl, J., Teskač, K., Caddeo, C., Abramović, Z., & Šentjurc, M. (2009). Improvements of cellular stress response on resveratrol in liposomes. *European Journal of Pharmaceutics and Biopharmaceutics, 73*, 253–259.

Kumar, M. N. R. (2000). A review of chitin and chitosan applications. *Reactive and Functional Polymers, 46*, 1–27.

Larrauri, J. A., Ruperez, P., & Saura-Calixto, F. (1997). Effect of drying temperature on the stability of polyphenols and antioxidant activity of red grape pomace peels. *Journal of Agricultural and Food Chemistry, 45*, 1390–1393.

Li, B., Du, W., Jin, J., & Du, Q. (2012). Preservation of (−)-epigallocatechin-3-gallate antioxidant properties loaded in heat treated β-lactoglobulin nanoparticles. *Journal of Agricultural and Food Chemistry, 60*, 3477–3484.

Li, H., Zhao, X., Ma, Y., Zhai, G., Li, L., & Lou, H. (2009a). Enhancement of gastrointestinal absorption of quercetin by solid lipid nanoparticles. *Journal of Controlled Release, 133*, 238–244.

Li, M., Zahi, M. R., Yuan, Q., Tian, F., & Liang, H. (2015a). Preparation and stability of astaxanthin solid lipid nanoparticles based on stearic acid. *European Journal of Lipid Science and Technology, 118*(4), 592–602.

Li, X., Anton, N., Arpagaus, C., Belleteix, F., & Vandamme, T. F. (2010). Nanoparticles by spray drying using innovative new technology: The Büchi Nano Spray Dryer B-90. *Journal of Controlled Release, 147*, 304–310.

Li, Y., Lim, L. T., & Kakuda, Y. (2009b). Electrospun zein fibers as carriers to stabilize (−)-epigallocatechin gallate. *Journal of Food Science, 74*, C233–C240.

Li, Z., Jiang, H., Xu, C., & Gu, L. (2015b). A review: Using nanoparticles to enhance absorption and bioavailability of phenolic phytochemicals. *Food Hydrocolloids, 43*, 153–164.

López-Rubio, A., & Lagaron, J. M. (2012). Whey protein capsules obtained through electrospraying for the encapsulation of bioactives. *Innovative Food Science & Emerging Technologies, 13*, 200–206.

Madrigal-Carballo, S., Lim, S., Rodriguez, G., Vila, A. O., Krueger, C. G., Gunasekaran, S., & Reed, J. D. (2010). Biopolymer coating of soybean lecithin liposomes via layer-by-layer self-assembly as novel delivery system for ellagic acid. *Journal of Functional Foods*, *2*, 99–106.

Mahdavi, S. A., Jafari, S. M., Ghorbani, M., & Assadpoor, E. (2014). Spray-drying microencapsulation of anthocyanins by natural biopolymers: A review. *Drying Technology*, *32*, 509–518.

Manzocco, L., Calligaris, S., Mastrocola, D., Nicoli, M. C., & Lerici, C. R. (2000). Review of non-enzymatic browning and antioxidant capacity in processed foods. *Trends in Food Science & Technology*, *11*, 340–346.

Mao, L., Xu, D., Yang, J., Yuan, F., Gao, Y., & Zhao, J. (2009). Effects of small and large molecule emulsifiers on the characteristics of b-carotene nanoemulsions prepared by high pressure homogenization. *Food Technology and Biotechnology*, *47*, 336–342.

Mayer, S., Weiss, J., & Mcclements, D. J. (2013). Vitamin E-enriched nanoemulsions formed by emulsion phase inversion: Factors influencing droplet size and stability. *Journal of Colloid and Interface Science*, *402*, 122–130.

Mcclements, D. J. (2015). *Food emulsions: Principles, practices, and techniques*. New York: CRC Press.

Mcclements, D. J., & Li, Y. (2010). Review of in vitro digestion models for rapid screening of emulsion-based systems. *Food & Function*, *1*, 32–59.

Mirhosseini, H., Tan, C. P., Hamid, N. S., & Yusof, S. (2008). Effect of arabic gum, xanthan gum and orange oil contents on ζ-potential, conductivity, stability, size index and pH of orange beverage emulsion. *Colloids and Surfaces A: Physicochemical and Engineering Aspects*, *315*, 47–56.

Misra, K., Dhillon, G. S., Brar, S. K., & Verma, M. (2014). *Antioxidants. Biotransformation of Waste Biomass into High Value Biochemicals*. New York: Springer.

Mitri, K., Shegokar, R., Gohla, S., Anselmi, C., & Muller, R. H. (2011). Lipid nanocarriers for dermal delivery of lutein: Preparation, characterization, stability and performance. *International Journal of Pharmaceutics*, *414*, 267–275.

Moeiniafshari, A.-A., Zarrabi, A., & Bordbar, A.-K. (2015). Exploring the interaction of naringenin with bovine beta-casein nanoparticles using spectroscopy. *Food Hydrocolloids*, *51*, 1–6.

Mohammadi, A., Jafari, S. M., Assadpour, E., & Esfanjani, A. F. (2016a). Nano-encapsulation of olive leaf phenolic compounds through WPC–pectin complexes and evaluating their release rate. *International Journal of Biological Macromolecules*, *82*, 816–822.

Mohammadi, A., Jafari, S. M., Esfanjani, A. F., & Akhavan, S. (2016b). Application of nano-encapsulated olive leaf extract in controlling the oxidative stability of soybean oil. *Food Chemistry*, *190*, 513–519.

Mozafari, M. (2010). *Nanoliposomes: Preparation and analysis. Liposomes: Methods and protocols, volume 1: Pharmaceutical nanocarriers*, Springer, Berlin, 29–50.

Mozafari, M., Johnson, C., Hatziantoniou, S., & Demetzos, C. (2008). Nanoliposomes and their applications in food nanotechnology. *Journal of Liposome Research*, *18*, 309–327.

Muèller, R. H., Maèder, K., & Gohla, S. (2000). Solid lipid nanoparticles (SLN) for controlled drug delivery–A review of the state of the art. *European Journal of Pharmaceutics and Biopharmaceutics*, *50*, 161–177.

Muller, R., & Lucks, J. (1996). Arzneistoffträger aus festen Lipidteilchen. *Feste Lipidnanosphären (SLN), European Patent*, 605497.

Muller, R., Radtke, M., & Wissing, S. (2002). Nanostructured lipid matrices for improved microencapsulation of drugs. *International Journal of Pharmaceutics*, *242*, 121–128.

Neo, Y. P., Ray, S., Jin, J., Gizdavic-Nikolaidis, M., Nieuwoudt, M. K., Liu, D., & Quek, S. Y. (2013). Encapsulation of food grade antioxidant in natural biopolymer by electrospinning technique: A physicochemical study based on zein–gallic acid system. *Food Chemistry*, *136*, 1013–1021.

Neves, A. R., Lucio, M., Martins, S., Lima, J., & Reis, S. (2013). Novel resveratrol nanodelivery systems based on lipid nanoparticles to enhance its oral bioavailability. *International Journal of Nanomedicine*, *8*, 177–187.

Nitta, S. K., & Numata, K. (2013). Biopolymer-based nanoparticles for drug/gene delivery and tissue engineering. *International Journal of Molecular Sciences*, *14*, 1629–1654.

Ostertag, F., Weiss, J., & Mcclements, D. J. (2012). Low-energy formation of edible nanoemulsions: Factors influencing droplet size produced by emulsion phase inversion. *Journal of Colloid and Interface Science*, *388*, 95–102.

Palafox-Carlos, H., Ayala-Zavala, J. F., & González-Aguilar, G. A. (2011). The role of dietary fiber in the bioaccessibility and bioavailability of fruit and vegetable antioxidants. *Journal of Food Science*, *76*, R6–R15.

Perez-Masiá, R., López-Nicolás, R., Periago, M. J., Ros, G., Lagaron, J. M., & López-Rubio, A. (2015). Encapsulation of folic acid in food hydrocolloids through nanospray drying and electrospraying for nutraceutical applications. *Food Chemistry*, *168*, 124–133.

Pinho, E., Grootveld, M., Soares, G., & Henriques, M. (2014). Cyclodextrins as encapsulation agents for plant bioactive compounds. *Carbohydrate Polymers*, *101*, 121–135.

Priprem, A., Watanatorn, J., Sutthiparinyanont, S., Phachonpai, W., & Muchimapura, S. (2008). Anxiety and cognitive effects of quercetin liposomes in rats. *Nanomedicine: Nanotechnology, Biology and Medicine*, *4*, 70–78.

Qi, C., Chen, Y., Huang, J. H., Jin, Q. Z., & Wang, X. G. (2012). Preparation and characterization of catalase-loaded solid lipid nanoparticles based on soybean phosphatidylcholine. *Journal of the Science of Food and Agriculture*, *92*, 787–793.

Qian, C., Decker, E. A., Xiao, H., & Mcclements, D. J. (2012). Physical and chemical stability of β-carotene-enriched nanoemulsions: Influence of pH, ionic strength, temperature, and emulsifier type. *Food Chemistry*, *132*, 1221–1229.

Rafiee, Z., Jafari, S., Alami, M., & Khomeiri, M. (2012). Antioxidant effect of microwave-assisted extracts of olive leaves on sunflower oil. *Journal of Agricultural Science and Technology*, *14*, 1497–1509.

Rahmanian, N., Jafari, S. M., & Wani, T. A. (2015). Bioactive profile, dehydration, extraction and application of the bioactive components of olive leaves. *Trends in Food Science & Technology*, *42*, 150–172.

Ru, Q., Yu, H., & Huang, Q. (2010). Encapsulation of epigallocatechin-3-gallate (EGCG) using oil-in-water (O/W) submicrometer emulsions stabilized by ι-carrageenan and β-lactoglobulin. *Journal of Agricultural and Food Chemistry*, *58*, 10373–10381.

Sabaghi, M., Maghsoudlou, Y., & Habibi, P. (2015a). Enhancing structural properties and antioxidant activity of kefiran films by chitosan addition. *Food Structure*, *5*, 66–71.

Sabaghi, M., Maghsoudlou, Y., Khomeiri, M., & Ziaiifar, A. M. (2015b). Active edible coating from chitosan incorporating green tea extract as an antioxidant and antifungal on fresh walnut kernel. *Postharvest Biology and Technology*, *110*, 224–228.

Sáiz-Abajo, M.-J., González-Ferrero, C., Moreno-Ruiz, A., Romo-Hualde, A., & González-Navarro, C. J. (2013). Thermal protection of β-carotene in re-assembled casein micelles during different processing technologies applied in food industry. *Food Chemistry*, *138*, 1581–1587.

Samad, A., Sultana, Y., & Aqil, M. (2007). Liposomal drug delivery systems: An update review. *Current Drug Delivery*, *4*, 297–305.

Sanguansri, P., & Augustin, M. A. (2006). Nanoscale materials development—A food industry perspective. *Trends in Food Science & Technology*, *17*, 547–556.

Sari, T., Mann, B., Kumar, R., Singh, R., Sharma, R., Bhardwaj, M., & Athira, S. (2015). Preparation and characterization of nanoemulsion encapsulating curcumin. *Food Hydrocolloids*, *43*, 540–546.

Saura-Calixto, F., Pérez-Jiménez, J., & Goñi, I. (2009). Dietary fiber and associated antioxidants in fruit and vegetables. In L. A. de la Rosa, E. Alvarez-Parrilla, & G. A. González-Aguilar (Eds.), *Fruit and vegetable phytochemicals: Chemistry, nutritional value, and stability* (pp. 223–234). Oxford, UK: Wiley-Blackwell.

Scartezzini, P., & Speroni, E. (2000). Review on some plants of Indian traditional medicine with antioxidant activity. *Journal of Ethnopharmacology*, *71*, 23–43.

Sessa, M., Balestrieri, M. L., Ferrari, G., Servillo, L., Castaldo, D., D'onofrio, N., ... Tsao, R. (2014). Bioavailability of encapsulated resveratrol into nanoemulsion-based delivery systems. *Food Chemistry*, *147*, 42–50.

Shahidi, F., Janitha, P., & Wanasundara, P. (1992). Phenolic antioxidants. *Critical reviews in Food Science & Nutrition*, *32*, 67–103.

Shalaby, E. A., & Shanab, S. M. (2013). Antioxidant compounds, assays of determination and mode of action. *African Journal of Pharmacy and Pharmacology*, *7*, 528–539.

Shangguan, M., Lu, Y., Qi, J., Han, J., Tian, Z., Xie, Y., ... Wu, W. (2014). Binary lipids-based nanostructured lipid carriers for improved oral bioavailability of silymarin. *Journal of Biomaterials Applications*, *28*, 887–896.

Shpigelman, A., Israeli, G., & Livney, Y. D. (2010). Thermally-induced protein–polyphenol co-assemblies: beta lactoglobulin-based nanocomplexes as protective nanovehicles for EGCG. *Food Hydrocolloids*, *24*, 735–743.

Siegrist, M., Cousin, M.-E., Kastenholz, H., & Wiek, A. (2007). Public acceptance of nanotechnology foods and food packaging: The influence of affect and trust. *Appetite*, *49*, 459–466.

Siro, I., Kapolna, E., Kapolna, B., & Lugasi, A. (2008). Functional food. Product development, marketing and consumer acceptance—A review. *Appetite*, *51*, 456–467.

Souto, E., Wissing, S., Barbosa, C., & Muller, R. (2004). Development of a controlled release formulation based on SLN and NLC for topical clotrimazole delivery. *International Journal of Pharmaceutics*, *278*, 71–77.

Souza, V. B., Almeida, S. M., Spinelli, L. S., & Mansur, C. R. (2011). Stability of orange oil/water nanoemulsions prepared by the PIT method. *Journal of Nanoscience and Nanotechnology*, *11*, 2237–2243.

Sozer, N., & Kokini, J. L. (2009). Nanotechnology and its applications in the food sector. *Trends in Biotechnology*, *27*, 82–89.

Sun, J., Bi, C., Chan, H. M., Sun, S., Zhang, Q., & Zheng, Y. (2013). Curcumin-loaded solid lipid nanoparticles have prolonged in vitro antitumour activity, cellular uptake and improved in vivo bioavailability. *Colloids and Surfaces B: Biointerfaces*, *111*, 367–375.

Sun, M., Nie, S., Pan, X., Zhang, R., Fan, Z., & Wang, S. (2014). Quercetin-nanostructured lipid carriers: Characteristics and anti-breast cancer activities in vitro. *Colloids and Surfaces B: Biointerfaces*, *113*, 15–24.

Sundar, S., Kundu, J., & Kundu, S. C. (2016). Biopolymeric nanoparticles. *Science and Technology of Advanced Materials*, *11*(1), 1–13.

Taghvaei, M., & Jafari, S. M. (2015). Application and stability of natural antioxidants in edible oils in order to substitute synthetic additives. *Journal of Food Science and Technology*, *52*, 1272–1282.

Tang, D.-W., Yu, S.-H., Ho, Y.-C., Huang, B.-Q., Tsai, G.-J., Hsieh, H.-Y., ... Mi, F.-L. (2013). Characterization of tea catechins-loaded nanoparticles prepared from chitosan and an edible polypeptide. *Food Hydrocolloids*, *30*, 33–41.

Tapia-Hernández, J. A., Torres-Chavez, P. I., Ramirez-Wong, B., Rascon-Chu, A., Plascencia-Jatomea, M., Barreras-Urbina, C. G., ... Rodríguez-Felix, F. (2015). Micro- and nanoparticles by electrospray: advances and applications in foods. *Journal of Agricultural and Food Chemistry*, *63*(19), 4699–4707.

Teeranachaideekul, V., Souto, E. B., Junyaprasert, V. B., & Muller, R. H. (2007). Cetyl palmitate-based NLC for topical delivery of Coenzyme Q 10—Development, physicochemical characterization and in vitro release studies. *European Journal of Pharmaceutics and Biopharmaceutics*, *67*, 141–148.

Teixeira, B. N., Ozdemir, N., Hill, L. E., & Gomes, C. L. (2013). Synthesis and characterization of nano-encapsulated black pepper oleoresin using hydroxypropyl beta-cyclodextrin for antioxidant and antimicrobial applications. *Journal of Food Science*, *78*, N1913–N1920.

Tiyaboonchai, W., Tungpradit, W., & Plianbangchang, P. (2007). Formulation and characterization of curcuminoids loaded solid lipid nanoparticles. *International Journal of Pharmaceutics*, *337*, 299–306.

Tubesha, Z., Bakar, Z. A., & Ismail, M. (2013). Characterization and stability evaluation of thymoquinone nanoemulsions prepared by high-pressure homogenization. *Journal of Nanomaterials*, *2013*, 126.

Ünlusayin, M., Hădărugă, N. G., Rusu, G., Gruia, A. T., Păunescu, V., & Hădărugă, D. I. (2016). Nano-encapsulation competitiveness of omega-3 fatty acids and correlations of thermal analysis and Karl Fischer water titration for European anchovy (*Engraulis encrasicolus* L.) oil/β-cyclodextrin complexes. *LWT-Food Science and Technology*, *68*, 135–144.

Wang, X., Jiang, Y., Wang, Y.-W., Huang, M.-T., Ho, C.-T., & Huang, Q. (2008). Enhancing anti-inflammation activity of curcumin through O/W nanoemulsions. *Food Chemistry*, *108*, 419–424.

Woods, K. K., & Selling, G. W. (2007). Improved tensile strength of zein films using glyoxal as a crosslinking reagent. *Journal of Biobased Materials and Bioenergy*, *1*, 282–288.

Wu, Y., Luo, Y., & Wang, Q. (2012). Antioxidant and antimicrobial properties of essential oils encapsulated in zein nanoparticles prepared by liquid–liquid dispersion method. *LWT-Food Science and Technology*, *48*, 283–290.

Xia, S., Xu, S., & Zhang, X. (2006). Optimization in the preparation of coenzyme Q10 nanoliposomes. *Journal of Agricultural and Food Chemistry*, *54*, 6358–6366.

Yadav, A., Lomash, V., Samim, M., & Flora, S. J. (2012). Curcumin encapsulated in chitosan nanoparticles: a novel strategy for the treatment of arsenic toxicity. *Chemico-biological interactions*, *199*, 49–61.

Yallapu, M. M., Jaggi, M., & Chauhan, S. C. (2010). β-Cyclodextrin-curcumin self-assembly enhances curcumin delivery in prostate cancer cells. *Colloids and Surfaces B: Biointerfaces*, *79*, 113–125.

Yi, J., Li, Y., Zhong, F., & Yokoyama, W. (2014). The physicochemical stability and in vitro bioaccessibility of beta-carotene in oil-in-water sodium caseinate emulsions. *Food Hydrocolloids*, *35*, 19–27.

Yuan, C., Jin, Z., Xu, X., Zhuang, H., & Shen, W. (2008). Preparation and stability of the inclusion complex of astaxanthin with hydroxypropyl-β-cyclodextrin. *Food Chemistry*, *109*, 264–268.

Zhang, Y., Yang, Y., Tang, K., Hu, X., & Zou, G. (2008). Physicochemical characterization and antioxidant activity of quercetin-loaded chitosan nanoparticles. *Journal of Applied Polymer Science*, *107*, 891–897.

Zou, T., Li, Z., Percival, S. S., Bonard, S., & Gu, L. (2012). Fabrication, characterization, and cytotoxicity evaluation of cranberry procyanidins-zein nanoparticles. *Food Hydrocolloids*, *27*, 293–300.

Chapter 3

Nanoencapsulation of Fish Oil and Essential Fatty Acids

İsmail Tontul[1,2], Emrah Eroglu[1] and Ayhan Topuz[1]
[1]*Akdeniz University, Antalya, Turkey,* [2]*Necmettin Erbakan University, Konya, Turkey*

3.1 INTRODUCTION TO ESSENTIAL FATTY ACIDS

Fatty acids are the simplest and basic unit of lipids and are generally classified into two groups, which are saturated and unsaturated fatty acids. They consist of a carbon chain ($-CH_2-$), which ends with methyl ($-CH_3$) and carboxylic acid ($-COOH$) groups. Fatty acids are required for lots of biological functions which are vital for normal function of the body and good health.

The fatty acids differ in the number of carbons in their chain and the number and presence of carbon-to-carbon double bonds. According to the distance between methyl end and the double bond, they are classified as omega-9, omega-6, and omega-3 fatty acids. Oleic acid is the most important omega-9 fatty acid. The richest source of oleic acid is olive oil but significant amount is found in almost all fatty foods. Linoleic and arachidonic acids are the most important omega-6 fatty acids. They can be used in the transformation of higher fatty acids in human body by some elongation and desaturation reactions. These fatty acids are dominant in human body because of high content of omega-6 fatty acids in our diet. α-Linolenic acid (ALA), eicosapentaenoic acid (EPA), and docosahexaenoic acid (DHA) are the most important omega-3 fatty acids. These fatty acids are essential for humans since our body cannot produce these fatty acids due to lack of specific desaturase enzymes. Therefore, they must be taken by diet. However, human body can transform these fatty acids to each other by several steps of elongation, desaturation, and oxidation reactions (Benatti, Peluso, Nicolai, & Calvani, 2004). The rate of these transformations depends on gender, age, and physiological status. In general, low transformation rates are shown from ALA to EPA or DHA in humans (Walker, Decker, & McClements, 2015a). Therefore, the best way to increase plasma levels of these fatty acids is to consume the foods that contain high amount of them. In different studies, it is suggested to increase omega-3/omega-6 ratio in our diet (Simolopoulos, 1989). Different international organizations recomended to get

Nanoencapsulation of Food Bioactive Ingredients. DOI: http://dx.doi.org/10.1016/B978-0-12-809740-3.00003-9

more than 1% of energy from long chain polyunsaturated fatty acids (PUFAs; Force, 1992; Sugano, 1996).

3.1.1 Source of Essential Fatty Acids

Fish oils are the richest source of EPA and DHA. However, linolenic acids are generally found in the seeds, nuts, and plant oils. The EPA, DHA, and ALA content of different fishes, seeds, nuts, and plants are presented in Table 3.1.

3.1.2 Health Benefits of Essential Fatty Acids

Consuming fish oil and PUFAs are believed to have different health benefits throughout the world. The pioneer researches on the effect of the fish and fish

TABLE 3.1 **The Natural Sources of Essential Fatty Acids (g/100 g) (Raper, Cronin, & Exler, 1992; Tontul & Topuz, 2015)**

Source	ALA	EPA	DHA
Fishes			
Salmon	n.r.	0.3	0.5
Atlantic mackerel	n.r.	0.9	1.4
Atlantic herring	n.r.	0.7	0.9
Whitefish species	n.r.	0.3	0.9
Swordfish	n.r.	0.1	0.5
Smelt, rainbow	n.r.	0.3	0.4
Seed and Plant Oils			
Flaxseed oil	56.0	n.r.	n.r.
Perilla seed oil	56.8	n.r.	n.r.
Sacha inchi oil	50.4	n.r.	n.r.
Chia oil	52.0–63.4	n.r.	n.r.
Camelina oil	33.7–40.3	n.r.	n.r.
Purslane oil	31.4–43.7	n.r.	n.r.
Lingonberry seed oil	25.2–27.3	n.r.	n.r.
Walnut	10.6–16.2	n.r.	n.r.
Hemp seed oil	5.0–8.7	n.r.	n.r.

n.r. not reported,
ALA: α-Linolenic acid, EPA: Eicosapentaenoic acid, DHA: Docosahexaenoic acid

oil consuming on the health was conducted to determine why Greenland Eskimos had the lowest heart disease in 1970s (Sidhu, 2003). Since then effect of fish oils and PUFA consumption on different health problems was studied.

3.1.2.1 Coronary Heart Disease

Coronary heart disease is generally shown in the patients with high levels of plasma cholesterol, very low-density lipoproteins (VLDL), low-density lipoproteins (LDL), triglycerides, and hypertension. On the other hand, these patients generally have low levels of high-density lipoproteins (HDL). Several researches showed that diets with high level of PUFAs tend to lower plasma cholesterol, VLDL, LDL, and triglycerides (Goh, Jumpsen, Ryan, & Clandinin, 1997). Additionally, higher HDL levels in plasma were also determined in different studies (Damsgaard et al., 2006; Ramprasath, Eyal, Zchut, & Jones, 2013; Sanders & Hinds, 1992). However, there are also several papers that reported no linkage between fish oils and PUFAs consumption, and plasma lipoproteins (Ramprasath et al., 2013). Therefore, conflicting results reported about fish oil and PUFAs consumption on plasma lipoproteins level thereby coronary heart disease. The differences between studies could be related to source of oil, type of fatty acids, doses, and test subject (animal, human, blood, etc.).

There could be also different effects of fish oils and PUFAs that resulted in low coronary heart disease. Indeed, Bouwens et al. (2009) showed intake of PUFAs changed the expression of 1040 genes in human blood and decreased expression of genes involved in inflammatory and atherogenesis-related pathways. Additionally, improved plaque stability, vascular permeability, and decreased endothelial activation thereby lower cardiovascular risk was determined (Cawood et al., 2010; Schiano et al., 2008). Consumption of PUFAs has been also effective in anticoagulation therapies by dose escalation of the drugs which is very important for the patients who have low response to these drugs (Lev et al., 2010).

3.1.2.2 Rheumatoid Arthritis

Rheumatoid arthritis (RA) also has a low incidence in Greenland Eskimos. Therefore, some studies were conducted to determine the linkage between fish oils and PUFAs consumption and RA. Intake of fish oils and PUFAs showed alleviation of the symptoms and a reduction of neutrophil chemiluminescence (Belch & Muir, 1998; Magaro et al., 1988). However, Cleland, James, and Proudman (2003) suggest high consumption of fish oils and PUFAs to see these effects.

3.1.2.3 Brain Development and Function

Human nervous system contains substantial amount of DHA which could have some specific functions (Ruxton, Reed, Simpson, & Millington, 2004).

Yehuda, Rabinovitz, and Mostofsky (1999) claimed that omega-3 PUFA provide fluidity in neuronal membranes and regulation of transmitters. Additionally, consumption of PUFA increases the visual acuity.

3.1.2.4 Cancer

Some studies showed that omega-6 PUFAs can cause breast cancer invasion and metastasis via eicosanoid production. Increasing consumption of omega-3 PUFAs may suppress these effects (Rose & Connolly, 1990). Other researchers determined omega-3 PUFAs may retard the prostate cancer progression (Karmali et al., 1987; Rose & Cohen, 1988).

3.1.2.5 Mental Health

Richardson and Ross (2000) claimed fatty acid status had an effect on mental status, and imbalances can cause some disorders such as attention deficit hyperactivity disorder (ADHD), dyslexia, dyspraxia, and autism.

Stevens et al. (1995) observed low level of EPA and DHA in the plasma of children with ADHD compared to normal children. Additionally, a poor omega-3 PUFA status could be related to high risk of depression. Indeed, different researchers determined significant correlation between fish consumption and depression. (Grosso et al., 2016; Hibbeln, 1998; Smith et al., 2014; Timonen et al., 2004).

3.1.2.6 Infant Health

Increasing omega-3 PUFA intake during pregnancy not only enhances maternal PUFA levels but also benefits the fetus (Connor, Lowensohn, & Hatcher, 1996). After birth, the fatty acid status of the mother continues to impact her baby by breastfeeding. Breast milk is a natural rich source of DHA (Crawford, Hassam, & Stevens, 1981). High fish oil consumption during pregnancy increases visual acuity and development score in infants (Daniels, Longnecker, Rowland, Golding, & Team, 2004; Jørgensen, Hernell, Hughes, & Michaelsen, 2001).

3.1.3 Stability of the Fish Oil and Essential Fatty Acids

Oxidation is a big problem in fish oil and essential fatty acids since these oils are sensitive to oxidation. Oxidation not only deteriorates the taste and aroma of the oils but also leads to formation of hazardous compounds. Therefore, the mechanism and prevention method of oxidation must be known.

Lipid oxidation is a complex sequence of chemical reactions, which occurs between lipid and oxygen. Oxidation generally starts with the formation of free radicals; molecules that have unpaired electrons. The pathway of oxidation can be described in three general steps: Initiation, propagation, and termination. In the initiation step, a hydrogen from fatty acid is abstracted to form alkyl radical. Then, oxidation propagates by addition of oxygen to the alkyl radical. The combination of alkyl radical with another radical results in the formation of a covalent bond. The other radical on the oxygen remains free and peroxyl radical (LOO•) is formed. When the peroxyl radical is formed, oxidation promotes by the abstraction of a hydrogen from another molecule. At the last step, termination occurs between two radicals by combination to form nonradical species (McClements & Decker, 2008).

Lipid oxidation rates are affected from the oxygen concentration, temperature, surface area, water activity, and oil type. Among these factors oil type is very important for fish oils since these oils have high content of PUFAs. PUFAs are more susceptible to oxidation since C–H covalent bonds in unsaturated fatty acids are weaker (McClements & Decker, 2008).

During oxidation reactions, fatty acids decompose into small volatile molecules that cause off-flavor also known as rancidity. Additionally, free radicals formed during oxidation can cause toxic effects on the human body. Therefore, oxidation must be suppressed or retarded in fish oils and essential fatty acids. Nanoencapsulation is one of the techniques that is used to protect sensitive oils.

3.1.4 Nanoencapsulation of Fish Oil and Essential Fatty Acids

Encapsulation is a technology for physical packaging of sensitive oils in a wall material for different purposes. According to the size of obtained capsules, this technology is classified as microencapsulation or nanoencapsulation. In nanoencapsulation, the size of these physical capsules is lower than 1 μm while it was previously bigger up to 1 mm.

There are many different approaches in microencapsulation of fish oil and essential fatty acids. Among these techniques, spray-drying, freeze-drying, and coacervation are the most common techniques in fish oil and essential fatty acid microencapsulation. Some of the studies on microencapsulation of fish oil and essential oil are summarized in Table 3.2.

Nanocapsules have lower size thereby, much broader surface area than microcapsules. The lower size of the nanocapsules increased the transparency of the solution when utilized in clear liquids such as beverages and sauces. The main aims of the nanoencapsulations are:

- Protection of oils against oxidation
- Reducing the reactions between molecules and surrounding medium
- Controlled release of the oils
- Higher surface area for absorption of oils
- Improved solubility of the bioactive oils
- High intracellular uptake
- Pass along the smallest body fenestrations
- Formulating transparent oil products
- Reduction in the quantity of utilized core-shell material
- Reinforced physical stability against coalescence and gravitational separations (Katouzian & Jafari, 2016).

TABLE 3.2 Microencapsulation of Fish Oil and Essential Fatty Acids

Core Material	Wall Materials	Key Findings	Reference
Spray-Drying			
Menhaden oil	Maltodextrin (MD) + fish gelatin and/or κ-carragenan	MD + fish gelatin as wall material provided the highest microencapsulation efficiency. The release of fish oil from the microcapsules showed similar trend regardless of the wall material combination. In first 3 h a quick release was shown by burst effect, then it became steady	(Mehrad, Shabanpour, Jafari, & Pourashouri, 2015)
Cod liver oil or omega-3-enriched oil	Four different wall material combination tested [MD + fish gelatin, MD + chitosan, MD + fish gelatin + chitosan and MD + fish gelatin + Transglutaminase (TGase)]	Cross-linking of wall materials by TGase limited the burst effect during release. Microencapsulation of fish oil increased the stability of the oils during 60-day storage in terms of peroxide value, p-anisidine value, and thiobarbituric acid reactive substances. Among the wall material combinations, MD + chitosan and MD + fish gelatin + TGase provided the highest stability	(Pourashouri et al., 2014b)
Cod liver oil or omega-3-enriched oil	Four different wall material combination tested (MD + fish gelatin, MD + chitosan, MD + fish gelatin + chitosan and MD + fish gelatin + TGase)	The physiochemical properties of the microcapsules prepared with different wall material combinations were determined. Using fish gelatin in wall material combination resulted in higher microencapsulation efficiency in comparison to the chitosan. Cross-linking of the gelatin and maltodextrin with transglutaminase increased the efficiency. Scanning electron microscopy revealed that microcapsules prepared with MD + fish gelatin and MD + fish gelatin + TGase had less shrinkage and cracks than others	(Pourashouri et al., 2014a)

Anchovy oil	Gum Arabic and/or hemicellulose	Using hemicellulose as wall materials resulted in not only fewer dents and cracks in microcapsules but also higher microencapsulation efficiency	(Tatar, Tunç, Dervisoglu, Cekmecelioglu, & Kahyaoglu, 2014)
Grass carp oil	Gum Arabic/casein/beta-cyclodextrin mixtures	Wall material concentration, core-to-wall ratio, emulsifying temperature, and air inlet temperature was optimized according to emulsion viscosity, emulsion stability, encapsulation efficiency, and yield. Optimum conditions were determined as wall materials concentration of 35%, core-to-wall ratio of 3:7, emulsifying temperature of 55°C, and air inlet temperature of 200°C	(Li, Xiong, Wang, Regenstein, & Liu, 2015)
Freeze-Drying			
Sandeel oil	Sodium caseinate and lactose	Initial oil droplet size was not affected microencapsulation efficiency and storage stability of the microcapsules. However, fast freezing using liquid nitrogen prior to the freeze-drying provided longer shelf life	(Heinzelmann, Franke, Velasco, & Marquez-Ruiz, 2000)
Tilapia oil	Trehalose, gelatin, sucrose, xanthan gum	Inlet air temperature, solid content, drying air flow rate, and atomizing pressure were optimized in tilapia oil microencapsulation by spray-drying. The optimum conditions were determined as 121°C inlet air temperature, 0.65 m^3/min drying air flow rate and 100 kPa atomization pressure which provided higher than 90% of microencapsulation efficiency. Trehalose was more effective against to the oxidation	(Huang et al., 2014)
Fish oil	Hydrolyzed soy protein isolate and maltodextrin conjugate	Hydrolization of soy protein isolate and maltodextrin improved the oxidative and thermal stability of obtained microcapsules	(Zhang et al., 2014)

(Continued)

TABLE 3.2 (Continued)

Core Material	Wall Materials	Key Findings	Reference
α-linolenic acid rich oil	Gum ghatti, gum arabic, and soy protein isolate	The ratio between gum and soy protein isolate and wall-to-core, and homogenization speed was optimized to achieve maximum microencapsulation efficiency. The optimum conditions were determined as Gum:SPI ratio of 0.32, wall-to-core ratio of 2 and homogenization speed of 10,000 rpm	(Naik, Meda, & Lele, 2014)
Coacervation			
Fortified tuna oil	Gelatin and sodium hexametaphosphate (SHMP). TGase as cross-linker	High encapsulation efficiency and yield (>88) was reported. The size of the microcapsules varied in the range of 50–80 μm. About 15,000 rpm was the best homogenization speed for generating emulsions prior to complex coacervation	(Wang, Vongsvivut, Adhikari, & Barrow, 2015)
		FTIR results indicated that there was no chemical bonding between the oil and the wall material and chemical changes for gelation probably due to interaction with SHMP	
Tuna oil	Gelatin and sodium hexametaphosphate. TGase as cross-linker	pH 4.7 was determined as the optimum pH for coacervation	(Wang, Adhikari, & Barrow, 2014)
		Among different tested gelatin/SHMP ratio (5 to 45) the ratio of 15 provided highest yield. There was no precipitation of coacervates when the gelatin/SHMP ratio was either lower than 7.5 or higher than 35	
		Oxidative stability of the noncross-linked (39.32 h) and cross-linked microcapsules (40.16 h) were found to be similar but higher than bulk oil	

Tuna oil	Whey protein isolate and gum Arabic (GA)	The effect of spray-drying and freeze-drying on oxidative stability of coacervates were determined. Oxidative stability index of spray-dried microcapsules was found to be higher than freeze-dried samples	(Eratte et al., 2015)
Omega-3 rich tuna oil	Whey protein isolate and GA	The optimal conditions were determined as pH of 3.75 and whey protein isolate and gum Arabic ratio of 3:1. Spray-drying of coacervates resulted in more stable microcapsules	(Eratte, Wang, Dowling, Barrow, & Adhikari, 2014)
Fish oil	Gelatin and GA	Increasing wall to core ratio in the formulas until a certain level increased the microencapsulation efficiency from 17% to 92%	(Tamjidi, Nasirpour, & Shahedi, 2013)
Omega-3 ethyl ester	Soy protein isolate (SPI) and GA	Wall material concentration, wall-to-core material ratio, and the concentration of cross-linker in the microencapsulation was optimized. The lower core material concentration and the higher SPI:GA ratio resulted in the higher microencapsulation yield. Microencapsulation efficiency was not modeled adequately with a mathematical model	(de Conto, Grosso, & Gonçalves, 2013)
		The conditions of 1.5:1.0 SPI:GA, 1.0:1.0 wall:core and 6.0 UA of TG/g was determined as optimum	

(*Continued*)

TABLE 3.2 (Continued)

Core Material	Wall Materials	Key Findings	Reference
		conditions and resulted in 25 g of EPA + DHA in 100 g of microcapsules	
Other Techniques			
DHA	Whey protein isolate	Physicochemical properties and storage stability of the microcapsules produced by spray-freeze-drying (SFD) compared to those produced by spray-drying and freeze-drying. Although, microencapsulation by SFD resulted in lower microencapsulation efficiency, the storage stability of these microcapsules was found to be higher than the samples produced with other techniques	(Karthik & Anandharamakrishnan, 2013)
Tuna oil	lecithin and chitosan	Microencapsulation of tuna oil by layer by layer technique increased the stability of the oils. Addition of Ethylenediaminetetraacetic acid (EDTA) or EDTA and mixed tocopherols on the stability of microencapsulated oils was also tested. EDTA was found to be more effective	(Klinkesorn, Sophanodora, Chinachoti, McClements, & Decker, 2005)

Nanoencapsulation can be achieved by different techniques which can be classified as bottom-up and top-down approaches (Fig. 3.1). The nanoencapsulation techniques used in fish oil and essential fatty acids are summarized in Table 3.3. In the top-down approaches, the size of the system is precisely decreased. On the other hand, materials are constructed by the self-assembly and self-organization of molecules in bottom-up approaches.

The morphology and the size of the particles can be various according to the applied encapsulation method (Fig. 3.2). The details and the applications of these techniques in fish oil and essential fatty acids are discussed in the following subsections.

3.2 LIPID FORMULATION NANOENCAPSULATION OF FISH OIL AND ESSENTIAL FATTY ACIDS

3.2.1 Encapsulation by Nanoemulsions

An emulsion is defined as the mixture of two immiscible liquids in which one is dispersed in the other. In case the droplet size of the dispersed phase ranges between 50 and 1000 nm, the emulsion can be called as nanoemulsion

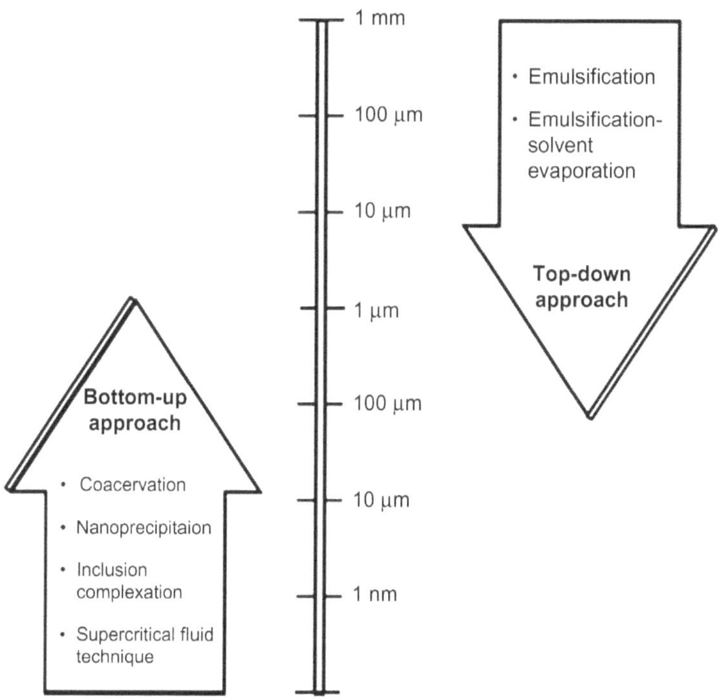

FIGURE 3.1 Top-down and bottom-up approaches in nanoencapsulation techniques (Ezhilarasi et al. 2013).

TABLE 3.3 Nanoencapsultaion Techniques Used for Fish Oil and Essential Fatty Acids

Nanoencapsulation Technique	Methods	Specifications and Brief Results
Nanoemulsions	High energy (microfluidization, emulsification, high-pressure homogenization) and low-energy (self-emulsification, phase inversion, and emulsion inversion point) approaches	High-energy approaches need expensive equipment but lower amount of surfactant than low-energy approaches. Scale up of the low-energy approaches is harder than high-energy approaches. High-energy approaches results in generally higher oxidation products soon after emulsification in comparison to the low-energy approaches due to the applied mechanical force
Nanoliposomes	Mechanical (extrusion, ultrasonication, high-pressure homogenization, microfluidization) and nonmechanical (reversed-phase evaporation and depletion of mixed detergent-lipid micelles)	It is a suitable technique for large-scale production. Among different methods, Mozafari method was an effective, fast, scalable, and safe method in nanoliposome production from essential fatty acids
Solid lipid nanoparticles	Hot homogenization method or cold homogenization method	High encapsulation efficiency, no need for organic solvents, easiness of the scale-up and sterilization, flexible control of release, possibility of delayed release, and the effective protection of the core material against environmental factors. However, release of the oils take longer time than other techniques and expulsion of oil during storage can observe during storage
Nanostructured lipid carriers	Hot homogenization method or cold homogenization method	Nanostructured lipid carriers are developed to overcome the disadvantages of the solid lipid nanoparticles
Entrapment in caseins	Complexation	Caseins have the ability to host water insoluble

(*Continued*)

TABLE 3.3 (Continued)

Nanoencapsulation Technique	Methods	Specifications and Brief Results
		compounds, thereby, increase its solubility and bioavailability. A casein molecule can bind up to four molecules of DHA
Entrapment in β-cyclodextrin	Complexation	Cyclodextrins are good options for nanoencapsulation of bioactive oils due to their nonpolar molecular cavity
Electrospinning and electrospraying		Bioactive oils are entrapped into wall materials using electrical forces. Only molecular cohesion in the raw material is variable in these techniques
Nanospray-drying		Spray-drying is a process which transforms a liquid feed into solid particles using hot drying medium
Coacervation		Coacervation is the complexation of two or more oppositely charged biopolymers. The technique is needed emulsification prior to the encapsulation. Cross-linking of the coacervates increased the stability and mechanical strength of the nanocapsules
Inclusion complexation		Inclusion complexation refers to the encapsulation of bioactive oils in a cavity-bearing material by chemical forces

(Anandharamakrishnan, 2014). However, real nanoemulsions must have droplet size smaller than 100 nm (Mason, Wilking, Meleson, Chang, & Graves, 2006; Rodríguez, Martín, Ruiz, & Clares, 2016).

Nanoemulsions have some benefits and challenges. The most important benefits of the nanoemulsions are increased stability against creaming,

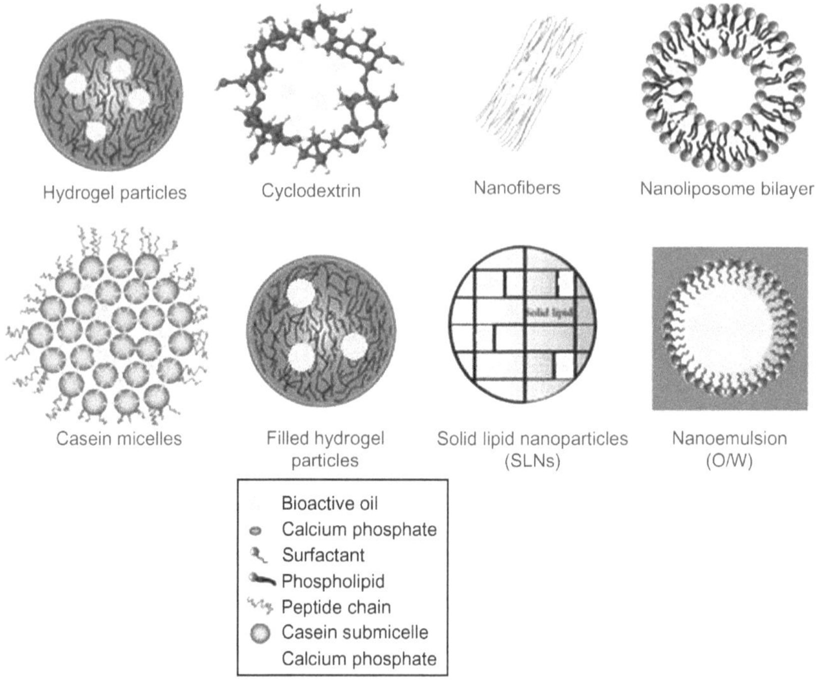

FIGURE 3.2 Schematics of nanoencapsulation systems. Reorganized from Katouzian and Jafari (2016).

sedimentation, and coalescence. This high stability is mainly related to two different reasons. First, the creaming or sedimentation velocity is proportional to the square of the particle size according to the Stoke's law. Second, the motion of the small droplets is dominated by the Brownian motion more than gravitational forces. In addition to the stability, nanoemulsions are generally transparent or slightly turbid which is very important for some products (Mason et al., 2006). However, they are thermodynamically unstable due to their susceptibility to Ostwald ripening compared to conventional emulsions. In an emulsion, Ostwald ripening occurs by diffusion of dispersed phase from small droplets to the larger droplets which results in droplet growth. However, Ostwald ripening can be prevented by careful selection of oil phase or using ripening inhibitors.

Nanoemulsions have been used as encapsulation systems to protect and control release for food and pharmaceutical bioactive compounds. Therefore, it is one of the most promising techniques for the encapsulation of fish oil and essential fatty acids. The bioavailability of nanoemulsified oils is generally greater than bulk oil. Previous reports suggested different mechanisms for higher bioavailability of nanoemulsions. For example, the broader surface area of the nanoemulsions increases the enzyme activity at the interface and leads to faster digestion (Acosta, 2009; McClements, 2011). Additionally, smaller droplets can penetrate through the mucus layer of the epithelium cells of small intestine which allows longer time for the digestion and absorption of the bioactive compounds. Moreover, smaller droplet size increases the solubility of the oils in aqueous media which provides better absorption of the oils.

Nanoemulsions can be produced by different emulsification techniques classified in high- and low-energy approaches (Fig. 3.3). In the high-energy approaches, the dispersed phase is disrupted using a mechanical device which applies intensive forces to the emulsion. Whereas, in low-energy approaches, emulsification occurs itself by altering environmental conditions such as temperature and composition of mixture (Walker et al., 2015a).

Nanoemulsification in high-energy approaches is achieved by high-pressure homogenization, microfluidization, and ultrasonication. The droplet size of the nanoemulsion depends on design of homogenizer, operation conditions (pressure, duration, temperature, flow rate, number of passes or cycles, etc.), environmental conditions, emulsion composition, and physicochemical properties of the liquids (Anandharamakrishnan, 2014).

Self-emulsification (spontaneous emulsification), phase inversion, and emulsion inversion point (EIP) methods are the low-energy approaches using in nanoemulsion production. These methods are more simple and economic than the high-energy methods. The droplet size of the emulsion is related to the composition of phases, type and amount of surfactant, environmental

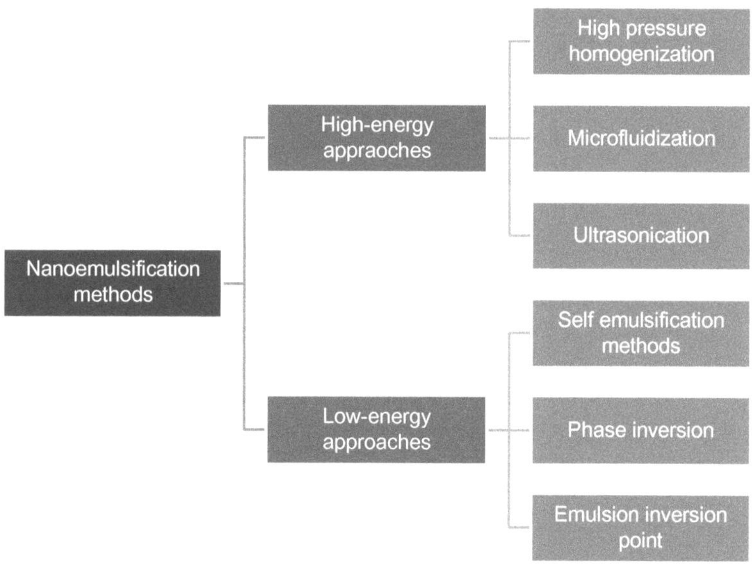

FIGURE 3.3 Clasification of the nanoemulsification methods (Walker et al., 2015a).

factors, and mixing conditions. In self-emulsification method, continuous phase is added on to the dispersed phase containing surfactant and emulsification occurs due to the rapid diffusion of the surfactant and/or solvent to the continuous phase. In phase inversion methods, nanoemulsification occurs by either changing temperature or composition of the conventional emulsion with regard to the properties of the surfactant. However, in EIP method, nanoemulsion is formed by adding continuous phase which leads to the catastrophic phase inversion (Walker et al., 2015a). In literature, there are many studies on nanoemulsification of fish oil and related products by using either low- or high-energy emulsification methods as summarized below.

Esquerdo, Dotto, and Pinto (2015) tried to obtain fish oil nanoemulsions using chitosan and Tween 80 as emulsifiers by mechanical stirring at 10,000 rpm for 10–20 min. In this study, chitosan coated spherical fish oil droplets in monomodal particle size distribution was achieved in droplet size between 332 and 488 nm. No creaming or phase separation occurred in the emulsion during 30 days storage. Additionally, the obtained emulsion was lyophilized to increase the stability of the fish oil.

Gulotta, Saberi, Nicoli, and McClements (2014) studied the effect of surfactant-to-oil ratio, addition of different oils [medium chain triglycerides (MCT) or lemon oil added because plain fish oil did not yield nanoemulsion in preliminary tests] to fish oil, and cosolvent composition on the formation and stability of the fish oil nanoemulsion formed by self-emulsification method. According to their results, increasing content of fish oil in MCT or lemon oil mixture resulted in bigger droplets. Therefore, fish oil must be lower than 40% in fish oil-MCT mixture and 50% in fish oil-lemon oil mixture to form a nanoemulsion with droplet sizes lower than 200 nm. Additionally, the surfactant-to-oil ratio must be higher than 1 to produce a fine nanoemulsion. The effect of ethanol, polypropylene glycol and glycerol was also evaluated as cosolvent in different ratios. Although replacing water with ethanol and polypropylene glycol did not affect the droplet size of the emulsion, addition of 40% glycerol decreased the droplet size to 50 nm which increased the transparency of the emulsion. The obtained emulsion using best conditions was stable in terms of the droplet size at both 4 and 20°C during storage for 30 days. However, storage at 37°C caused a linear increase in droplet size during the storage and it was measured as 350 nm at the end of storage

In another similar study, physical and chemical stability of the fish oil-lemon oil nanoemulsions were determined by spontaneous emulsification (Walker, Decker, & McClements, 2015b). Increasing the surfactant-to-oil ratio increased the physical stability of the nanoemulsions. Lipid oxidation products [peroxides and thiobarbituric acid reactive substances (TBARS)] was tracked during storage of the emulsions with different droplet sizes and it was found that droplet size of the emulsion did not affect the oxidation. Additionally, spontaneous emulsification and microfluidization methods were compared in terms of lipid oxidation products. Although, no significant difference was observed after emulsification, the secondary oxidation products (TBARS) was significantly higher in the microfluidized nanoemulsions after storage at 55°C for 14 days. The lower oxidation in the emulsion produced by spontaneous emulsification was explained by the high concentration of surfactant which formed a physical barrier between oxidants and lipids (Silvestre, Chaiyasit, Brannan, McClements, & Decker, 2000; Waraho, McClements, & Decker, 2011).

Ostertag, Weiss, and McClements (2012) studied the effect of oil type (MCT, limonene, olive oil, mineral oil, sesame oil, grape seed oil, peanut oil, canola oil and orange oil), surfactant type, surfactant-to-oil ratio, and initial surfactant location on the particle size distribution of the nanoemulsion formed by emulsion phase inversion method. Similar to the results of Gulotta et al. (2014) who also used a low-energy approach, a high surfactant-to-oil ratio (>0.65) was required to produce nanoemulsion with low mean droplet size; however, much higher surfactant-to-oil ratio was needed to decrease size distribution range. The authors hypothesized that physicochemical properties and the mean droplet size could be depended on each other. However, the results did not confirm this hypothesis and no linkage between physicochemical properties and mean droplet size was determined. The lowest droplet size was obtained by using MCT as oil which was followed by flavor oil (limonene and orange) and long chain triglycerides (olive, canola, peanut, sesame, grape), respectively. The mixture of three surfactants (Tween 20, 80, and 85) and Tween 80 alone provided the lowest droplet size in the emulsions. Additionally, the location of the surfactant was determined to be critical in emulsion phase inversion method. According to their results, the higher portion of the surfactant must be in the organic phase. Finally, emulsion phase inversion method was compared with microfluidization (a high-energy approach) in different surfactant-to-oil ratios. In microfluidization method, nanoemulsion could be produced even in very low surfactant-to-oil ratio (0.1); however, in emulsion phase inversion methods, high surfactant-to-oil ratio (2.5) was needed to produce nanoemulsion with similar mean particle size.

Uluata, McClements, and Decker (2015) compared the effectiveness of the two natural (lecithin and quillaja saponin) and two synthetic [Tween 80 and sodium dodecyl sulfate (SDS)] surfactants in fish oil nanoemulsification by microfluidization method. The smallest droplet size was obtained by Tween 80 which was followed by lecithin, SDS, and quillaja saponin, respectively. The effect of pH and NaCl concentration was also studied. pH had a negative effect on nanoemulsion prepared with lecithin. Additionally, increasing the NaCl content resulted in bigger droplet size with all surfactants, except SDS. Emulsion prepared with Tween 80 and lecithin was less stable to thermal treatments. In terms of the photosensitizer-promoted oxidative stability, quillaja saponins provided the highest stability. The authors explained this phenomenon with the absorption of the light which generates reactive oxygen species generated at 355 nm. This explanation was confirmed by the photosensitizer-promoted oxidative stability test presence of the rose Bengal. Additionally, the higher antioxidant activity, 10-fold higher than the lecithin in terms of oxygen radical absorbance capacity (ORAC) analysis, of the quillaja saponins was also effective in the lowest oxidation.

Komaiko, Sastrosubroto, and McClements (2016) tried to use four different sunflower phospholipids in fish oil nanoemulsion fabrication using microfluidization. Three of the tested sunflower phospholipids resulted in smaller droplets at high surfactant-to-oil ratio. The other one acts differently because of the high viscosity of the emulsion at higher levels. The smallest droplets were produced at high phosphatidylcholine levels and surfactant-to-oil ratios. This study showed that sunflower phospholipids are good and natural alternatives for synthetic surfactants in nanoemulsion production.

Lane, Li, Smith, and Derbyshire (2016) produced vegetarian nanoemulsions from flaxseed oil and DHA-rich algal oil using soy lecithin and/or Tween 40 as emulsifier by ultrasonication. Tween 40 alone and in combination with soy lecithin was more effective in size reduction than soy lecithin alone. The smallest droplet size obtained in 10 min ultrasonication. For all surfactants, emulsion prepared with algal oil resulted in smaller droplet size than that of flaxseed oil.

In a comprehensive study, Habib, Amr, and Hamadneh (2012) determined the effect of solvent (ethyl acetate or acetone), aqueous phase/organic phase ratio (1/5 or 1/2), and stabilizer type (pluronic-F68, tween 20, gelatin) on the ALA nanoemulsions. Their results showed that an emulsion with the smallest droplet size was obtained using the conditions of acetone, 1/5 aqueous phase/organic phase ratio, and tween 20.

Nejadmansouri, Hosseini, Niakosari, Yousefi, and Golmakani (2016) studied the effect of whey protein isolate (WPI)/fish oil ratio, pH, and storage temperature on the physicochemical properties of nanoemulsions. Increasing the WPI/fish oil ratio and pH resulted in lower particle size. The oxidative stability of the nanoemulsions was found to be higher than coarse emulsion. Storage for 28 days did not affect the particle size of emulsions. However, slight increase in viscosity was determined during storage.

3.2.2 Encapsulation by Nanoliposomes

Liposomes are vesicles formed using membrane-like phospholipids in aqueous medium. These phospholipids have a polar head and hydrophobic hydrocarbon tails. Owing to these properties, they can be used in large-scale production for the encapsulation of hydrophilic, hyprophobic, and amphiphilic materials using natural ingredients (Fathi, Martín, & McClements, 2014). As a carrier of the encapsulant, liposomes provide stability in environmental factors and targeted and controlled release of the materials.

The liposomes are classified according to their lamellae number and size. Liposomes that have only one bilayer are called as unilamellar vesicles. This type of liposome is generally suitable for water soluble materials. On the other hand, liposomes can be formed from a number of bilayers which are called as multilamellar vesicles. However, if many small nonconcentric vesicles are formed within a single bilayer, then this type of liposome is named as multivesicular liposome. Multilamellar and multivesicular liposomes are generally shown in the entrapment of the lipid-soluble materials (Anandharamakrishnan, 2014). According to their size, liposomes are categorized as small and large vesicles. Large vesicles diameters can reach up to few micrometers. On the contrary, small vesicles have diameters smaller than 100 nm. Therefore, they are called as nanoliposomes. This type of small liposomes is mostly applied in encapsulation of sensitive materials.

Mechanical and nonmechanical methods can be used for the production of nanoliposomes. Some examples of mechanical methods are extrusion, ultrasonication, high-pressure homogenization, and microfluidization. Reversed-phase evaporation and depletion of mixed detergent-lipid micelles are nonmechanical methods used in nanoliposome production (Fathi, Mozafari, & Mohebbi, 2012). The detailed description of these methods can be found in literature (Gregoradis, 1993; Mozafari, 2010; Taylor, Weiss, Davidson, & Bruce, 2005). So far, different technologies have been applied to stabilize the obtained liposomes such as freeze-drying, spray-drying, freezing, and supercritical fluid (SCF) technologies. Among these technologies, freeze-drying is the most promising one for liposomes containing heat sensitive compounds (Chen, Han, Cai, & Tang, 2010).

Different nanoliposome researches on fish oils and essential fatty acids have been carried out to increase their stability against oxidation reactions and increase their bioavailability. Some examples of these researches are listed below.

Sahari, Moghimi, Hadian, Barzegar, and Mohammadi (2016) produced multilamellar vesicles of both DHA and EPA with or without α-tocopherol using dipalmitoyl phosphocholine as membrane. The obtained nanoliposomes were determined in the range of 73–107 nm. The physical and chemical stability of the obtained nanoliposomes was also tested during storage for 90 days at different temperatures. Generally, nanoliposomes were found to be stable in terms of particle size, especially in lower storage temperatures. On the other hand, high content of the EPA and DHA was degraded during storage of the nanoliposomes. Addition of α-tocopherol limited the degradation of EPA and DHA which was much clear in the samples stored at 4°C for 90 days.

Rasti, Jinap, Mozafari, and Yazid (2012) compared the oxidative and physical stability of the EPA + DHA nanoliposomes produced by two different methods namely conventional (Bangham thin-film hydration using organic solvents) and Mozafari (without organic solvents) methods. The nanoliposomes produced by Mozafari method showed highest oxidative stability during storage for 10 months at 4°C. Additionally, smaller liposomes were found to be more stable in terms of physical and chemical properties. The authors concluded, Mozafari method was an effective, fast, scalable, and safe method in nanoliposome production from essential fatty acids.

Rasti, Jinap, Mozafari, and Abd-Manap (2014) optimized the conditions for nanoliposome production by Mozafari method using Box–Behnken design. Mixing rate (600–1000 rpm), mixing time (30–60 min), and ultrasonication time (10–20 min) were chosen as independent variables to produce vesicles with the maximum entrapment efficiency and the smallest size. Both of these dependent variables fitted well to the quadratic models. Increasing mixing time generally increased the entrapment efficiency of the nanoliposomes. On the other hand, entrapment efficiency increased until a certain level of mixing rate. Higher level of mixing rate resulted in lower efficiencies. The size of the vesicles generally decreased with increasing all three variables. Using 795 rpm mixing for 60 min and ultrasonication for 10 min resulted in vesicles with an entrapment efficiency of 100% and 81.4 nm average diameter.

Hadian, Sahari, Moghimi, and Barzegar (2014) compared some physical properties of EPA and DHA-loaded nanoliposomes produced by different homogenization methods (extrusion, ultrasonic irradiation, bath sonication, probe sonication, and combined probe and bath sonication). Probe and combined probe and bath sonication resulted in the smallest and the most efficient vesicles. Similar level (103–107 μg/kg) of secondary oxidation products (propanal, pentanal, hexanal, and heptanal) were formed in probe sonicated and extruded liposomes. However, probe and bath-sonicated samples had higher level of these products. Considering high loading of DHA and EPA and low oxidation, probe sonication was advised for future studies.

Kubo, Sekine, and Saito (2003) tested resistance of DHA-loaded nanoliposomes produced using two different types of phosphatidylethanolamine (1,2-dipalmitoyl-3-sn-phosphatidylcholine and 1,2-dioleoyl-3-sn-phosphatidylcholine) to *in vivo* lipid peroxidation. According to their results, liposomes formed with phosphatidylethanolamine protected the DHA from lipid oxidation.

Ghorbanzade, Jafari, Akhavan, and Hadavi (2017) produced fish oil-loaded nanoliposomes and applied them in fortifying yoghurt. According to the results of this study, yoghurt fortified with nanoencapsulated fish oil had similar sensorial characteristics as control samples. DHA and EPA after 21 days of storage was found to be higher in yoghurt fortified with nanoencapsulated fish oil than yoghurt fortified with free fish oil.

3.2.3 Encapsulation by Solid Lipid Nanoparticles

SLNs are aqueous colloidal dispersions in nanosized solid lipid shells. High encapsulation efficiency, no need for organic solvents, easiness of the scale-up and sterilization, flexible control of release, possibility of delayed release, and the effective protection of the core material against environmental factors are the unique advantages of the SLNs in comparison to nanoemulsions and nanoliposomes (Saupe & Rades, 2006). On the other hand, release of the core material in SLNs takes longer time than other techniques due to longer degradation period of the solid matrix (Müller, Dingler, Schneppe, & Gohla, 2000).

The production of SLNs is achieved by either hot homogenization method or cold homogenization method in food processing. In the hot homogenization, the lipid is melted approximately 10°C above the melting point. Then the core material and the surfactant which is heated to same temperature is added into the melted lipid. The mixture is homogenized using a high-pressure homogenization at the controlled temperature. SLNs are produced by cooling of the emulsion which leads to recrystallization (Fathi et al., 2012). However, this method is not suitable for the heat-sensitive compounds. Therefore, cold homogenization must be applied for this type of products. In cold homogenization method, first core materials are added into the melted lipid. After cooling and solidification of the lipid, it is grounded by a mill. The obtained nanoparticles are dispersed in a cold surfactant and homogenized using a high-pressure homogenizer. However, special care must be taken to prevent temperature rising during homogenization (Fathi et al., 2012).

Only a study on production of SLNs containing fish oils was found in the literature. Salminen, Gömmel, Leuenberger, and Weiss (2016) produced SLNs from fish oil using tristearin as carrier and quillaja extract alone or in combination with lecithin (high- or low-melting point) as surfactant. Physical and chemical stability was tested during storage at dark for 51 days. The particle size and polydispersity index of the SLNs were found to be stable except quillaja-low-melting lecithin. On the other hand, in terms of chemical stability, SLNs produced with quillaja-hot-melting lecithin combination was found to be more protective than those of quillaja-low-melting lecithin and quillaja alone.

3.2.4 Encapsulation by Nanostructured Lipid Carriers

Nanostructured lipid carriers (NLCs) are considered as developed SLNs to overcome the problems of SLNs such as low-loading capacity and possibility of expulsion during storage (Fathi et al., 2012). It was first developed by Radtke and Müller (2001). NLCs are produced by similar methods as SLNs but use a mixture of different types of lipids (solid lipids and liquid lipids). The obtained NLCs have higher loading capacity and low expulsion risk.

Similar to the SLNs, NLCs can also be prepared by cold or hot high-pressure homogenization. In recent years, several researches have successfully prepared NLCs loaded with fish oils as core materials. In one of these studies, Zhu, Zhuang, Luan, Sun, and Cao (2015) produced NLCs with high load of krill oil using palm stearin and lecithin by hot homogenization method. They optimized the content of the krill oil and lecithin by central composite design using mean diameter and polydispersity index. According to the results of optimization, usage of 65% krill oil and 1.1% lecithin resulted in the lowest mean diameter and polydispersity index. The obtained NLCs were found to be stable under high-speed centrifugation. Also NLCs were found to be more stable in low-temperature longtime treatment than high-temperature short time application. Chemical stability during storage was also tested and EPA, DHA and astaxanthin content of the NLCs were determined. The NLCs protected the krill oil especially at 4°C and room temperature.

Effectiveness of low-melting lecithin and high-melting lecithin on physical and chemical stability of fish oil-loaded NLCs were evaluated by Salminen, Helgason, Kristinsson, Kristbergsson, and Weiss (2013). NLCs produced with both lecithins were stable during 50 days of storage in terms of their mean diameter. However, high-melting lecithin was found to be more protective against lipid oxidation.

Salminen, Aulbach, Leuenberger, Tedeschi, and Weiss (2014) compared the fish oil-loaded NLCs and nanoemulsions in terms of physical and chemical stability. The content of the fish oil and surfactant was similar in two systems but tristearin was also used in the production of NLCs. Using same surfactant concentration, NLCs yielded in the smaller particles than nanoemulsions. Additionally, NLCs maintained their size throughout the storage. On the other hand, aggregation was observed for the nanoemulsions during storage. NLCs were also more protective against lipid oxidation and had 72%, 53%, and 57% lower content of lipid hydroperoxides, propanal, and hexanal than nanoemulsions, respectively.

Averina, Müller, Popov, and Radnaeva (2011) produced PUFA-loaded NLCs from fish oil and pine seed oil using Tween 80 and Poloxamer 188 as surfactant. According to the results of the study, Poloxamer 188 was a suitable surfactant for both oils in terms of physical stability. Fatty acid compositions of the NLCs slightly changed even at 40°C after storage for 90 days.

Wang et al. (2014) prepared NLCs from microalgal oils using stearic acid and Poloxamer 188 by microfluidization method. The mean size of the carriers was determined as 300–350 nm. High encapsulation efficiency (>88.8%) was obtained even in high oil concentrations. Higher pH values than 8.0 and high storage temperatures caused increase in both particle size and polydispersity index of the NLCs.

Zhu et al. (2015) studied the effectiveness of the fish oil-loaded NLCs as carrier of the water insoluble ingredients. They first produced stable fish oil-loaded NLCs using glyceryl monostearate and Tween 80 by ultrasonication. Then fluorescein isothiocyanate-loaded NLCs were produced using obtained conditions. Finally, loaded NLCs were tested using cytometric assays. The results showed that NLCs are not only effective to deliver fish oil but also have potential to increase cellular uptake of water insoluble bioactives.

3.3 NATURAL NANOCARRIERS FOR NANOENCAPSULATION OF FISH OIL AND ESSENTIAL FATTY ACIDS

3.3.1 Nanoencapsulation by Caseins

Caseins are the major milk proteins, natural nanosystems for better and more effective transport of dairy minerals and proteins. There are four main types of caseins (α_{s1}, α_{s2}, β, and κ) and their average diameter is 150 nm. Caseins have the ability to host water insoluble compounds which improves their solubilization and bioavailability. Therefore, casein micelles are used for nanoencapsulation for several compounds in food technology. The studies on encapsulation of food compounds using casein micelles are focused on fat-soluble vitamins which are out of the scope of the current chapter. Only a study on encapsulation of DHA in reassembled casein micelles was found in literature.

Zimet, Rosenberg, and Livney (2011) reported that a casein molecule binds three or four molecules of DHA and the obtained micelles had a diameter 288.9 nm. However, the average diameter decreased to 50–60 nm by the addition of calcium and phosphate. Pasteurization at 74°C for 20 s did not significantly change the particle size of micelles. The concentration of DHA–casein micelles was observed during storage at 4°C for 37 days and compared with control samples. Results indicated that DHA content of the micelles did not differ statistically throughout the storage period while drastic degradation was observed in control samples.

3.3.2 Nanoencapsulation by Cyclodextrins

Cyclodextrins are cyclic oligosaccharides that can be obtained by enzymatic conversion of starch. There are three types of cyclodextrins namely α-, β- and γ-cyclodextrin, which are 6-, 7-, and 8-membered, respectively. Cyclodextrins are excellent options for nanoencapsulation of fatty acids and bioactive oils or masking odors and off-flavors due to their apolar molecular cavity. There are several studies on the inclusion of fish oil within cyclodextrins which are summarized below.

Choi, Ruktanonchai, Min, Chun, and Soottitantawat (2010) prepared β-cyclodextrin (BCD) and fish oil complexes at different ratios. The obtained complexes had particle size in the range of 278–714 nm and encapsulation efficiency higher than 84%. The release characteristic of encapsulated fish oil in four different continuous phases (water, fish sauce, 15% NaCl, and

25% NaCl) was also investigated. The fastest release was observed in the fish sauce which was followed by 25% NaCl and water.

Na, Kim, Kim, and Lee (2011) produced fish oil-loaded nanocapsules using cyclodextrins (β and γ), whey protein concentrate, and xanthan gum. The smallest particles and the highest encapsulation efficiency were obtained using γ-cyclodextrin. The most stable fish oil-loaded nanocapsules providing low peroxide value was obtained using γ-cyclodextrin, whey protein concentrate, and xanthan gum combination at the ratio of 0.8:0.2:0.5, respectively.

Hădăruga et al. (2016) produced BCD and salmon oil complexes at different ratios. The best ratio according to the hydratability and thermal properties was determined at the ratio of 3:1.

Ünlüsayin et al. (2016) prepared successfully anchovy oil BCD complexes using crystallization and kneading methods at two different oil to BCD ratios. In their study, the ratio of saturated and monounsaturated fatty acids in nanocomplex was found to be about 2.5-fold higher than raw anchovy oil. The authors explained this phenomenon by binding capacity of BCD could be related to the structure of fatty acids and claimed that the saturated and monounsaturated fatty acids could be more suitable for encapsulation in BCD.

3.4 EQUIPMENT BASED-NANOENCAPSULATION OF FISH OIL AND ESSENTIAL FATTY ACIDS

Nanoencapsulation of fish oil and essential fatty acids by electrospinning, electrospraying, and nanospray-drying is discussed in this section. Among these technologies, electrospinning and electrospraying may be considered as sister technologies since only molecular cohesion in the raw material is variable. In the electrospinning technology, the jet from Taylor cone is stabilized by high concentration or molecular chain entanglement and formed jet produces a nanofiber. On the other hand, in the electrospraying technology, the jet is broken to produce fine spherical nanoparticles.

3.4.1 Nanoencapsulation by Electrospinning

Electrospinning is a new technology to produce nanosized polymer fibers using electrostatic forces (Fig. 3.4A). In this technique, a liquid polymer droplet with a very low volume is formed using a needle tip. When the formed droplet is exposed to an electric field, it stretches toward the nearest lower potential point. The structure formed during the stretching of the polymer is called as Taylor cone (Fig. 3.4D). The formation of the Taylor cone is actualizing above a critical value when the electrical forces overcome the surface tension of the droplets. Electrospinning has many advantages in nanofiber production such as scale-up opportunities, reproducibility, processability of obtained fibers, and controlling the fiber

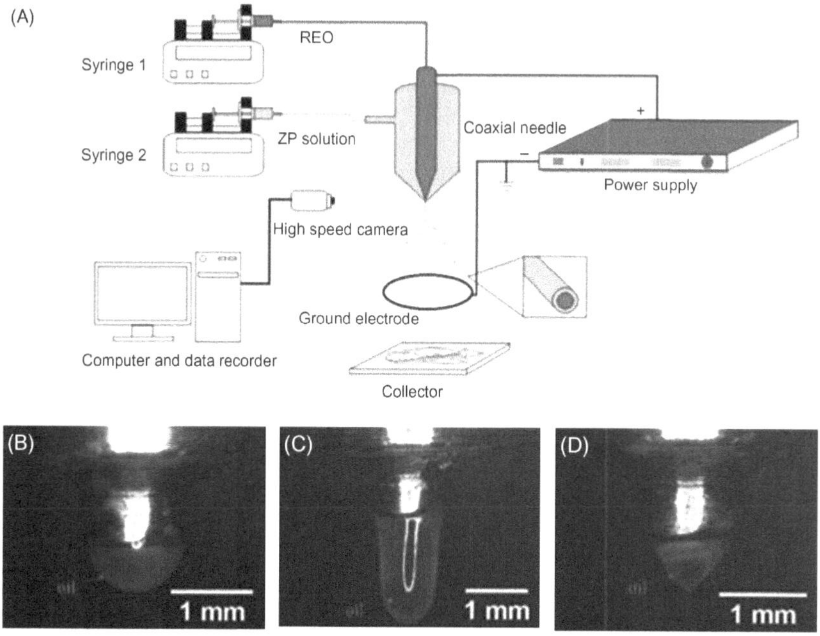

FIGURE 3.4 Schematic of electrospinning process (A). Variations in jet formation (B) dripping, (C) transition (D) Taylor cone jet (REO: Core oil, ZP: Wall material) (Yao, Chang, Ahmad, & Li, 2016).

diameters. The properties of the solution (molecular weight and viscosity, surface tension, pH, conductivity, etc.), process parameters (voltage, feed rate, needle diameter, distance between needle and collector, etc.), and environmental factors (humidity, temperature, pressure, etc.) are effective on the obtained fibers. Detailed information on the effect of these parameters on obtained fibers can be found in the literature (Schiffman & Schauer, 2008).

García-Moreno et al. (2016) produced nanofibers of fish oil emulsions (formed by either WPI or fish protein hydrolysate) in a synthetic polymer (polyvinyl alcohol) by electrospinning. They studied the effect of polymer concentration (7.5–10.5%) and fish oil concentration (1.5–2.5%) on the physical and chemical properties of obtained nanofibers. According to their results, high concentration of the polymer (10.5%) was needed regardless of the emulsifier type to produce a smooth nanofiber. The higher fish oil in the solution resulted in the thicker nanofibers. Surprising results were obtained in the oxidative stability analyses. The encapsulated oils oxidized much faster than unprotected oils. The authors explained this poor stability by catalyzing of the trace metals presented in the polymer. One possible explanation of the poor stability could be the increasing surface area of the oils which enable oxygen permeability into the nanofiber structure.

In two similar studies, fish oil encapsulation in zein nanofibers was achieved by electrospinning and oxidative stability (Moomand & Lim, 2014a) and properties of the oil under simulated *in vitro* conditions (Moomand & Lim, 2014b) were determined. In both studies, different solvents (isopropanol and ethanol), zein concentration, and fish oil loading were studied. According to the result of the oxidative stability test, nanofibers produced using isopropanol and 30% fish oil loading provided the most stable fish oil nanofibers (Moomand & Lim, 2014a). Additionally, isopropanol enabled greater swelling and erosion rates under gastrointestinal fluids (Moomand & Lim, 2014b).

3.4.2 Nanoencapsulation by Electrospraying

Electrospraying technique can be used for different purposes such as film coating, chocolate processing, stabilization of food ingredients, drug delivery, and especially encapsulation of bioactive compounds (Ghorani & Tucker, 2015).

In a study by Torres-Giner, Martinez-Abad, Ocio, and Lagaron (2010), DHA-loaded nanocapsules were produced using zein prolamine as the coating material by electrospraying technique. Increasing zein/DHA ratio resulted in lower oxidation of DHA. Additionally, effect of storage humidity and temperature on oxidation kinetics was also studied. According to the results, storing at dry conditions and higher temperatures accelerates the oxidation of the oil. The activation energy of bulk DHA and encapsulated DHA was determined as 61.97 and 71.03 kJ/mole, respectively.

In another study, Gómez-Mascaraque and López-Rubio (2016) produced ALA-loaded capsules using gelatin, whey protein concentrate, or soy protein isolate as carrier materials using electrospraying technique. The properties of the obtained capsules were compared with the capsules produced by spray-drying. The obtained encapsulation efficiency was determined as 40, 61, and 67% for gelatin, whey protein concentrate, and soy protein isolate, respectively. The most stable capsules were produced using whey protein concentrate which was followed by soy protein isolate. It must be noted that, rapid oil oxidation in the capsules produced using gelatin was observed. The authors explained this rapid oxidation with the presence of acetic acid in these capsules.

3.4.3 Nanoencapsulation by Nanospray Dryer

Spray-drying is a process which transforms a liquid feed into solid particles using hot drying medium. It is a well-established technique and one of the oldest and the most widely used techniques for the microencapsulation of bioactive ingredients. Although spray-drying is generally used for microencapsulation purposes, nanoparticles can be obtained by adapting conditions.

Nanoencapsulation by spray-drying is generally achieved by spray-drying of nanoemulsions. Additionally, recently some companies developed spray dryers that produce submicron particles. There are no studies on fish oils and other essential fatty acids using nanospray-drying technique. In a previous study, (Jafari, Assadpoor, Bhandari, & He, 2008) spray-dried fish oil nanoemulsions were prepared by microfluidization technique. However, the obtained particles had a diameter of 25–32 μm.

Nassar, Rom, Nyska, and Benita (2009) prepared a medicinal mixture contained argan oil and spray dried using hydroxypropylmethylcellulose as wall material. The average diameter of the nanocapsules was reported in the range of 350–400 nm.

3.5 BIOPOLYMER-BASED NANOENCAPSULATION OF FISH OIL AND ESSENTIAL FATTY ACIDS

Coacervation and inclusion complexation are used in micro/nanoencapsulation. Coacervation technique involves one or more oppositely charged polyelectrolyte solutions which form a shell around the core material by electrostatic interactions, hydrogen bonding, hydrophobic interactions, and polarization-induced attractive interactions. Micro/nanoencapsulation by coacervation is achieved in five basic steps namely dissolution, emulsification, coacervation, gelation, and hardening/rinsing/filtering/drying (Xiao, Liu, Zhu, Zhou, & Niu, 2014). Many factors including the biopolymer type (molar mass, flexibility, and charge), pH, ionic strength, concentration, and the ratio of the biopolymers affect the physicochemical properties of the coacervates (Ezhilarasi, Karthik, Chhanwal, & Anandharamakrishnan, 2013). Different biopolymers such as gelatin, Arabic gum, chitosan, WPI, and maltodextrin can be used as wall materials. Generally, cross-linking is applied using glutaraldehyde, tannic acid, or transglutaminase to increase mechanical properties of nanocapsules. Finally, the wet coacervates can be dried using different drying techniques.

Fish oil coacervates were successfully prepared using electrostatic interactions between sodium caseinate and gum Arabic. The obtained nanoparticles had 232 nm size and 78.9% encapsulation efficiency. The bioaccessibility of the EPA and DHA in enriched fruit juice was determined as 36% and 56%, respectively (Ilyasoglu & El, 2014). Similarly, krill oil was nanoencapsulated using tripolyphosphate and chitosan by coacervation technique. The stability of the nanoencapsulated krill oil was much better than bulk oil during storage at high temperatures (Haider, Majeed, Williams, Safdar, & Zhong, 2017). Jafarpour, Esfahani, and Jafarpour (2016) optimized coacervation process of omega-3 fatty acids using gelatin and gum Arabic as polymers. The optimum conditions were determined as 33% gelatin, 30% omega-3 fatty acids, and 20,000 rpm homogenization speed by Box–Behnken design. In another similar study, optimum conditions were calculated as 1% sodium casseinate, 0.45% pectin, and pH 4.1 (Bahrani, Ghanbarzadeh, Hamishekar, & Sowti khiyabani, 2013).

Inclusion complexation refers to the encapsulation of bioactive compounds in a cavity-bearing material through hydrogen bonding, Van der Waals force, or an entropy-driven hydrophobic effect. Although this technique is mainly used in the encapsulation of volatile organic molecules (essential oils and vitamins); it can be used for other core materials. However, only few particular molecular compounds like BCD and β-lactogloglobulin are suitable for encapsulation through this method (Ezhilarasi et al., 2013). Zimet and Livney (2009) produced DHA-loaded β-Lactoglobulin nanocomplexes with low methoxyl pectin. The nanocomplexes provided good protection against oxidation of DHA during an accelerated stability test.

Zein is the storage protein mainly present in the corn kernel endosperms and formed from water-insoluble prolamines. It is highly soluble in alcohols and its odorless and tasteless character is an advantage in encapsulation studies (Pereira, Valentão, & Andrade, 2014). Zein has been used as a nanoencapsulation material for fish oil and essential fatty acids in a few studies listed below.

Zhong, Tian, and Zivanovic (2009) produced fish oil-loaded nanocapsules in zein using precipitation (desolvation) technique by decreasing alcohol concentration in stock solution. The particle size of the zein particles is measured in the range of 350–450 nm. The chemical stability of the obtained fish oil-loaded nanocapsules was found to be good, especially when the zein: oil ratio was higher than 4:1.

Soltani and Madadlou (2015) produced fish oil-loaded zein nanoparticles by solvent evaporation method. The obtained nanoparticles had an average particle size of 83 nm. These authors explained the lower particle size observed in this study by differences in the alcohol removal procedure.

Tavakolipour, Bagheri, and Madadlou (2015) cross-linked the pomegranate seed oil-loaded zein nanoparticles by citric acid and determined the physicochemical properties of the noncross-linked and cross-linked nanoparticles. According to the results of this study, the particle size of the cross-linked nanoparticles was lower than the noncross-linked counterpart. Fourier transform infrared spectroscopy (FTIR) analysis clearly showed the occurrence of chemical linkage between citric acid and zein. Cross-linking not only increased the thermal stability of the oil but also decreased the release of the oil from particles in gastric digestion.

Sponton, Perez, Carrara, and Santiago (2015) tried preparation of ovalbumin-linoleic acid (LA) complexes. After obtaining ovalbumin nanoparticles at 100 nm size, binding experiments of LA was carried out by mixing. The fluorescence and turbidity assays showed the interaction between ovalbumin and LA. Therefore, LA nanocomplexes were successfully prepared and can be used in several industries.

3.6 OTHER NANOENCAPSULATION TECHNOLOGIES FOR FISH OIL AND ESSENTIAL FATTY ACIDS

Bejrapha, Min, Surassmo, and Choi (2010) compared the effect of conventional freeze-drying and vacuum freeze-drying of fish oil nanoemulsions on the properties obtained nanocapsules. The particle size of the nanocapsules was ranged 262–353 nm after drying. Oxidation onset temperature of the nanocapsules produced using conventional freeze-drying was higher than that of vacuum freeze-drying.

The effect of freeze-drying conditions on nanoencapsulated fish oil emulsions prepared by emulsion diffusion method using BCD and poly(ε-caprolactone) (PCL) has also been studied. According to the results of this study, fish oil:PCL yielded in smaller particles than fish oil:BCD. Additionally, PCL was more effective according to oil loading, leakage, and encapsulation efficiency. However, PCL particles were broken during freeze-drying (Choi, Briançon, Andrieu, Min, & Fessi, 2004).

3.7 RELEASE, CHARACTERIZATION, AND SAFETY OF NANOENCAPSULATED FISH OIL AND ESSENTIAL FATTY ACID

3.7.1 Controlled Release of Fatty Acids From Nanocapsules

Providing controlled release is one of the main aims of the encapsulation of bioactive oils. Controlled release has two different means which are delayed and sustained release. In the delayed release, the time and place of the release is modified to protect the bioactive oil from the adverse conditions such as gastrointestinal fluids or to deliver the bioactive oils into the right action or absorption site (Averina & Allémann, 2013). On the other hand, in the sustained release, the release duration of bioactive oils is prolonged to increase bioavailability and nutritive properties. The release characteristics of the nanoparticles are governed generally by the properties of the wall materials (Rodríguez et al., 2016).

Different release mechanisms such as diffusion, dissolution, swelling, degradation/erosion, external forces, or combinations are proposed to be involved in the release of nanoencapsulated bioactive oils (Siegel & Rathbone, 2012). Among these mechanisms, diffusion is the most common one especially for the water-soluble matrixes. The driving force of diffusion is the concentration gradient, thereby molecular displacement occurs from the high-concentration regions to the low-concentration regions. Dissolution of the wall material used in the nanoencapsulation of the bioactive oils is very important since other mechanisms can involve after dissolution of the coating material. It must be noted that, in certain types of nanocapsules, bioactive oils may not be exclusively located in the matrix but also at the surface. In such particles, diffusion could occur before dissolution of the coating material (Rodríguez et al., 2016; Siegel & Rathbone, 2012). Swelling is the water uptake of nanocapsule matrix resulting in volume increases. Swelling increases macromolecular mobility after a certain level. Erosion is the result of the release medium to the nanocapsule matrix. However, breakdown of the polymer chain linkage might occur simultaneously to erosion. Certain conditions of release media (such as temperature, pH, humidity, etc.) may also contribute to the release of the bioactive oils (Siegel & Rathbone, 2012).

3.7.2 Instrumental Analysis and Characterization of Nanocapsules Containing Fish Oil

Characterization of the nanocapsules containing fish oil and essential fatty acids can be classified in two different approaches namely physical and chemical characterization. In physical characterization, particle size, zeta potential, thermal properties, and morphology of the nanocapsules are evaluated. Additionally, moisture content, fatty acids profile, and chemical stability are generally determined to characterize chemical properties of the nanocapsules.

3.7.2.1 Physical Characterization Methods

Particle size and its distribution is one of the most important physical characteristics of the nanocapsules since it directly affects the stability of the bioactive oils. Although there are many different methods for particle size determination such as image analysis, electrozone particle counters, sieve analysis, and sedimentation, particle size of the bioactive oil-loaded nanocapsules is determined by laser diffraction method (Dodds, 2013). This technique does not measure directly the size of particles but determines diffraction pattern formed by passing laser light through the dispersion. In this method, light adsorption and refractive index of the media is needed since these values are used in particle size calculations. This method can be used for wet and dry samples using both bench type and online measurement instruments. The diffraction of the light when passing through the particle or droplet is inversely proportional to the particle size. Therefore, the diffraction pattern is measured using wide-angle and backscattering detectors. The obtained signals are used to calculate the particle size using Mie theory (Dodds, 2013). Detailed information about calculations using Mie theory can be found in Wriedt (2012). The particle size of the nanocapsules are generally reported as D_{10}, D_{50}, D_{90}, $D_{[3,2]}$, $D_{[4,3]}$, and span. D_{10}, D_{50}, and D_{90} refers that 10%, 50%, and 90% of the particles/droplets are lower than this value, respectively. $D_{[4,3]}$ (De Brouckere mean diameter), is the mean diameter of a sphere with the same volume with particles/droplets and $D_{[3,2]}$ (Sauter mean diameter) is the mean diameter of a sphere with the same surface area (Karlović et al., 2015). Span is generally used to determine the spread of the distribution and calculate from the ratio of $D_{90}-D_{10}$ to D_{50}.

Zeta potential is the magnitude of the electrostatic repulsion or attraction between particles and directly affects the stability in nanocapsule suspensions. If the particles in suspension have large zeta potential, they will tend to repel each other which decrease the stability. There are several factors affecting the zeta potential of the particles such as pH, conductivity, and concentration. The zeta potential of the suspensions are based on electrophoresis and streaming potential measurements (Li, 2004).

Thermal properties which show the physical stability of the nanocapsules are generally determined using differential scanning calorimetry (DSC) or thermogravimetric analysis (TGA). The energy changes in nanocapsules during cooling or heating are observed during DSC analysis. The physical and chemical changes in the samples result in changes in the differential heat flow (Biliaderis, 1983). Melting point, glass transition temperature, and lipid oxidation points can be determined by DSC analysis (Gordon, Marom, & Magdassi, 2015). On the other hand, weight loss rate which shows degradation of wall and core materials during heating is recorded during TGA analysis (Gómez-Mascaraque & López-Rubio, 2016; Hădăruga et al., 2016).

Morphology of the nanocapsules can be determined using different techniques such as light microscopy (Jafari, He, & Bhandari, 2007; Joos, Weiss, & McClements, 2014), scanning electron microscopy (SEM) (Moomand & Lim, 2014a), transmission electron microscopy (TEM) (Henna Lu, Nielsen, Timm-Heinrich, & Jacobsen, 2011), or atomic force microscopy. Light microscopy is used generally for nanoemulsions, while SEM can be used for all types of nanocapsules. Atomic force microscopy is used generally for determination of surface properties of nanocapsules.

3.7.2.2 Chemical Characterization Methods

FTIR is used for the determination of the chemical interactions between bioactive oil and wall materials (Esquerdo et al., 2015).

Chemical stability of the bioactive oils is determined generally during shelf life or oxidation test by monitoring fatty acid profile and oxidation products. Fatty acids profile determination is important for the bioactive oils since these oils generally have high content of sensitive fatty acids to the environmental factors. It is generally determined using gas chromatography after a suitable methylation process. Oxidation products such as peroxides, anisidine, volatiles (hexanal, nonanal, etc.), TBARS, etc. are monitored to observe degree of oxidation. The amount of these products in nanoencapsulated bioactive oils is generally compared with the amount of bulk oils during oxidation test (Gökmen et al., 2011; Gómez-Mascaraque & López-Rubio, 2016; Moomand & Lim, 2014a).

3.7.2.3 Other Characterization Methods

In a few studies, different properties of the nanocapsulated bioactive oils have also been tested. For instance, Choi et al. (2010) evaluated the release of the bioactive oils from BCD nanocomplexes in different temperatures and humidities. Moomand and Lim (2014b) studied the fate of nanoencapsulated bioactive oils during *in vitro* gastrointestinal digestion. Dey, Ghosh, Ghosh, Koley, and Dhar (2012) compared the intestinal absorption of EPA–DHA-loaded nanoemulsions and conventional emulsions in rats. Similarly, adsorption in Caco-2 human intestinal cells was also tested (Heo, Kim, Pan, & Kim, 2016). Lane, Li, Smith, and Derbyshire (2014) applied blood testing to subjects who consumed either free or nanoencapsulated fish oil.

Additionally, integration of nanoencapsulated bioactive oils into different food products such as yoghurt, fruit juice, and bread was also reported (Gökmen et al., 2011; Ilyasoglu & El, 2014; Lane et al., 2014; Lane et al., 2016).

3.7.3 Safety and Regulatory Issues of Fish Oil-Loaded Nanocapsules

Nanoencapsulated ingredients have some safety, environmental, ethical, and regulatory issues due to the lack of knowledge of these materials on human health and the environment (EFSA, 2009). Although there is no clear report on the undesirable effect of nanoencapsulated bioactive oils, there are several researches on pharmaceutical lipid nanoparticles. The conclusions of these researches raised some issues. These issues are increased bioavailability of bioactive compounds that are toxic at high levels, direct absorption of lipid nanoparticles, interference to gastrointestinal function, and wall materials. Decreasing the particle size of bioactive compounds generally increases the bioavailability of them. Although this is an advantage of the nanoencapsulation, it causes some problems for the materials that are toxic at high levels. One of the main aim of the nanoencapsulation of bioactive oils is incorporation of these oils into the food products. When an enriched food/beverage with these nanoencapsulated bioactive oils is consumed regularly in large volumes, it can cause some problems. Nanosized particles can directly be absorbed through the epithelium cells in the mouth, esophagus, stomach, and intestine. Therefore, they may be metabolized, excreted, or accumulated within certain tissues. On the other hand, absorption to the epithelium cells before intestine where the lipids normally digested can metabolize differently. Additionally, wall materials and surfactant used in the production of nanoencapsulated bioactive oils can also cause some health problems (McClements, 2013). There is need for detailed multidisciplinary studies to address these issues.

3.8 CONCLUSION AND FURTHER REMARKS

Nanoencapsulation increases the stability, controlled release, and bioavailability of the fish oil and essential oil. It can be achieved by different techniques such as emulsification, nanoliposomes, lipid carriers, and equipment-based encapsulation. Although some of these techniques (nanoemulsions, nanoliposomes, SLNs) highly applied in nanoencapsulation of fish oil and essential fatty acids, others (biopolymer complexes, electrospinning and electrospraying, etc.) are still in infancy. Therefore, a significant progress is still needed in this concept. Additionally, cross-linking of the nanoparticles to increase mechanical stability and release characteristics of the cross-linked nanoparticles must be studied. Finally, and the most importantly, effectiveness and safety issues of the nanoencapsulated fish oil and essential fatty acids must be determined by *in vivo* and *in vitro* studies.

REFERENCES

Acosta, E. (2009). Bioavailability of nanoparticles in nutrient and nutraceutical delivery. *Current Opinion in Colloid & Interface Science*, *14*(1), 3–15.

Anandharamakrishnan, C. (2014). *Techniques for nanoencapsulation of food ingredients*. New York: Springer.

Averina, E., & Allémann, E. (2013). Encapsulation of alimentary bioactive oils of the Baikal Lake area into pH-sensitive micro- and nanoparticles. *LWT—Food Science and Technology*, *53*(1), 271–277. Available from http://dx.doi.org/10.1016/j.lwt.2013.01.020.

Averina, E., Müller, R., Popov, D., & Radnaeva, L. (2011). Physical and chemical stability of nanostructured lipid drug carriers (NLC) based on natural lipids from Baikal region (Siberia, Russia). *Die Pharmazie-An International Journal of Pharmaceutical Sciences*, *66*(5), 348–356.

Bahrani, S., Ghanbarzadeh, B., Hamishekar, H., & Sowti khiyabani, M. (2013). Nanoencapsulation of omega-3 fatty acids using caseinate-pectin based complexes: FTIR, DSC, particle size, and encapsulation efficiency. *Iranian Journal of Nutrition Sciences & Food Technology*, *8*(3), 1–15.

Bejrapha, P., Min, S. G., Surassmo, S., & Choi, M. J. (2010). Physicothermal properties of freeze-dried fish oil nanocapsules frozen under different conditions. *Drying Technology*, *28*(4), 481–489. Available from http://dx.doi.org/10.1080/07373931003613684.

Belch, J., & Muir, A. (1998). n-6 and n-3 essential fatty acids in rheumatoid arthritis and other rheumatic conditions. *Proceedings of the Nutrition Society*, *57*(04), 563–569.

Benatti, P., Peluso, G., Nicolai, R., & Calvani, M. (2004). Polyunsaturated fatty acids: Biochemical, nutritional and epigenetic properties. *Journal of the American College of Nutrition*, *23*(4), 281–302. Available from http://dx.doi.org/10.1080/07315724.2004.10719371.

Biliaderis, C. G. (1983). Differential scanning calorimetry in food research—A review. *Food Chemistry*, *10*(4), 239–265. Available from http://dx.doi.org/10.1016/0308-8146(83)90081-X.

Bouwens, M., van de Rest, O., Dellschaft, N., Bromhaar, M. G., de Groot, L. C., Geleijnse, J. M., & Afman, L. A. (2009). Fish-oil supplementation induces antiinflammatory gene expression profiles in human blood mononuclear cells. *The American journal of clinical nutrition*, *90*(2), 415–424.

Cawood, A. L., Ding, R., Napper, F. L., Young, R. H., Williams, J. A., Ward, M. J., & Ye, S. (2010). Eicosapentaenoic acid (EPA) from highly concentrated n−3 fatty acid ethyl esters is incorporated into advanced atherosclerotic plaques and higher plaque EPA is associated with decreased plaque inflammation and increased stability. *Atherosclerosis*, *212*(1), 252−259.

Chen, C., Han, D., Cai, C., & Tang, X. (2010). An overview of liposome lyophilization and its future potential. *Journal of Controlled Release*, *142*(3), 299−311.

Choi, M. J., Briançon, S., Andrieu, J., Min, S. G., & Fessi, H. (2004). Effect of freeze-drying process conditions on the stability of nanoparticles. *Drying Technology*, *22*(1−2), 335−346. Available from http://dx.doi.org/10.1081/DRT-120028238.

Choi, M. J., Ruktanonchai, U., Min, S. G., Chun, J. Y., & Soottitantawat, A. (2010). Physical characteristics of fish oil encapsulated by β-cyclodextrin using an aggregation method or polycaprolactone using an emulsion−diffusion method. *Food Chemistry*, *119*(4), 1694−1703. Available from http://dx.doi.org/10.1016/j.foodchem.2009.09.052.

Cleland, L. G., James, M. J., & Proudman, S. M. (2003). The role of fish oils in the treatment of rheumatoid arthritis. *Drugs*, *63*(9), 845−853.

Connor, W. E., Lowensohn, R., & Hatcher, L. (1996). Increased docosahexaenoic acid levels in human newborn infants by administration of sardines and fish oil during pregnancy. *Lipids*, *31*(1), S183−S187.

Crawford, M., Hassam, A., & Stevens, P. A. (1981). Essential fatty acid requirements in pregnancy and lactation with special reference to brain development. *Progress in Lipid Research*, *20*, 31−40.

Damsgaard, C. T., Schack-Nielsen, L., Michaelsen, K. F., Fruekilde, M. B., Hels, O., & Lauritzen, L. (2006). Fish oil affects blood pressure and the plasma lipid profile in healthy Danish infants. *The Journal of Nutrition*, *136*(1), 94−99.

Daniels, J. L., Longnecker, M. P., Rowland, A. S., Golding, J., & Team, A. S. (2004). Fish intake during pregnancy and early cognitive development of offspring. *Epidemiology*, *15*(4), 394−402.

de Conto, L. C., Grosso, C. R. F., & Gonçalves, L. A. G. (2013). Chemometry as applied to the production of omega-3 microcapsules by complex coacervation with soy protein isolate and gum Arabic. *LWT—Food Science and Technology*, *53*(1), 218−224. Available from http://dx.doi.org/10.1016/j.lwt.2013.02.017.

Dey, T. k, Ghosh, S., Ghosh, M., Koley, H., & Dhar, P. (2012). Comparative study of gastrointestinal absorption of EPA & DHA rich fish oil from nano and conventional emulsion formulation in rats. *Food Research International*, *49*(1), 72−79. Available from http://dx.doi.org/10.1016/j.foodres.2012.07.056.

Dodds, J. (2013). Techniques to Analyse Particle Size of Food Powders. In B. Bhandari, N. Bansal, M. Zhang, & P. Schuck, (Eds.), *Handbook of food powders* (pp. 309−338). Woodhead Publishing, Oxford.

EFSA (2009). Scientific opinion on 'the potential risks arising from nanoscience and nanotechnologies on food and feed safety', scientific opinion of the scientific committee, adopted on 10 February 2009. *The EFSA Journal*, *958*, 1−39.

Eratte, D., McKnight, S., Gengenbach, T. R., Dowling, K., Barrow, C. J., & Adhikari, B. P. (2015). Co-encapsulation and characterisation of omega-3 fatty acids and probiotic bacteria in whey protein isolate−gum Arabic complex coacervates. *Journal of Functional Foods*, *19* (Part B), 882−892. Available from http://dx.doi.org/10.1016/j.jff.2015.01.037.

Eratte, D., Wang, B., Dowling, K., Barrow, C. J., & Adhikari, B. P. (2014). Complex coacervation with whey protein isolate and gum arabic for the microencapsulation of omega-3 rich tuna oil. *Food and Function*, *5*(11), 2743–2750. Available from http://dx.doi.org/10.1039/c4fo00296b.

Esquerdo, V. M., Dotto, G. L., & Pinto, L. A. A. (2015). Preparation of nanoemulsions containing unsaturated fatty acid concentrate-chitosan capsules. *Journal of Colloid and Interface Science*, *445*, 137–142. Available from http://dx.doi.org/10.1016/j.jcis.2014.12.094.

Ezhilarasi, P. N., Karthik, P., Chhanwal, N., & Anandharamakrishnan, C. (2013). Nanoencapsulation techniques for food bioactive components: A review. *Food and Bioprocess Technology*, *6*(3), 628–647. Available from http://dx.doi.org/10.1007/s11947-012-0944-0.

Fathi, M., Martín, T., & McClements, D. J. (2014). Nanoencapsulation of food ingredients using carbohydrate based delivery systems. *Trends in Food Science and Technology*, *39*(1), 18–39. Available from http://dx.doi.org/10.1016/j.tifs.2014.06.007.

Fathi, M., Mozafari, M. R., & Mohebbi, M. (2012). Nanoencapsulation of food ingredients using lipid based delivery systems. *Trends in Food Science and Technology*, *23*(1), 13–27. Available from http://dx.doi.org/10.1016/j.tifs.2011.08.003.

Force, B.N.F.T. (1992). Unsaturated fatty acids—nutritional and physiological significance. Chapman & Hall, London.

García-Moreno, P. J., Stephansen, K., Van Der Kruijs, J., Guadix, A., Guadix, E. M., Chronakis, I. S., & Jacobsen, C. (2016). Encapsulation of fish oil in nanofibers by emulsion electrospinning: Physical characterization and oxidative stability. *Journal of Food Engineering*, *183*, 39–49. Available from http://dx.doi.org/10.1016/j.jfoodeng.2016.03.015.

Ghorani, B., & Tucker, N. (2015). Fundamentals of electrospinning as a novel delivery vehicle for bioactive compounds in food nanotechnology. *Food Hydrocolloids*, *51*, 227–240. Available from http://dx.doi.org/10.1016/j.foodhyd.2015.05.024.

Ghorbanzade, T., Jafari, S. M., Akhavan, S., & Hadavi, R. (2017). Nano-encapsulation of fish oil in nano-liposomes and its application in fortification of yogurt. *Food Chemistry*, *216*, 146–152. Available from http://dx.doi.org/10.1016/j.foodchem.2016.08.022.

Goh, Y. K., Jumpsen, J. A., Ryan, E. A., & Clandinin, M. T. (1997). Effect of omega 3 fatty acid on plasma lipids, cholesterol and lipoprotein fatty acid content in NIDDM patients. *Diabetologia*, *40*(1), 45–52.

Gökmen, V., Mogol, B. A., Lumaga, R. B., Fogliano, V., Kaplun, Z., & Shimoni, E. (2011). Development of functional bread containing nanoencapsulated omega-3 fatty acids. *Journal of Food Engineering*, *105*(4), 585–591. Available from http://dx.doi.org/10.1016/j.jfoodeng.2011.03.021.

Gómez-Mascaraque, L. G., & López-Rubio, A. (2016). Protein-based emulsion electrosprayed micro- and submicroparticles for the encapsulation and stabilization of thermosensitive hydrophobic bioactives. *Journal of Colloid and Interface Science*, *465*, 259–270. Available from http://dx.doi.org/10.1016/j.jcis.2015.11.061.

Gordon, V., Marom, G., & Magdassi, S. (2015). Formation of hydrophilic nanofibers from nanoemulsions through electrospinning. *International Journal of Pharmaceutics*, *478*(1), 172–179. Available from http://dx.doi.org/10.1016/j.ijpharm.2014.11.038.

Gregoradis, G. (1993). Liposome preparation and related techniques. *Liposome Technology*, *1*, 1–63.

Grosso, G., Micek, A., Marventano, S., Castellano, S., Mistretta, A., Pajak, A., & Galvano, F. (2016). Dietary n-3 PUFA, fish consumption and depression: A systematic review and meta-analysis of observational studies. *Journal of Affective Disorders*, *205*, 269–281. Available from http://dx.doi.org/10.1016/j.jad.2016.08.011.

Gulotta, A., Saberi, A. H., Nicoli, M. C., & McClements, D. J. (2014). Nanoemulsion-based delivery systems for polyunsaturated (ω-3) oils: Formation using a spontaneous emulsification method. *Journal of Agricultural and Food Chemistry*, *62*(7), 1720–1725. Available from http://dx.doi.org/10.1021/jf4054808.

Habib, S. M., Amr, A. S., & Hamadneh, I. M. (2012). Nanoencapsulation of alpha-linolenic acid with modified emulsion diffusion method. *Journal of the American Oil Chemists' Society*, *89*(4), 695–703.

Hădăruga, D. I., Ünlüsayin, M., Gruia, A. T., Birău, C., Rusu, G., & Hădăruga, N. G. (2016). Thermal and oxidative stability of Atlantic salmon oil (*Salmo salar* L.) and complexation with β-cyclodextrin. *Beilstein Journal of Organic Chemistry*, *12*, 179–191. Available from http://dx.doi.org/10.3762/bjoc.12.20.

Hadian, Z., Sahari, M. A., Moghimi, H. R., & Barzegar, M. (2014). Formulation, characterization and optimization of liposomes containing eicosapentaenoic and docosahexaenoic acids; A methodology approach. *Iranian Journal of Pharmaceutical Research*, *13*(2), 393–404.

Haider, J., Majeed, H., Williams, P., Safdar, W., & Zhong, F. (2017). Formation of chitosan nanoparticles to encapsulate krill oil (*Euphausia superba*) for application as a dietary supplement. *Food Hydrocolloids*, *63*, 27–34.

Heinzelmann, K., Franke, K., Velasco, J., & Marquez-Ruiz, G. (2000). Microencapsulation of fish oil by freeze-drying techniques and influence of process parameters on oxidative stability during storage. *European Food Research and Technology*, *211*(4), 234–239.

Henna Lu, F. S., Nielsen, N. S., Timm-Heinrich, M., & Jacobsen, C. (2011). Oxidative stability of marine phospholipids in the liposomal form and their applications. *Lipids*, *46*(1), 3–23. Available from http://dx.doi.org/10.1007/s11745-010-3496-y.

Heo, W., Kim, J. H., Pan, J. H., & Kim, Y. J. (2016). Lecithin-based nano-emulsification improves the bioavailability of conjugated linoleic acid. *Journal of Agricultural and Food Chemistry*, *64*(6), 1355–1360. Available from http://dx.doi.org/10.1021/acs.jafc.5b05397.

Hibbeln, J. R. (1998). Fish consumption and major depression. *Lancet*, *351*(9110), 1213.

Huang, H., Hao, S., Li, L., Yang, X., Cen, J., Lin, W., & Wei, Y. (2014). Influence of emulsion composition and spray-drying conditions on microencapsulation of tilapia oil. *Journal of food science and technology*, *51*(9), 2148–2154.

Ilyasoglu, H., & El, S. N. (2014). Nanoencapsulation of EPA/DHA with sodium caseinate–gum arabic complex and its usage in the enrichment of fruit juice. *LWT-Food Science and Technology*, *56*(2), 461–468.

Jafari, S. M., Assadpoor, E., Bhandari, B., & He, Y. (2008). Nano-particle encapsulation of fish oil by spray drying. *Food Research International*, *41*(2), 172–183. Available from http://dx.doi.org/10.1016/j.foodres.2007.11.002.

Jafari, S. M., He, Y., & Bhandari, B. (2007). Optimization of nano-emulsions production by microfluidization. *European Food Research and Technology*, *225*(5–6), 733–741. Available from http://dx.doi.org/10.1007/s00217-006-0476-9.

Jafarpour, S. A., Esfahani, R., & Jafari, S. M. (2016). Efficiency evaluation of nanoencapsulation of omega-3 fatty acids with gelatin-Arabic gum complex using coaservation technique. *Iranian Scientific Fisheries Journal*, *25*(2), 29–42.

Joos, A., Weiss, J., & McClements, D. J. (2014). Fabrication of lipophilic nanoparticles by spontaneous emulsification: Stabilization by cosurfactants. *Food Biophysics*, *10*(1), 83–93. Available from http://dx.doi.org/10.1007/s11483-014-9364-1.

Jørgensen, M. H., Hernell, O., Hughes, E. L., & Michaelsen, K. F. (2001). Is there a relation between docosahexaenoic acid concentration in mothers' milk and visual development in term infants? *Journal of pediatric gastroenterology and nutrition*, *32*(3), 293–296.

Karlović, S., Bosiljkov, T., Brnčić, M., Semenski, D., Dujmić, F., Tripalo, B., & Ježek, D. (2015). Reducing fat globules particle-size in goat milk: Ultrasound and high hydrostatic pressures approach. *Chemical and Biochemical Engineering Quarterly*, *28*(4), 499–507.

Karmali, R. A., Reichel, P., Cohen, L. A., Terano, T., Hirai, A., Tamura, Y., & Yoshida, S. (1987). The effects of dietary ω-3 fatty acids on the DU-145 transplantable human prostatic tumor. *Anticancer Research*, *7*(6), 1173–1180.

Karthik, P., & Anandharamakrishnan, C. (2013). Microencapsulation of docosahexaenoic acid by spray-freeze-drying method and comparison of its stability with spray-drying and freeze-drying methods. *Food and Bioprocess Technology*, *6*(10), 2780–2790.

Katouzian, I., & Jafari, S. M. (2016). Nano-encapsulation as a promising approach for targeted delivery and controlled release of vitamins. *Trends in Food Science & Technology*, *53*, 34–48. Available from http://dx.doi.org/10.1016/j.tifs.2016.05.002.

Klinkesorn, U., Sophanodora, P., Chinachoti, P., McClements, D. J., & Decker, E. A. (2005). Increasing the oxidative stability of liquid and dried tuna oil-in-water emulsions with electrostatic layer-by-layer deposition technology. *Journal of Agricultural and Food Chemistry*, *53*(11), 4561–4566. Available from http://dx.doi.org/10.1021/jf0479158.

Komaiko, J., Sastrosubroto, A., & McClements, D. J. (2016). Encapsulation of ω-3 fatty acids in nanoemulsion-based delivery systems fabricated from natural emulsifiers: Sunflower phospholipids. *Food Chemistry*, *203*, 331–339. Available from http://dx.doi.org/10.1016/j.foodchem.2016.02.080.

Kubo, K., Sekine, S., & Saito, M. (2003). Docosahexaenoic acid-containing phosphatidylethanolamine in the external layer of liposomes protects docosahexaenoic acid from 2,2′-azobis(2-aminopropane)dihydrochloride-mediated lipid peroxidation. *Archives of Biochemistry and Biophysics*, *410*(1), 141–148. Available from http://dx.doi.org/10.1016/S0003-9861(02)00679-3.

Lane, K. E., Li, W., Smith, C. J., & Derbyshire, E. J. (2016). The development of vegetarian omega-3 oil in water nanoemulsions suitable for integration into functional food products. *Journal of Functional Foods*, *23*, 306–314. Available from http://dx.doi.org/10.1016/j.jff.2016.02.043.

Lane, K. E., Li, W., Smith, C., & Derbyshire, E. (2014). The bioavailability of an omega-3-rich algal oil is improved by nanoemulsion technology using yogurt as a food vehicle. *International Journal of Food Science & Technology*, *49*(5), 1264–1271. Available from http://dx.doi.org/10.1111/ijfs.12455.

Lev, E. I., Solodky, A., Harel, N., Mager, A., Brosh, D., Assali, A., & Kornowski, R. (2010). Treatment of aspirin-resistant patients with omega-3 fatty acids versus aspirin dose escalation. *Journal of the American College of Cardiology*, *55*(2), 114–121.

Li, D. (2004). *Microfluidic methods for measuring zeta potential.* In: *Electrokinetics in microfluidics,* Volume 2 (pp. 617–640). Academic Press, Amsterdam.

Li, J., Xiong, S., Wang, F., Regenstein, J. M., & Liu, R. (2015). Optimization of microencapsulation of fish oil with gum Arabic/casein/beta-cyclodextrin mixtures by spray drying. *Journal of Food Science*, *80*(7), C1445–C1452.

Magaro, M., Altomonte, L., Zoli, A., Mirone, L., De Sole, P., Di Mario, G., & Oradei, A. (1988). Influence of diet with different lipid composition on neutrophil chemiluminescence and disease activity in patients with rheumatoid arthritis. *Annals of the Rheumatic Diseases*, *47*(10), 793–796. Available from http://dx.doi.org/10.1136/ard.47.10.793.

Mason, T. G., Wilking, J. N., Meleson, K., Chang, C. B., & Graves, S. M. (2006). Nanoemulsions: formation, structure, and physical properties. *Journal of Physics: Condensed Matter*, *18*(41), R635.

McClements, D. J. (2011). Edible nanoemulsions: Fabrication, properties, and functional performance. *Soft Matter*, *7*(6), 2297–2316.

McClements, D. J. (2013). Edible lipid nanoparticles: Digestion, absorption, and potential toxicity. *Progress in lipid research*, *52*(4), 409–423. Available from http://dx.doi.org/10.1016/j.plipres.2013.04.008.

McClements, D. J., & Decker, E. A. (2008). Lipids. In S. Damodaran, K. L. Parkin, & O. R. Fennema, (Eds.), *Fennema's food chemistry, Fourth Edition* (pp. 155–216). Boca Raton: CRC Press.

Mehrad, B., Shabanpour, B., Jafari, S., & Pourashouri, P. (2015). Characterization of dried fish oil from Menhaden encapsulated by spray drying. *AACL Bioflux*, *8*(1), 57–69.

Moomand, K., & Lim, L. T. (2014a). Oxidative stability of encapsulated fish oil in electrospun zein fibres. *Food Research International*, *62*, 523–532. Available from http://dx.doi.org/10.1016/j.foodres.2014.03.054.

Moomand, K., & Lim, L. T. (2014b). Properties of encapsulated fish oil in electrospun zein fibres under simulated in vitro conditions. Food and Bioprocess. *Technology*, *8*(2), 431–444. Available from http://dx.doi.org/10.1007/s11947-014-1414-7.

Mozafari, M. (2010). Nanoliposomes: Preparation and analysis. In V. Weissig (Ed.), *Liposomes: Methods and protocols, volume 1: Pharmaceutical nanocarriers* (pp. 29–50). Glendale, AZ: Humana Press.

Müller, R. H., Dingler, A., Schneppe, T., & Gohla, S. (2000). Large-scale production of solid lipid nanoparticles (SLN) and nanosuspensions. In D. L. Wise, (Ed.), *Handbook of pharmaceutical controlled release technology* (pp. 377–392). New York: CRC Press.

Na, H. S., Kim, J. N., Kim, J. M., & Lee, K. Y. (2011). Encapsulation of fish oil using cyclodextrin and whey protein concentrate. *Biotechnology and Bioprocess Engineering*, *16*(6), 1077–1082. Available from http://dx.doi.org/10.1007/s12257-011-0099-2.

Naik, A., Meda, V., & Lele, S. (2014). Freeze drying for microencapsulation of α-linolenic acid rich oil: A functional ingredient from Lepidium sativum seeds. *European Journal of Lipid Science and Technology*, *116*(7), 837–846.

Nassar, T., Rom, A., Nyska, A., & Benita, S. (2009). Novel double coated nanocapsules for intestinal delivery and enhanced oral bioavailability of tacrolimus, a P-gp substrate drug. *Journal of Controlled Release*, *133*(1), 77–84. Available from http://dx.doi.org/10.1016/j.jconrel.2008.08.021.

Nejadmansouri, M., Hosseini, S. M. H., Niakosari, M., Yousefi, G. H., & Golmakani, M. T. (2016). Physicochemical properties and storage stability of ultrasound-mediated WPI-stabilized fish oil nanoemulsions. *Food Hydrocolloids*, *61*, 801–811. Available from http://dx.doi.org/10.1016/j.foodhyd.2016.07.011.

Ostertag, F., Weiss, J., & McClements, D. J. (2012). Low-energy formation of edible nanoemulsions: Factors influencing droplet size produced by emulsion phase inversion. *Journal of Colloid and Interface Science*, *388*(1), 95–102. Available from http://dx.doi.org/10.1016/j.jcis.2012.07.089.

Pereira, D. M., Valentão, P., & Andrade, P. B. (2014). Nano-and microdelivery systems for marine bioactive lipids. *Marine Drugs*, *12*(12), 6014–6027. Available from http://dx.doi.org/10.3390/md12126014.

Pourashouri, P., Shabanpour, B., Razavi, S. H., Jafari, S. M., Shabani, A., & Aubourg, S. P. (2014a). Impact of wall materials on physicochemical properties of microencapsulated fish oil by spray drying. *Food and Bioprocess Technology*, *7*(8), 2354–2365.

Pourashouri, P., Shabanpour, B., Razavi, S. H., Jafari, S. M., Shabani, A., & Aubourg, S. P. (2014b). Oxidative stability of spray-dried microencapsulated fish oils with different wall materials. *Journal of Aquatic Food Product Technology*, *23*(6), 567–578.

Radtke, M., & Müller, R. H. (2001). Nanostructured lipid drug carriers. *New Drugs*, *2*, 48–52.

Ramprasath, V. R., Eyal, I., Zchut, S., & Jones, P. J. (2013). Enhanced increase of omega-3 index in healthy individuals with response to 4-week n-3 fatty acid supplementation from krill oil versus fish oil. *Lipids in Health and Disease*, *12*(1), 1.

Raper, N. R., Cronin, F., & Exler, J. (1992). Omega-3 fatty acid content of the US food supply. *Journal of the American College of Nutrition*, *11*(3), 304–308.

Rasti, B., Jinap, S., Mozafari, M. R., & Abd-Manap, M. Y. (2014). Optimization on preparation condition of polyunsaturated fatty acids nanoliposome prepared by Mozafari method. *Journal of Liposome Research*, *24*(2), 99–105. Available from http://dx.doi.org/10.3109/08982104.2013.839702.

Rasti, B., Jinap, S., Mozafari, M. R., & Yazid, A. M. (2012). Comparative study of the oxidative and physical stability of liposomal and nanoliposomal polyunsaturated fatty acids prepared with conventional and Mozafari methods. *Food Chemistry*, *135*(4), 2761–2770. Available from http://dx.doi.org/10.1016/j.foodchem.2012.07.016.

Richardson, A. J., & Ross, M. (2000). Fatty acid metabolism in neurodevelopmental disorder: A new perspective on associations between attention-deficit/hyperactivity disorder, dyslexia, dyspraxia and the autistic spectrum. *Prostaglandins, Leukotrienes and Essential Fatty Acids (PLEFA)*, *63*(1), 1–9.

Rodríguez, J., Martín, M. J., Ruiz, M. A., & Clares, B. (2016). Current encapsulation strategies for bioactive oils: From alimentary to pharmaceutical perspectives. *Food Research International*, *83*, 41–59. Available from http://dx.doi.org/10.1016/j.foodres.2016.01.032.

Rose, D. P., & Cohen, L. A. (1988). Effects of dietary menhaden oil and retinyl acetate on the growth of DU 145 human prostatic adenocarcinoma cells transplanted into athymic nude mice. *Carcinogenesis*, *9*(4), 603–605. Available from http://dx.doi.org/10.1093/carcin/9.4.603.

Rose, D. P., & Connolly, J. M. (1990). Effects of fatty acids and inhibitors of eicosanoid synthesis on the growth of a human breast cancer cell line in culture. *Cancer Research*, *50*(22), 7139–7144.

Ruxton, C., Reed, S. C., Simpson, M., & Millington, K. (2004). The health benefits of omega-3 polyunsaturated fatty acids: A review of the evidence. *Journal of Human Nutrition and Dietetics*, *17*(5), 449–459.

Sahari, M. A., Moghimi, H. R., Hadian, Z., Barzegar, M., & Mohammadi, A. (2016). Improved physical stability of docosahexaenoic acid and eicosapentaenoic acid encapsulated using nanoliposome containing α-tocopherol. *International Journal of Food Science & Technology*, *51*(5), 1075–1086. Available from http://dx.doi.org/10.1111/ijfs.13068.

Salminen, H., Aulbach, S., Leuenberger, B. H., Tedeschi, C., & Weiss, J. (2014). Influence of surfactant composition on physical and oxidative stability of Quillaja saponin-stabilized lipid particles with encapsulated ω-3 fish oil. *Colloids and Surfaces B: Biointerfaces*, *122*, 46–55. Available from http://dx.doi.org/10.1016/j.colsurfb.2014.06.045.

Salminen, H., Gömmel, C., Leuenberger, B. H., & Weiss, J. (2016). Influence of encapsulated functional lipids on crystal structure and chemical stability in solid lipid nanoparticles: Towards bioactive-based design of delivery systems. *Food Chemistry*, *190*, 928–937. Available from http://dx.doi.org/10.1016/j.foodchem.2015.06.054.

Salminen, H., Helgason, T., Kristinsson, B., Kristbergsson, K., & Weiss, J. (2013). Formation of solid shell nanoparticles with liquid ω-3 fatty acid core. *Food Chemistry*, *141*(3), 2934–2943. Available from http://dx.doi.org/10.1016/j.foodchem.2013.05.120.

Sanders, T. A. B., & Hinds, A. (1992). The influence of a fish oil high in docosahexaenoic acid on plasma lipoprotein and vitamin E concentrations and haemostatic function in healthy male volunteers. *British Journal of Nutrition, 68*(1), 163–173. Available from http://dx.doi.org/10.1079/BJN19920074.

Saupe, A., & Rades, T. (2006). Solid lipid nanoparticles. In M. R. Mozafari, (Ed.), *Nanocarrier technologies: Frontiers of nanotherapy* (pp. 41–50). Dordrecht: Springer.

Schiano, V., Laurenzano, E., Brevetti, G., De Maio, J. I., Lanero, S., Scopacasa, F., & Chiariello, M. (2008). Omega-3 polyunsaturated fatty acid in peripheral arterial disease: Effect on lipid pattern, disease severity, inflammation profile, and endothelial function. *Clinical Nutrition, 27*(2), 241–247.

Schiffman, J. D., & Schauer, C. L. (2008). A review: Electrospinning of biopolymer nanofibers and their applications. *Polymer Reviews, 48*(2), 317–352. Available from http://dx.doi.org/10.1080/15583720802022182.

Sidhu, K. S. (2003). Health benefits and potential risks related to consumption of fish or fish oil. *Regulatory Toxicology and Pharmacology, 38*(3), 336–344. Available from http://dx.doi.org/10.1016/j.yrtph.2003.07.002.

Siegel, R. A., & Rathbone, M. J. (2012). Overview of controlled release mechanisms. In J. Siepmann, R. A. Siegel, & M. J. Rathbone (Eds.), *Fundamentals and applications of controlled release drug delivery*. Boston, MA: Springer, 19–4.

Silvestre, M. P. C., Chaiyasit, W., Brannan, R. G., McClements, D. J., & Decker, E. A. (2000). Ability of surfactant headgroup size to alter lipid and antioxidant oxidation in oil-in-water emulsions. *Journal of Agricultural and Food Chemistry, 48*(6), 2057–2061. Available from http://dx.doi.org/10.1021/jf9911621.

Simolopoulos, A. (1989). Summary of NATO advanced research workshop on dietery n-3 and n-6 fatty acids: Biologycal effects nutritional essentiality. *Journal of Nutrion, 199*, 512–528.

Smith, K. J., Sanderson, K., McNaughton, S. A., Gall, S. L., Dwyer, T., & Venn, A. J. (2014). Longitudinal associations between fish consumption and depression in young adults. *American Journal of Epidemiology, 179*(10), 1228–1235. Available from http://dx.doi.org/10.1093/aje/kwu050.

Soltani, S., & Madadlou, A. (2015). Gelation characteristics of the sugar beet pectin solution charged with fish oil-loaded zein nanoparticles. *Food Hydrocolloids, 43*, 664–669. Available from http://dx.doi.org/10.1016/j.foodhyd.2014.07.030.

Sponton, O. E., Perez, A. A., Carrara, C. R., & Santiago, L. G. (2015). Linoleic acid binding properties of ovalbumin nanoparticles. *Colloids and Surfaces B: Biointerfaces, 128*, 219–226. Available from http://dx.doi.org/10.1016/j.colsurfb.2015.01.037.

Stevens, L. J., Zentall, S. S., Deck, J. L., Abate, M. L., Watkins, B. A., Lipp, S. R., & Burgess, J. R. (1995). Essential fatty acid metabolism in boys with attention-deficit hyperactivity disorder. *The American journal of clinical nutrition, 62*(4), 761–768.

Sugano, M. (1996). Characteristics of fats in Japanese diets and current recommendations. *Lipids, 31*(1), S283–S286.

Tamjidi, F., Nasirpour, A., & Shahedi, M. (2013). Mixture design approach for evaluation of fish oil microencapsulation in gelatin-acacia gum coacervates. *International Journal of Polymeric Materials and Polymeric Biomaterials, 62*(8), 444–449. Available from http://dx.doi.org/10.1080/00914037.2012.719138.

Tatar, F., Tunç, M. T., Dervisoglu, M., Cekmecelioglu, D., & Kahyaoglu, T. (2014). Evaluation of hemicellulose as a coating material with gum arabic for food microencapsulation. *Food Research International, 57*, 168–175. Available from http://dx.doi.org/10.1016/j.foodres.2014.01.022.

Tavakolipour, H., Bagheri, L., & Madadlou, A. (2015). Pomegranate seed oil-loaded particles of the zein cross-linked with citric acid. *Journal of Food Process Engineering*, *38*(1), 49–56. Available from http://dx.doi.org/10.1111/jfpe.12125.

Taylor, T. M., Weiss, J., Davidson, P. M., & Bruce, B. D. (2005). Liposomal nanocapsules in food science and agriculture. *Critical Reviews in Food Science and Nutrition*, *45*(7–8), 587–605.

Timonen, M., Horrobin, D., Jokelainen, J., Laitinen, J., Herva, A., & Räsänen, P. (2004). Fish consumption and depression: The Northern Finland 1966 birth cohort study. *Journal of Affective Disorders*, *82*(3), 447–452. Available from http://dx.doi.org/10.1016/j.jad.2004.02.002.

Tontul, I., & Topuz, A. (2015). Microencapsulation of plant oils rich in alpha-linolenic acid: Effect of processing parameters. In L. M. C. Sagis, (Ed.), *Microencapsulation and microspheres for food applications* (pp. 253–270). Amsterdam: Academic Press.

Torres-Giner, S., Martinez-Abad, A., Ocio, M. J., & Lagaron, J. M. (2010). Stabilization of a nutraceutical omega-3 fatty acid by encapsulation in ultrathin electrosprayed zein prolamine. *Journal of Food Science*, *75*(6), N69–N79. Available from http://dx.doi.org/10.1111/j.1750-3841.2010.01678.x.

Uluata, S., McClements, D. J., & Decker, E. A. (2015). Physical stability, autoxidation, and photosensitized oxidation of ω-3 oils in nanoemulsions prepared with natural and synthetic surfactants. *Journal of Agricultural and Food Chemistry*, *63*(42), 9333–9340. Available from http://dx.doi.org/10.1021/acs.jafc.5b03572.

Ünlüsayin, M., Hădărugă, N. G., Rusu, G., Gruia, A. T., Păunescu, V., & Hădărugă, D. I. (2016). Nano-encapsulation competitiveness of omega-3 fatty acids and correlations of thermal analysis and Karl Fischer water titration for European anchovy (*Engraulis encrasicolus* L.) oil/β-cyclodextrin complexes. *LWT—Food Science and Technology*, *68*, 135–144. Available from http://dx.doi.org/10.1016/j.lwt.2015.12.017.

Walker, R. M., Decker, E. A., & McClements, D. J. (2015b). Physical and oxidative stability of fish oil nanoemulsions produced by spontaneous emulsification: Effect of surfactant concentration and particle size. *Journal of Food Engineering*, *164*, 10–20. Available from http://dx.doi.org/10.1016/j.jfoodeng.2015.04.028.

Walker, R., Decker, E. A., & McClements, D. J. (2015a). Development of food-grade nanoemulsions and emulsions for delivery of omega-3 fatty acids: Opportunities and obstacles in the food industry. *Food & function*, *6*(1), 41–54.

Wang, B., Adhikari, B., & Barrow, C. J. (2014). Optimisation of the microencapsulation of tuna oil in gelatin-sodium hexametaphosphate using complex coacervation. *Food Chemistry*, *158*, 358–365. Available from http://dx.doi.org/10.1016/j.foodchem.2014.02.135.

Wang, B., Vongsvivut, J., Adhikari, B., & Barrow, C. J. (2015). Microencapsulation of tuna oil fortified with the multiple lipophilic ingredients vitamins A, D3, E, K2, curcumin and coenzyme Q10. *Journal of Functional Foods*, *19*, 893–901. Available from http://dx.doi.org/10.1016/j.jff.2015.03.027.

Wang, J. L., Dong, X. Y., Wei, F., Zhong, J., Liu, B., Yao, M. H., & Chen, H. (2014). Preparation and characterization of novel lipid carriers containing microalgae oil for food applications. *Journal of Food Science*, *79*(2), E169–E177. Available from http://dx.doi.org/10.1111/1750-3841.12334.

Waraho, T., McClements, D. J., & Decker, E. A. (2011). Mechanisms of lipid oxidation in food dispersions. *Trends in Food Science & Technology*, *22*(1), 3–13. Available from http://dx.doi.org/10.1016/j.tifs.2010.11.003.

Wriedt, T. (2012). Mie theory: A review. In W. Hergert, & T. Wriedt, (Eds.), *The Mie theory* (pp. 53–71). Berlin: Springer.

Xiao, Z., Liu, W., Zhu, G., Zhou, R., & Niu, Y. (2014). A review of the preparation and application of flavour and essential oils microcapsules based on complex coacervation technology. *Journal of the Science of Food and Agriculture, 94*(8), 1482–1494. Available from http://dx.doi.org/10.1002/jsfa.6491.

Yao, Z. C., Chang, M. W., Ahmad, Z., & Li, J. S. (2016). Encapsulation of rose hip seed oil into fibrous zein films for ambient and on demand food preservation via coaxial electrospinning. *Journal of Food Engineering, 191*, 115–123. Available from http://dx.doi.org/10.1016/j.jfoodeng.2016.07.012.

Yehuda, S., Rabinovitz, S., & Mostofsky, D. I. (1999). Essential fatty acids are mediators of brain biochemistry and cognitive functions. *Journal of Neuroscience Research, 56*(6), 565–570.

Zhang, Y., Tan, C., Abbas, S., Eric, K., Zhang, X., Xia, S., & Jia, C. (2014). The effect of soy protein structural modification on emulsion properties and oxidative stability of fish oil microcapsules. *Colloids and Surfaces B: Biointerfaces, 120*, 63–70. Available from http://dx.doi.org/10.1016/j.colsurfb.2014.05.006.

Zhong, Q., Tian, H., & Zivanovic, S. (2009). Encapsulation of fish oil in solid zein particles by liquid-liquid dispersion. *Journal of Food Processing and Preservation, 33*(2), 255–270. Available from http://dx.doi.org/10.1111/j.1745-4549.2009.00390.x.

Zhu, J., Zhuang, P., Luan, L., Sun, Q., & Cao, F. (2015). Preparation and characterization of novel nanocarriers containing krill oil for food application. *Journal of Functional Foods, 19*, 902–912. Available from http://dx.doi.org/10.1016/j.jff.2015.06.017.

Zhu, Q. Y., Guissi, F., Yang, R. Y., Wang, Q., Wang, K., Chen, D., & Gu, Y. Q. (2015). Preparation of deep sea fish oil-based Nanostructured lipid carriers with enhanced cellular uptake. *Journal of Nanoscience and Nanotechnology, 15*(12), 9539–9547. Available from http://dx.doi.org/10.1166/jnn.2015.10880.

Zimet, P., & Livney, Y. D. (2009). Beta-lactoglobulin and its nanocomplexes with pectin as vehicles for ω-3 polyunsaturated fatty acids. *Food Hydrocolloids, 23*(4), 1120–1126.

Zimet, P., Rosenberg, D., & Livney, Y. D. (2011). Re-assembled casein micelles and casein nanoparticles as nano-vehicles for ω-3 polyunsaturated fatty acids. *Food Hydrocolloids, 25*(5), 1270–1276. Available from http://dx.doi.org/10.1016/j.foodhyd.2010.11.025.

Chapter 4

Nanoencapsulation of Vitamins

Iman Katouzian[1,2] and Seid Mahdi Jafari[1]

[1]*Gorgan University of Agricultural Sciences and Natural Resources, Gorgan, Iran,*
[2]*Nano-encapsulation in the Food, Nutraceutical, and Pharmaceutical Industries Group (NFNPIG), Universal Scientific Education and Research Network (USERN), Tehran, Iran*

4.1 INTRODUCTION

Today, the hot market of functional foods has advanced the usage of colloidal delivery systems, especially the micro- and nanoencapsulation techniques by which the challenges during food enrichment processes are overcome. The encapsulation procedures prevent the interaction of bioactive compounds with the external medium as they are entrapped in a shell of an organic or inorganic material. The engineered capsules can release their content within the determined zone of the body for nutraceutical or pharmaceutical purposes.

Industrial food manufacturing processes require the addition of different flavors, colors, preservatives, etc. to provide palatable products for the consumers as well as extending the shelf life of the products. Moreover, specific micronutrients and beneficial microelements are usually added to the foods to fabricate value-added and enriched food products. Nevertheless, incorporating these bioactive compounds is often faced with some problems that limit the implementation of these nutrients, for instance, the storage period, inconsistency with the food network, and the undesirable sensory properties that some bioactive compounds induce (Velikov & Pelan, 2008). In addition, the challenges in the manufacturing processes in combination with the controlled release and bioavailability issues make fortification a complex process (Gleeson, Ryan, & Brayden, 2016; Grumezescu, 2016). Hence, micro- and, specifically, nanoencapsulation systems are novel and proper candidates for achieving the goal of food fortification (Assadpour, Maghsoudlou, Jafari, Ghorbani, & Aalami, 2016a; Esfanjani & Jafari, 2016; Ghorbanzade, Jafari, Akhavan, & Hadavi, 2017). By implementing these nanostructures, the uptake of nutrients and especially vitamins is elevated in the gastrointestinal (GI) tract. The release profile of the vitamins is regulated more by the nanoparticle size than other physical properties; therefore, scholars

Nanoencapsulation of Food Bioactive Ingredients. DOI: http://dx.doi.org/10.1016/B978-0-12-809740-3.00004-0

are seeking to decrease and keep the size of particles in the nanoscale for better absorption in the epithelial cells (Katouzian & Jafari, 2016).

In this chapter, first, the vital biological functions of vitamins plus their deficiency consequences are reviewed. Next, the versatile types of micro- and nanocarriers are studied together with their production methods for the encapsulation and delivery of vitamins.

4.2 NUTRITIONAL PROPERTIES AND CLASSIFICATION OF VITAMINS

Vitamins are essential parts of the human diet as they regulate several biochemical reactions in the human body. Thirteen vitamins are recognized at present that have inevitable roles in the physiological reactions and are categorized into two classes considering their solubility. The lipophilic vitamins include vitamins A, E, K, and D plus the carotenoids exhibiting the functional properties of vitamin A (Eitenmiller, Landen Jr, & Ye, 2016). Moreover, the hydrophilic vitamins contain vitamin C, the class of B vitamins which includes thiamin (B_1), riboflavin (B_2), nicotinic acid (B_3), pyridoxine (B_6), pantothenic acid (B_5), folate and cyanocobalamin (B_{12}), which interact in significant body functions. This type of classification indeed reflects the bioavailability of these bioactive agents since the solubility rate is attributed to the uptake and absorption in the intestine. Also, the solubility profile represents the vitamin distribution in different types of foods (Sebrell & Harris, 2014; Teleki, Hitzfeld, & Eggersdorfer, 2013).

Based on the structure of different vitamins, they reveal different biological activities and the structure of these compounds are termed vitamers. The vitamin precursors (provitamins) are organic compounds that do not display the properties of vitamins themselves but can be converted to vitamins by the metabolism process (Failla, 2011). It is obvious that vitamins must be a part of human diet since most of the vitamin types are not synthesized within the body except vitamin B_2 and vitamin D (Bikle, 2014; Cooper, Murrell, Roane, & Harirforoosh, 2015). The formation of vitamin D is conducted in the presence of sunlight; furthermore, the formation of vitamin B_3 relies on the adequate content of tryptophan in proteins that triggers its synthesis. Plants are able to produce all sorts of vitamins except cyanocobalamin and therefore should be incorporated into the human diet (Fink, Fink, & Mikesky, 2013).

Some of the crucial functions of vitamins in the physiological system include boosting the immune system and eyesight, supporting the dermal health, and cancer prevention (vitamin A); empowering the immunity against microorganisms, alleviate anxiety and depression, reduce stroke chances, and soothing PMS (premenstrual syndrome) (vitamin B-complex); strengthening the immune system, healing common cold, keeping the skin healthy, healing wounds, controlling the cholesterol content and blood sugar concentration, reducing disorders related to the nervous system (vitamin C) (Hickey, 2009); preventing cancer and cardiovascular diseases as well as developing healthy and strong bones and teeth (vitamin D); restraining neural diseases, such as, Alzheimer and other abnormalities, elevating the stamina, and avoiding skin disorders (vitamin E); formation of coagulates, signal propagation in nerves, and empowering the bone structure (vitamin K) (Zempleni, Suttie, Gregory, & Stover, 2013).

4.2.1 Vitamin Deficiency Outcomes

Usually, the group of B vitamins is employed in the catabolic reactions to produce the required energy for living cells in the form of coenzymes that are needed for the enzyme function. A coenzyme comprises a protein (apoenzyme) binding to the vitamin. The vitamin moiety of the coenzyme induces the affinity for the enzyme and the substrate. When the consumed vitamin is not modified to the coenzyme form, it does not take part in metabolic reactions, thus it is deposited in the body tissues and the metabolism route is changed.

Lack of some B-group vitamins may cause disorders with definite symptoms, while the deficiency of other vitamins has a less detrimental impact on health compared to the B-group vitamins. Nevertheless, the lack of other vitamins in animal species probably leads to peculiar signs. To sum up, the reasons for vitamin deficiency may include one or a combination of factors mentioned below.

- Insufficient ingestion
- Limited absorption
- Excess elimination
- Enhanced requirement
- Higher rate of catabolism
- Capacity of the body to keep vitamins, for instance, thiamin can be preserved for only 2 weeks, while cyanocobalamin is stored in the liver for so long (Eitenmiller, et al., 2016).

4.2.2 Dosage of Vitamin Consumption

Metabolic reactions satisfy the needs of our body all the time and vitamins are needed for maintaining this process. The Food and Nutrition Center of Medicine in the United States has specified the sufficient amount of nutrient consumption in order to maintain the essential body functions in an individual. The recommended dietary allowance (RDA) determines the average intake extent of a nutrient, which supports the vital body functions in healthy people plus a specific life phase and gender group. Indeed, the RDA emanates from an estimated average requirement (EAR) that is the assessment of nutrient intake in which the chance of insufficiency is 50%. The RDA definition has been applied to the vitamins; however, there is not enough information about biotin and pantothenic acid to calculate their EAR and instead the reference adequate intake (AI) is exerted. In this way, the AI is obtained experimentally from the intake extent or the average nutrient intake by groups of healthy people (Bolzetta, et al., 2015).

4.2.3 Vitamin Fortification in Food Products

Many terms are used to outline the increased level of vitamins in food products, such as enrichment, standardization, and nitrification (Chalupa & Schroeder, 2016; Dasgupta, Ranjan, Mundra, Ramalingam, & Kumar, 2016). The addition of vitamins, fully or in part is required to compensate for the vitamin loss occurred in the processing stage. Some examples of vitamin fortification include vitamins A and D applied to skimmed milk powder, vitamin D incorporated in evaporated milk, vitamin B addition to the flour, and adding vitamin C to instant potato (Adolphe, Whiting, & Dahl, 2009; Gupta, 2015; Hilliam, 1996; Murphy, et al., 2001).

Fortification with vitamin is usually applied to the foods that are proper vehicles for the implemented vitamins. Moreover, the fortified food product must not necessarily have the used vitamin naturally. As an example, in many countries margarine is fortified with vitamin A when it is used as an alternative for butter (Solon, et al., 1996). Also, margarine contains higher levels of vitamin D in comparison to butter as it enhances the public health standards (Calvo & Whiting, 2013). Since 1998, almost all the cereal products have been fortified with folic acid in the United States but this nitrification procedure is voluntary in the United Kingdom (Assadpour, et al., 2016a).

Enrichment can be defined as the deliberate increase in the natural level of an agent to make a product beneficial and interesting for consumers. The addition of compounds to compensate for the loss of vitamins during the processing conditions is called standardization. Accordingly, milk and butter experience seasonal modifications in their vitamin A and D levels that are subsequently added to the formulations to keep the vitamin levels normal. Finally, nitrification is referred to the improvement of vitamins in fabricated foods that are purchased as meal substitutes (Ottaway, 2012).

In addition to the nutritional properties, vitamins impart other functional features in food products. As an illustration, β-carotene induces a red-orange color when it is added to the food products. Currently, it has been applied to margarine, pasta, cakes, etc. Besides, vitamin E and C (ascorbyl palmitate type) can be implemented as antioxidants to stabilize fats and oils. Ascorbic acid is known for its reduction potential and it is responsible for the color of the cured meat, ham, and bacon during the fabrication process (Pearson & Gillett, 2012; Richards & Tatiyaborworntham, 2015).

4.2.4 Vitamin Retention under Food Processing Conditions

New technologies in the fabrication of food products are oriented in a way to inhibit the destruction of naturally occurring vitamins as much as possible. Factors, such as temperature, light, moisture content, pH, water activity, degradative enzymes plus trace metals like copper and iron, affect the degradation of vitamins. Ball (2005) has explored these factors in detail. Also, Ryley, et al. (1990) have provided the mathematical functions for determining the level of vitamin retention during food processing operations.

Food processing operations destroy the food-borne pathogens and degenerative enzymes, which would prolong the shelf life of food products. Moreover, food packaging plays a key role in the storage of food products following the production stages; thus, employing appropriate packaging materials enable better preservation conditions. Good manufacturing practices (GMP) standards are executed to ensure that food products are consistently fabricated and controlled. In the end, the distribution stage is considered in very sensitive food products.

The factors, which influence the destruction of vitamins, are mentioned above. The stability of vitamins is not similar for all of them as water-soluble vitamins are normally lost during leaching that includes the washing, blanching, and cooking operations. Furthermore, vitamin C is rapidly oxidized during the processing and cooking activities. Vitamin B_1 (thiamin) is susceptible to high-temperature in neutral and alkaline food networks and is labile in the air. Vitamin B_2 (riboflavin) is decomposed by light, while niacin and vitamin B_6 are resistant types under processing and storage conditions. Vitamins A and E are susceptible to oxidation, which occurs by means of heat, light, metal ions, etc. The structure of vitamin K is dramatically modified by fluorescent and visible light. Finally, vitamin D is not considerably influenced by the storage and processing circumstances.

It is true that not all the content of vitamins can be stored during processing conditions but each food commodity has an abundant source of vitamin that should be considered in choosing the food preparation conditions. For instance, the loss of vitamin C in milk is not normally considered since this commodity is not a great source of vitamin C compared with other food products like fruit juices. Moreover, the natural changes of the vitamin extent in the raw food may have a more dramatic effect on the vitamin content than the individual processing.

In order to overcome these undesirable conditions and enhance the preservation of vitamins in cuisines during food processing and storage stages, encapsulation techniques have revealed great potential in the preservation and targeted delivery of vitamins. This process involves the entrapment of a bioactive compound (vitamin) or a mixture of bioactive elements by another compound called "wall material." The entrapped nutrient is basically a liquid but other physical states can also be implemented. The encapsulation

TABLE 4.1 Importance of Using the Micro/nanoencapsulation Processes for Vitamins and Their Key Features

Microencapsulation Benefits	Nanoencapsulation Advantages
Protect the vitamin from external environment	Faster dissociation
Controlled release of vitamins	Higher surface area compared to mass proportion
Improved flow properties	High intracellular uptake
Reduce overages	Pass along the smallest body fenestrations
Measuring the precise level of vitamin delivery	Enable precision targeting
Form light-scattering vitamin solutions	Reduce the reactions between vitamins and other molecules plus surrounding medium
Being cost-effective, especially for spray drying method	Formulating optically transparent vitamin solutions
Undesirable flavor of some vitamins are masked	Reduction in the quantity of utilized core-shell material
Enriching the food products with a complex of vitamins	Rendering long shelf life coated vitamins
	Reinforced physical stability against coalescence and gravitational separations

Source: Reprinted with permission from Katouzian, I., & Jafari, S.M. (2016). Nano-encapsulation as a promising approach for targeted delivery and controlled release of vitamins. Trends in Food Science & Technology, 53, 34–48.

techniques are classified into microencapsulation and nanoencapsulation methods which are outlined in this chapter. According to Katouzian and Jafari (2016), nanoencapsulation is a promising approach for the retention of vitamins in food products and have more benefits compared with microencapsulation techniques. Table 4.1 points out the advantages of implementing the encapsulation techniques in the vitamin sector. Moreover, nanoencapsulation is more complicated compared with the conventional microencapsulation procedures, having its own exclusive features.

4.3 MICROENCAPSULATION OF VITAMINS

Microencapsulation techniques enable a sustained and targeted release in the right place of the body as well as protecting the vitamins against destructive conditions either in the body or the ambient (Khazaei, Jafari, Ghorbani, & Kakhki, 2014; Mahdavi, Jafari, Ghorbani, & Assadpoor, 2014; Pourashouri, et al., 2014). Fig. 4.1 displays the types of microcapsules used for the microencapsulation of vitamins.

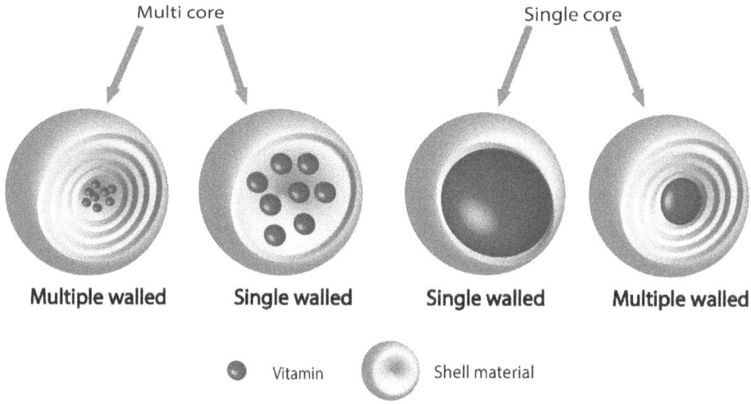

FIGURE 4.1 Microcapsule forms applied in vitamin encapsulation. *Source: Reprinted with permission from Katouzian, I., & Jafari, S.M. (2016). Nano-encapsulation as a promising approach for targeted delivery and controlled release of vitamins. Trends in Food Science & Technology, 53, 34–48.*

According to Gibbs (1999), the microcapsule size varies between 5 μm and 300 μm. The microcapsules may have a continuous payload phase encircled by a continuous shell or may have an irregular morphology with several tiny core particles dispersed in a network of shell material. Normally, the hydrophobic core is surrounded by a hydrophilic wall material and a hydrophilic core is surrounded by a hydrophobic wall material (De Prisco & Mauriello, 2016). The wall materials are responsible for protecting the vitamins against light, oxygen, and other factors that lead to the degradation of vitamins (Katouzian & Jafari, 2016; Wang, Vongsvivut, Adhikari, & Barrow, 2015).

The common microencapsulation techniques applied to the vitamins are explained briefly below. Furthermore, Table 4.2 summarizes the studies being conducted for the microencapsulation of vitamins.

4.3.1 Spray Drying

It is a common encapsulation method that yields powders in different particle sizes (Mehrad, Shabanpour, Jafari, & Pourashouri, 2015; Rajabi, Ghorbani, Jafari, Mahoonak, & Rajabzadeh, 2015). Spray drying is usually employed for entrapping water-insoluble vitamins at an industrial level (Desai & H. J. Park, 2005; Gharsallaoui, Roudaut, Chambin, Voilley, & Saurel, 2007). In this method, structures like maltodextrin and gum Acacia are dissolved and then the oily material is added during blending. Subsequently, the mixture is homogenized to obtain an emulsion and ultimately fine powders are achieved via the spray drying procedure (Jafari, Assadpoor, He, & Bhandari, 2008; Jafari, He, & Bhandari, 2007). The vitamins which are encapsulated

TABLE 4.2 Microencapsulation Techniques for Different Vitamins

Microencapsulation Technique	Wall Material	Vitamin	Purpose	Reference
Spray drying	Granules of rice starch and gum arabic	Vitamin C	Examining the stability of ascorbic acid and determining the size distribution of microcapsules	Trindade & Grosso, 2000
Spray drying	Trypolyphosphate cross-linked chitosan microspheres	Vitamin C	Investigating the release rate and stability in the capsule	Desai & Park, 2005
Liposome	Egg phosphatidylcholine, cholesterol, DL-α-tocopherol	Vitamin C	Comparing the half-life of pure vitamin C and capsulated one	Kirby et al., 1991
Emulsion technique	Starch, glycerin of vegetable origin, carrageenan, disodium phosphate, medium chain triglycerides	Vitamin E	Bioavailability of vitamin E in fortified breakfast cereals	Leonard et al., 2004
Spray drying	Chitosan/ethyl cellulose	Vitamin D_2	Morphology and release properties of the microcapsules	Shi & Tan, 2002
Emulsion technique	$mPEG_{5000}$-b-p(HPMAm-lac_2), a thermosensitive block copolymer	Vitamin K	Evaluating the influence of bile acids on the oral bioavailability	Van Hasselt et al., 2008
Emulsion technique	A-axial 5,6-dimethylbenzimidazole ligand with cucurbit-7-uril	Vitamin B_{12}	Stabilizing cob (III) almins such as CNCbl and AdoCbl with the suggested capsules	Wang et al., 2009
Fluidized bed coating	Gelatin	Vitamin C	Encapsulation efficiency and microdiet effect on larval shrimps	Xie et al., 2010
Coacervation	Gelatin and gum arabic	Vitamin A	Effect of process variables on the encapsulation process	Junyaprasert et al., 2001
Spray drying	Starch and β-cyclodextrin	Vitamin C	Analyzing the encapsulation efficiency and the degradation of ascorbic acid	Uddin et al., 2001
Spray cooling	Fully hydrogenated palm fat and 1% lecithin	Vitamin A	Food fortification to combat health problems in developing countries	Wegmüller et al., 2006

Source: Reprinted with permission from Katouzian, I., & Jafari, S.M. (2016). Nano-encapsulation as a promising approach for targeted delivery and controlled release of vitamins. Trends in Food Science & Technology, 53, 34–48.

by this method are typically used in tablets. In this type, the tablet features are influenced by the type of the applied matrix (Schrooyen, De Ruiter, & De Kruif, 2000).

In a study, Shi and Tan (2002) used a complex of chitosan–ethyl cellulose to encapsulate vitamin D_2 and analyzed the morphology and release profile of the formulated microcapsules. It was suggested that the microcapsules were stable in the intestine juice by conducting in vitro assays. Besides, Desobry, Netto, and Labuza (1999) showed that the combination of 25DE maltodextrin with galactose, lactose, or glucose prolong the shelf life of encased *trans*-β-carotene, while maltodextrin alone cannot provide this condition.

Spray drying technique has also been implemented for the water-soluble vitamins. As an example, the extracted vitamin C from camu-camu fruit (found in Amazon which is rich in vitamin C) was encapsulated by spray drying (Dib Taxi, De Menezes, Santos, & Grosso, 2003). The retention rate for vitamin C was about 15% and the optimized temperature was 150°C. Vitamin C is greatly used in the food and nutraceutical industries due to its high antioxidant ability, which is very sensitive to oxidation and other ambient factors (Kirby, Whittle, Rigby, Coxon, & Law, 1991). In another investigation, Desai & H. Park (2005) exerted the tripolyphosphate cross-linked chitosan microspheres as the shell material for encasing vitamin C. Consequently, the cross-linking factor agent affected the particle size in the range of 6.1–9 μm.

4.3.2 Spray Chilling and Cooling

The methods comprise the diffusion of vitamins in a molten fat or wax medium. Accordingly, the mixture is atomized via the high-temperature nozzles and discharged into a vessel at ambient temperature. The melting point of the microcapsules lies in the range of 45–122°C. Typically, at refrigerated temperatures, materials tend to melt at 32–42°C (Risch & Reineccius, 1995). The fabricated microcapsules are hydrophobic and will be molten at high temperatures. In this way, the fat crystallization step in these methods should be carefully monitored. Wegmüller, Zimmermann, Bühr, Windhab, and Hurrell (2006) formulated microcapsules consisting iodine, vitamin A, and iron plus hydrogenated palm fat through spray cooling. During 6 months of storage, only 12% loss of vitamins was observed in these formulated microcapsules.

4.3.3 Emulsion Techniques

In this method, vitamins are dispersed in a nonmiscible liquid phase, serving as the core material. Subsequently, modifications are carried out to initiate the formation of shells around the dispersed vitamins in the solution. The most common type of emulsions in microencapsulation is the O/W form (Kondo, 1979). Wang, MacGillivray, and Macartney (2009) encapsulated some types of cob (III) alamins along with 5,6-dimethylbenzimidazole and cucurbituril. The complex of cucurbituril and vitamin B_{12} exhibited the photochemical and electrochemical features of cob (III) alamins. Also, Leonard, Good, Gugger, and Traber (2004) incorporated vitamin E microcapsules in breakfast cereals and compared its bioavailability to the supplements including vitamin E. As a result, the bioavailability of microcapsules was enhanced significantly when it was used in fortified foods. Van Hasselt, et al. (2009) employed polymeric micelles to entrap vitamin K and evaluated its absorption in the bile duct medium of sham rats. In effect, the free bile influenced the absorption of formulated microcapsules and the uptake by pinocytosis route was detected to be inconsiderable.

4.3.4 Fluidized Bed Coating

Suspension coating is another term used for this technique. The solid materials are suspended in either hot or cold upward flow of air. Following this process, solid materials are sprayed on the top of the atomized particles that might be molten or dispersed in an evaporable solvent (Risch & Reineccius, 1995). The wall material may have versatile compositions like starch and its derivatives, dextrins, emulsifiers, lipids, and proteins (Shahidi & Han, 1993). The application of this technique is mostly in formulating nutritional supplements with a complex of vitamins and minerals. Besides, it has been implemented in some food products, such as puddings, fillings, and sauces (Risch & Reineccius, 1995). In a study, Xie, et al. (2010) prepared microcapsules incorporating vitamin C and gelatin as the wall material. They fed shrimps (*Penaeus japonicas*) with this diet and observed that wet weight of shrimps increased to 300% during 30 days following hatching. Moreover, the encapsulation efficiency was about 88.2%.

4.3.5 Liposome Carriers

These types of carriers have different forms; they can be single- or multilayered and water-soluble and insoluble vitamins can be incorporated in these structures. Kirby, et al. (1991) used this method to encapsulate vitamin C. Phospholipids, lecithin, and cholesterol form stable liposomes. A typical way of fabricating liposomes is employing the hydration–rehydration process that yields stable liposomes.

4.3.6 Coacervation

In this case, both the polymeric solution and core material are in a container and polymeric material is decomposed gradually, depositing around the core material to form a uniform layer Gibbs, et al. (1999). Indeed, it involves the desolvation of the gelling protein plus the emulsification process of the nutrient into the protein network. The coating process occurs as a result of controlled physical blending. The solidification process takes place due to the thermal, desolvation, and cross-linking procedures. At last, the formulated microcapsules are acquired by centrifugation and filtration processes. The coacervation technique can either be simple or complex. The simple type includes only one colloidal solution like gelatin, while the other type needs a secondary oppositely charged water-soluble compound, such as gelatin and gum arabic or gelatin together with polysaccharides Gibbs, et al. (1999). Junyaprasert, Mitrevej, Sinchaipanid, Boonme, and Wurster (2001) microencapsulated vitamin A palmitate by implementing the complex coacervation method and used gelatin and gum Acacia as the coating materials.

4.4 NANOENCAPSULATION OF VITAMINS

Microcapsules have a diameter of 3–800 μm, while nanoparticles are between 1 nm and 100 nm and above the size of 100 nm they are called submicron particles (Lamouroux, 2016). Nanoencapsulation is the entrapment of nutrients at the nanoscale (Lopez-Rubio, Gavara, & Lagaron, 2006). The delivery of any type of nutrient scales directly with the particles size. Hence, nanoencapsulation provides a much higher absorbance and delivery of nutrients compared with the microscale delivery (Katouzian & Jafari, 2016). Food-grade nanovehicles are commonly used for the encapsulation of vitamins. Cyclodextrins, albumin, alginate, gelatin, collagen, chitosan, whey proteins, pectin, etc. are natural biopolymeric compounds that are employed in the nanodelivery systems (Grumezescu, 2016). A tremendous development has been observed in engineering novel nanodelivery systems to obtain the best results in food, pharmaceutical, and nutraceutical sciences (Benshitrit, Levi, Tal, Shimoni, & Lesmes, 2012; He & Hwang, 2016; Peters, et al., 2016).

Here, the common nanocarriers for the preservation and delivery of vitamins are discussed in detail. Before moving to the description part of nanovehicles, their structure has been presented in Fig. 4.2 in order to have a background of these technologies. Also, the nanoencapsulation systems and techniques applied to hydrophilic and lipophilic vitamins are presented in Tables 4.3 and 4.4, respectively.

4.4.1 Nanoemulsions

These types of systems are synthesized by either low-energy or high-energy techniques like phase inversion temperature and microfluidization, respectively. The size of the nanodroplets in these structures normally varies between 50 nm to 1000 nm (Assadpour, Maghsoudlou, Jafari,

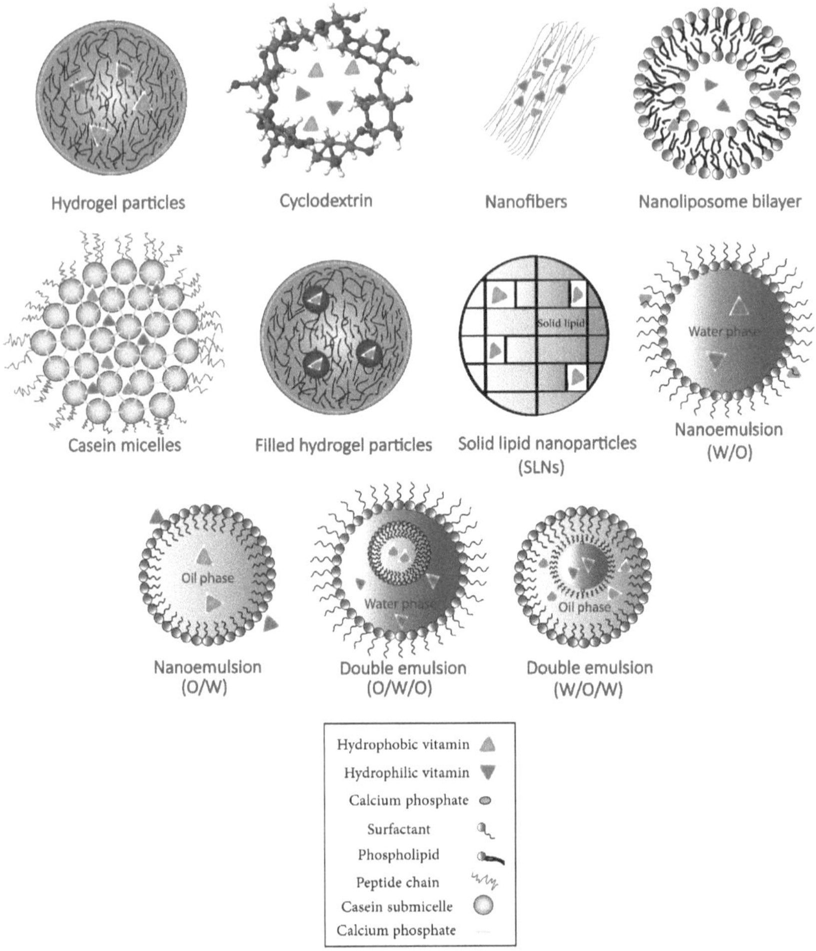

FIGURE 4.2 Types of nanostructured delivery systems applied in vitamin encapsulation. *Source: Adapted from Katouzian, I., & Jafari, S.M. (2016). Nano-encapsulation as a promising approach for targeted delivery and controlled release of vitamins. Trends in Food Science & Technology, 53, 34–48.*

Ghorbani, & Aalami, 2016b; Silva, Cerqueira, & Vicente, 2012). Nanoemulsions containing vitamins can be used in their original state or in their powder form obtained via spray drying process, which is more stable (Esfanjani, et al., 2016; Assadpour & Jafari, 2017).

As an example, Cho, Seo, Yim, and Lee (2013) noted that the nanoencapsulation of a type of vitamin B called thiamine dilauryl sulfate (TDS), by using lecithin as the external phase, inhibited the spore germination of *Fusarium oxypansum* f. sp. Moreover, this formulation was effective in stalling the mycelial growth.

TABLE 4.3 Nanoencapsulation Structures for Water-Soluble Vitamins

Nanoencapsulation Technique	Nanovehicle Type	Wall Material Ingredients	Water-soluble Vitamin Type	Goal of Entrapping	Reference
Nanoemulsification (spontaneous)	Nanoemulsion	Maltodextrin–whey protein	Folic acid (vitamin B_9)	Exerting a low-energy method to encapsulate folic acid	Assadpour et al., 2016a
Nanoemulsification	Nano double emulsion	Pectin–whey protein	Folic acid (vitamin B_9)	Creating a double protection method to encapsulate folic acid	Assadpour et al., 2016b
Nano emulsification	Nano double emulsion–spray drying	Pectin–whey protein	Folic acid (vitamin B_9)	Producing powder particles containing encapsulated folic acid	Assadpour et al., 2017; Assadpour and Jafari, 2017
Ultrasonic irradiation	Nanoliposome	L-α-Phosphatidylcholine, Cholesterol and egg yolk lecithin	Vitamin B_{12}	Fabricating multi/unilamellar food-grade nanoliposomes to encase three different vitamins	Bochicchio et al., 2016
Self-assembly	Nanoliposome	Positive chitosan and negative sodium alginate	Vitamin C	Evaluating the delivery of chitosan and sodium alginate	Liu et al., 2016
High speed homogenization	Nanoliposome	Phosphatidylcholine	Vitamin B_1	Prepare and characterize nanoliposomes containing thiamine	Fathima et al., 2016
Double-emulsion-dynamic high-pressure microfluidization and double emulsion	Nanoliposome	Soybean phospholipids and cholesterol	Vitamin C	Investigating the release profile of vitamin C incorporated in nanoliposomes	Li et al., 2015

(Continued)

TABLE 4.3 (Continued)

Nanoencapsulation Technique	Nanovehicle Type	Wall Material Ingredients	Water-soluble Vitamin Type	Goal of Entrapping	Reference
Ionotropic gelation	Nanohydrogel	Alginate/chitosan nanoparticles	Vitamin B_2	Evaluating encapsulation and controlled release of vitamin B_2 considering the wall materials	Azevedo et al., 2014
Coacervation and aggregation	Biopolymer nanoparticle	β-lactoglobulin (BLG) and lactoferrin (LF) coassemblies	Vitamin B_9	Fabricating a novel biocarrier for vitamin B9	Chapeau et al., 2016
Ultracentrifugation and freeze drying	Biopolymer nanoparticle	Double layered chitosan–zein complex	Vitamin C	Protecting polar drug from degradation and delayed the release in the stomach	Dong & Wang, 2016
Coacervation	Biopolymer nanoparticle	Casein nanoparticles	Folic acid (Vitamin B_9)	Evaluating the oral bioavailability through in vitro and in vivo studies	Penalva et al., 2015
Ionotropic gelation	Biopolymer nanoparticle	Chitosan and cyclodextrin	Vitamin C	Delivering ascorbic acid (AA) to marine and freshwater organisms	Jiménez-Fernández et al., 2014
Ionotropic gelation	Biopolymer nanoparticle	Chitosan	Vitamin C	Extending shelf life and delivery of vitamin C	Alishahi et al., 2011
Self-assembly	Nanohydrogel	Bovine serum albumin (BSA), and *Citrus* peel pectin	Vitamin C	Formulating a novel nanohydrogel system for food and non-food delivery applications	Peng et al., 2016

Bio-crosslinking	Nanohydrogel	Anionic bovine serum albumin (BSA) and pullulan	Vitamin B_6	Fabricating nanogels by employing organic crosslinkers	Tsuchido et al., 2015
High-pressure extrusion	Nano-organogel	Isocarb20 and L-ascorbic acid	Vitamin C	Creating a biocompatible delivery system to solubilize and shield bioactives	Lo Nostro et al., 2007
Electrospinning	Nanofiber mats	Cellulose acetate	Vitamins B_2 and C	Investigating the antibacterial property and slow release of vitamins	Agarwal et al., 2016
Electrospraying and nanospray drying	Nanocapsule	Whey protein concentrate (WPC) and a commercial resistant starch	Folic acid (vitamin B_9)	Increasing the thermal resistance of folic acid	Pérez-Masiá, et al., 2015
Electrospinning	Nanofiber	Polycaprolactone nanofiber	Vitamin B_{12}	Investigating water-soluble vitamin delivery with hydrophobic polymer nanofibers for transdermal applications	Madhaiyan et al., 2013
Electrospinning	Nanofiber	Electrospun polyacrylonitrile nanofibers	Vitamin C	Fabricating core-shell nanofibers encapsulating vitamins for photoprotection	Wu et al., 2011
Inclusion complexation	Cyclodextrins	ß-Cyclodextrin–co-guanidine	Vitamin B_2	Improving the physicochemical properties of vitamin B_2	Heydari et al., 2016

(Continued)

TABLE 4.3 (Continued)

Nanoencapsulation Technique	Nanovehicle Type	Wall Material Ingredients	Water-soluble Vitamin Type	Goal of Entrapping	Reference
Supercritical coinjection process and production of particles from gas-saturated solutions (PGSS)	SLN	Fully hydrogenated canola oil	Vitamin B_2	Introducing a novel and efficient method for the preparation of encapsulation systems	Couto et al., 2017
Hot homogenization	SLN	Compritol and Tween 80	Vitamin B_{12}	Analyzing the cancer treatment potential and possible therapeutic applications of synthetized SLNs	Genç et al., 2015
High pressure homogenization	Nanoemulsion	$W_1/O/W_2$ double emulsions with four different lipid sources	Vitamin B_2	Using this process as functional healthier fat-food ingredients	Bou et al., 2014
Solvent evaporation method	Nanoemulsion	Lecithin	Thiamine dilauryl sulfate (TDS), a vitamin B derivative	Inhibiting spore germination of *Fusarium oxysporum* f. sp. *raphani* using TDS in nanocapsules	Cho et al., 2013

TABLE 4.4 Nanoencapsulation Structures for Water-Insoluble Vitamins

Nanoencapsulation Technique	Nanovehicle Type	Wall Material	Water-insoluble Vitamin type	Goal of Entrapping	Reference
Thin-film evaporation	Nanoliposome coated by chitosan (chitosome)	Egg yolk phospholipid with Tween 80 and chitosan	β-carotene	Developing a novel structure for an efficient delivery of β-carotene	Tan et al., 2016
Thin-film hydration–sonication	Nanoliposome	Lecithin–cholesterol	Vitamin A palmitate	Protecting vitamin A from light, oxygen and chemical degradation	Pezeshky et al., 2016
Ultrasonic irradiation	Nanoliposome	L-α-Phosphatidylcholine, cholesterol, and egg yolk lecithin	Vitamin E and D_2	Fabricating multi/unilamellar food-grade nanoliposomes to encase three different vitamins	Bochicchio et al., 2016
Electrospraying	Electrospun layer	Chitosan	Vitamin E	Generating functional coating layers for packaging and medical applications	Stoleru et al., 2016
Electrospinning	Nanofiber	Gelatin	Vitamin A palmitate and E TPGS	Fabricating wound-healing dressings	Li et al., 2016

(*Continued*)

TABLE 4.4 (Continued)

Nanoencapsulation Technique	Nanovehicle Type	Wall Material	Water-insoluble Vitamin type	Goal of Entrapping	Reference
Electrospinning	Nanofibrous mat	Silk fibroin (SF) nanofibrous mats	Vitamin E	Fabrication and viewing the skin benefit of vitamin E loaded with these nanofibers	Sheng et al., 2013
Electrospinning	Nanofiber	Electrospun polyacrylonitrile nanofibers	Vitamin E	Fabricating core-shell nanofibers encapsulating vitamins for photoprotection	Wu et al., 2011
Saturated aqueous vacuum drying	Cyclodextrin	β-cyclodextrin	Vitamin D_3	Increase the stability and bioavailability of vitamin D_3	Liu and Zhang, 2016
Freeze drying	Cyclodextrin	β-cyclodextrin	Vitamin A palmitate	Increase the water solubility of vitamin A palmitate and its stability against external factors	Vilanova et al., 2015
Inclusion complexation	Cyclodextrin	Dextran nanoparticles	Vitamin D	Encapsulating vitamin D to increase its regulation of body weight effect	Soares et al., 2012
Hot homogenization	SLN	Tripalmitin	Vitamin D_2 (ergocalciferol)	Increasing the stability of vitamin D_2 to enrich milk and margarine	Patel and Martin-Gonzalez, 2012
Melt-emulsification	NLC	Rambutan seed fat and Tween 20	Vitamin E	Extract the seed fat of Rambutan for developing NLCs	Uraiwan and Satirapipathkul, 2016

Hot hemogenation	NLC	Liquid oil (octyloctanoate) and solid lipid (precirol)	Vitamin A palmitate	Investigate surfactant concentration effect on physicochemical properties of prepared formulation	Pezeshki et al., 2014
pH treatment and ultrasonication	Biopolymer nanoparticle	Commercial soy protein isolate (SPI) and canola oil	Vitamin D_3	Analyzing the functional properties of soluble nanosized soy protein aggregates	Lee et al., 2016
Nanoprecipitation	Biopolymer nanoparticle	Potato proteins	Vitamin D_3	Utilizing potato proteins as natural nanovehicles for the encapsulation of vitamin D_3	David and Livney, 2016
Supercritical fluid extraction of emulsions (SFEE)	Biopolymer nanoparticle	Polycaprolactone and Tween 80	Vitamin E	Evaluate the stability of the formulated nanocapsules over storage periods	Prieto and Calvo, 2016
Sonication and freeze drying	Biopolymer nanoparticle	High amylose starch	Vitamin D_3	Fortification of milk using starch nanoparticles	Hasanvand et al., 2015
Girox method	Biopolymer nanoparticle	Whey protein isolate (WPI) nanoparticles	Vitamin D_3	Studying the stability of vitamin D_3	Abbasi et al., 2014
Wash-out method	Nanoemulsion	Edible mustard oil with Tween-80	Vitamin E	Employing a simple and low-energy method to formulate nanoemulsions with vitamin E	Dasgupta et al., 2016
Sonication and pH shifting	Nanoemulsion	Pea protein	Vitamin D	Test the efficacy of the prepared nanoemulsion for micronutrient delivery	Almajwal et al., 2016

(*Continued*)

TABLE 4.4 (Continued)

Nanoencapsulation Technique	Nanovehicle Type	Wall Material	Water-insoluble Vitamin type	Goal of Entrapping	Reference
Spontaneous emulsification	Nanoemulsion	Alpha-tocopherol (aTOC), ethanol, and Tween 80	Vitamin K_1	Prepare a cost-effective topical formulation by low-energy method	Campani et al., 2016
Emulsion phase inversion	Nanoemulsion	Octenyl succinic anhydride (OSA) modified starches	Vitamin E	Investigating physicochemical stability and thermal degradation of vitamin E	Hategekimana et al., 2015
Spontaneous emulsification	Nanoemulsion	Medium chain triglyceride oil (MCT)	Vitamin D	Investigating particle size and stability of vitamin D	Guttoff et al., 2015
High pressure homogenization	Nanoemulsion	Carrier oil (MCT, corn oil, fish oil, mineral oil or orange oil)	Vitamin D_3	Emulsifying and stabilizing capacities of natural surfactants	Ozturk et al., 2014
Wash out method	Nanoemulsion	Canola oil and Span80	Vitamin E acetate	Developing a practical HPLC method to estimate vitamin E	Morais et al., 2014
Spontaneous emulsification	Nanoemulsion	Medium chain triglyceride oil (MCT)	Vitamin E	Studying the influence of cosolvents on formation and stability of vitamin E	Saberi et al., 2013

Also in the field of natural surfactants, Ozturk, Argin, Ozilgen, and McClements (2014) entrapped vitamin D_3 in O/W emulsions by employing quillaja saponin as a natural surfactant. Their results suggested that the bioavailability of vitamin D_3 declined in the following order: corn oil $\approx$ fish oil > orange oil > mineral oil > medium-chain triglycerides (MCT). In addition, long-chain triglycerides were considered as the most effective compounds that boost the vitamin bioaccessibility.

Double emulsions are another type of delivery systems for food bioactive ingredients such as vitamins (Assadpour, et al., 2016a, 2016b; Esfanjani, Jafari, Assadpoor, & Mohammadi, 2015; Mohammadi, Jafari, Assadpour, & Esfanjani, 2016; Mohammadi, Jafari, Esfanjani, & Akhavan, 2016). Bou, Cofrades, and Jiménez-Colmenero (2014) fabricated $W_1/O/W_2$ double emulsions (chia oil, sunflower oil, olive oil, or rendered pork backfat) and incorporated riboflavin in this formulation. Retention of riboflavin was significantly high in chia oil in the beginning of the encapsulation process; nevertheless, the double emulsions in rendered pork backfat exhibited a better protection after 8 days of storage. All in all, double emulsions are resistant structures, which can withstand the stresses of food processing stages and ingestion.

Hategekimana, Chamba, Shoemaker, Majeed, and Zhong (2015) and Hategekimana, Masamba, Ma, and Zhong (2015) produced vitamin E-loaded nanocapsules composed of *octenyl succinic anhydride* starches as the external phase. The emulsification characteristics were advanced by the high level of replacement, low molecular weight and low interfacial tension. Besides, the Weibull model was the appropriate model for the release profile of vitamin E. Altogether, low molecular weights displayed the best results and generated stable nanocapsules that are of great interest in the beverage sector.

Saberi, Fang, and McClements (2013) produced vitamin E-bearing nanoemulsions via spontaneous emulsification method. This procedure involves the formation of tiny oil droplets upon the titration of an oil/surfactant admixture in an aqueous solution. The highest transparency and smallest droplet diameter ($d < 50$ nm) were obtained by utilizing 30% propylene glycol (PG) or 20% ethanol, respectively. However, at high temperatures, the Ostwald ripening process modified the physicochemical properties of the prepared nanoemulsions. Concentrated nanoemulsions rendered an irreversible increase in turbidity when exposed to heat (53°C) for the formulation of 30% PG at 38°C for the one containing 20% ethanol. In regard to the diluted agents, high thermal stability was detected at 75.5°C for both systems. The release profile of lipid-soluble active vitamin E acetate from oil/water nanoemulsions was suggested by Morais and Burgess (2014), prepared by a low-energy emulsification technique. Nanoemulsions comprised canola oil, cremophorRH40, and Span80. Dialysis sac and reverse dialysis sac approaches were applied to measure the vitamin E release. In conclusion, micellar solubilization increased vitamin E transport into the buffer solution and the external aqueous phase of the nanoemulsion was not modified in the presence of micelles. Guttoff, Saberi, and McClements (2015) synthesized vitamin D-enriched nanoemulsions through spontaneous emulsification. They analyzed the impact of vitamin D and MCT for surfactant to oil ratio, surfactant type (Tween 20, 40, 60, 80, and 85) plus the stirring properties on the particle size of the yield. Results depicted that tiny droplets with diameter < 200 nm were generated by employing Tween at high rotational speed (800 rpm). The prepared nanoemulsions were susceptible to high temperatures ($T > 80°C$). Nonetheless, the thermal resistance depends on the exerted cosurfactant and can be enhanced by choosing the appropriate agent (sodium dodecyl sulphate).

Furthermore, Almajwal, Abulmeaty, and Andrade (2016) prepared vitamin D-nanoemulsions (VDN) by employing sonication and pH modifications of pea protein. Healthy male albino rats were divided into different groups and fed with VDN and control emulsions (some lacking in vitamin D). After the treatments, the femur bones were isolated for the histopathological assays. After all, VDN treatment exhibited dramatic changes in calcium, parathyroid hormone (PTH), alkaline phosphatase, and phosphorus levels. Thus, nanoemulsions seem to be an effective delivery system for the fortification of foodstuffs with vitamin D.

Campani, et al. (2016) incorporated vitamin K_1 (VK1) in nanoemulsions for topical purposes in order to overcome the problems existing in the semisolid formulations. They implemented a low-energy and simple approach to formulate nanoemulsions. Subsequently, the possibility of nebulization without dramatic chemical alterations was tested on the nanoemulsions. In essence, after the permeation tests made on skin, nanoemulsions presented a favorable quality to be employed in commercial spray formulations for the topical absorption of VK1.

4.4.2 Nanohydrogels and Nano-Organogels

Nanohydrogels are composed of a linkage of polymeric chains, present in the form of colloidal gels with high capacity of water absorbance. Some of the advantages of employing these structures as nanocarriers include broad range of applications and exhibiting a high similarity to the natural tissues according to their high water capacity. Both naturally occurring and fabricated materials are included in the category of hydrogels (Sharma et al., 2017; Fuciños, et al., 2014). Moreover, the nano-organogels are defined as semisolid structures involving an organic lipid structure, which is fixed by a three-dimensional matrix comprising self-assembled intertwined fibers with gelation properties (Da Pieve, Calligaris, Nicoli, & Marangoni, 2010; Wambura, 2009).

Peng, et al. (2016) formulated a novel nanohydrogel consisting of the naturally occurring polymers of citrus peel pectin and bovine serum albumin (BSA) via the self-assembly technique. They nanoencapsualted vitamin C

(VC) in this network. The encapsulation efficiency of VC was approximately 65.31% and the in vitro release mechanisms involved diffusion, swelling, and erosion. More importantly, the stability of the prepared nanohyrdogel systems was 73.95% after storage during 10 weeks.

Also, Tsuchido, Sasaki, Sawada, and Akiyoshi (2015) employed zinc ions for preparing strong protein nanogels made of BSA along with vitamin B_6-bearing pullulan (VBBP) acting as bio-crosslinkers. Hence, the attachment of BSA to VBBP was increased considerably by exerting the zinc ions in spite of the electrostatic repulsions between the two chemical structures.

Lo Nostro, et al. (2007) used the combination of 2-octyl-dodecanoyl-6-*O*-ascorbid acid and L-ascorbic acid to fabricate nano-organogels. This surfactant compound created viscous nano-organogels and formed water-induced organogels upon the presence of water in the medium. To sum up, this formulation can be applied to shield and stabilize the sensible and valuable bioactive compounds such as vitamins.

4.4.3 Solid Lipid Nanoparticles and Nanostructured Lipid Carriers (SLNs and NLCs)

The SLNs and NLCs are basically composed of lipids as the inner phase (solid or solid and liquid), surface active materials (emulsifiers), and water. The nanoparticles are formed by congealing. The vitamins are entrapped in a solid state lipid that has a high stability. However in NLCs, the matrix is a combination of solid and liquid lipids so that it can accommodate more bioactive compounds due to the increased space between the solid state lipids (Aditya & Ko, 2015; Ekambaram, Sathali, & Priyanka, 2012; Livney, 2015).

Couto, Alvarez, and Temelli (2017) incorporated vitamin B_2 in the SLN systems made of fully hydrogenated canola oil. Four variables (pressure, vitamin concentration, molecular weight, and stabilizer concentration) were studied according to their effects on encapsulation efficiency and loading efficiency. All in all, the optimal circumstances were pressure of 15 MPa, 2% vitamin concentration, 5% polyethylene glycol (PEG) with a molecular weight of 35,000 Da and the experiments revealed that high concentration of water-soluble compounds can be incorporated in SLN networks.

Genç, Kutlu, and Güney (2015) successfully incorporated vitamin B_{12} in the matrix of an SLN system and analyzed the cytotoxicity of the entrapped vitamin on two control cell lines (H-*R*as 5RP7 and NIH/3T3). Consequently, the results revealed the enhanced anticarcinogenic effect of prepared SLN system in comparison with the free vitamin form. Also, the characterization tests showed that the size of vitamin B_{12}-loaded SLN structures was about 200 nm that raised the reaction rates and elevated the release efficiency.

Uraiwan and Satirapipathkul (2016) used the fat from Rambutan seed (*Nephelium lappaceum*) to fabricate NLCs together with Tween 20 and vitamin E as emulsifier and core material, respectively. The aim was to detect the stability of nanoparticles in accordance with the employed surfactant level. As a result, 5% (w/w) of Tween 20 was considered as the optimal concentration and formed a stable NLC system with an average particle diameter of 139.43 ± 1.15 nm along with a polydispersity index of about 0.165 ± 0.017.

Pezeshki, Ghanbarzadeh, Mohammadi, Fathollahi, and Hamishehkar (2014) prepared vitamin A palmitate-loaded NLC structures via the hot homogenization procedure. Precoil was implemented as the solid lipid, octyloctanoate as the liquid phase, and poloxamer as the surfactant. To sum up, the lowest particle size was obtained in the presence of 6% poloxamer, also the encapsulation efficiency was about 98.5% and the shelf life of the formulation was 2 months at 25°C.

4.4.4 Biopolymer Nanoparticles

Currently, the application of biopolymer compounds and their combinations in the nanoencapsulation of vitamins has attracted the attention of food scientists. Some of these organic polymers include gelatin, proteins in milk, chitosan/starch nanocrystals, and combinations of these edible nanostructures, which are synthesized by several techniques.

For instance, Prieto and Calvo (2016) utilized the supercritical fluid extraction of emulsions to incorporate vitamin E into polycaprolactone nanocapsules. The formulation revealed high encapsulation efficiency (90%) and the particle sizes between 8 nm and 276 nm. By using CO_2 with the flow rate of 7.2 kg h^{-1} kg along with the conditions 8 MPa and 313 K, the concentration of the solvent decreased to 50 ppm. As a final point, the yielded nanocapsules showed high stability during 6–12 months of storage and remained unchanged. In another study, Abbasi, Emam-Djomeh, Mousavi, and Davoodi (2014) exerted whey protein isolates to encase vitamin D_3 and tested the stability of this system in 7 days. Regarding the ultimate results, nanoparticles had a higher level of retaining vitamin D compared with the control sample. Oxygen was not able to diffuse into the nanoparticle network due to its dense structure in the presence of calcium. These nanoparticles were suggested to be useful in the beverage industry.

Penalva, et al. (2015) coated folic acid by using casein nanoparticles as the wall material and the mixture was dried via spray drying method. The average sizes of the fabricated nanoparticles were around 150 nm; in addition, the folic acid value was about 25 μg/mg in the yield. Concerning the in vitro release profile, folic acid was resistant against the GI circumstances and was released in the intestinal zone. Considering the in vivo studies, animal models were gavaged using this vitamin. Overall, in the animal models fed with casein nanoparticles, the bioavailability value was approximately 50%–52% higher than the traditional solution. At the same time, the bioavailability and release rate of the nanovehicles was constant after being exposed to high hydrostatic pressure treatment. Also, Hasanvand, Fathi,

Bassiri, Javanmard, and Abbaszadeh (2015) nanoencapsualted vitamin D_3 by exerting starch nanoparticles. The engineered nanocarriers were incorporated in milk as an abundant source of calcium. On the whole, these novel nanovehicles improved the bioavailability of vitamin D_3 and masked the unwanted aftertaste in the fortified yield.

Dong and Wang (2016) nanoencapsulated vitamin C by using a double-layer chitosan–zein complex in which, firstly, vitamin C was incorporated in chitosan by sodium tripolyphosphate as the cross-linker. Subsequently, the formulation was coated by zein. In conclusion, this novel structure prolonged the release of vitamin C in gastric fluid conditions and enhanced the oxidation stability. Chapeau, et al. (2016) implemented β-lactoglobulin (BLG) and lactoferrin (LF) coassemblies to envelope vitamin B_9 (B9). The coassemblies were obtained by coacervation and aggregation and analyzed by compiling screening maps. Taken together, B9–LF–BLG coacervates depicted a considerable encapsulation efficiency for vitamin B_9 ($\approx$1 mg B9/g protein), presenting the great potential of food-grade formulations to be applied as nanocarriers in different sectors and for various applications.

Jiménez-Fernández, et al. (2014) fabricated chitosan-based nanoparticles to deliver vitamin C to the intended marine animals. Zebrafish liver cell-line was the model for in vitro tests. Besides, the in vivo studies were performed on fish and rotifers to assess the practical utilization of this nanoencapsulation system. The antioxidant property of the engineered system was much higher in comparison to the nonloaded nanoparticles. In postmetamorphic larvae of *S. senegalensis*, nanoparticles entered the epithelial tissue of the intestine after 2 h. Also, in rotifers with the diet of vitamin C nanoparticles, the content of ascorbic acid elevated nearly twofold in comparison to control groups. In another study, Alishahi, et al. (2011) prepared chitosan nanoparticles to prolong the storage and boost the targeted delivery of vitamin C. The release process was pH-dependent and burst release occurred at 0.1 M phosphate buffer solution (PBS, pH 7.4), while a relatively low rate of release was detected at 0.1 M HCl. In short, the shelf life of vitamin C was enhanced by utilizing this nanocarrier complex and the in vivo release profile in the intestine of the sample trout was akin to the in vitro test.

Lee, et al. (2016) selected commercial soy protein isolate and canola oil as natural nanovehicles to shield vitamin D_3 against undesirable circumstances (especially under UV rays) by using ultrasonication for 5 min. On average, the encapsulation of this delivery system was nearly 73.5%, which underlines the efficiency of this food-grade transporting system in the food and pharmaceutical industries.

4.4.5 Nanoliposomes

A single or several lipid or phospholipid bilayers constitute nanoliposomes and may contain biological molecules like carbohydrates or proteins in their network. The hydrophobic–hydrophilic interactions between water and phospholipid lead to the formation of nanoliposomes. Both the water-soluble and water-insoluble vitamins can be incorporated in these nanovehicles (Mozafari, et al., 2008). For example, Pezeshky, Ghanbarzadeh, Hamishehkar, Moghadam, and Babazadeh (2016) prepared vitamin A palmitate-loaded nanoliposomes composed of different ratios of lecithin–cholesterol blend using the thin-film-hydration sonication method. The results suggested that at high levels of cholesterol, the encapsulation efficiency of vitamin A palmitate was reduced, while by implementing lecithin–cholesterol mixture (50/10 mg), the encapsulation efficiency reached to 15.8% and the size of nanoparticles was nearly 76 nm.

Liu, et al. (2016) synthetized vitamin C-bearing multilayered nanoliposomes by depositing positive chitosan and negative sodium alginate on the surface of the anionic nanoliposomes, respectively, and applied it to mandarin juice. The coated structure of nanoliposomes modified the surface characteristics of the nanoliposomes and lead to a reduction in the rate of lipid peroxidation and vitamin C protection after 90 days of storage; moreover, no significant organoleptic changes were observed in the fortified samples. In another study, Li, et al. (2015) synthetized complex nanoliposomes loaded with vitamin C and medium-chained fatty acids via the double-emulsion-dynamic high-pressure microfluidization and double emulsion techniques. The complex nanocarriers prepared via the double-emulsion-dynamic high pressure microfluidization method depicted higher encapsulation efficiency (nearly 64%) and stability for 90 days of storage at 4°C compared with the other method. Altogether, the double-emulsion-dynamic high-pressure microfluidization was found to be an effective approach to encapsulate both water-soluble and insoluble bioactive compounds.

Fathima, Fathima, Abhishek, and Khanum (2016) nanoencapsulated thiamin by using nanoliposomes formulated through phosphatidylcholine as an edible wall material. Thiamin is a labile compound and by implementing the nanoliposomes, it was possible to store it for more than 3 months at unequal temperatures. Moreover, the encapsulation efficiency was 97% and the size of the nanocarriers was around 150 nm.

In a study, Tan, Feng, Zhang, Xia, and Xia (2016) synthetized nanoliposomes with a layer of phospholipid–chitosan complex (chitosomes), composed of carotenoids, lycopene, β-carotene, lutein, and canthaxanthin. This complex system was prepared via layer self-assembly deposition technique. In brief, the biopolymer network shields lutein and β-carotene to a greater extent compared with canthaxanthin and lycopene. The stability of the formulated biopolymer nanoparticles can be related to the arrangement of free lipids at the water-soluble heads and the nonpolar membrane. Besides, Bochicchio, Barba, Grassi, and Lamberti (2016) loaded a set of vitamins (vitamin E, vitamin B_{12}, and vitamin D_2) by implementing multilamellar large vesicles (MLVs) and small unilamellar vesicles (SUVs) of nanoliposomes. All in all, the highest encapsulation efficiency values were ascribed to MLVs (between 72% and 95%) and SUVs (between 56% and 76%).

4.4.6 Cyclodextrins

These cyclic structures are originated from starch and include 6, 7, and 8 glucopyranose groups typically known as α, β, and γ cyclodextrins. The internal cavities of these rings are about 0.6, 0.8, and 0.1 nm in diameter. Cyclodextrins are generated from the enzymatic conversion of various starches. Corn and potato starch are abundantly used for the production of cyclodextrins. Also, maize and wheat starches are great sources of amylose so that the produced cyclodextrins contain less amylopectin in their structure (Li, Chen, Gu, Chen, & Wu, 2014; Marques, 2010; Singh, Bharti, Madan, & Hiremath, 2010).

Heydari, Doostan, Khoshnood, and Sheibani (2016) incorporated vitamin B_2 into cationic β-cyclodextrin structures to enhance the stability of vitamin B_2 plus the controlled release of the entrapped vitamin. Results from the in vitro tests presented that by using this engineered system, initially a burst release occurs and after that, a slower release rate is observed. Liu and Zhang (2016) fabricated a host-guest structure by employing a β-cyclodextrin molecule to wrap vitamin D_3 via the saturated aqueous vacuum drying technique. Through running orthogonal experiments, the optimum proportion of β-cyclodextrin to vitamin D_3 was about 15:1 and the uptake of vitamin D_3 enhanced dramatically in relation to the original form of the vitamin.

In another study, Vilanova and Solans (2015) prepared β-cyclodextrin molecules loaded with vitamin A palmitate without exerting organic solvents. They produced this system to enrich the edible solutions and enhance the stability of vitamin A palmitate in these cage molecules. Altogether, the surface activity of this intricate system depicted its possible application in emulsions as efficient stabilizers.

4.4.7 Nanofibers and Electrosprayed Nanoparticles

Nanofibers are generated via the electrospinning technique. In this method, a high-voltage electrical field is applied to the droplets from the polymer solution and as a result the droplets are stretched due to the electrostatic repulsion that dominates the surface tension. Subsequently, the droplets will constitute a conical shape called "Taylor Cone" and as the field strength increases, a fluid jet is formed next to the spinneret tip that is stretched toward the collector. Rapid evaporation occurs under the electrical field and the yield is nonwoven nanoscale fibers (Bhushani & Anandharamakrishnan, 2014; Rieger & Schiffman, 2014). Electrospraying is another method in which nanoparticles are obtained by applying electrical forces. The difference in the content of the solutions justifies the difference between electrospraying and electrospinning.

Accordingly, for low-concentrated solutions, the jet is destabilized because of varicose and ultrafine particles are obtained. On the contrary, if the concentration is high, the elongated droplets form nanofibers by whipping instability procedure (Bhushani & Anandharamakrishnan, 2014).

Pérez-Masiá, et al. (2015) exerted the electrospraying and nanospray drying approaches to encapsulate folic acid by using whey protein concentrate (WPC) and resistant starch as the components of the wall material. The thermal resistance and encapsulation efficiency of enveloped folic acid were evaluated under different conditions during storage. Altogether, the WPC network enhanced the encapsulation efficiency due to its interactions with folic acid and both techniques generated micro, submicron, and nanoscale particles, while electrospraying produced smaller capsules along with controllable size distribution. In another study, Wu, Branford-White, Yu, Chatterton, and Zhu (2011) produced polyacrylonitrile nanofibers to encase vitamin C and E. They used the coaxial electrospinning method to encapsulate the intended vitamins and obtained desirable delivery profiles that are proper for targeted delivery networks.

Stoleru, et al. (2016) employed the electrospinning technique to fabricate a bioactive dual coating composed of chitosan–vitamin E compounds that exhibited both biocidal and antioxidant properties. Also, vitamin E modified the rheological properties of the synthesized layer and caused a fluid-like state as its level was increased. Agarwal, et al. (2016) formulated cellulose acetate (CA) nanofiber mats loaded with vitamins B_2 and C, and ZnO nanoparticles. The vitamin-loaded nanofibers revealed slow and controlled release compared to CA films that exhibited burst release, which can be implemented in oral delivery formulations. Li, et al. (2016) produced gelatin nanofibers loaded with vitamin A palmitate and E to be applied as wound-healing dressings. Whey protein isolate, zein, and soy protein isolate were employed in the structure of wall materials. By the addition of vitamins into the nanocarriers, the size of the fibers decreased; furthermore, the nanofibers loaded with vitamin A or E individually revealed a prolonged release for more than 60 h. When both vitamins were incorporated in the system, similar liberation properties were obtained. Finally, the in vivo assays suggested that these engineered dressings are proper candidates for wound-healing purposes in relation to the traditional methods.

Sheng, et al. (2013) fabricated silk fibroin nanofibers loaded with vitamin E. The entrapment of vitamin E in silk nanofibers protected skin fibroblast cells against oxidative stress caused by tert-butyl hydroperoxide. Overall, this vitamin delivery system has a great potential in skin care plus tissue regeneration and related subjects. Madhaiyan, Sridhar, Sundarrajan, Venugopal, and Ramakrishna (2013) synthesized vitamin B_{12}-bearing polycaprolactone nanofibers and tested it for transdermal delivery. The morphology of the nanofibers was detected by the SEM apparatus; moreover, the pore size measurements and mechanical properties of the nanofibers were determined by running the related tests. The fibers were exposed to plasma and gradually made hydrophilic in order to heighten the vitamin release. Finally, by considering release rate in PBS buffer performed in the in vitro medium, the vitamin-loaded system was considered as an applicable formulation to be used in transdermal patches.

4.5 CONCLUSIONS AND FUTURE TRENDS

Nanovehicles can be dissociated and absorbed faster due to their miniature scale, specifically concerning the nutrients that exhibit low bioavailability. Studies have suggested that the properties of materials are modified when transformed into nanoscale but the chemical structure of the treated compounds also play an important role in their functionality. However, the safety of using nanocarriers is still a challenging question and therefore the toxicity experiments and the delivery of nanoencapsulated vitamins need to be studied in detail. All the methods used for the nanoencapsulation of vitamins have their own pros and cons, for example, disadvantages of freeze drying and spray drying are related to the high operation costs and modifications to retain the nanoscale, respectively. Recently, most of the nanoencapsulation studies, including the retention of vitamins, are centered on employing green procedures like solvent-free approaches (Prieto & Calvo, 2016; Rao & Khanum, 2016; Zhu, Wang, Zhang, & Mei, 2016), low-energy preparation methods to decline the usage of surfactant and exclude the cosurfactant (Assadpour, et al., 2016a,b; Dasgupta, et al., 2016; Komaiko & McClements, 2016; Zhang, et al., 2016), exerting edible coatings in the nanoformulations (Liu, et al., 2016; Richardson, Barrett, & O'Conner, 2016; Varona, Martín, & Cocero, 2016), and using mathematical and numerical methods to optimize the formulation and fabrication processes such as particle swarm optimization, Levenberg–Marquardt algorithms in artificial neural networks, genetic algorithm optimization, etc. (Moradi Bidhendi, et al., 2016; Tengse, Priya, & Kumar, 2017; Wari & Zhu, 2016; Wu, 2016) for nanoencapsulation techniques plus predicting their release profiles by novel methods including Monte Carlo and cellular automata techniques (Eibinger, Zahel, Ganner, Plank, & Nidetzky, 2016; Eltayeb, Stride, & Edirisinghe, 2015; Guerrisi, Giordano, Marabotti, Raiconi, & Tagliaferri, 2016).

REFERENCES

Abbasi, A., Emam-Djomeh, Z., Mousavi, M. A. E., & Davoodi, D. (2014). Stability of vitamin D 3 encapsulated in nanoparticles of whey protein isolate. *Food Chemistry*, *143*, 379–383.

Aditya, N., & Ko, S. (2015). Solid lipid nanoparticles (SLNs): Delivery vehicles for food bioactives. *RSC Advances*, *5*, 30902–30911.

Adolphe, J. L., Whiting, S. J., & Dahl, W. J. (2009). Vitamin fortification of pureed foods for long-term care residents. *Canadian Journal of Dietetic Practice and Research*, *70*, 143–150.

Agarwal, A., Jeengar, A., Bhowmick, M., Samanta, K. K., Satyamurthy, P., D'Souza, C., & Vigneshwaran, N. (2016). Performance Characteristics of Electrospun Cellulose Acetate Nanofiber Mat Embedded with Nano-Zno/Vitamins. *International Journal of Nanotechnology and Application (IJNA) ISSN (P)*, 2277–4777.

Alishahi, A., Mirvaghefi, A., Tehrani, M., Farahmand, H., Shojaosadati, S., Dorkoosh, F., & Elsabee, M. Z. (2011). Shelf life and delivery enhancement of vitamin C using chitosan nanoparticles. *Food Chemistry*, *126*, 935–940.

Almajwal, A. M., Abulmeaty, M., & Andrade, J. (2016). Efficacy of a novel food fortification system to combat vitamin D deficiency in rats. *The FASEB Journal*, *30*, lb267–lb267.

Assadpour, E., & Jafari, S. M. (2017). Spray drying of folic acid within nano-emulsions; optimization by Taguchi approach. *Drying Technology*. Available from http://dx.doi.org/10.1080/07373937.2016.1242016

Assadpour, E., Jafari, S.-M., & Maghsoudlou, Y. (2017). Evaluation of folic acid release from spray dried powder particles of pectin-whey protein nano-capsules. *International Journal of Biological Macromolecules*, *95*, 238–247.

Assadpour, E., Maghsoudlou, Y., Jafari, S.-M., Ghorbani, M., & Aalami, M. (2016a). Optimization of folic acid nano-emulsification and encapsulation by maltodextrin-whey protein double emulsions. *International Journal of Biological Macromolecules*, *86*, 197–207.

Assadpour, E., Maghsoudlou, Y., Jafari, S. M., Ghorbani, M., & Aalami, M. (2016b). Evaluation of folic acid nano-encapsulation by double emulsions. *Food and Bioprocess Technology*, *9* (12), 2024–2032.

Azevedo, M. A., Bourbon, A. I., Vicente, A. A., & Cerqueira, M. A. (2014). Alginate/chitosan nanoparticles for encapsulation and controlled release of vitamin B2. *International Journal of Biological Macromolecules*, *71*, 141–146.

Ball, G.F. (2005). Vitamins in foods: Analysis, bioavailability, and stability. CRC Press, Boca Raton, FL.

Benshitrit, R. C., Levi, C. S., Tal, S. L., Shimoni, E., & Lesmes, U. (2012). Development of oral food-grade delivery systems: current knowledge and future challenges. *Food & function*, *3*, 10–21.

Bhushani, J. A., & Anandharamakrishnan, C. (2014). Electrospinning and electrospraying techniques: Potential food based applications. *Trends in Food Science & Technology*, *38*, 21–33.

Bikle, D. D. (2014). Vitamin D metabolism, mechanism of action, and clinical applications. *Chemistry & Biology*, *21*, 319–329.

Bochicchio, S., Barba, A. A., Grassi, G., & Lamberti, G. (2016). Vitamin delivery: Carriers based on nanoliposomes produced via ultrasonic irradiation. *LWT-Food Science and Technology*, *69*, 9–16.

Bolzetta, F., Veronese, N., De Rui, M., Berton, L., Toffanello, E. D., Carraro, S., ... Manzato, E. (2015). Are the recommended dietary allowances for vitamins appropriate for elderly people? *Journal of the Academy of Nutrition and Dietetics*, *115*, 1789–1797.

Bou, R., Cofrades, S., & Jiménez-Colmenero, F. (2014). Physicochemical properties and riboflavin encapsulation in double emulsions with different lipid sources. *LWT-Food Science and Technology*, *59*, 621–628.

Calvo, M. S., & Whiting, S. J. (2013). Survey of current vitamin D food fortification practices in the United States and Canada. *The Journal of Steroid Biochemistry and Molecular Biology*, *136*, 211–213.

Campani, V., Biondi, M., Mayol, L., Cilurzo, F., Pitaro, M., & De Rosa, G. (2016). Development of nanoemulsions for topical delivery of vitamin K1. *International Journal of Pharmaceutics*, *511*, 170–177.

Chalupa, W.F., & Schroeder, G.M. (2016). Method and apparatus for vitamin d enhancement in mushrooms. In: Google Patents.

Chapeau, A.-L., Tavares, G. M., Hamon, P., Croguennec, T., Poncelet, D., & Bouhallab, S. (2016). Spontaneous co-assembly of lactoferrin and β-lactoglobulin as a promising biocarrier for vitamin B9. *Food Hydrocolloids*, *57*, 280–290.

Cho, J. S., Seo, Y. C., Yim, T. B., & Lee, H. Y. (2013). Effect of nanoencapsulated vitamin B1 derivative on inhibition of both mycelial growth and spore germination of *Fusarium oxysporum* f. sp. *raphani*. *International Journal of Molecular Sciences*, *14*, 4283–4297.

Cooper, D. L., Murrell, D. E., Roane, D. S., & Harirforoosh, S. (2015). Effects of formulation design on niacin therapeutics: Mechanism of action, metabolism, and drug delivery. *International Journal of Pharmaceutics*, *490*, 55–64.

Couto, R., Alvarez, V., & Temelli, F. (2017). Encapsulation of Vitamin B2 in solid lipid nanoparticles using supercritical CO_2. *The Journal of Supercritical Fluids*, *120*, 432–442.

Da Pieve, S., Calligaris, S., Nicoli, M. C., & Marangoni, A. G. (2010). Shear nanostructuring of monoglyceride organogels. *Food Biophysics*, *5*, 211–217.

Dasgupta, N., Ranjan, S., Mundra, S., Ramalingam, C., & Kumar, A. (2016). Fabrication of food grade vitamin E nanoemulsion by low energy approach, characterization and its application. *International Journal of Food Properties*, *19*, 700–708.

David, S., & Livney, Y. D. (2016). Potato protein based nanovehicles for health promoting hydrophobic bioactives in clear beverages. *Food Hydrocolloids*, *57*, 229–235.

De Prisco, A., & Mauriello, G. (2016). Probiotication of foods: A focus on microencapsulation tool. *Trends in Food Science & Technology*, *48*, 27–39.

Desai, K., & Park, H. (2005). Encapsulation of vitamin C in tripolyphosphate cross-linked chitosan microspheres by spray drying. *Journal of Microencapsulation*, *22*, 179–192.

Desobry, S., Netto, F., & Labuza, T. (1999). Influence of maltodextrin systems at an equivalent 25de on encapsulated β-carotene loss during storage. *Journal of Food Processing and Preservation*, *23*, 39–55.

Dib Taxi, C., De Menezes, H., Santos, A., & Grosso, C. (2003). Study of the microencapsulation of camu-camu (*Myrciaria dubia*) juice. *Journal of Microencapsulation*, *20*, 443–448.

Dong, F., & Wang, Y. (2016). Encapsulation of vitamin C by a double-layer zein/chitosan structure with improved stability and controlled release. *International Journal of Nanomedicine and Nanosurgery*, 2.

Eibinger, M., Zahel, T., Ganner, T., Plank, H., & Nidetzky, B. (2016). Cellular automata modeling depicts degradation of cellulosic material by a cellulase system with single-molecule resolution. *Biotechnology for biofuels*, *9*, 1.

Eitenmiller, R.R., Landen Jr, W., & Ye, L. (2016). Vitamin analysis for the health and food sciences. CRC press, London.

Ekambaram, P., Sathali, A. A. H., & Priyanka, K. (2012). Solid lipid nanoparticles: A review. *Scientific Reviews and Chemical Communications*, *2*, 80–102.

Eltayeb, M., Stride, E., & Edirisinghe, M. (2015). Preparation, characterization and release kinetics of ethylcellulose nanoparticles encapsulating ethylvanillin as a model functional component. *Journal of Functional Foods*, *14*, 726–735.

Esfanjani, A. F., & Jafari, S. M. (2016). Biopolymer nano-particles and natural nano-carriers for nano-encapsulation of phenolic compounds. *Colloids and Surfaces B: Biointerfaces*, *146*, 532–543.

Esfanjani, A. F., Jafari, S. M., Assadpoor, E., & Mohammadi, A. (2015). Nano-encapsulation of saffron extract through double-layered multiple emulsions of pectin and whey protein concentrate. *Journal of Food Engineering*, *165*, 149–155.

Failla, M. (2011). Bioavailability and efficacy of vitamins and provitamins in foods. In *Abstracts of Papers of the American Chemical Society* (Vol. 241): Amer Chemical Soc, Washington, DC, USA.

Fathima, S. J., Fathima, I., Abhishek, V., & Khanum, F. (2016). Phosphatidylcholine, an edible carrier for nanoencapsulation of unstable thiamine. *Food Chemistry*, *197*, 562–570.

Fink, H.H., Fink, W.S.L.H.H., & Mikesky, A.E. (2013). Practical applications in sports nutrition. Jones & Bartlett Publishers, Sudbury, MA.

Fuciños, C., Fuciños, P., Míguez, M., Katime, I., Pastrana, L. M., & Rúa, M. L. (2014). Temperature-and pH-sensitive nanohydrogels of poly (N-Isopropylacrylamide) for food packaging applications: modelling the swelling-collapse behaviour. *PloS One*, *9*, e87190.

Genç, L., Kutlu, H. M., & Güney, G. (2015). Vitamin B12-loaded solid lipid nanoparticles as a drug carrier in cancer therapy. *Pharmaceutical Development and Technology*, *20*, 337–344.

Gharsallaoui, A., Roudaut, G., Chambin, O., Voilley, A., & Saurel, R. (2007). Applications of spray-drying in microencapsulation of food ingredients: An overview. *Food Research International*, *40*, 1107–1121.

Ghorbanzade, T., Jafari, S. M., Akhavan, S., & Hadavi, R. (2017). Nano-encapsulation of fish oil in nano-liposomes and its application in fortification of yogurt. *Food Chemistry*, *216*, 146–152.

Gibbs, B. F., Kermasha, S., Alli, I., & Mulligan, C. N. (1999). Encapsulation in the food industry: A review. *International Journal of Food Sciences and Nutrition*, *50*, 213–224.

Gleeson, J. P., Ryan, S. M., & Brayden, D. J. (2016). Oral delivery strategies for nutraceuticals: Delivery vehicles and absorption enhancers. *Trends in Food Science & Technology*, *53*, 90–101.

Grumezescu, A. (2016). Nutraceuticals: nanotechnology in the food industry, volume 4. *Nanotechnology*.

Guerrisi G., Giordano D., Marabotti A., Raiconi G., and Tagliaferri R., A cellular automata to simulate the growth and death of a cell culture, 2016, 2167–9843.

Gupta, A. (2015). Fortification of foods with vitamin D in India: Strategies targeted at children. *Journal of the American College of Nutrition*, *34*, 263–272.

Guttoff, M., Saberi, A. H., & McClements, D. J. (2015). Formation of vitamin D nanoemulsion-based delivery systems by spontaneous emulsification: factors affecting particle size and stability. *Food Chemistry*, *171*, 117–122.

Hasanvand, E., Fathi, M., Bassiri, A., Javanmard, M., & Abbaszadeh, R. (2015). Novel starch based nanocarrier for vitamin D fortification of milk: Production and characterization. *Food and Bioproducts Processing*, *96*, 264–277.

Hategekimana, J., Chamba, M. V., Shoemaker, C. F., Majeed, H., & Zhong, F. (2015). Vitamin E nanoemulsions by emulsion phase inversion: Effect of environmental stress and long-term storage on stability and degradation in different carrier oil types. *Colloids and Surfaces A: Physicochemical and Engineering Aspects*, *483*, 70–80.

Hategekimana, J., Masamba, K. G., Ma, J., & Zhong, F. (2015). Encapsulation of vitamin E: Effect of physicochemical properties of wall material on retention and stability. *Carbohydrate Polymers*, *124*, 172–179.

He, X., & Hwang, H.-M. (2016). Nanotechnology in food science: Functionality, applicability, and safety assessment. *Journal of Food and Drug Analysis*, *24*, 671–681.

Heydari, A., Doostan, F., Khoshnood, H., & Sheibani, H. (2016). Water-soluble cationic poly (β-cyclodextrin-co-guanidine) as a controlled vitamin B 2 delivery carrier. *RSC Advances*, *6*, 33267–33278.

Hickey, S. (2009). *Vitamin C: the real story: The remarkable and controversial healing factor: Easyread comfort edition*. readhowyouwant.com.

Hilliam, M. (1996). Functional foods: the Western consumer viewpoint. *Nutrition Reviews*, *54*, S189.

Jafari, S. M. (2017). An overview of nano-encapsulation techniques and their classification. In S. M. Jafari (Ed.), *Nanoencapsulation technologies for the food and nutraceutical Industries* (Chapter 1, pp. 1–34). Amsterdam: Elsevier.

Jafari, S. M., Assadpoor, E., He, Y., & Bhandari, B. (2008). Encapsulation efficiency of food flavours and oils during spray drying. *Drying Technology*, *26*, 816–835.

Jafari, S. M., He, Y., & Bhandari, B. (2007). Role of powder particle size on the encapsulation efficiency of oils during spray drying. *Drying Technology*, *25*, 1081–1089.

Jiménez-Fernández, E., Ruyra, A., Roher, N., Zuasti, E., Infante, C., & Fernández-Díaz, C. (2014). Nanoparticles as a novel delivery system for vitamin C administration in aquaculture. *Aquaculture*, *432*, 426–433.

Junyaprasert, V. B., Mitrevej, A., Sinchaipanid, N., Boonme, P., & Wurster, D. E. (2001). Effect of process variables on the microencapsulation of vitamin A palmitate by gelatin-acacia coacervation. *Drug Development and Industrial Pharmacy*, *27*, 561–566.

Katouzian, I., & Jafari, S. M. (2016). Nano-encapsulation as a promising approach for targeted delivery and controlled release of vitamins. *Trends in Food Science & Technology*, *53*, 34–48.

Khazaei, K. M., Jafari, S., Ghorbani, M., & Kakhki, A. H. (2014). Application of maltodextrin and gum arabic in microencapsulation of saffron petal's anthocyanins and evaluating their storage stability and color. *Carbohydrate Polymers*, *105*, 57–62.

Kirby, C., Whittle, C., Rigby, N., Coxon, D., & Law, B. (1991). Stabilization of ascorbic acid by microencapsulation in liposomes. *International Journal of Food Science & Technology*, *26*, 437–449.

Komaiko, J. S., & McClements, D. J. (2016). Formation of food-grade nanoemulsions using low-energy preparation methods: A review of available methods. *Comprehensive Reviews in Food Science and Food Safety*, *15*, 331–352.

Kondo A, Van Valkenburg JW. (1979). Microcapsule processing and technology. Marcel Dekker, New York.

Lamouroux, E. (2016). Nanoparticles. By Raz Jelinek. De Gruyter, 2015. Pp. 283. Price EUR 69.95. ISBN 9783110330021. *Acta Crystallographica Section B: Structural Science, Crystal Engineering and Materials*, *72*, 288–289.

Lee, H., Yildiz, G., dos Santos, L., Jiang, S., Andrade, J., Engeseth, N., & Feng, H. (2016). Soy protein nano-aggregates with improved functional properties prepared by sequential pH treatment and ultrasonication. *Food Hydrocolloids*, *55*, 200–209.

Leonard, S. W., Good, C. K., Gugger, E. T., & Traber, M. G. (2004). Vitamin E bioavailability from fortified breakfast cereal is greater than that from encapsulated supplements. *The American Journal of Clinical Nutrition*, *79*, 86–92.

Li, H., Wang, M., Williams, G. R., Wu, J., Sun, X., Lv, Y., & Zhu, L.-M. (2016). Electrospun gelatin nanofibers loaded with vitamins A and E as antibacterial wound dressing materials. *RSC Advances*, *6*, 50267–50277.

Li, T., Yang, S., Liu, W., Liu, C., Liu, W., Zheng, H., & Tong, G. (2015). Preparation and characterization of nanoscale complex liposomes containing medium-chain fatty acids and vitamin C. *International Journal of Food Properties*, *18*, 113–124.

Li, Z., Chen, S., Gu, Z., Chen, J., & Wu, J. (2014). Alpha-cyclodextrin: Enzymatic production and food applications. *Trends in Food Science & Technology*, *35*, 151–160.

Liu, W., Tian, M., Kong, Y., Lu, J., Li, N., & Han, J. (2016). Multilayered vitamin C nanoliposomes by self-assembly of alginate and chitosan: Long-term stability and feasibility application in mandarin juice. *LWT-Food Science and Technology*, *75*, 608–615.

Liu, Y., & Zhang, H. (2016). Study of VD3-β-clodextrin inclusion complex. *Journal of Geoscience and Environment Protection*, *4*, 163.

Livney, Y. D. (2015). Nanostructured delivery systems in food: Latest developments and potential future directions. *Current Opinion in Food Science*, *3*, 125–135.

Lo Nostro, P., Ramsch, R., Fratini, E., Lagi, M., Ridi, F., Carretti, E., & Baglioni, P. (2007). Organogels from a vitamin C-based surfactant. *The Journal of Physical Chemistry B*, *111*, 11714–11721.

Lopez-Rubio, A., Gavara, R., & Lagaron, J. M. (2006). Bioactive packaging: Turning foods into healthier foods through biomaterials. *Trends in Food Science & Technology*, *17*, 567–575.

Madhaiyan, K., Sridhar, R., Sundarrajan, S., Venugopal, J. R., & Ramakrishna, S. (2013). Vitamin B 12 loaded polycaprolactone nanofibers: a novel transdermal route for the water soluble energy supplement delivery. *International Journal of Pharmaceutics*, *444*, 70–76.

Mahdavi, S. A., Jafari, S. M., Ghorbani, M., & Assadpoor, E. (2014). Spray-drying microencapsulation of anthocyanins by natural biopolymers: A review. *Drying Technology*, *32*, 509–518.

Marques, H. M. C. (2010). A review on cyclodextrin encapsulation of essential oils and volatiles. *Flavour and Fragrance Journal*, *25*, 313–326.

Mehrad, B., Shabanpour, B., Jafari, S., & Pourashouri, P. (2015). Characterization of dried fish oil from Menhaden encapsulated by spray drying. *AACL Bioflux*, *8*, 57–69.

Mohammadi, A., Jafari, S. M., Assadpour, E., & Esfanjani, A. F. (2016). Nano-encapsulation of olive leaf phenolic compounds through WPC–pectin complexes and evaluating their release rate. *International Journal of Biological Macromolecules*, *82*, 816–822.

Mohammadi, A., Jafari, S. M., Esfanjani, A. F., & Akhavan, S. (2016). Application of nano-encapsulated olive leaf extract in controlling the oxidative stability of soybean oil. *Food chemistry*, *190*, 513–519.

Moradi Bidhendi, S., Zolfagharian, H., Mohamadpour Dounighi, N., Saraei, F., Khaki, P., & Inanlou, F. (2016). Design and evaluate alginate nanoparticles as a protein delivery system. *Archives of Razi Institute*, *68*, 139–146.

Morais, J. M., & Burgess, D. J. (2014). In vitro release testing methods for vitamin E nanoemulsions. *International Journal of Pharmaceutics*, *475*, 393–400.

Mozafari, M. R., Khosravi-Darani, K., Borazan, G. G., Cui, J., Pardakhty, A., & Yurdugul, S. (2008). Encapsulation of food ingredients using nanoliposome technology. *International Journal of Food Properties*, *11*, 833–844.

Murphy, S., Whited, L., Rosenberry, L., Hammond, B., Bandler, D., & Boor, K. (2001). Fluid milk vitamin fortification compliance in New York State. *Journal of Dairy Science*, *84*, 2813–2820.

Ottaway, P.B. (2012). The technology of vitamins in food. Springer Science & Business Media, Glascow.

Ozturk, B., Argin, S., Ozilgen, M., & McClements, D. J. (2014). Formation and stabilization of nanoemulsion-based vitamin E delivery systems using natural surfactants: Quillaja saponin and lecithin. *Journal of Food Engineering*, *142*, 57–63.

Patel, M. R., & San Martin-Gonzalez, M. F. (2012). Characterization of ergocalciferol loaded solid lipid nanoparticles. *Journal of Food Science*, *77*(1), N8–N13.

Pearson, A.M., & Gillett, T.A. (2012). Processed meats. Springer, Ireland.

Penalva, R., Esparza, I., Agüeros, M., Gonzalez-Navarro, C. J., Gonzalez-Ferrero, C., & Irache, J. M. (2015). Casein nanoparticles as carriers for the oral delivery of folic acid. *Food Hydrocolloids*, *44*, 399–406.

Peng, H., Chen, S., Luo, M., Ning, F., Zhu, X.-M., & Xiong, H. (2016). Preparation and self-assembly mechanism of bovine serum albumin-citrus peel pectin conjugated hydrogel: a potential delivery system for vitamin C. *Journal of Agricultural and Food Chemistry, 64*, 7377–7384.

Pérez-Masiá, R., López-Nicolás, R., Periago, M. J., Ros, G., Lagaron, J. M., & López-Rubio, A. (2015). Encapsulation of folic acid in food hydrocolloids through nanospray drying and electrospraying for nutraceutical applications. *Food Chemistry, 168*, 124–133.

Peters, R. J., Bouwmeester, H., Gottardo, S., Amenta, V., Arena, M., Brandhoff, P., & Pesudo, L. Q. (2016). Nanomaterials for products and application in agriculture, feed and food. *Trends in Food Science & Technology, 54*, 155–164.

Pezeshki, A., Ghanbarzadeh, B., Mohammadi, M., Fathollahi, I., & Hamishehkar, H. (2014). Encapsulation of vitamin A palmitate in nanostructured lipid carrier (NLC)-effect of surfactant concentration on the formulation properties. *Advanced Pharmaceutical Bulletin, 4*, 563.

Pezeshky, A., Ghanbarzadeh, B., Hamishehkar, H., Moghadam, M., & Babazadeh, A. (2016). Vitamin A palmitate-bearing nanoliposomes: Preparation and characterization. *Food Bioscience, 13*, 49–55.

Pourashouri, P., Shabanpour, B., Razavi, S. H., Jafari, S. M., Shabani, A., & Aubourg, S. P. (2014). Impact of wall materials on physicochemical properties of microencapsulated fish oil by spray drying. *Food and Bioprocess Technology, 7*, 2354–2365.

Prieto, C., & Calvo, L. (2016). Supercritical fluid extraction of emulsions to nanoencapsulate vitamin E in polycaprolactone. *The Journal of Supercritical Fluids, 119*, 274–282.

Rajabi, H., Ghorbani, M., Jafari, S. M., Mahoonak, A. S., & Rajabzadeh, G. (2015). Retention of saffron bioactive components by spray drying encapsulation using maltodextrin, gum Arabic and gelatin as wall materials. *Food Hydrocolloids, 51*, 327–337.

Rao, P. J., & Khanum, H. (2016). A green chemistry approach for nanoencapsulation of bioactive compound—Curcumin. *LWT-Food Science and Technology, 65*, 695–702.

Richards, M., & Tatiyaborworntham, N. (2015). Oxygen scavenging stabilizes color in meats in low oxygen. In: Google Patents.

Richardson, M.J., Barrett, A.H., & O'Conner, L. (2016). Applications of Films and Coatings for Special Missions.

Rieger, K. A., & Schiffman, J. D. (2014). Electrospinning an essential oil: Cinnamaldehyde enhances the antimicrobial efficacy of chitosan/poly (ethylene oxide) nanofibers. *Carbohydrate Polymers, 113*, 561–568.

Risch, S.J., & Reineccius, G. (1995). Encapsulation and controlled release of food ingredients. In *ACS symposium series (USA)*: American Chemical Society.

Ryley, J., Abdelkader, Z., Guevara, L., Lappo, B., Lamb, J., Manan, F., & Race, P. (1990). The prediction of vitamin retention in food-processing. *Journal of Micronutrient Analysis, 7*, 329–347.

Saberi, A. H., Fang, Y., & McClements, D. J. (2013). Fabrication of vitamin E-enriched nanoemulsions by spontaneous emulsification: Effect of propylene glycol and ethanol on formation, stability, and properties. *Food Research International, 54*, 812–820.

Schrooyen, P., De Ruiter, G., & De Kruif, C. (2000). Spray-dried orange oil emulsions: influence of cold water dispersible matrices on retention and shelf life. In *Proceedings of the International Symposium on the Controlled Release of BioactiveMaterials: Deerfield, Controlled Release Society* (pp. 1317–1318).

Sebrell, W.H., & Harris, R.S. (2014). The vitamins: chemistry, physiology, pathology, methods. Academic Press, New York.

Shahidi, F., & Han, X. Q. (1993). Encapsulation of food ingredients. *Critical Reviews in Food Science & Nutrition, 33*, 501–547.

Sharma, G., Naushad, M., Ala'a, H., Kumar, A., Khan, M. R., Kalia, S., & Sharma, A. (2017). Fabrication and characterization of chitosan-crosslinked-poly (alginic acid) nanohydrogel for adsorptive removal of Cr (VI) metal ion from aqueous medium. *International Journal of Biological Macromolecules*, *95*, 484–493.

Sheng, X., Fan, L., He, C., Zhang, K., Mo, X., & Wang, H. (2013). Vitamin E-loaded silk fibroin nanofibrous mats fabricated by green process for skin care application. *International Journal of Biological Macromolecules*, *56*, 49–56.

Shi, X.-Y., & Tan, T.-W. (2002). Preparation of chitosan/ethylcellulose complex microcapsule and its application in controlled release of Vitamin D2. *Biomaterials*, *23*, 4469–4473.

Silva, H. D., Cerqueira, M. Â., & Vicente, A. A. (2012). Nanoemulsions for food applications: development and characterization. *Food and Bioprocess Technology*, *5*, 854–867.

Singh, R., Bharti, N., Madan, J., & Hiremath, S. (2010). Characterization of cyclodextrin inclusion complexes—A review. *Journal of Pharmaceutical Science and Technology*, *2*, 171–183.

Soares, M. J., Murhadi, L. L., Kurpad, A. V., Chan She Ping-Delfos, W. L., & Piers, L. S. (2012). Mechanistic roles for calcium and vitamin D in the regulation of body weight. *Obesity Reviews*, *13*(7), 592–605.

Solon, F., Solon, M., Mehansho, H., West, K., Jr, Sarol, J., Perfecto, C., & Wasantwisut, E. (1996). Evaluation of the effect of vitamin A-fortified margarine on the vitamin A status of preschool Filipino children. *European Journal of Clinical Nutrition*, *50*, 720–723.

Soukoulis, C., & Bohn, T. (2015). A comprehensive overview on the micro-and nano-technological encapsulation advances for enhancing the chemical stability and bioavailability of carotenoids. *Critical Reviews in Food Science and Nutrition*, . Available from http://dx.doi.org/10.1080/10408398.2014.971353

Stoleru, E., Munteanu, S. B., Dumitriu, R. P., Coroaba, A., Drobotă, M., Zemljic, L. F., & Vasile, C. (2016). Polyethylene materials with multifunctional surface properties by electrospraying chitosan/vitamin E formulation destined to biomedical and food packaging applications. *Iranian Polymer Journal*, *25*, 295–307.

Tan, C., Feng, B., Zhang, X., Xia, W., & Xia, S. (2016). Biopolymer-coated liposomes by electrostatic adsorption of chitosan (chitosomes) as novel delivery systems for carotenoids. *Food Hydrocolloids*, *52*, 774–784.

Teleki, A., Hitzfeld, A., & Eggersdorfer, M. (2013). 100 Years of vitamins: The science of formulation is the key to functionality. *KONA Powder and Particle Journal*, *30*, 144–163.

Tengse, D.D., Priya, B., & Kumar, P.A.R. Optimization for encapsulation of green tea (Camelliasinensis L.) extract by spray drying technology. *Journal of Food Measurement and Characterization*,11, 2017, 85–92.

Trindade, M. A., & Grosso, C. R. F. (2000). The stability of ascorbic acid microencapsulated in granules of rice starch and in gum arabic. *Journal of Microencapsulation*, *17*(2), 169–176.

Tsuchido, Y., Sasaki, Y., Sawada, S.-i, & Akiyoshi, K. (2015). Protein nanogelation using vitamin B6-bearing pullulan: effect of zinc ions. *Polymer Journal*, *47*, 201–205.

Uddin, M. S., Hawlader, M. N., & Zhu, H. J. (2001). Microencapsulation of ascorbic acid: Effect of process variables on product characteristics. *Journal of Microencapsulation*, *18*(2), 199–209.

Uraiwan, K., & Satirapipathkul, C. (2016). The entrapment of vitamin E in nanostructured lipid carriers of rambutan seed fat for cosmeceutical uses. *Key Engineering Materials*, *675*, 77–80.

van Hasselt, P. M., de Koning, T. J., Kvist, N., de Vries, E., Lundin, C. R., ... Verkade, H. J. (2008). Prevention of Vitamin K deficiency bleeding in breastfed infants: Lessons from the Dutch and Danish biliary atresia registries. *Pediatrics*, *121*(4), e857–e863.

Van Hasselt, P., Janssens, G., Slot, T., Van der Ham, M., Minderhoud, T., Talelli, M., & Van Nostrum, C. (2009). The influence of bile acids on the oral bioavailability of vitamin K encapsulated in polymeric micelles. *Journal of Controlled Release*, *133*, 161–168.

Varona, S., Martín, Á., & Cocero, M. J. (2016). Encapsulation of edible active compounds using supercritical fluids, Encapsulation and Controlled Release Technologies in Food Systems. *Second Edition*, 16–40 , ISBN: 9780470277881.

Velikov, K. P., & Pelan, E. (2008). Colloidal delivery systems for micronutrients and nutraceuticals. *Soft Matter*, *4*, 1964–1980.

Vilanova, N., & Solans, C. (2015). Vitamin A palmitate–β-cyclodextrin inclusion complexes: characterization, protection and emulsification properties. *Food Chemistry*, *175*, 529–535.

Wambura, P. (2009). Vaccination of chickens using raw rice coated with novel trehalose nano-organogels containing Newcastle disease (strain I-2) vaccine. *Tropical Animal Health and Production*, *41*, 797–802.

Wang, B., Vongsvivut, J., Adhikari, B., & Barrow, C. J. (2015). Microencapsulation of tuna oil fortified with the multiple lipophilic ingredients vitamins A, D3, E, K2, curcumin and coenzyme Q 10. *Journal of Functional Foods*, *19*, 893–901.

Wang, R., MacGillivray, B. C., & Macartney, D. H. (2009). Stabilization of the base-off forms of vitamin B12 and coenzyme B12 by encapsulation of the α-axial 5, 6-dimethylbenzimidazole ligand with cucurbit [7] uril. *Dalton Transactions*, 3584–3589.

Wari, E., & Zhu, W. (2016). A survey on metaheuristics for optimization in food manufacturing industry. *Applied Soft Computing*, *46*, 328–343.

Wegmüller, R., Zimmermann, M. B., Bühr, V. G., Windhab, E. J., & Hurrell, R. F. (2006). Development, stability, and sensory testing of microcapsules containing iron, iodine, and vitamin A for use in food fortification. *Journal of Food Science*, *71*, S181–S187.

Wu, X. Y. (2016). Strategies for optimizing polymer-lipid hybrid nanoparticle-mediated drug delivery. *Expert Opinion on Drug Delivery*, *13*, 609–612.

Wu, X.-M., Branford-White, C. J., Yu, D.-G., Chatterton, N. P., & Zhu, L.-M. (2011). Preparation of core-shell PAN nanofibers encapsulated α-tocopherol acetate and ascorbic acid 2-phosphate for photoprotection. *Colloids and Surfaces B: Biointerfaces*, *82*, 247–252.

Xie, Z., Wang, F., Liu, H., Guo, S., Zhu, A., & Niu, H. (2010). Gelatin-walled microencapsulated diet for larval shrimp (*Penaeus japonicus* Bate) manufactured using the fluidized bed coating process. *Aquaculture Research*, *42*, 65–73.

Zempleni, J., Suttie, J. W., Gregory, J. F., III, & Stover, P. J. (2013). *Handbook of vitamins*. Boca Raton, FL: CRC Press.

Zhang, R., Zhang, Z., Kumosani, T., Khoja, S., Abualnaja, K. O., & McClements, D. J. (2016). Encapsulation of β-carotene in nanoemulsion-based delivery systems formed by spontaneous emulsification: Influence of lipid composition on stability and bioaccessibility. *Food Biophysics*, *11*, 154–164.

Zhu, B., Wang, J.-R., Zhang, Q., & Mei, X. (2016). Greener solid-state synthesis: Stereoselective [2 + 2] photodimerization of vitamin K3 controlled by halogen bonding. *Crystal Engineering Communication*, *18*, 6327–6330.

Chapter 5

Nanoencapsulation of Food Antimicrobial Agents and Essential Oils

Yue Zhang[1,*], Huaiqiong Chen[2,*] and Kang Pan[3]
[1]*University of Nebraska—Lincoln, Lincoln, NE, United States,* [2]*Texas Tech University, Lubbock, TX, United States,* [3]*Kellogg Company, Battle Creek, MI, United States*

5.1 INTRODUCTION

Food antimicrobials are chemicals that stop or inhibit the growth of microorganisms, and are often used in combination with other preservation procedures to preserve a food (Davidson et al., 2013b). Nowadays, consumers demand for foods that are convenient to store and use and yet are "fresher," "more natural," and "minimally processed." It has forced the food industry either modifying the existing preservation technologies to meet the requirements of consumers, or researching or applying new alternatives to replace the traditional synthetic antimicrobial substances. Moreover, the release of synthetic antimicrobial agents to environment such as sources of drinking water has aroused public concerns about the risks of ingesting the contaminated water and the chronic presence of antimicrobial agents may promote antibiotic resistance (Kolpin et al., 2002). Using naturally occurring antimicrobials such as antimicrobial peptides, plant-based essential oils, bacteriocins can be a solution to address these concerns. However, many of these natural antimicrobials are not as effective as the synthetic additives, being more expensive, unstable, or water insoluble; therefore, additional procedures should be developed and applied to effectively protect the antimicrobials against chemical degradation, improve their solubility and bioactivity. Furthermore, recent emergence of foodborne disease outbreaks related to *Escherichia coli* O157:H7, *Salmonella Saintpaul*, and *Listeria monocytogenes*, etc. has made microbial safety a high priority that more effective intervention systems are needed to preserve the natural antimicrobials stability, especially during food processing and storage.

---

* Yue Zhang and Huaiqiong Chen have contributed equally to this chapter.

Nanoencapsulation of Food Bioactive Ingredients. DOI: http://dx.doi.org/10.1016/B978-0-12-809740-3.00005-2

The use of nanoencapsulation systems may provide the protection of antimicrobial agents against degradation, enhance the bioavailability and the target delivery of antimicrobial agents, and thereby decrease the amount of antimicrobials required for effective food preservation. In addition, some nanoencapsulation systems are helpful to incorporate antimicrobial agents in food packaging materials. Numerous benefits offered by nanoencapsulation make it among the most promising alternative strategies to ensure food safety. The scope of this chapter is to provide a background of the two classes of antimicrobial compounds (traditional and natural antimicrobial agents), current nanoencapsulation techniques applied on antimicrobial agents as well as their characterizations and recent research achievements, along with their state-of-the-art applications in food systems.

5.2 TRADITIONAL AND NATURAL FOOD ANTIMICROBIAL AGENTS

Antimicrobial agents have been investigated for their effectiveness to kill or inhibit growth of microorganisms in the matrix and on the surface of foods. This has been done with the aim to increase food safety for the consumers, as well as to increase the shelf life of food products. Based on their origins, they can be simply classified as traditional chemical preservatives and natural antimicrobials.

5.2.1 Traditional Chemical Preservative Agents

According to definition of Davidson et al. (2013b), antimicrobials are classified as traditional ones when they (1) have been used by many years, (2) are approved by many countries for inclusion as preservatives in foods, or (3) are produced in large scale by synthetic processes. Commonly used traditional chemical preservatives include acetates, benzoates, lactates, sulfites, nitrites, fatty acids and esters, etc. (Jay et al., 2005). Among these traditional chemical preservatives, one of the most widely studied antimicrobials are organic acids, such as lactic, acetic, and sorbic acids, which will be discussed in this following part (Table 5.1).

As a group of chemicals, organic acids are considered to be any organic carboxylic acid of the general structure R-COOH, including fatty acids and amino acids. Initially, organic acids are used as food additives, and these organic acids have a pK_a, the pH at which the acid is half dissociated, between 3 and 5, are usually found to have antimicrobial activity.

The mode of action of organic acids and their derivatives is mainly due to the consumption of ATPase. As depicted in Fig. 5.1, at pH lower than their pK_a, more organic acids in the undissociated form exist, which is better able to penetrate the microbial membrane due to the lipophilic property. This undissociated form is prior to dissociate into the free proton and acid anion,

TABLE 5.1 The Structures and Properties of Commonly Used Organic Acids (Davidson and Zivanovic, 2003)

Compound	Structure	pK_a	Microbial Target	Limitations in Food
Acetic acid	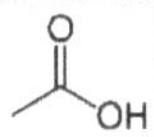	4.71	Yeasts, bacteria	Not exceed GMP[a] (21CFR184.1005)
Lactic acid	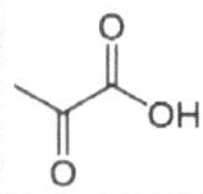	3.79	Bacteria	Not exceed GMP (21CFR184.1061)
Benzoic acid	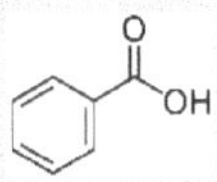	4.19	Yeasts, molds	0.1% (21CFR184.1733)
Sorbic acid	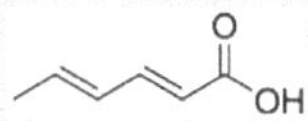	4.75	Yeasts, molds	Not limited (21CFR182.3089)

[a]*GMP: Good manufacturing practice.*

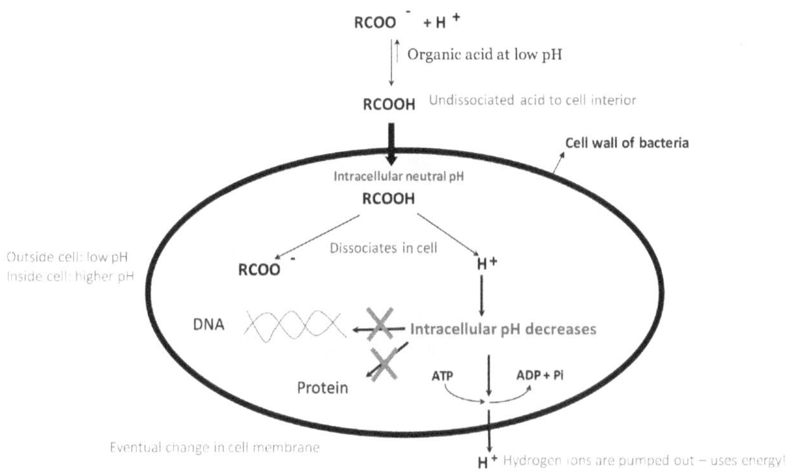

FIGURE 5.1 Mode of action of organic acids and their derivatives. Scheme was modified from Davidson et al. (2013b).

since the interior cytoplasmic pH of bacteria cell is usually neutral. This dissociation causes the acidification of the cell interior. In order to maintain the cytoplasmic pH level to protect compounds such as structural proteins, enzymes, nucleic acids, and phospholipid, bacteria cells have to consume their ATP to extrude the liberated protons to the outer environment. The bacteria cells are finally too exhausted to proliferate, resulting in some degree of bacteriostasis. Derivatives of some weak organic acids likely have similar mechanisms of microbial inhibition.

Acetic acid is the primary component of vinegar. Acetic acid and its salts are considered as some of the oldest food antimicrobials. In general, they are more effective against yeasts and bacteria than against molds. Their activity in foods varies in different studies. The application of 150 mM acetic acid to raw chicken breast meat resulted in 1−2log (CFU/ml) reduction of *S. typhimurium* and *E. coli* O157:H7, while there was no significant effect on the levels of *L. monocytogenes* (Over et al., 2009) for 12 days storage at 4°C. While Carpenter et al. (2011) stated that 2% acetic acid washes only lowered recoverable numbers of pathogens by 0.6 to 1log/cm^2 for *E. coli* O157:H7 on beef plate, *Salmonella* on chicken skin and pork belly, and *L. monocytogenes* on turkey roll. In another study, populations of *E. coli* O157:H7 and *Salmonella* on beef trim exposed to 2% or 4% acetic acid were 2.0−2.5log (CFU/g) less than populations on controls after acid application (Harris et al., 2006). Sodium acetate and sodium diacetate (SDA), the salt forms of acetic acid, are approved at levels not to exceed 0.25% of the product formulation (9 CFR 424.21). Fresh salmon slices treated by dipping in 2.5% (w/v) aqueous solution of sodium acetate have 4−7 days more shelf life than that of the control (Ibrahim Sallam, 2007). Frankfurters with 0.25% SDA inhibited *L. monocytogenes* on surfaces over 40 days of storage at 10°C and the levels of the pathogen were 2.5 log (CFU/cm^2) less than controls at the end of storage (Barmpalia et al., 2004).

Lactic acid is synthesized naturally by lactic acid bacteria during fermentation. Similar to acetic acid, lactates are the salt forms of lactic acid, both of which show similar activities. It was found that dipping of salmon slices in aqueous solutions (2.5%) of the sodium lactate (SL) was efficient against the proliferation of various categories of spoilage microorganisms (Ibrahim Sallam, 2007). Combination of SL (2.5%) and SDA (0.2%) was bacteriostatic to *L. monocytogenes* and bactericidal to *S. enteritidis* after 20 days at 5°C and 10°C, which was more effective than the salts applied alone in beef emulsions samples of 79% moisture (Mbandi and Shelef, 2001). These authors also reported the enhanced antimicrobial effect of combination of SL (2.5%) and SDA (0.2%) on ready-to-eat meat (Mbandi and Shelef, 2002).

Benzoic acid occurs naturally in cranberries, plums, prunes, cinnamon, cloves, and most berries. Benzoates, the sodium salt of benzoic acid, is much more soluble (66.0 g/100 ml at 20°C) in water than benzoic acid (0.27% at 18°C) and is much more preferred to use in many cases (Chipley, 2005). It is a commonly used food additive preservative that is listed among the "generally recognized as safe" (GRAS) compounds by the United States Food and Drug Administration (FDA), and can be present in foods at a concentration up to 0.1%. Both of them exhibited inhibitory activity against a wide range of microorganisms, especially fungi, yeasts, and molds (Chipley, 2005). *E. coli* 0157:H7 was reduced from 5.2log CGU/ml to 0.3log CFU/ml and 1.4log CFU/ml by 0.1% sodium benzoate and potassium benzoate, respectively, at 8°C after 14 days of storage. At 25°C, the population of *E. coli* 0157:H7 was reduced by 4.8 CFU/ml and 4.0 CFU/ml by 0.1% sodium benzoate and potassium benzoate, respectively, at the end of storage (Ceylan et al., 2004). Incorporation of 0.1% sodium benzoate in mango juice reduced the time to half compared to the control group to achieve 3-log reduction of heat-resistant mold *Neosartorya fischeri* ATCC 200957 (Rajashekhara et al., 1998).

Sorbic acid is a *trans-trans* unsaturated monocarboxylic fatty acid that is slightly soluble in water (0.15 g per 100 ml at 20°C), while its salt form is significantly more water soluble (potassium sorbate as 58.2 g per 100 ml at 20°C) (Stopforth et al., 2005). Sorbic acid and sorbates inhibit a wide range of microorganisms, including *Acinetobacter, Aeromonas, Alicyclobacillus acido terrestris, Bacillus, Campylobacter, E. coli* O157:H7, *Lactobacillus, L. innocua, Pseudomonas, Salmonella, Staphylococcus, Vibrio*, and *Y. enterocolitica* (Davidson et al., 2013b).

5.2.2 Natural Antimicrobial Agents

Based on the origins, natural antimicrobials can be further classified as animal, plant, and microbial sources. There activity against bacteria and fungi was summarized in Table 5.2.

5.2.2.1 Animal Sources

5.2.2.1.1 Chitosan

Chitosan, a linear polysaccharide, is the *N*-deacetylation product of chitin, the byproduct of shellfish processing. Chitosan comprises a series of polymers with differing ratios of glucosamine and *N*-acetyl glucosamine units via β (1–4) linkage. Chitosan has been reported to have antimicrobial activities against foodborne fungi and bacteria, with minimum inhibition concentration (MIC) varying from 0.01% to 5%, depending on the source of chitosan, charge density, the molecular weight, degree of deacetylation, pH, temperature, ionic strength, concentration, derivatives of chitosan, complex with certain materials, type of microorganisms, and food components (Hosseinnejad and Jafari, 2016; Kong et al., 2010). Up to now, four models have been proposed to explain the antimicrobial activity of chitosan (Goy et al., 2009; Hosseinnejad and Jafari, 2016; Kong et al., 2010): (1) the electrostatic interaction between positively charged chitosan molecules and negatively charged microbial cell membranes, causing the change of membrane permeability and hydrolysis of peptidoglycans in the microorganism wall. This is the most acceptable one. (2) The penetration of chitosan into

TABLE 5.2 Antimicrobial Spectrum of Selected Natural Antimicrobial Agents

Sources	Compound	Bacteria Inhibited		Fungi Inhibited
		Gram-positive Bacteria	**Gram-negative Bacteria**	
Animal	Chitosan	*L. monocytogenes* *Bacillus cereus* *Staphylococcus aureus*	*E. coli* *E. coli* O157:H7 *Salmonella typhimurium*	*Zygosaccharomyces bailii*
	Lysozyme	*B. stearothermophilus* *B. cereus* *Lactobacillus spp.* *L. monocytogenes*	*C. jejuni* *E. coli* *E. coli* O157:H7 *Pseudomonas aeruginosa* *S. enteritidis* *Y. enterocolitica*	*Aspergillus* *Candida* *Fusarium* *Graminearum* *Paecilomyces* *Penicillium* *Saccharomyces*
	Lactoferrin	*B. subtilis* *B. stearothermophilus,* *L. monocytogenes*	*E. coli* *E. coli* O157:H7 *S. enteritidis*	
Plant	Basil oil	*Shigella sonnei* *Bacillus* *Enterococcus* *Staphylococcus*	*S. flexneri* *E. coli* *S. enteritidis* *Pseudomonas* *Vibrio parahaemolyticus*	*Aspergillus* *Candida* *Geotrichum candidum*
	Thyme oil	*Shigella sonnei* *S. aureus* *L. monocytogenes* Lactic acid bacteria	*S. flexneri* *E. coli* *E. coli* O157:H7 *S.* enteritidis *S. typhimurium* *Y. enterocolitica*	*Botrytis cinerea* *C. albicans* *P. digitatum* *Rhizopus stolonifer*
	Clove bud oil	*L. monocytogenes* *S. epidermidis* *S. aureus*	*C.jejuni* *E. coli* *S. enteritidis*	*C. albicans* *Trichophyton mentagrophytes*
	Oregano oil	*L. monocytogenes* *B. subtilis* *S. aureus* (methicillin-resistant) *S. aureus*	*E. coli* *E. coli* O157:H7 *P. aeruginosa* *S. enterica* *S. enteritidis* *S. typhimurium*	*C. albicans* *P. digitatum* *S. cerevisiae*
	Tea tree oil	*B. cereus,* *S. aureus* (methicillin resistant); *L. monocytogenes* *B. subtilis*	*Acinetobacter baumannii* *E. coli* *E. coli* O157:H7 *S. flexneri* *S. choleraesuis*	*A. flavus* *C. albicans* *Cryptococcus neoformans*
	Rosemary extract	*Streptococcus agalactiae* *L. monocytogenes,* *Lactococcus lactis* *S. aureus*	*P. putida* *E. coli* *Klebsiella pneumoniae*	*Aspergillus* *Saccharomyces cerevisiae* *C. albicans*

(*Continued*)

TABLE 5.2 (Continued)

Sources	Compound	Bacteria Inhibited		Fungi Inhibited
		Gram-positive Bacteria	Gram-negative Bacteria	
		B. megaterium *B. subtilis* *Enterococcus faecalis*	*Proteus mirabilis*	*P. pastoris*
	Arrowroot tea extract	*L. monocytogenes* *S. aureus*	*E. coli* O157:H7 *S. enterica*	
	Grapefruit seed extract	*B. subtilis* *Micrococcus flavus* *S. aureus* *Serratia marcescens*	*E. coli* *P. mirabilis*	*C. maltose*
	Olive leaf extract	*S. aureus* *B. cereus* *B. subtilis* *E. faecalis* *Kocuria rhizophila* *Lactobacillus acidophilus* *L. casei* *L. spp.* *L. innocua* *L. monocytogenes* *Micrococcus luteus* *S. aureus* (methicillin resistant) *S. capitis* *S. epidermidis* *S. hominis*	*E. coli* *Acinetobacter calcoaceticus* *C. jejuni* *Helicobacter pylori* *K. pneumoniae* *P. aeruginosa* *S. enterica* *Serratia marcescens*	*C. albicans* *C. glabrata* *C. parapsilosis* *C. oleophila* *Schizosaccharomyces pombe*
Microbial	Nisin	*Listeria* *Staphylococcus* *Bacillus* *Clostridium* *Enterococcus* *Lactococcus* Spores formed by bacteria such as *Bacillus* and *Clostridium*	No effect	
	Natamycin	No effect	No effect	*Penicillium candidum* *A. flavus* *S. cerevisiae* *Byssochlamys nivea* *Hemerocallis fulva* *Zygosaccharomyces bailii*

the nuclei of microorganisms results in the binding of chitosan with microbial DNA and thereof prevents the RNA transformation and protein synthesis. (3) The chelation of chitosan with metals, the suppression of spores and binding to essential nutrients inhibit the microbial growth. (4) chitosan forms a polymer membrane on the surface of the cell. This membrane could prevent nutrients from entering the cell or block oxygen from the cells which can inhibit the growth of aerobic bacteria. Chitosan is a great polymer candidate for biofilm formation. Therefore, it has been widely used for biofilm formation. The films consisting of chitosan have been evaluated to extend the shelf life of ground meat (Dehnad et al., 2014), fish fillets (Kakaei and Shahbazi, 2016), fruits (Jongsri et al., 2016), and vegetables (Bilbao-Sainz et al., 2016).

5.2.2.1.2 Lysozyme

Lysozyme is an enzyme that is often isolated from hen egg, as well as other natural sources, such as certain vegetables, insects, plants, fungi, human colostrum, and milks (Larson, 2005). The antimicrobial activity of lysozyme relies on the ability of hydrolyzing the β-1,4 linkage between C1 of N-acetylmuramic acid and C4 of N-acetyl glucosamine present in peptidoglycan. Gram-positive bacteria are very susceptible to lysozyme, because their cell wall is made up of 90% of peptidoglycan; while in Gram-negative bacteria, peptidoglycan counts only for 5–10% of the cell wall and lies beneath the outer membrane of the cell envelope, making them not susceptible to lysozyme (Barbiroli et al., 2012). However, the sensitivity of lysozyme to gram-negative bacteria can be increased by adding chelators, such as ethylenediaminetetraacetic acid to bind with Ca^{2+} and Mg^{2+} to destroy the integrity of lipopolysaccharide (LPS) layer (Davidson and Zivanovic, 2003). The activity of lysozyme toward Gram-negative bacteria can also be improved by lacroferrin (Barbiroli et al., 2012), which is introduced below.

5.2.2.1.3 Lactoferrin

Lactoferrin is an innate defense protein and a component of milk. Lactoferrin is an 80 kDa whey glycoprotein that can bind two ferric ions for each protein molecule. Because of its ability to reversibly bind ions, lactoferrin is naturally a mixture of iron-free (apo-lactoferrin) and of forms containing one or two iron atoms (holo-lactoferrin) (González-Chávez et al., 2009). The antibacterial activity of lactoferrin is due to two different and unrelated mechanisms, one based on iron deprivation that inhibits bacterial growth, and one related to the large cationic patches present on the lactoferrin surface (Jenssen and Hancock, 2009). These cationic patches allow direct interaction with the anionic Lipid A, a component of the LPS of Gram-negative bacteria, thus altering the outer membrane permeability and resulting in the release of LPS. As a natural component of milk, lactoferrin satisfies all regulatory requirements, and has been used directly in an antimicrobial spray for the treatment of beef carcasses (Taylor et al., 2004). It has been found that combinations of various antimicrobial compounds such as lysozyme (Barbiroli et al., 2012), nisin (Murdock et al., 2007), and fluconazole (Lupetti et al., 2003) could enhance the antimicrobial activity of lactoferrin in a synergistic manner.

5.2.2.2 Plant Sources

5.2.2.2.1 Essential Oils

Essential oils are the aromatic volatile liquids or semiliquids obtained from whole plant or plant parts such as flower, seeds, buds, leaves, bark, fruits, roots, and resins (Burt, 2004). They are also called as volatile or ethereal oils due to their volatility (Burt, 2004). The use of essential oils or their main components as antimicrobial agents are gaining interest because they have natural sources and could be classified as organic. Recently, many studies on composition and antimicrobial activity of essential oil from various sources have been published. Up to now, FDA has listed more than 70 essential oils and their components as GRAS additives (FDA, 1998).

Antimicrobial activities of essential oils have been demonstrated against a wide variety of microorganisms, including Gram-positive and Gram-negative bacteria as well as fungi. Basil and thyme oils and their major components inhibited the growth of *Shigella sonnei* and *S. flexneri* in both agar medium and food model matrix lettuce (Bagamboula et al., 2004). Clove bud oil demonstrated activities to bacteria *L. monocytogenes*, *Staphylococcus epidermidis*, *Campylobacter jejuni*, *S. enteritidis*, *E. coli*, *Staphylococcus aureus* and fungi *Candida albicans* and *Trichophyton mentagrophytes* (Chaieb et al., 2007). Oregano essential oil was effective against *Pseudomonas aeruginosa* and *E. coli* (Bozin et al., 2006). Tea tree oil showed a broad-spectrum activity against more than 27 strains of bacteria, including *Acinetobacter baumannii, Bacillus cereus, S. aureus* (methicillin resistant), and 24 strains of fungi, such as *Aspergillus flavus*, *C. albicans*, *Cryptococcus neoformans* (Carson et al., 2006). More information about the researches on biological activities including antibacterial activities of essential oils can be found in review articles such as Burt Bakkali et al. (2008); (Burt, 2004).

5.2.2.2.2 Plant Extracts

Plant extracts usually contain phenolic compounds such as phenols and flavonoids (Cowan, 1999), which are potential antimicrobial agents. A lot of extracts from different plants have been investigated with their antimicrobial activity against bacteria. The extracts of several edible Asian plants were screened for their antibacterial activity against *Bacillus cereus*, *Staphylococcus aureus*, *L. monocytogenes*, *E. coli,* and *S. infantis* (Alzoreky and Nakahara, 2003). About 16 out of 26 extracts were found to be effective against these bacteria, with MIC ranging from 0. 165 to 2.64 mg/mL. The antimicrobial activities of rosemary extracts with different solvent were compared. Methanol extract with higher polyphenol content showed lower MICs against bacteria than water extract (Moreno et al., 2006). Antimicrobial activity of water-soluble arrowroot tea extract was evaluated against *E. coli* O157:H7, *S. enterica* Serotype Enteritidis, *L. monocytogenes*, and *S. aureus* in ground beef and mushroom soup. At 6% concentration, the extract could inhibit 1.5log for pathogens inoculated on beef after 7 days storage. The extract was more effective in liquid food. At 1%, it could reduce the pathogen population by at least 2.3 log CFU/ml in mushroom soup after storage (Kim and Fung, 2004). The antimicrobial efficacy of six grapefruit seed extracts was tested, and five out of six extracts showed a high growth inhibiting activity against the test germs including *Bacillus subtilis*, *Micrococcus flavus*, *Staphylococcus aureus*, *Serratia marcescens*, *E. coli*, *Proteus mirabilis*, and *Candida maltose* (Von Woedtke et al., 1999). The methanol extract of the leaves of *Calpurnia aurea* exhibited broad spectrum of antibacterial activities, with MIC ranging from 31.25 to 125 mg/mL against seven tested pathogens (Umer et al., 2013). Olive leaves extracts have been found to be effective against a wide spectrum of bacteria and fungi, such as *B. cereus, L. monocytogenes, E. coli*, and *C. oleophila* (Rahmanian et al., 2015). The antimicrobial activities are related to their terpenes content and phenolic compounds that are well reported to possess a good antimicrobial activity. Even though many plant extracts have shown antimicrobial activities, the effective concentration is usually high and the dark colors of raw extract is one big challenge of applying them directly in foods.

5.2.2.3 Microbial Sources

Bacteria produce many compounds that are active against other bacteria, which can be harnessed to inhibit the growth of potential spoilage or pathogenic microorganisms. Both Gram-negative and Gram-positive bacteria produce bacteriocins. They are proteinaceous antibacterial compounds, which constitute a heterologous subgroup of ribosomally synthesized antimicrobial peptides (Tiwari et al., 2009).

5.2.2.3.1 Nisin

Nisin is the most widely used bacteriocin. It is a cationic peptide produced by certain strains of *Lactococcus lactis* ssp. *lactis* (Bonev et al., 2000; Jozala et al., 2009). It has 34 amino acids and a molecular weight of around 3.5 kDa (Abdullah et al., 2010). The three lysine residues contribute to its positive charges that make nisin active against gram-positive bacteria such as *Listeria* and *Staphylococcus* as well as the outgrowth of spores of *Bacillus* and *Clostridium* (Jozala et al., 2009). Currently, nisin is the only bacteriocin that is recognized as safe for food applications in over 50 countries (Thomas and Delves-Broughton, 2005). In the United States, nisin has been listed as a GRAS food preservative by the FDA since 1988 and has been widely used in cheese, sausages, and ready-to-eat meat (Jozala et al., 2009). The main antimicrobial mechanism is the formation of pore complexes by eight nisin units with four Lipid II units (Breukink and de Kruijff, 2006). Since Lipid II is the transporter of peptidoglycan precursor subunits for the cellular wall synthesis, the complexes formation causes the synthesis of cell wall interrupted. Due to this mechanism, it has a limited spectrum of activity. It does not inhibit Gram-negative bacteria or fungi.

5.2.2.3.2 Natamycin

Natamycin is an antifungal produced by *Streptomyces natalensis* that is effective against nearly all molds and yeasts but has little or no effect on bacteria. It is the only antifungal medication approved by FDA. It acts by binding with ergosterol, which is an essential component in fungal cell wall, and blocks fungal growth (Roller, 2003). Several characteristics of natamycin make it an ideal broad-spectrum antifungal bio-preservative for foods and beverages. It is a safe compound for consumption; is effective at low concentrations; has no converse effect on quality of foods, and has prolonged antimicrobial activity specifically on food surfaces, such as cheese and sausages (Stark and Roller, 2003). In addition, natamycin can be used for inhibition of spoilage fungi but not hampering the bacterial for fermentation (Davidson et al., 2013a).

5.2.3 Problems and Opportunities of Natural Antimicrobials

Natural antimicrobials are gaining interest by consumers for their uses as alternatives to traditional antimicrobial treatments due to their "green" nature. There is a big market of using natural antimicrobials in foods. However, there are many limitations in the application of natural antimicrobials in foods. It requires further research on their antimicrobial efficacy, consumer acceptability, and cost. There are several challenges of using naturally derived antimicrobials in food, including:

1. Hydrophobic antimicrobials have low solubility in the aqueous phase, at which concentration they are not able to function their bioactivities.
2. Essential oils, which have been investigated widely in the past several years as natural antimicrobials, are very volatile, and therefore a large amount may be lost during food processing, causing the increased cost.
3. Some sensitive compounds degrade during the manufacturing, storage, transportation, and utilization, such as thermal processing, UV radiation, pH adjustment, and enzyme digestion, resulted in the loss of antimicrobial activity.
4. Further, the addition of some natural substances can lead to adverse changes in the sensory properties of foods, such as herbs and essential oils. When applied at the concentrations needed to achieve the desired

level of antimicrobial potency, these antimicrobials can adversely affect the organoleptic properties of food beyond its threshold and consumer acceptance.

Even though many of the natural antimicrobials are categorized as GRAS for food application, their use in commercial applications requires regulatory approval. Natural antimicrobials can provide a tremendous opportunity for advancing the field of food preservation and safety; however, additional research is needed to optimize their applications.

The challenge of using these natural antimicrobials as preserving agents may be overcome to some extent by nanoencapsulation instead of applying them as free antimicrobials directly to the food. By loading antimicrobials into nanoparticles/formulations, the physical properties such as dispersibility, dispersion stability, turbidity, viscosity can be significantly improved, and hence the bioactivities including antimicrobial activity can be promoted in contrast to the free antimicrobials (Blanco-Padilla et al., 2014). Nanoencapsulation may also be helpful to protect the encapsulated antimicrobials from degradation during processing and storage, mask unpleasant flavors, and release the antimicrobials at a controlled rate. Moreover, some designed nanocarriers can trigger the release of antimicrobials according to the environment surrounding the food systems (pH, temperature, etc.). These functions are typically enabled by incorporating the encapsulated antimicrobials into fibers, or coatings/films.

5.3 NANOENCAPSULATION OF FOOD ANTIMICROBIAL AGENTS

Many techniques have been developed for the fabrication of nanoencapsulated antimicrobials. These methods can be classified into two main categories according to whether the formulation requires a high mechanical energy (top-down method) or can be achieved directly without applying mechanical energy (bottom-up method). They can also be classified based on encapsulating materials (such as lipid, biopolymer, and inorganic material) (Jafari, 2017). The most important methods applied for antimicrobial agents are summarized below (as shown in Table 5.3).

5.3.1 Nanoencapsulation of Antimicrobials through Lipid-Formulation Techniques

5.3.1.1 Nanoemulsions

Among numerous nanoencapsulation techniques for antimicrobial agents, emulsions are particularly suitable for food applications due to abundant food compatible emulsion ingredients and scalable top-down approaches such as high-pressure homogenization for preparation (McClements, 2015).

TABLE 5.3 Nanoencapsulation Techniques for Food Antimicrobial Agents and their Target Food Applications

Technique		Advantages	Antimicrobial Agents
Lipid-formulation	Nanoemulsions	Small droplet size Transparent/translucent systems suitable to use in beverages Rapid absorption Improved antimicrobial activity	Essential oils including peppermint oil (Liang et al., 2012), clove bud oil (Zhang et al., 2014a), thyme oil (Wu et al., 2014; Xue et al., 2015), sago oil (Moghimi et al., 2016)
			Essential oil components such as thymol, eugenol (Ma et al., 2016), carvacrol (Chang et al., 2013)
			Nystatin (Campos et al., 2012)
	Nanoliposomes	Capability to encapsulate both hydrophilic and hydrophobic antimicrobial agents either in their cavity or in their bilayer High drug carrying capacity and readily tunable formulation	Antimicrobial peptides such as nisin (Colas et al., 2007; TAYLOR et al., 2008)
			Anethum graveolens essential oil (Ortan et al., 2009)
			Rose essential oil (Wen et al., 2011)
			Artemisia arborescens L. essential oil (Sinico et al., 2005)
			Daptomycin (Li et al., 2013)
	Solid lipid nanoparticles	Both lipophilic and hydrophilic compounds can be loaded into the solid matrix Possibility of large-scale production	Nisin (Prombutara et al., 2012) Clove extract (eugenol) (Cortés-Rojas et al., 2014)
Biopolymer based	Protein	Improved bioavailability and antimicrobial activity; low toxicity	Thymol (Bhavini et al., 2012; Pan et al., 2014a), carvacrol (Wu et al., 2012) Clove bud oil (Luo et al., 2014)
	Carbohydrate	Low toxicity and low cost High stability in a wide pH range	Clove bud oil (Luo et al., 2014) Lavandin essential oil (Varona et al., 2013a)

(*Continued*)

TABLE 5.3 (Continued)

Technique		Advantages	Antimicrobial Agents
		Good biodegradability	Essential oil components including d-limonene (Donsì et al., 2011), cinnamaldehyde, eugenol (Hill et al., 2013) Chitosan with benzoyl peroxide(Friedman et al., 2013) Nisin (Chopra et al., 2014)
Equipment based	Electrospinning and electrospraying	Large surface area, amendable size, and morphology Capability to carry heat-sensitive compounds Possibility of large-scale production	Bacteriocin (Heunis et al., 2010) Peppermint oil (Ghayempour and Mortazavi, 2014)
	Nanospray-drying	Preparation of nano-size ultrafine powder Good morphological properties and redispersibility	Eugenol (Hu et al., 2016) Peppermint oil (Wang et al., 2016)
Nanocomposites		Improved mechanical and barrier properties	Carvacrol (Tunç and Duman, 2011) Rosemary essential oil (Abdollahi et al., 2012) Clove, coriander, caraway, marjoram, cinnamon, and cumin essential oils (Alboofetileh et al., 2014)

(A)

(B)

(C)

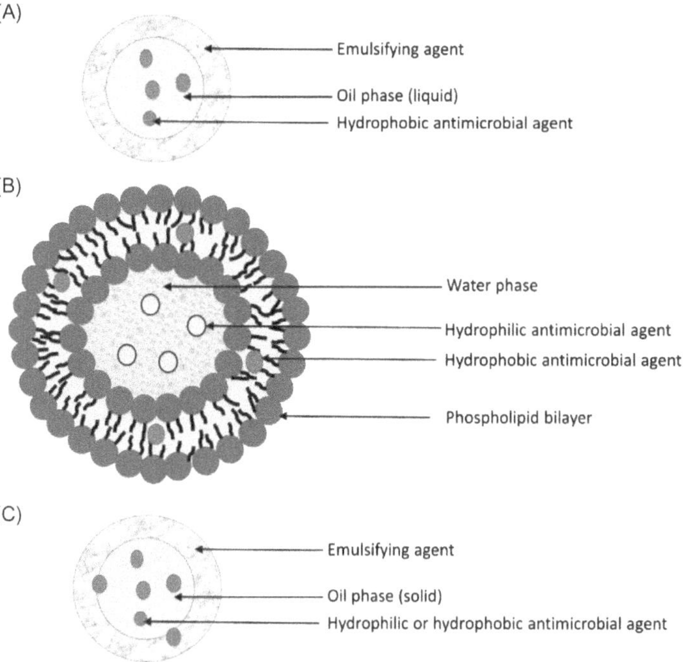

FIGURE 5.2 Schematic illustration of three lipid formulations for antimicrobial agent delivery: (A) nanoemulsion, (B) nanoliposome, and (C) solid lipid nanoparticle.

Nanoemulsions are a specific type of colloidal dispersion containing oil, water, and an emulsifier with a remarkable small droplet size, usually covering the size range of 20–200 nm (Fig. 5.2A) (Solans et al., 2003). Depending on the nature of hydrophobicity or hydrophilicity of the antimicrobial agents, either O/W or W/O emulsions can be applied to stabilize oil-soluble or water-soluble agents, respectively. Nanoemulsions are nonequilibrium systems with a spontaneous tendency to separate into the constituent phases, while microemusions, which may appear to be similar to nanoemulsions in composition and structure, are equilibrium systems (Pan and Zhong, 2016b). Nanoemulsions can be prepared through both top-down (high-energy,) and bottom-up (low-energy) methods, while microemulsions are produced by low-energy methods only (Galanakis, 2015). The high-energy emulsification methods require a large external force by mechanical devices like high-pressure homogenization, microfludization, or ultrasonication to be applied in order to disrupt and intermingle the oil and water phases (McClements, 2012). High-pressure and high shear homogenizations are two of the most popular methods to prepare food emulsions (Pan and Zhong, 2016b).

Antimicrobial agents (e.g., nystatin, peppermint oil) can be predissolved in commonly used oils (e.g., triacylglycerol, Labrafac lipophile, soybean oil) and then emulsified in the aqueous phase with predissolved emulsifiers using high-energy methods (Campos et al., 2012; Liang et al., 2012). Nanoemulsions of essential oils and their components (e.g., thyme oil, eugenol) can also be formed by homogenizing them in the aqueous solutions directly, showing similar or enhanced antimicrobial properties than free essential oils (Ma et al., 2016; Wu et al., 2014; Xue et al., 2015). Compared to high-energy approaches, low-energy emulsification methods rely on the spontaneous assembling of oil droplets within mixed surfactant–oil–water systems when the conditions are altered such as phase inversion and spontaneous emulsification methods (McClements, 2012). Phase-inversion methods utilize the interfacial behavior of surfactant which are a function of composition or temperature (Pan and Zhong, 2016b). For phase-inversion composition method, essential oil microemulsions with and without other oil phase can be formed above a critical surfactant-to-oil ratio by simple mixing (Ma and Zhong, 2015; Zhang et al., 2014a). The phase-inversion temperature method applies a heating step to induce the phase inversion from crude O/W to W/O, followed by another phase inversion from W/O to O/W during rapid cooling. (Pan and Zhong, 2016b) Additionally, a self-emulsifying technique based on the deprotonation of essential oils in alkaline condition and self-assemble property of emulsion during neutralization has also be demonstrated as a successful method to form essential oil nanoemulsions (Luo et al., 2014; Zhang et al., 2016).

5.3.1.2 Nanoliposomes

Liposomes are spherical lipid vesicles with an aqueous core and amphiphilic lipid bilayer, as shown in Fig. 5.2B (Zhang et al., 2010). As nanoscale lipid vesicles, nanoliposomes have been extensively studied as antimicrobial delivery systems because of their excellent physical properties such as high drug carrying capacity and readily tunable formulation properties (Thamphiwatana et al., 2013). Their unique bilayer structure can make them easily enter into microbes by fusing with bacterial membranes. As antimicrobial drug delivery carriers, nanolipsomes also showed advantages like increasing bactericidal efficacy, enhancing infection site concentrations. and decreasing the potential toxicity of antimicrobial agents (Halwani et al., 2008; Mugabe et al., 2006; Sachetelli et al., 2000). To encapsulate lipid-soluble antimicrobials and essential oils into liposomes, these compounds are predissolved with the phospholipids together. Then a sufficient amount of energy such as sonication, homogenization, shaking, or heating is supplied to phospholipids placed in water. Organized and closed bilayer vesicles with entrapped hydrophobic antimicrobials can be obtained during the process (Reza Mozafari et al., 2008). A variety of liposome techniques including sonication, thin lipid film hydration, freeze-thaw cycles, and extrusion, etc. and various lipids have been employed for the encapsulation. Although nanoliposomes may be limited by their poor stability, short half life, as well as leakage and fusion, these weaknesses can be overcome by coating with other polymers or binding tiny charged nanoparticles to the surfaces (Zhang and Granick, 2006). Colas et al. prepared nanoliposome encapsulated nisin Z using different lipids [Phospholipon 90 H, Phospholipon 100 H, dipalmitoyl-phosphatidylcholine (DPPC), stearylamine (SA), dicetyl phosphate (DCP) and cholesterol] via membrane extrusion method with an encapsulation efficacy up to 54% (Colas et al., 2007). The liposome-encapsulated nisin completely inhibited the growth of *E.coli* O157:H7 at concentrations below it reported necessary for unencapsulated nisin (Taylor et al., 2008). A daptomycin-loaded nanoliposome with smaller mean particle size (55.4 nm) and a higher encapsulation efficiency (87.85%) using dry film dispersion method was reported by Li et al. (Li et al., 2013). A very large range of sizes have been obtained for essential oil-loaded liposomes. To achieve essential oil-encapsulated small unilamellar vesicle (SUV) with size between 20 and 100 nm, thin film hydration was usually employed with sonification together (Ortan et al., 2009; Sinico et al., 2005). A novel rapid expansion of supercritical solution process was applied to encapsulate rose essential oil and achieved an entrapment efficiency of 89.46% and average particle size of 94 nm, respectively (Wen et al., 2011).

5.3.1.3 Solid Lipid Nanoparticles

Solid lipid nanoparticle (SLN) is another delivery platform instead of nanoemulsions and nanoliposomes that has attracted much attention since it was introduced in 1991. SLNs are mainly made from solid lipids including fatty acids (e.g., palmitic acid), triglycerides (e.g., trilaurin), steroids (e.g., cholesterol), partial glycerides (e.g., glyceryl monosterate), and waxes (Fig. 5.2C) (Zhang et al., 2010). Surfactants are also needed to stabilize the lipid dispersions. The basic production methods for SLN include the high-pressure homogenization, cold precipitation, spray-drying, high-shearing homogenization, and ultrasonication (MuÈller et al., 2000). Although most studies considered SLNs ideal carriers for cosmetic and pharmaceutical applications, their unique properties including the composition of physiological compounds, the possibility for loading both lipophilic and hydrophilic compounds into the solid matrix and large-scale production have also attracted the attention to use them as carriers of food antimicrobials (MuÈller et al., 2000). Nisin was entrapped in the SLN by high-pressure homogenization and showed a sustained release to inhibit the growth of *L. monocytogenes* and *L. plantarum* over at least 15–20 days (Prombutara et al., 2012). Clove extract (eugenol) was encapsulated in Compritol-based solid lipid carriers by spray-drying and showed a high retention in the formulation (Cortés-Rojas et al., 2014).

5.3.2 Biopolymer-Based Nanoencapsulation of Food Antimicrobial Agents

Biopolymers have found tremendous applications in nanoencapsulation of antimicrobial agents. Various sources serve as origins of biopolymers, which could be either animal or plant based, in their native form or being modified for specific goals such as controlled or target delivery. Commonly used biopolymers for nanoencapsulation include protein-based materials like dairy, plant proteins and carbohydrate-based biopolymers like starch and gums.

5.3.2.1 Protein-Based Biopolymers

Proteins are biopolymers comprising amino acids, adopting different molecular structures determined by amino acids sequence and environmental factors like pH, ionic strength, heat, pressure, etc. Proteins from various sources could be drastically different in terms of emulsification capability, solubility, stability, and their impacts on texture, flavor, color of food matrix that they will be incorporated in. To function as an encapsulation material for antimicrobial agents, only proteins have balanced hydrophobic/hydrophilic portions/amino acids sequence that can be used to minimize the overall free energy between dispersed phase (antimicrobial agents) and continuous phase (solvent) that generally have significant polarity differences.

Whey proteins have been used as an effective encapsulant for various hydrophobic antimicrobial agents such as essential oils for preparation of stable dispersions in aqueous systems like milk. By forming conjugation with maltodextrin, whey-maltodextrin conjugate was able to encapsulate and stabilize thymol, inhibiting growth of *Escherichia coli* O157:H7, and *Listeria monocytogenes* in both reduced fat milk and apple cider while not affecting the transparency of the beverage. (Bhavini et al., 2012) The anti-*listerial* activity can also be improved significantly by utilizing the synergistic effect of thymol and propylene glycol after being encapsulated in whey-maltodextrin conjugate, resulting in much faster and improved bactericidal effect in skim, reduced and whole fat milk, compared with free thymol applied at same levels (Pan et al., 2014a).

Caseins are dominant dairy protein, comprising 80% of the total bovine milk protein, with the rest of being mainly whey and other minor globular proteins. Different from whey protein which is globular protein with defined three-dimensional structure, caseins are considered to have random coil spatial structure with flexible chains in aqueous solution. There are four types of caseins forming micelles in their native state in milk. After being depleted of calcium ions, which serve as bridges for keeping the integrity of casein micelles, caseins can be transformed into caseinate. Because of the balanced hydrophobic and hydrophilic amino acid portions, casein or caseinate has found valuable uses as a natural biopolymer for nanoencapsulation. Through direct high shear homogenization of thymol mixed in caseinate solution, nanoencapsulated thymol showed transparent appearance and significantly improved anti-*listerial* activity in milk of different fat levels (Pan et al., 2014a).

Zein is a group of prolamine proteins insoluble in water. Zein has been used extensively for nanoencapsulation of bioactive compounds due to its high portion of hydrophobic amino acid residues (>50%) that provides more binding sites for hydrophobic molecules, as well as its superb film formation property and resistance to water enables its applications in antimicrobial food packaging. Liquid-liquid dispersion or antisolvent method (which will be covered in later sections) is commonly used for preparation of zein nanoparticles. Spherical particles are formed with hydrophobic antimicrobial agents like essential oils entrapped into lipophilic chamber of zein nanoparticles (Parris et al., 2005). To improve the water redispersibility of antimicrobial encapsulated in zein nanoparticles, additional layers of coatings could be applied with material of better water binding capacity such as sodium caseinate and chitosan (Chen and Zhong, 2014; Zhang et al., 2014b). However, in cases where organic solvent is not desired, dispersibility of zein nanoparticles could also be improved by "green" method without involvement of organic solvent by utilizing the pH-dependent self-assembly properties of biopolymers (Pan and Zhong, 2016a).

5.3.2.2 Carbohydrate-Based Biopolymers

Carbohydrate-based biopolymers, depending upon the number of types of consisting unit monosaccharide, can be classified as homo-polysaccharide or hetero-polysaccharide. Polysaccharides also differ from one another with different degrees of polymerization (number of unit monosaccharide), molecular weight, hydrophobicity/hydrophilicity, electrostatic charges, gelation properties, and viscosity due to different chemical properties of the unit monosaccharides.

Starch is one of the most abundant storage polysaccharides in nature composed of single monosaccharide glucose linked by α-d-(1−4) and α-d-(1−6) linkages. Due to its high hydrophilicity, starches are normally needed to be hydrophobically modified into octenyl succinic anhydride (OSA) starch, acetylated starch, propyl starch, and PEGylated starch, etc. to provide hydrophobic cavity for lipophilic compounds. OSA starch-encapsulated lavandin essential oil demonstrated enhanced antibacterial activity against *E. coli*, *Staphylococcus aureus*, and *Bacillus cereus* due to the protection and controlled release of the essential oil (Varona et al., 2013b). D-limonene nanoencapsulated in modified starch was shown to possess significantly improved MIC against three microorganisms (*Saccharomyces cerevisiae, Escherichia coli, and Lactobacillus delbruecki*) due to the reduced particle size compared with pure D-limonene (Donsi et al., 2011).

Chitosan [(1-4)-2-amino-2-deoxy-β-D-glucan] is a polysaccharide produced by partially N-deacetylation of the natural polymer chitin and is positively charged in neutral to slight acidic pH. Because of its ideal properties like biodegradability, low cost, and nontoxicity, chitosan has been used extensively as a carrier for nanoencapsulation. Derivatives of chitosan have high efficacy against bacteria and fungi with their quaternary ammonium groups, by targeting at the cytoplasmic membrane of bacterial cells with their cationic side groups. Chitosan–alginate-based nanoparticles either with or without other antimicrobial agents exhibited superior antimicrobial activity against *Propionibacterium acnes in vitro* and showed low toxicity when cultured with eukaryotic cells (Friedman et al., 2013). When loaded with other natural food antimicrobial agents, the efficacy of chitosan-based antimicrobial nanoparticles can be significantly improved. Nisin-loaded chitosan–carageenan nanoparticles prepared through ionic complexation method demonstrated better antimicrobial activity both *in vitro* and in modeled tomato juice (Chopra et al., 2014).

Cyclodextrin is derived from starch through enzyme modification. Three types of cyclodextrin (α-, β-, γ-cyclodextrin) are commonly used for nanoencapsulation based on the size of their lipophilic cavity (6, 8, 10 Å, respectively), (Szejtli, 1998) which provide compartment for hydrophobic molecules through inclusion complexation and hydrophilic surface for dispersion in aqueous systems. The complex formed between β-cyclodextrin and essential oils showed significantly improved antimicrobial activity against different microorganisms and decreased MICs of antimicrobial agents, beyond their ability of flavor masking (Hill et al., 2013).

Gums such as pectin, alginate, and gum Arabic have also been used as carbohydrate-based biopolymers for encapsulation of antimicrobial agents. Essential oil-loaded alginate/cashew gum nanoparticles prepared through spray-drying demonstrated controlled release mechanism for over 30 h to release 95% of oil encapsulated (de Oliveira et al., 2014). Calcium alginate was used to nanoencapsulate lactoferrin to achieve a sustained release and provide protection against potential harsh conditions (Raei et al., 2015). Through self-emulsification of clove bud oil, gum arabic along with other biopolymers like whey protein concentrate, lecithin were able to form nanoparticles with clear appearance in aqueous systems and demonstrated significantly increased antimicrobial activity against *Escherichia coli*, *Listeria monocytogenes*, and *Salmonella*, indicating the effectiveness of nanoencapsulation for improved antimicrobial formulations (Luo et al., 2014).

5.3.3 Equipment-Based Nanoencapsulation of Food Antimicrobials

Electrospinning and electrospraying are two modes of electrohydrodynamic processes that utilize electrically charged jet to spin or spray polymer solution for production of fibers or particles (Bhushani and Anandharamakrishnan, 2014). The concentrations of polymer solution for two techniques are different. Fine droplets are formed at low solution concentration by jet due to varicose instability while elongation takes place at high concentration by stabilized jet due to whipping instability mechanism (Bhushani and Anandharamakrishnan, 2014). Because of the facile, cost effective, and flexible properties, the prospective of using these two techniques in food preservation and application is promising. Antimicrobial agents can also be directly entrapped into active ingredients/wall materials by these simple, one-step processes to obtain nanoencapsulated antimicrobials in dried form. The advantages of large area to volume ratio, amendable size and morphology, suitability in encapsulating heat-sensitive active compounds further facilitate the nanoencapsulation. The electrospinning technique has been used to prepare antimicrobial nanofibers to encapsulate the bacteriocins from probiotic bacteria to achieve sustained release during food processing and storage (Heunis et al., 2010). A water-insoluble fiber mat of blending zein with natural antimicrobial chitosan was formed by using electrospinning (Torres-Giner et al., 2009). The fibers had antimicrobial performance and could be of great interest in active food packaging applications. The electrospraying can be used for formation of edible antimicrobial coating on food materials directly. Peppermint essential oil was successfully encapsulated into an alginate biopolymer using the electrospray method and the essential oil was not degraded after encapsulation (Ghayempour and Mortazavi, 2014).

Nanospray-drying offers a new, simple approach for nanoparticle productions suitable for encapsulation applications. By equipping with a vibrating mesh spray technology and an electrostatic particle collector, fine droplets can be generated and fine particles can be then collected (Lee et al., 2011). Nano-size ultrafine powder of eugenol nanoemulsions was prepared by this technology, which exhibited better morphological properties and redispersibility than freeze drying (Hu et al., 2016). Although the utilization of this novel technology on the encapsulation of natural antimicrobial agents is currently limited, it has attracted the interest of research groups in food fields due to the favorable properties of the antimicrobial delivery systems such as narrow size distribution, ease of scale-up and high yield, etc. (Gu et al., 2015).

5.3.4 Nanocomposites of Food Antimicrobial Agents

Recently, nanocomposites consisting of inorganic nanolayers of layered silicate have evoked intense research interest to be used as antimicrobial carriers. Their applications have been exploring. The nanoclays with and without modification can be used as the nanofiller to prepare hybrid polymer nanocomposite material. These materials usually have improved mechanical, thermal, electrical, barrier, and optical properties and hence have the potential to be used in active antimicrobial food packaging applications. Montmorillonite nanoclay, consisting of two tetrahedral sheets of silica fused to an edge-shared octahedral sheet of alumina, was incorporated into alginate or chitosan and formed bio-nanocomposites with encapsulated essential oils (Alboofetileh et al., 2014; Abdollahi et al., 2012). The obtained nanocomposite films showed increased mechanical properties and antimicrobial activities. Carvacrol was employed in the methyl cellulose-based nanocomposites and produced a film with enhanced antimicrobial activity and a controlled release of carvacrol during storage (Tunç and Duman, 2011). The antimicrobial nanocomposites can also be deposited onto the material surface as an additional layer by various methods (Nikiforov et al., 2016).

5.3.5 Other Nanoencapsulation Technologies for Food Antimicrobial Agents

Other nanotechnology strategies such as nanogels have also been studied to encapsulate antimicrobial agents. Nanogels including nanohydrogels and nanoorganogels, three-dimensional networks formed via covalent linkages or self-assembly, are promising carriers for natural antimicrobials due to their high-loading capacity, high stability, and unique release properties. Encapsulation of antimicrobial agents such as *Mentha piperita* essential oils within a nanogel has other benefits such as controlled and sustained release of a centain amount of oils from the carrier for a longer time (Beyki et al., 2014). Essential oils encapsulated in chitosan–cinnamic acid nanogel showed high antifungal activity under nonsealed condition. Nanolaminates, nanotubes formed from globular proteins and polysaccharides also showed the potential for the food applications with encapsulated antimicrobial agents (Weiss et al., 2006; Graveland-Bikker and De Kruif, 2006).

5.4 RELEASE, CHARACTERIZATION, AND SAFETY OF NANOENCAPSULATED FOOD ANTIMICROBIAL AGENTS

5.4.1 Controlled Release of Antimicrobial Agents from Nanocapsules

Delivering the bioactive compounds (e.g., antimicrobial agents) precisely and safely at a right period of time to the targeted system to achieve the controlled release is a yardstick in the design and development of nanocapsules. The release is conducted by one or a combination of three different mechanisms including (1) diffusion, (2) erosion, and (3) swelling (Fathi et al., 2012). For lipid-based nanoencapsulation, the first one is more likely to occur, while the latter two are more related to hydrophilic-based (proteins and polysaccharides) nanocarriers. Fickian diffusion principle (Equation 5.1) is used to model the release from a structure of core–shell or matrix types. However, many factors including the interactions between encapsulated compound with capsule, the infusion of water and other cosolutes, etc. can change the structure of capsule during the release process (Zhong et al., 2015). This can cause the formation of "channels" inside particles that change release characteristics, and hence modified Fickian or non-Fickian diffusion models are usually used to evaluate the release kinetics from nanocapsules.

The transport of thymol and carvacrol from zein nanoparticles was studied using Korsmeyer–Peppas model that thyme oil exhibited a combination of Fickian and non-Fickian diffusion mechanism while carvacrol followed type II super transport where the diffusion rate is much higher than that of polymer relaxation (da Rosa et al., 2015). When the shell curvature radius is typically much bigger than encapsulated molecules in a core–shell type structure, the Fick's law for a planar geometry can be applied (Equation 5.2) (Zhong et al., 2015). This equation is also widely used to determine the release kinetics of encapsulated antimicrobial agents from food packaging and coatings (Choi et al., 2005; Del Nobile et al., 2008).

$$\frac{\partial c}{\partial t} = D\frac{1}{r^2}\frac{\partial}{\partial r}\left(r^2\frac{\partial c}{\partial r}\right) \tag{5.1}$$

$$\left(\frac{\partial c}{\partial t}\right)_x = D\left(\frac{\partial^2 c}{\partial x^2}\right)_t, \tag{5.2}$$

where c is the concentration of encapsulated molecule at a radial distance r from the sphere center, t is time, D is the diffusion coefficient of the molecule in the matrix, and x is the space coordinate in the direction of transport.

Triggered release is the burst release of an cncapsulated compound from nanosystems responding to various internal and external changes, such as pH, specific enzymes, temperature, ultrasound, and light, etc. (Esser-Kahn et al., 2011). Allyl isothiocyanate, a naturally occurring antimicrobial compound found in cruciferae, was encapsulated in soy protein and poly(lactic acid) electrospun fibers. The release of allyl isothiocyanate was negligible under dry conditions, but was triggered at 75% RH (Vega-Lugo and Lim, 2009). The relative humidity in-package of fresh-cut tomatoes has also been demonstrated as a trigger to release antimicrobial garlic oil from β-cyclodextrin capsules to reduce the microbial growth (Ayala-Zavala and González-Aguilar, 2010). The pH variations associated with growth of bacteria such as *S. epidermidis* was used as an internal pH trigger to release antimicrobial agents from hydrogels (Pavlukhina et al., 2010).

5.4.2 Instrumental Analysis and Characterization of Antimicrobial-Loaded Nanocapsules

5.4.2.1 *Physicochemical Properties*

It is considered that the effects of nanomaterials on antimicrobial activity are largely related to their physicochemical properties such as size, shape, structure, state of aggregation, dispersion and sorption, etc. (Magnuson et al.,

2011). The equipment like transmission electron microscopy (TEM), scanning electron microscopy (SEM), and atomic force microscopy (AFM) are the most commonly used techniques to image the nanocapsules (Pan and Zhong, 2016b). Particle dimension and size distribution can be determined by scattering techniques with the light source of laser light, X-rays, or neutrons. Among these techniques, laser light scattering is the most readily available method for size characterization (Hassellöv et al., 2008). Surface properties such as surface charge are important characteristics of colloidal nanoparticle in aqueous solutions because they determine types and magnitudes of interparticle interactions (Ninham, 1999). Zeta potential and surface hydrophobicity, the two most commonly measured parameters, are used to determine the significance of either electrostatic interactions or hydrophobic interactions on the functionality and stability of nanocapsules. Zeta potential can be measured by electrophoresis and laser techniques, while the latter is usually determined by the binding of fluorescence probes or hydrophobic chromatography (Azevedo et al., 2014; Pan and Zhong, 2016b). The surface composition can also be analyzed by X-ray photoelectron spectroscopy or microscopy coupled with infrared spectroscopy (Jafari et al., 2008a).

5.4.2.2 Encapsulation Efficiency and Loading Capacity

Encapsulation efficiency is usually defined as the percentage of a molecule encapsulated in the nanocapsule with respect to the total amount used (Equation 5.3). For nanoemulsions and liposomes, encapsulation efficiency is expected to be high because of the insolubility of the encapsulated lipophilic molecule in water solution, while some systems may exhibit low efficiency due to the overall low yield during manufacture such as the volatile compounds losses significantly during spray-drying (Jafari et al., 2008b). The encapsulation efficiency can be affected by molecular structures, pH, ionic strength, and cosolutes, etc. All these four factors can be summarized as four groups: (1) properties of capsule materials; (2) properties of core materials/encapsulated molecules; (3) characterizations of nanocapsules in aqueous solution; and (4) conditions of the preparation into dry form (powder, fiber, etc.). The methods used to determine encapsulation efficiency may be different for different delivery systems and encapsulated molecules. The loading capacity may also be determined as another important parameter for nanocapsules (Equation 5.4). This concept may not be applicable for nanoemulsions and nanoliposomes in aqueous solutions (Amidi et al., 2006).

$$\text{Encapsulation efficiency (\%)} = \frac{\text{Mass of a molecule enpsulated in a nanocapsule}}{\text{Total mass of the molecule used before encapsulation}} \tag{5.3}$$

$$\text{Loading capacity (\%)} = \frac{\text{Mass of a molcculc enpsulated in a nanocapsule}}{\text{Total mass of the nanocapsule in dry form}}. \tag{5.4}$$

5.4.2.3 In Vitro *and* In Vivo *Antimicrobial Activity Evaluation*

The antimicrobial activity of nanocapsules is a key indicator to evaluate the ability of nanocapsules to incorporate and release antimicrobial agents, and the applicability of nanocapsules in real food applications. Therefore, qualitative and quantitative measurements are needed to characterize the antimicrobial activity *in vitro* and *in vivo*. Many diffusion and dilution methods such as disk diffusion, agar dilution, and microbroth dilution methods have been applied to compare the MIC of antimicrobial agents with and without nanoencapsulation *in vitro*. The latter one is considered more sensitive than other two methods, and hence can be more appropriate for evaluating the antimicrobial effect of nanocapsules in aqueous solution (Klančnik et al., 2010). The minimum bactericidal concentration (MBC) of nanocapsules such as nanoemulsions, nanoliposomes can be further determined by using the microbroth dilution method (Alipour et al., 2009; Zhang et al., 2016). The inactivation kinetics of microorganisms in the presence of encapsulated antimicrobial agents are usually determined in comparison to free antimicrobial agents that characterize the sustained release of nanocapsules in aqueous base (Donsì et al., 2011). Disk diffusion method is usually used to qualitatively evaluate the nanocapsules with fiber structure such as electrospun nanofibers/mat, nanocapsule-incorporated films/coatings that the nanocomposite films/fibers can be punched into discs and placed on the plates with inoculated bacteria (Abdollahi et al., 2012; Torres-Giner et al., 2009). The disk diffusion method can also be used to evaluate antimicrobial effect of antimicrobial nanogels by impregnating paper disks within gel-forming solutions (Farag and Mohamed, 2012). The *in vivo* efficacy of nanoencapsulated antimicrobial agents is assessed by different methods according to their potential applications in different food systems. Nanoemulsions, nanoliposomes, rehydrated nanoparticles with encapsulated essential oils can be directly added into food systems such as milk as alternative preservatives to test their effects on the growth kinetics of bacteria *in vivo* (Chen et al., 2015; da Silva Malheiros et al., 2010; Xue et al., 2015). These antimicrobial nanocapsules can also be diluted in water and used to wash inoculated vegetables and fruits as alternative washing solutions (Ayala-Zavala and González-Aguilar, 2010; Zhang et al., 2016). For the use in food packaging and coating applications, the assessment of nanocapsules-incorporated coatings or films can be done by immersing the inoculated food systems into the film-forming solutions, spraying the film-form solutions on the surface of food, or packaging the food with films (Huang et al., 2012) (Brasil et al.,

2012; Zhang et al., 2015). Since sustained release of antimicrobial agents is a significant benefit of nanoencapsulation, *in vivo* efficacy of nanocapsules on shelf life of food products is usually investigated. Above methods can be applied on different food systems first and then the food systems under practical storage conditions are monitored during a storage period (Brasil et al., 2012; Zhang et al., 2015).

5.4.3 Safety of Nanoencapsulated Food Antimicrobial Agents

Apart from the advantages of nanotechnology in food industry, the lack of knowledge about the impacts of nano-sized materials on human health and the environment has raised some safety concerns (Cushen et al., 2012). Although many foods contain nanoscale components that have been consumed safely for a long time such as milk, there is a scientific consensus that the toxicity of formulated/engineered nanostructures could be different and hence create new risks to human health and environment (Taylor, 2014). And also, there are some concerns on utilization of very small lipid droplets in foods since they may change the extent or route of absorption of lipophilic components (McClements, 2011). There is also a risk of incorporating natural antimicrobial agents into food applications. For example, some botanicals may have no significant history of use as food ingredients and typical toxicological information such as acceptable daily intake (ADI) or no observed adverse effect level (NOEL) may not be available for them (Negi, 2012). Therefore, for the successful exploitation of the nanoencapsulated natural antimicrobials as food preservatives, further studies on both the antimicrobial agents and nanocapsules are clearly needed.

5.5 APPLICATIONS OF NANOENCAPSULATED FOOD ANTIMICROBIAL AGENTS

Promising applications by using nanoencapsulation technology to enhance the efficacy of antimicrobials in foods are being developed. The purposes of using nanoencapsulation method include the following four reasons:

1. To stabilize the volatile antimicrobials, for example, essential oils, against evaporation during processing
2. To reduce the interaction of antimicrobials with food matrices
3. To control the release rate of antimicrobials in food matrices, so as to prolong the exposure of microbes to the antimicrobials
4. To improve the solubility of antimicrobials in undesirable foods, so as to broaden the application of antimicrobials.

5.5.1 Aqueous Food Systems

The two most consumed aqueous food systems can be classified as carbohydrate-based products and dairy-based products. Carbohydrate-based products, such as fruit juices and soft drinks, usually have some amount of carbohydrate, but no or very little amount of proteins and lipids. The problems of applying antimicrobials for carbohydrate-based products are incorporating of hydrophobic compound into the system and the effect of acidic pH on the stability of antimicrobials. With regard todairy-based products, due to the large amount of proteins, and fat compositions in some products, the interaction of antimicrobials with these complicated food compounds and the stability of compounds during pasteurization are challenges as well. Nanoencapsulation has been applied to overcome these challenges. Researchers at the University of Tennessee-Knoxville found essential oils after encapsulation were more efficient at preventing the growth of *E. coli* O157:H7 and *L. monocytogenes* in milk of different fat levels than adding the oils directly (Chen et al., 2015; Pan et al., 2014b; Shah et al., 2012; Shah et al., 2013; Xue et al., 2013). It was explained as the improved dispersibility and faster mass transfer of essential oils to the aqueous phase after nanoencapsulation (Pan et al., 2014b). In addition, the encapsulation masks the strong aromatic odors of essential oils in some extent (Chen et al., 2015). Nisin and thymol oil coencapsulated by zein showed better reduction than free antimicrobials at the 4 h time point in 2% fat-reduced milk (Xiao et al., 2011). The terpenes nanocapsules were tested in orange and pear juices, inoculated with *L. delbrueckii* (Donsì et al., 2011). The addition of the nanoencapsulated terpenes at 1.0 g/l was able to delay the microbial growth. At 5.0 g/l, the addition of nanoencapsulated terpen completely inactivate the microorganisms while minimally altering the organoleptic properties of the fruit juices (Donsì et al., 2011). In some situations, the application of antimicrobials after encapsulation in food systems have a verse antimicrobial effect than free antimicrobials. The addition of micelle-encapsulated eugenol to milk resulted to be less or as inhibitory as unencapsulated eugenol (Gaysinsky et al., 2007). Thyme oil encapsulated in nisin-saccharide also showed less effect than free thyme oil and free nisin combinations (Chen et al., 2014). Therefore, biopolymers used to encapsulate antimicrobials need to be chosen carefully so that the partitioning of antimicrobials with bacteria is favored rather with capsules and food matrix components.

5.5.2 Solid Food Matrix

Antimicrobials after nanoencapsulation are also applied in solid food matrix to improve the homogeneous mixing and prolong the shelf life. One of the studies that researchers are interested is the application of liposome encapsulated antimicrobials in solids food models. The encapsulation of pediocin AcH within liposomes resulted in a 27.5% and 28.9% average increase in the recovery of pediocin activity in slurries of heated beef muscle and tallow, respectively, compared to otherwise similar beef slurries containing free pediocin (Degnan and Luchansky, 1992). In a similar investigation, the authors evaluated liposomes as pediocin AcH vectors in slurries of nonfat dry milk, butterfat, beef muscle tissue, or beef tallow (prepared in deionized water, at final concentration of 10%, w/v), and a 29%–62% increase in pediocin recovery was reported, compared to free pediocin AcH (Degnan et al., 1993). Since nisin is approved for food use, a great amount of investigation is directed toward encapsulation and application of this antimicrobial peptide. Laridi et al. (2003) investigated the stability and the effects of nisin Z entrapped into liposomes during Cheddar cheese manufacture. Encapsulated nisin Z lowered the viable counts of lactococci and did not severely disturb Cheddar cheese fermentation. It was stable throughout the Cheddar cheese-making temperature cycle. Benech, Kheadr, Lacroix et al. (2002) and Benech et al. (2003) added liposome-encapsulated nisin to cheese milk, evaluating the antimicrobial activity, physicochemical property, and sensory attributes during 6 months of Cheddar cheese ripening (Benech et al., 2002, 2003). After 6 months, less than 10 CFU/g of *L. monocytogenes* were found and 90% of the initial nisin activity was recovered in nisin-loaded liposomes, whereas ~103 CFU/g of *L. monocytogenes* and only 12% recovery of nisin activity were reported in cheeses manufactured with a nisin-producing mixed starter culture. This was correlated with a high activity and stability of encapsulated nisin compared to nisin produced by the nisinogenic strain (Benech et al., 2002). Further study showed that liposome-encapsulated nisin did not affect cheese proteolysis, rheology, and sensory characteristics, while the proteolysis and lipolysis were increased in the incorporation of a nisinogenic strain into cheese starter culture (Benech et al., 2003).

5.5.3 Fresh Produces and Vegetables

Foodborne disease outbreaks linked to consumption of fresh fruits and vegetables are increasing (Sivapalasingam et al., 2004). Due to the variation of surface properties of fruits and vegetables, chlorine is not effective in controlling all the pathogens that may attach to the produces or vegetables. Meanwhile, it is preferred that antimicrobials could function from the harvesting to folk. Encapsulated antimicrobials have been widely studied to enhance the safety issue of fresh produces and vegetables. Chitosan coatings with allyl isothiocyanate and nisin resulted in a 4.8 log CFU/g reduction of *Salmonella*, and it completely inactivated mold and yeast on cantaloupe at day 1 and no regrowth occurred even up to 14 days of storage (Chen et al., 2012). Cinnamon bark oil was incorporated into the alginate films and used to coat cantaloupes. Results showed that coatings with cinnamon bark oil enhanced the quality as well as the microbiological safety (Zhang et al., 2015). In authors' another study, thyme oil emulsion was used to wash cantaloupes, and similar trend was obtained. The inoculated pathogen population was gradually reduced during storage at room temperature after the emulsion rinsing (Zhang et al., 2016). *Trans*-cinnamaldehyde preencapsulated in beta-cyclodextrin was dispersed into the multilayer coating solutions

(Sipahi et al., 2013). Fresh-cut watermelon samples were then coated using the layer-by-layer dipping technique and stored at 4°C for 15 days. Coated samples maintained their quality for 13–15 days compared to uncoated fruits (7 days). Limonene, oregano essential oil, red thyme, peppermint oil, and lemongrass oil were individually incorporated into modified chitosan films as coating materials for strawberries. Formulations based on modified chitosan containing limonene provided a better preservation than other formulations, reducing the decay level. On day 8, 45% of strawberries coating containing limonene had some level of decay while the percentage was 65%, 75%, and 90% of decay for chitosan itself coating, chitosan with peppermint oil coating and uncoated strawberries, respectively (Vu et al., 2011). A total of 0.05% oregano oil emulsified by Tween 80 showed up to 3.44, 2.31, and 3.05log CFU/g reductions in *L. monocytogenes*, *S. Typhimurium*, and *E. coli O157:H7*, respectively, on lettuce after 3 days storage at 4°C (Bhargava et al., 2015). On spinach leaves, a significant inhibition of inoculated *E. coli* O157:H7 and *S. Saintpaul* was achieved by washing with carvacrol on eugenol emulsified by sodium dodecyl sulfate (SDS) or CytoGuard LA 20, with no difference observed between spray and immersion application (Ruengvisesh et al., 2015).

5.6 CONCLUSION

In conclusion, nanoencapsulation has great promise as a novel strategy to deliver antimicrobial agents. Numerous encapsulation methods have been studied to fabricate appropriate systems for different food applications. The advantages such as enhanced bioavailability, solubility, and stability of antimicrobial agents have increased the focus of research and development efforts on real food systems and samples. Meanwhile, the nanoencapsulated antimicrobials have showed great effectiveness in controlling growth of pathogenic microorganisms and spoilage in foodstuff, and sustained release properties to prolong the shelf life. However, there are a number of challenges that need to be overcome before the nanoencapsulated antimicrobials can be more widely used in food industries.

REFERENCES

Abdollahi, M., Rezaei, M., & Farzi, G. (2012). A novel active bionanocomposite film incorporating rosemary essential oil and nanoclay into chitosan. *Journal of Food Engineering*, *111* (2), 343–350.

Abdullah, S. U., Badaruddin, M., Ali, R., & Riaz, M. N. (2010). Effect of elementary and advanced glycation products of nisin on its preservative efficacy and digestibility. *Food Chemistry*, *122*(4), 1043–1046.

Alboofetileh, M., Rezaei, M., Hosseini, H., & Abdollahi, M. (2014). Antimicrobial activity of alginate/clay nanocomposite films enriched with essential oils against three common foodborne pathogens. *Food Control*, *36*(1), 1–7.

Alipour, M., Suntres, Z. E., & Omri, A. (2009). Importance of DNase and alginate lyase for enhancing free and liposome encapsulated aminoglycoside activity against Pseudomonas aeruginosa. *Journal of Antimicrobial Chemotherapy*, dkp165.

Alzoreky, N. S., & Nakahara, K. (2003). Antibacterial activity of extracts from some edible plants commonly consumed in Asia. *International Journal of Food Microbiology*, *80*(3), 223–230.

Amidi, M., Romeijn, S. G., Borchard, G., Junginger, H. E., Hennink, W. E., & Jiskoot, W. (2006). Preparation and characterization of protein-loaded N-trimethyl chitosan nanoparticles as nasal delivery system. *Journal of Controlled Release*, *111*(1), 107–116.

Ayala-Zavala, J. F., & González-Aguilar, G. A. (2010). Optimizing the use of garlic oil as antimicrobial agent on fresh-cut tomato through a controlled release system. *Journal of Food Science*, *75*(7), M398–M405.

Azevedo, M. A., Bourbon, A. I., Vicente, A. A., & Cerqueira, M. A. (2014). Alginate/chitosan nanoparticles for encapsulation and controlled release of vitamin B 2. *International Journal of Biological Macromolecules*, *71*, 141–146.

Bagamboula, C. F., Uyttendaele, M., & Debevere, J. (2004). Inhibitory effect of thyme and basil essential oils, carvacrol, thymol, estragol, linalool and p-cymene towards Shigella sonnei and S-flexneri. *Food Microbiology*, *21*(1), 33–42.

Bakkali, F., Averbeck, S., Averbeck, D., & Waomar, M. (2008). Biological effects of essential oils—A review. *Food and Chemical Toxicology*, *46*(2), 446–475.

Barbiroli, A., Bonomi, F., Capretti, G., Iametti, S., Manzoni, M., Piergiovanni, L., & Rollini, M. (2012). Antimicrobial activity of lysozyme and lactoferrin incorporated in cellulose-based food packaging. *Food Control*, *26*(2), 387–392.

Barmpalia, I. M., Geornaras, I., Belk, K. E., Scanga, J. A., Kendall, P. A., Smith, G. C., & Sofos, J. N. (2004). Control of Listeria monocytogenes on Frankfurters with Antimicrobials in the Formulation and by Dipping in Organic Acid Solutions. *Journal of Food Protection*, *67*(11), 2456–2464.

Benech, R. O., Kheadr, E. E., Lacroix, C., & Fliss, I. (2002). Antibacterial activities of nisin Z encapsulated in liposomes or produced in situ by mixed culture during cheddar cheese ripening. *Applied and Environmental Microbiology*, *68*(11), 5607–5619.

Benech, R. O., Kheadr, E. E., Lacroix, C., & Fliss, I. (2003). Impact of nisin producing culture and liposome-encapsulated nisin on ripening of Lactobacillus added-Cheddar cheese. *Journal of Dairy Science*, *86*(6), 1895–1909.

Beyki, M., Zhaveh, S., Khalili, S. T., Rahmani-Cherati, T., Abollahi, A., Bayat, M., Tabatabaei, M., & Mohsenifar, A. (2014). Encapsulation of Mentha piperita essential oils in chitosan–cinnamic acid nanogel with enhanced antimicrobial activity against Aspergillus flavus. *Industrial Crops and Products*, *54*, 310–319.

Bhargava, K., Conti, D. S., da Rocha, S. R. P., & Zhang, Y. (2015). Application of an oregano oil nanoemulsion to the control of foodborne bacteria on fresh lettuce. *Food Microbiology*, *47*, 69–73.

Bhavini, S., Davidson, P. M., & Qixin, Z. (2012). Nanocapsular dispersion of thymol for enhanced dispersibility and increased antimicrobial effectiveness against *Escherichia coli* O157:H7 and *Listeria monocytogenes* in model food systems. *Applied and Environmental Microbiology*, *78*, 8448–8453.

Bhushani, J. A., & Anandharamakrishnan, C. (2014). Electrospinning and electrospraying techniques: Potential food based applications. *Trends in Food Science & Technology*, *38*(1), 21–33.

Bilbao-Sainz, C., Chiou, B. S., Valenzuela-Medina, D., Imam, S. H., Vega-Gálvez, A., & Orts, W. J. (2016). Biopolymer films to control fusarium dry rot and their application to preserve potato tubers. *Journal of Applied Polymer Science*, *133*, 40.

Blanco-Padilla, A., Soto, K. M., Hernández Iturriaga, M., & Mendoza, S. (2014). Food antimicrobials nanocarriers. *The Scientific World Journal, 2014*, 1–11.

Bonev, B. B., Chan, W. C., Bycroft, B. W., Roberts, G. C. K., & Watts, A. (2000). Interaction of the lantibiotic nisin with mixed lipid bilayers: A ^{31}P and ^{2}H NMR study. *Biochemistry, 39*(37), 11425–11433.

Bozin, B., Mimica-Dukic, N., Simin, N., & Anackov, G. (2006). Characterization of the volatile composition of essential oils of some lamiaceae spices and the antimicrobial and antioxidant activities of the entire oils. *Journal of Agricultural and Food Chemistry, 54*(5), 1822–1828.

Brasil, I., Gomes, C., Puerta-Gomez, A., Castell-Perez, M., & Moreira, R. (2012). Polysaccharide-based multilayered antimicrobial edible coating enhances quality of fresh-cut papaya. *LWT-Food Science and Technology, 47*(1), 39–45.

Breukink, E., & de Kruijff, B. (2006). Lipid II as a target for antibiotics. *Nature Reviews Drug Discovery, 5*(4), 321–332.

Burt, S. (2004). Essential oils: their antibacterial properties and potential applications in foods—A review. *International Journal of Food Microbiology, 94*(3), 223–253.

Campos, F. F., Calpena Campmany, A. C., Delgado, G. R., Serrano, O. L., & Naveros, B. C. (2012). Development and characterization of a novel nystatin-loaded nanoemulsion for the buccal treatment of candidosis: Ultrastructural effects and release studies. *Journal of Pharmaceutical Sciences, 101*(10), 3739–3752.

Carpenter, C. E., Smith, J. V., & Broadbent, J. R. (2011). Efficacy of washing meat surfaces with 2% levulinic, acetic, or lactic acid for pathogen decontamination and residual growth inhibition. *Meat Science, 88*(2), 256–260.

Carson, C. F., Hammer, K. A., & Riley, T. V. (2006). *Melaleuca alternifolia* (Tea Tree) oil: a review of antimicrobial and other medicinal properties. *Clinical Microbiology Reviews, 19*(1), 50–62.

Ceylan, E., Fung, D. Y. C., & Sabah, J. R. (2004). Antimicrobial Activity and Synergistic Effect of Cinnamon with Sodium Benzoate or Potassium Sorbate in Controlling Escherichia coli O157:H7 in Apple Juice. *Journal of Food Science, 69*(4), FMS102–FMS106.

Chaieb, K., Hajlaoui, H., Zmantar, T., Ben Kahla-Nakbi, A., Rouabhia, M., Mahdouani, K., & Bakhrouf, A. (2007). The chemical composition and biological activity of clove essential oil, *Eugenia caryophyllata* (*Syzigium aromaticum* L. myrtaceae): a short review. *Phytotherapy Research, 21*(6), 501–506.

Chang, Y., McLandsborough, L., & McClements, D. J. (2013). Physicochemical properties and antimicrobial efficacy of carvacrol nanoemulsions formed by spontaneous emulsification. *Journal of Agricultural and Food Chemistry, 61*(37), 8906–8913.

Chen, H., Zhang, Y., & Zhong, Q. (2015). Physical and antimicrobial properties of spray-dried zein–casein nanocapsules with co-encapsulated eugenol and thymol. *Journal of Food Engineering, 144*, 93–102.

Chen, H. Q., Davidson, P. M., & Zhong, Q. X. (2014). Antimicrobial properties of nisin after glycation with lactose, maltodextrin and dextran and the thyme oil emulsions prepared thereof. *International Journal of Food Microbiology, 191*, 75–81.

Chen, H. Q., & Zhong, Q. X. (2014). Processes improving the dispersibility of spray-dried zein nanoparticles using sodium caseinate. *Food Hydrocolloids, 35*, 358–366.

Chen, W., Jin, T. Z., Gurtler, J. B., Geveke, D. J., & Fan, X. (2012). Inactivation of Salmonella on whole cantaloupe by application of an antimicrobial coating containing chitosan and allyl isothiocyanate. *International Journal of Food Microbiology, 155*(3), 165–170.

Chipley, J. R. (2005). Sodium benzoate and benzoic acid. In P. M. Davidson, J. Sofos, & L. Branen (Eds.), *Antimicrobials in food* (3rd ed., pp. 11–38). Boca Raton, FL: CRC Press, Taylor & Francis Group, LLC.

Choi, J., Choi, W., Cha, D., Chinnan, M., Park, H. J., Lee, D., & Park, J. (2005). Diffusivity of potassium sorbate in κ-carrageenan based antimicrobial film. *LWT-Food Science and Technology*, *38*(4), 417–423.

Chopra, M., Kaur, P., Bernela, M., & Thakur, R. (2014). Surfactant assisted nisin loaded chitosan-carageenan nanocapsule synthesis for controlling food pathogens. *Food Control*, *37*, 158–164.

Colas, J. C., Shi, W., Rao, V. M., Omri, A., Mozafari, M. R., & Singh, H. (2007). Microscopical investigations of nisin-loaded nanoliposomes prepared by Mozafari method and their bacterial targeting. *Micron*, *38*(8), 841–847.

Cortés-Rojas, D. F., Souza, C. R., & Oliveira, W. P. (2014). Encapsulation of eugenol rich clove extract in solid lipid carriers. *Journal of Food Engineering*, *127*, 34–42.

Cowan, M. M. (1999). Plant Products as Antimicrobial Agents. *Clinical Microbiology Reviews*, *12*(4), 564–582.

Cushen, M., Kerry, J., Morris, M., Cruz-Romero, M., & Cummins, E. (2012). Nanotechnologies in the food industry—Recent developments, risks and regulation. *Trends in Food Science & Technology*, *24*(1), 30–46.

da Rosa, C. G., Maciel, M. Vd. O. B., de Carvalho, S. M., de Melo, A. P. Z., Jummes, B., da Silva, T., Martelli, S. M., Villetti, M. A., Bertoldi, F. C., & Barreto, P. L. M. (2015). Characterization and evaluation of physicochemical and antimicrobial properties of zein nanoparticles loaded with phenolics monoterpenes. *Colloids and Surfaces A: Physicochemical and Engineering Aspects*, *481*, 337–344.

da Silva Malheiros, P., Daroit, D. J., da Silveira, N. P., & Brandelli, A. (2010). Effect of nanovesicle-encapsulated nisin on growth of Listeria monocytogenes in milk. *Food Microbiology*, *27*(1), 175–178.

Davidson, P., & Zivanovic, S. (2003). The use of natural antimicrobials. *Food Preservation techniques*, 5–30.

Davidson, P. M., Critzer, F. J., & Taylor, T. M. (2013a). Naturally occurring antimicrobials for minimally processed foods. *Annual Review of Food Science and Technology*, *4*, 163–190.

Davidson, P. M., Taylor, T. M., & Schmidt, S. E. (2013b). Chemical preservatives and natural antimicrobial compounds. In M. P. Doyle, & R. L. Buchanan (Eds.), *Food microbiology: Fundamentals and frontiers* (pp. 765–801). Washington: American Society of Microbiology.

Degnan, A. J., Buyong, N., & Luchansky, J. B. (1993). Antilisterial Activity of Pediocin Ach in Model Food Systems in the Presence of an Emulsifier or Encapsulated within Liposomes. *International Journal of Food Microbiology*, *18*(2), 127–138.

Degnan, A. J., & Luchansky, J. B. (1992). Influence of Beef Tallow and Muscle on the Antilisterial Activity of Pediocin Ach and Liposome-Encapsulated Pediocin Ach. *Journal of Food Protection*, *55*(7), 552–554.

Dehnad, D., Mirzaei, H., Emam-Djomeh, Z., Jafari, S. M., & Dadashi, S. (2014). Thermal and antimicrobial properties of chitosan-nanocellulose films for extending shelf life of ground meat. *Carbohydrate Polymers*, *109*, 148–154.

Del Nobile, M., Conte, A., Incoronato, A., & Panza, O. (2008). Antimicrobial efficacy and release kinetics of thymol from zein films. *Journal of Food Engineering*, *89*(1), 57–63.

de Oliveira, E. F., Paula, H. C. B., & de Paula, R. C. M. (2014). Alginate/cashew gum nanoparticles for essential oil encapsulation. *Colloids and Surfaces B: Biointerfaces*, *113*, 146–151.

Donsi, F., Annunziata, M., Sessa, M., & Ferrari, G. (2011). Nanoencapsulation of essential oils to enhance their antimicrobial activity in foods. *LWT-Food Science and Technology*, *44*(9), 1908–1914.

Esser-Kahn, A. P., Odom, S. A., Sottos, N. R., White, S. R., & Moore, J. S. (2011). Triggered release from polymer capsules. *Macromolecules*, *44*(14), 5539–5553.

Farag, R. K., & Mohamed, R. R. (2012). Synthesis and characterization of carboxymethyl chitosan nanogels for swelling studies and antimicrobial activity. *Molecules*, *18*(1), 190–203.

Fathi, M., Mozafari, M., & Mohebbi, M. (2012). Nanoencapsulation of food ingredients using lipid based delivery systems. *Trends in Food Science & Technology*, *23*(1), 13–27.

Food and Drug Administration (FDA) (1998). Substances generally recognized as safe. In Fa. D. Administration (Ed.). (p. 26). Code of Federal Regulations.

Friedman, A. J., Phan, J., Schairer, D. O., Champer, J., Qin, M., Pirouz, A., Blecher-Paz, K., Oren, A., Liu, P. T., & Modlin, R. L. (2013). Antimicrobial and anti-inflammatory activity of chitosan–alginate nanoparticles: A targeted therapy for cutaneous pathogens. *Journal of investigative Dermatology*, *133*(5), 1231–1239.

Galanakis, C. M. (2015). *Food Waste Recovery: Processing Technologies and Industrial Techniques*. San Diego, CA: Academic Press.

Gaysinsky, S., Taylor, T. M., Davidson, P. M., Bruce, B. D., & Weiss, J. (2007). Antimicrobial efficacy of eugenol microemulsions in milk against *Listeria monocytogenes* and *Escherichia coli* O157: H7. *Journal of Food Protection*, *70*(11), 2631–2637.

Ghayempour, S., & Mortazavi, S. (2014). Antibacterial activity of peppermint fragrance micro–nanocapsules prepared with a new electrospraying method. *Journal of Essential Oil Research*, *26*(6), 492–498.

González-Chávez, S. A., Arévalo-Gallegos, S., & Rascón-Cruz, Q. (2009). Lactoferrin: structure, function and applications. *International Journal of Antimicrobial Agents*, *33*(4), 301, e301-301. e308.

Goy, R. C., Britto, Dd, & Assis, O. B. (2009). A review of the antimicrobial activity of chitosan. *Polímeros*, *19*(3), 241–247.

Graveland-Bikker, J., & De Kruif, C. (2006). Unique milk protein based nanotubes: food and nanotechnology meet. *Trends in Food Science & Technology*, *17*(5), 196–203.

Gu, B., Linehan, B., & Tseng, Y. C. (2015). Optimization of the Büchi B-90 spray drying process using central composite design for preparation of solid dispersions. *International Journal of Pharmaceutics*, *491*(1), 208–217.

Halwani, M., Yebio, B., Suntres, Z., Alipour, M., Azghani, A., & Omri, A. (2008). Co-encapsulation of gallium with gentamicin in liposomes enhances antimicrobial activity of gentamicin against Pseudomonas aeruginosa. *Journal of Antimicrobial Chemotherapy*, *62*(6), 1291–1297.

Harris, K., Miller, M. F., Loneragan, G. H., & Brashears, M. M. (2006). Validation of the use of organic acids and acidified sodium chlorite to reduce Escherichia coli O157 and Salmonella Typhimurium in beef trim and ground beef in a simulated processing environment. *Journal of Food Protection*, *69*(8), 1802–1807.

Hassellöv, M., Readman, J. W., Ranville, J. F., & Tiede, K. (2008). Nanoparticle analysis and characterization methodologies in environmental risk assessment of engineered nanoparticles. *Ecotoxicology*, *17*(5), 344–361.

Heunis, T., Botes, M., & Dicks, L. (2010). Encapsulation of Lactobacillus plantarum 423 and its bacteriocin in nanofibers. *Probiotics and Antimicrobial Proteins*, *2*(1), 46–51.

Hill, L. E., Gomes, C., & Taylor, T. M. (2013). Characterization, of beta-cyclodextrin inclusion complexes containing essential oils (trans-cinnamaldehyde, eugenol, cinnamon bark, and clove bud extracts) for antimicrobial delivery applications. *LWT-Food Science and Technology*, *51*(1), 86–93.

Hosseinnejad, M., & Jafari, S. M. (2016). Evaluation of different factors affecting antimicrobial properties of chitosan. *International Journal of Biological Macromolecules*, *85*, 467–475.

Hu, Q., Gerhard, H., Upadhyaya, I., Venkitanarayanan, K., & Luo, Y. (2016). Antimicrobial eugenol nanoemulsion prepared by gum arabic and lecithin and evaluation of drying technologies. *International Journal of Biological Macromolecules*, *87*, 130–140.

Huang, W., Xu, H., Xue, Y., Huang, R., Deng, H., & Pan, S. (2012). Layer-by-layer immobilization of lysozyme–chitosan–organic rectorite composites on electrospun nanofibrous mats for pork preservation. *Food Research International*, *48*(2), 784–791.

Ibrahim Sallam, K. (2007). Antimicrobial and antioxidant effects of sodium acetate, sodium lactate, and sodium citrate in refrigerated sliced salmon. *Food Control*, *18*(5), 566–575.

Jafari, S. M. (2017). An overview of nano-encapsulation techniques and their classification. In: *Nanoencapsulation Technologies for the Food and Nutraceutical Industries* (Chapter 1: pp. 1–34). Amsterdam: Elsevier.

Jafari, S. M., Assadpoor, E., Bhandari, B., & He, Y. (2008a). Nano-particle encapsulation of fish oil by spray drying. *Food Research International*, *41*(2), 172–183.

Jafari, S. M., Assadpoor, E., He, Y., & Bhandari, B. (2008b). Encapsulation efficiency of food flavours and oils during spray drying. *Drying Technology*, *26*(7), 816–835.

Jay, J. M., Loessner, M. J., & Golden, D. A. (2005). Food protection with chemicals, and by biocontrol. In J. M. Jay, M. J. Loessner, & D. A. Golden (Eds.), *Modern food microbiology* (7th ed., pp. 301–350). New York: Springer Science + Business Media, Inc.

Jenssen, H., & Hancock, R. E. (2009). Antimicrobial properties of lactoferrin. *Biochimie*, *91*(1), 19–29.

Jongsri, P., Wangsomboondee, T., Rojsitthisak, P., & Seraypheap, K. (2016). Effect of molecular weights of chitosan coating on postharvest quality and physicochemical characteristics of mango fruit. *LWT—Food Science and Technology*, *73*, 28–36.

Jozala, A. F., de Arauz, L. J., Mazzola, P. G., & Penna, T. C. V. (2009). Nisin biotechnological production and application: a review. *Trends in Food Science & Technology*, *20*(3–4), 146–154.

Kakaei, S., & Shahbazi, Y. (2016). Effect of chitosan-gelatin film incorporated with ethanolic red grape seed extract and Ziziphora clinopodioides essential oil on survival of Listeria monocytogenes and chemical, microbial and sensory properties of minced trout fillet. *LWT—Food Science and Technology*, *72*, 432–438.

Kim, S., & Fung, D. Y. C. (2004). Antibacterial Effect of Water-Soluble Arrowroot (Puerariae radix) Tea Extracts on Foodborne Pathogens in Ground Beef and Mushroom Soup. *Journal of Food Protection*, *67*(9), 1953–1956.

Klančnik, A., Piskernik, S., Jeršek, B., & Možina, S. S. (2010). Evaluation of diffusion and dilution methods to determine the antibacterial activity of plant extracts. *Journal of Microbiological Methods*, *81*(2), 121–126.

Kolpin, D. W., Furlong, E. T., Meyer, M. T., Thurman, E. M., Zaugg, S. D., Barber, L. B., & Buxton, H. T. (2002). Pharmaceuticals, hormones, and other organic wastewater contaminants in US streams, 1999-2000: A national reconnaissance. *Environmental Science & Technology*, *36*(6), 1202–1211.

Kong, M., Chen, X. G., Xing, K., & Park, H. J. (2010). Antimicrobial properties of chitosan and mode of action: A state of the art review. *International Journal of Food Microbiology*, *144* (1), 51–63.

Laridi, R., Kheadr, E. E., Benech, R. O., Vuillemard, J. C., Lacroix, C., & Fliss, I. (2003). Liposome encapsulated nisin Z: optimization, stability and release during milk fermentation. *International Dairy Journal*, *13*(4), 325–336.

Larson, E. A. Ja. A. E. (2005). Lysozyme. In P. M. Davidson, & A. L. Branen (Eds.), *Antimicrobials in Food* (3rd ed, pp. 361–387). Boca Raton, FL: CRC Press, Taylor & Francis Group, LLC.

Lee, S. H., Heng, D., Ng, W. K., Chan, H. K., & Tan, R. B. (2011). Nano spray drying: a novel method for preparing protein nanoparticles for protein therapy. *International Journal of Pharmaceutics*, *403*(1), 192–200.

Li, C., Zhang, X., Huang, X., Wang, X., Liao, G., & Chen, Z. (2013). Preparation and characterization of flexible nanoliposomes loaded with daptomycin, a novel antibiotic, for topical skin therapy. *International Journal Nanomedicine*, *8*(1), 1285–1292.

Liang, R., Xu, S., Shoemaker, C. F., Li, Y., Zhong, F., & Huang, Q. (2012). Physical and antimicrobial properties of peppermint oil nanoemulsions. *Journal of Agricultural and Food Chemistry*, *60*(30), 7548–7555.

Luo, Y., Zhang, Y., Pan, K., Critzer, F., Davidson, P. M., & Zhong, Q. (2014). Self-emulsification of alkaline-dissolved clove bud oil by whey protein, gum Arabic, lecithin, and their combinations. *Journal of Agricultural and Food Chemistry*, *62*(19), 4417–4424.

Lupetti, A., Paulusma-Annema, A., Welling, M. M., Dogterom-Ballering, H., Brouwer, C. P. J. M., Senesi, S., van Dissel, J. T., & Nibbering, P. H. (2003). Synergistic activity of the N-terminal peptide of human lactoferrin and fluconazole against Candida species. *Antimicrobial Agents and Chemotherapy*, *47*(1), 262–267.

Ma, Q., Davidson, P. M., & Zhong, Q. (2016). Nanoemulsions of thymol and eugenol co-emulsified by lauric arginate and lecithin. *Food Chemistry*, *206*, 167–173.

Ma, Q., & Zhong, Q. (2015). Incorporation of soybean oil improves the dilutability of essential oil microemulsions. *Food Research International*, *71*, 118–125.

Magnuson, B. A., Jonaitis, T. S., & Card, J. W. (2011). A Brief Review of the Occurrence, Use, and Safety of Food-Related Nanomaterials. *Journal of Food Science*, *76*(6), R126–R133.

Mbandi, E., & Shelef, L. A. (2001). Enhanced Inhibition of Listeria monocytogenes and Salmonella Enteritidis in Meat by Combinations of Sodium Lactate and Diacetate. *Journal of Food Protection*, *64*(5), 640–644.

Mbandi, E., & Shelef, L. A. (2002). Enhanced antimicrobial effects of combination of lactate and diacetate on Listeria monocytogenes and Salmonella spp. in beef bologna. *International Journal of Food Microbiology*, *76*(3), 191–198.

McClements, D. J. (2011). Edible nanoemulsions: fabrication, properties, and functional performance. *Soft Matter*, *7*(6), 2297–2316.

McClements, D. J. (2012). Nanoemulsions versus microemulsions: terminology, differences, and similarities. *Soft Matter*, *8*(6), 1719–1729.

McClements, D. J. (2015). *Food emulsions: principles, practices, and techniques* (2nd ed). Boca Raton, FL: CRC press.

Moghimi, R., Aliahmadi, A., McClements, D. J., & Rafati, H. (2016). Investigations of the effectiveness of nanoemulsions from sage oil as antibacterial agents on some food borne pathogens. *LWT-Food Science and Technology*, *71*, 69–76.

Moreno, S., Scheyer, T., Romano, C. S., & Vojnov, A. A. (2006). Antioxidant and antimicrobial activities of rosemary extracts linked to their polyphenol composition. *Free Radical Research*, *40*(2), 223–231.

MuÈller, R. H., MaÈder, K., & Gohla, S. (2000). Solid lipid nanoparticles (SLN) for controlled drug delivery—a review of the state of the art. *European Journal of Pharmaceutics and Biopharmaceutics*, *50*(1), 161—177.

Mugabe, C., Halwani, M., Azghani, A. O., Lafrenie, R. M., & Omri, A. (2006). Mechanism of enhanced activity of liposome-entrapped aminoglycosides against resistant strains of Pseudomonas aeruginosa. *Antimicrobial Agents and Chemotherapy*, *50*(6), 2016—2022.

Murdock, C. A., Cleveland, J., Matthews, K. R., & Chikindas, M. L. (2007). The synergistic effect of nisin and lactoferrin on the inhibition of Listeria monocytogenes and Escherichia coli O157: H7. *Letters in Applied Microbiology*, *44*(3), 255—261.

Negi, P. S. (2012). Plant extracts for the control of bacterial growth: Efficacy, stability and safety issues for food application. *International Journal of Food Microbiology*, *156*(1), 7—17.

Nikiforov, A., Deng, X., Xiong, Q., Cvelbar, U., DeGeyter, N., Morent, R., & Leys, C. (2016). Non-thermal plasma technology for the development of antimicrobial surfaces: a review. *Journal of Physics D: Applied Physics*, *49*(20), 204002—204009.

Ninham, B. (1999). On progress in forces since the DLVO theory. *Advances in Colloid and Interface Science*, *83*(1), 1—17.

Ortan, A., Câmpeanu, G., Dinu-Pirvu, C., & Popescu, L. (2009). Studies concerning the entrapment of Anethum graveolens essential oil in liposomes. *Romanian Biotechnology Letter*, *14*, 4411—4417.

Over, K. F., Hettiarachchy, N., Johnson, M. G., & Davis, B. (2009). Effect of Organic Acids and Plant Extracts on Escherichia coli O157:H7, Listeria monocytogenes, and Salmonella Typhimurium in Broth Culture Model and Chicken Meat Systems. *Journal of Food Science*, *74*(9), M515—M521.

Pan, K., Chen, H., Davidson, P. M., & Zhong, Q. (2014a). Thymol nanoencapsulated by sodium caseinate: physical and antilisterial properties. *Journal of Agricultural and Food Chemistry*, *62*, 1649—1657.

Pan, K., Chen, H. Q., Davidson, P. M., & Zhong, Q. X. (2014b). Thymol Nanoencapsulated by Sodium Caseinate: Physical and Antilisterial Properties. *Journal of Agricultural and Food Chemistry*, *62*(7), 1649—1657.

Pan, K., & Zhong, Q. (2016a). Low energy, organic solvent-free co-assembly of zein and caseinate to prepare stable dispersions. *Food Hydrocolloids*, *52*, 600—606.

Pan, K., & Zhong, Q. (2016b). Organic Nanoparticles in Foods: Fabrication, Characterization, and Utilization. *Annual Review of Food Science and Technology*, *7*, 245—266.

Parris, N., Cooke, P. H., & Hicks, K. B. (2005). Encapsulation of essential oils in zein nanospherical particles. *Journal of Agricultural and Food Chemistry*, *53*, 4788—4792.

Pavlukhina, S., Lu, Y., Patimetha, A., Libera, M., & Sukhishvili, S. (2010). Polymer multilayers with pH-triggered release of antibacterial agents. *Biomacromolecules*, *11*(12), 3448—3456.

Prombutara, P., Kulwatthanasal, Y., Supaka, N., Sramala, I., & Chareonpornwattana, S. (2012). Production of nisin-loaded solid lipid nanoparticles for sustained antimicrobial activity. *Food Control*, *24*(1), 184—190.

Raei, M., Rajabzadeh, G., Zibaei, S., Jafari, S. M., & Sani, A. M. (2015). Nano-encapsulation of isolated lactoferrin from camel milk by calcium alginate and evaluation of its release. *International Journal of Biological Macromolecules*, *79*, 669—673.

Rahmanian, N., Jafari, S. M., & Wani, T. A. (2015). Bioactive profile, dehydration, extraction and application of the bioactive components of olive leaves. *Trends in Food Science & Technology*, *42*(2), 150—172.

Rajashekhara, E., Suresh, E. R., & Ethiraj, S. (1998). Thermal death rate of ascospores of neosartorya fischeri ATCC 200957 in the presence of organic acids and preservatives in fruit juices. *Journal of Food Protection*, *61*(10), 1358–1362.

Reza Mozafari, M., Johnson, C., Hatziantoniou, S., & Demetzos, C. (2008). Nanoliposomes and their applications in food nanotechnology. *Journal of Liposome Research*, *18*(4), 309–327.

Roller, S. (2003). *Natural antimicrobials for the minimal processing of foods*. Cambridge, UK: Elsevier.

Ruengvisesh, S., Loquercio, A., Castell-Perez, E., & Taylor, T. M. (2015). Inhibition of bacterial pathogens in medium and on spinach leaf surfaces using plant-derived antimicrobials loaded in surfactant micelles. *Journal of Food Science*, *80*(11), M2522–M2529.

Sachetelli, S., Khalil, H., Chen, T., Beaulac, C., Sénéchal, S., & Lagacé, J. (2000). Demonstration of a fusion mechanism between a fluid bactericidal liposomal formulation and bacterial cells. *Biochimica et Biophysica Acta (BBA)-Biomembranes*, *1463*(2), 254–266.

Shah, B., Davidson, P. M., & Zhong, Q. (2012). Nanocapsular dispersion of thymol for enhanced dispersibility and increased antimicrobial effectiveness against *Escherichia coli* O157:H7 and *Listeria monocytogenes* in model food systems. *Applied and Environmental Microbiology*, *78*(23), 8448–8453.

Shah, B., Davidson, P. M., & Zhong, Q. (2013). Nanodispersed eugenol has improved antimicrobial activity against *Escherichia coli* O157:H7 and *Listeria monocytogenes* in bovine milk. *International Journal of Food Microbiology*, *161*(1), 53–59.

Sinico, C., De Logu, A., Lai, F., Valenti, D., Manconi, M., Loy, G., Bonsignore, L., & Fadda, A. M. (2005). Liposomal incorporation of Artemisia arborescens L. essential oil and in vitro antiviral activity. *European Journal of Pharmaceutics and Biopharmaceutics*, *59*(1), 161–168.

Sipahi, R. E., Castell-Perez, M. E., Moreira, R. G., Gomes, C., & Castillo, A. (2013). Improved multilayered antimicrobial alginate-based edible coating extends the shelf life of fresh-cut watermelon (Citrullus lanatus). *LWT—Food Science and Technology*, *51*(1), 9–15.

Sivapalasingam, S., Friedman, C. R., Cohen, L., & Tauxe, R. V. (2004). Fresh produce: A growing cause of outbreaks of foodborne illness in the United States, 1973 through 1997. *Journal of Food Protection*, *67*(10), 2342–2353.

Solans, C., Esquena, J., Forgiarini, A. M., Uson, N., Morales, D., Izquierdo, P., Azemar, N., & Garcia-Celma, M. J. (2003). Nano-emulsions: formation, properties, and applications. *Surfactant Science Series*, *109*, 525–554.

Stark, J., & Roller, S. (2003). Natamycin: An effective fungicide for food and beverages. In S. Roller (Ed.), *Natural antimicrobials for the minimal processing of foods* (pp. 82–97). Cambridge, UK: Woodhead Publishing.

Stopforth, J. D., Sofos, J. N., & Busta, F. F. (2005). 3 Sorbic Acid and Sorbates. In P. M. Davidson, J. N. Sofos, & A. L. Branen (Eds.), *Antimicrobials in Food* (pp. 49–90). New York: Taylor & Francis.

Szejtli, J. (1998). Introduction and general overview of cyclodextrin chemistry. *Chemical Reviews*, *98*, 1743–1754.

Taylor, M. (2014). *Handbook of natural antimicrobials for food safety and quality*. Cambridge, UK: Elsevier.

Taylor, S., Brock, J., Kruger, C., Berner, T., & Murphy, M. (2004). Safety determination for the use of bovine milk-derived lactoferrin as a component of an antimicrobial beef carcass spray. *Regulatory Toxicology and Pharmacology*, *39*(1), 12–24.

Taylor, T., Bruce, B. D., Weiss, J., & Davidson, P. M. (2008). Listeria Monocytogenes and Escherichia coli O157: H7 inhibition in vitro by liposome-encapsulated nisin and ethylene diaminetetraacetic acid. *Journal of Food Safety*, *28*(2), 183–197.

Thamphiwatana, S., Fu, V., Zhu, J., Lu, D., Gao, W., & Zhang, L. (2013). Nanoparticle-stabilized liposomes for pH-responsive gastric drug delivery. *Langmuir*, *29*(39), 12228–12233.

Thomas, L. V., & Delves-Broughton, J. (2005). Nisin. In P. M. Davidson, J. N. Sofos, & A. L. Branen (Eds.), *Antimicrobials in food* (3rd ed., pp. 237–274). Boca Raton, FL: CRC Press, Taylor & Francis Group, LLC.

Tiwari, B. K., Valdramidis, V. P., O'Donnell, C. P., Muthukumarappan, K., Bourke, P., & Cullen, P. (2009). Application of natural antimicrobials for food preservation. *Journal of Agricultural and Food Chemistry*, *57*(14), 5987–6000.

Torres-Giner, S., Ocio, M. J., & Lagaron, J. M. (2009). Novel antimicrobial ultrathin structures of zein/chitosan blends obtained by electrospinning. *Carbohydrate Polymers*, *77*(2), 261–266.

Tunç, S., & Duman, O. (2011). Preparation of active antimicrobial methyl cellulose/carvacrol/montmorillonite nanocomposite films and investigation of carvacrol release. *LWT-Food Science and Technology*, *44*(2), 465–472.

Umer, S., Tekewe, A., & Kebede, N. (2013). Antidiarrhoeal and antimicrobial activity of Calpurnia aurea leaf extract. *BMC Complementary and Alternative Medicine*, *13*(1), 1–5.

Varona, S., Rojo, S. R., Martín, Á., Cocero, M. J., Serra, A. T., Crespo, T., & Duarte, C. M. (2013a). Antimicrobial activity of lavandin essential oil formulations against three pathogenic food-borne bacteria. *Industrial Crops and Products*, *42*, 243–250.

Varona, S., Rojo, S. Ri, Martin, Á., Cocero, Mi. J., Serra, A. T., Crespo, T., & Duarte, C. M. M. (2013b). Antimicrobial activity of lavandin essential oil formulations against three pathogenic food-borne bacteria. *Industrial Crops and Products*, *42*, 243–250.

Vega-Lugo, A. C., & Lim, L. T. (2009). Controlled release of allyl isothiocyanate using soy protein and poly (lactic acid) electrospun fibers. *Food Research International*, *42*(8), 933–940.

Von Woedtke, T., Schlüter, B., Pflegel, P., Lindequist, U., & Jülich, W. D. (1999). Aspects of the antimicrobial efficacy of grapefruit seed extract and its relation to preservative substances contained. *Pharmazie*, *54*(6), 452–456.

Vu, K. D., Hollingsworth, R. G., Leroux, E., Salmieri, S., & Lacroix, M. (2011). Development of edible bioactive coating based on modified chitosan for increasing the shelf life of strawberries. *Food Research International*, *44*(1), 198–203.

Wang, T., Soyama, S., & Luo, Y. (2016). Development of a novel functional drink from all natural ingredients using nanotechnology. *LWT-Food Science and Technology*, *73*, 458–466.

Weiss, J., Takhistov, P., & McClements, D. J. (2006). Functional materials in food nanotechnology. *Journal of Food Science*, *71*(9), R107–R116.

Wen, Z., You, X., Jiang, L., Liu, B., Zheng, Z., Pu, Y., & Cheng, B. (2011). Liposomal incorporation of rose essential oil by a supercritical process. *Flavour and Fragrance Journal*, *26*(1), 27–33.

Wu, J. E., Lin, J., & Zhong, Q. (2014). Physical and antimicrobial characteristics of thyme oil emulsified with soluble soybean polysaccharide. *Food Hydrocolloids*, *39*, 144–150.

Wu, Y., Luo, Y., & Wang, Q. (2012). Antioxidant and antimicrobial properties of essential oils encapsulated in zein nanoparticles prepared by liquid–liquid dispersion method. *LWT-Food Science and Technology*, *48*(2), 283–290.

Xiao, D., Davidson, P. M., & Zhong, Q. (2011). Spray-dried zein capsules with coencapsulated nisin and thymol as antimicrobial delivery system for enhanced antilisterial properties. *Journal of Agricultural and Food Chemistry*, *59*(13), 7393–7404.

Xue, J., Davidson, P. M., & Zhong, Q. (2013). Thymol nanoemulsified by whey protein-maltodextrin conjugates: the enhanced emulsifying capacity and antilisterial properties in milk by propylene glycol. *Journal of Agricultural and Food Chemistry*, *61*(51), 12720–12726.

Xue, J., Davidson, P. M., & Zhong, Q. (2015). Antimicrobial activity of thyme oil co-nanoemulsified with sodium caseinate and lecithin. *International Journal of Food Microbiology*, *210*, 1–8.

Zhang, L., Critzer, F., Davidson, P. M., & Zhong, Q. (2014a). Formulating essential oil microemulsions as washing solutions for organic fresh produce production. *Food Chemistry*, *165*, 113–118.

Zhang, L., & Granick, S. (2006). How to stabilize phospholipid liposomes (using nanoparticles). *Nano letters*, *6*(4), 694–698.

Zhang, L., Pornpattananangkul, D., Hu, C. M., & Huang, C. M. (2010). Development of nanoparticles for antimicrobial drug delivery. *Current Medicinal Chemistry*, *17*(6), 585–594.

Zhang, Y., Ma, Q., Critzer, F., Davidson, P. M., & Zhong, Q. (2015). Effect of alginate coatings with cinnamon bark oil and soybean oil on quality and microbiological safety of cantaloupe. *International Journal of Food Microbiology*, *215*, 25–30.

Zhang, Y., Ma, Q., Critzer, F., Davidson, P. M., & Zhong, Q. (2016). Organic thyme oil emulsion as an alternative washing solution to enhance the microbial safety of organic cantaloupes. *Food Control*, *67*, 31–38.

Zhang, Y., Niu, Y., Luo, Y., Ge, M., Yang, T., Yu, L. L., & Wang, Q. (2014b). Fabrication, characterization and antimicrobial activities of thymol-loaded zein nanoparticles stabilized by sodium caseinate–chitosan hydrochloride double layers. *Food Chemistry*, *142*, 269–275.

Zhong, Q., Chen, H., Zhang, Y., Pan, K., & Wang, W. (2015). *Delivery systems for food applications, nanotechnology and functional foods*. John Wiley & Sons, Ltd, Chichester, UK., pp. 91–111.

Chapter 6

Nanoencapsulation of Natural Food Colorants

Sahar Akhavan and Seid Mahdi Jafari
Gorgan University of Agricultural Sciences and Natural Resources, Gorgan, Iran

6.1 INTRODUCTION

The color of food materials reflects their quality in most cases and generally predetermines the expectation of consumers and product acceptance. Consumers view the color of food products as a measure of quality as well as an indicator of fine processing. Hence, the food mass producers must assign a proportion of their production costs to pick and implement the acceptable and desirable synthetic/natural colorants in their food products. This is the reason why seeking a suitable food color and controlling their intensity/saturation is always an important step in the food production process. Moreover, the food processors try their best to prevent the alteration of the natural appearance of the raw materials (Khazaei, Jafari, Ghorbani, & Kakhki, 2014, 2016; Lauro & Francis, 2000, Vila, Chaud, & Balcão, 2015).

Historically, the early civilizations used colorants as a means to increase the appeal of human-made food products. Ancient reports point to the use of natural colorants for the fabrication of appetizing food products in which living species were primarily used as food colorants, followed by inorganics and synthetics. Nowadays, safety is an important public concern, so natural colorants are suitable alternatives to the synthetic dyes and their employment in food products has increased considerably. In fact, accurate and comprehensive regulations have been defined for the use of synthetic colorants, as they are classified into "concern level III" substances by the US Food and Drug Administration (FDA) (Delgado-Vargas & Paredes-López, 2002; Hendry & Houghton, 1996).

In general, color has a direct impact on the consumer's willingness to purchase the food products with the potential for commercialization. During processing, a substantial amount of color is lost, so synthetic or natural colors are added to make any food commodity attractive for consumers. Some

Nanoencapsulation of Food Bioactive Ingredients. DOI: http://dx.doi.org/10.1016/B978-0-12-809740-3.00006-4

reasons why color is added to food are (Rymbai, Sharma, & Srivastav, 2011; Vila et al., 2015):

1. To enhance the chroma of the food already present;
2. To replace color in the food, which is lost during processing;
3. To minimize batch-to-batch variations;
4. To supplement food with nutrients.

A certain type of the color additive should be chosen in regard to the intended application, thus many factors must then be considered in this selection, such as:

- Required color hue;
- Physicochemical properties of the processed food merchandise (e.g., oily or water-based product, pH);
- Physical state (e.g., liquid, solid, emulsion);
- Processing terms (like storage conditions whether the process requires heating or cooling).

Furthermore, the relevant legislations should also be taken into account. Overall, each country and even some states have their own regulations for the consumption of food colorants; besides, the application of color relies on the type of the utilized pigment. It is common to find product application formulations that are particular for one manufactured product: Spray dried powders are preferred for mass coloration, whereas oil-soluble colorants must be emulsified to be applied in citrus oils. As a result, it is strongly recommended to first determine the right colorant concerning our purpose so as to achieve the desired final chroma, solubility, physical form (liquid, solid powders, emulsions, etc.), pH, microbiological state plus the other considered features (Delgado-Vargas & Paredes-López, 2002; Socaciu, 2007). The study of natural colorants and the process of their extraction, encapsulation, and preservation have been extensively investigated due to the growing interest in eliminating synthetic colorants according to their possible toxicity.

6.2 CLASSIFICATION OF FOOD COLORANTS

Food colorants have been categorized on the basis of different systems as indicated in Table 6.1. In these classification structures, the same type of colorants could be arranged in different groups. However, today, the most important classification trend is based on the origin and legislations of the colorants (Delgado-Vargas, Jiménez, & Paredes-López, 2000).

Considering the origin of the naturally occurring food colorants, they are categorized into natural, synthetic, or inorganic. Natural food colors originate from a wide range of sources like vegetables, fruits, plants, minerals, and other edible natural sources. They impart color when added to foods or

TABLE 6.1 Different Systems for Classification of Food Colorants

Classification System	Groups	Examples
Origin of colorants	Natural	Carotenoids, anthocyanins,
	Synthetic	FD&C[a] colorants
	Inorganic	TiO_2
Structural characteristics	Tetrapyrrole derivatives	Chlorophylls and heme colors
	Isoprenoid derivatives	Carotenoids and iridoids
	N-heterocyclic compounds different from tetrapyrroles	Purines, pterins, flavins, phenazines, phenoxazines, and betalains
	Benzopyran derivatives (oxygenated heterocyclic compounds)	Anthocyanins and other flavonoid pigments.
	Quinones	Benzoquinone, naphthoquinone, anthraquinone
	Melanins	
Chemical structure of the colorants	Chromophores with conjugated systems	Carotenoids, anthocyanins, betalains, caramel, synthetic pigments and lakes.
	Metal-coordinated porphyrins	Myoglobin, chlorophyll, and their derivatives

[a]FD&C = Certifiable colorants permitted to be used for foods, drugs, and cosmetics.

beverages. Natural colors comprise a broad range of colorants indicating divergent solubility and stability properties. So, the chemical structure and stability of the intended natural colorants should be studied prior to usage in the food products in order to adapt them to the changing circumstances during processing, packaging, and distribution (Hendry & Houghton, 1996; Blanc, 2002).

Synthetic colorants generally have a negative impact on consumers, due to the possible deleterious effects of some synthetic pigments on human health, including allergic reactions, genotoxicity, and potential carcinogenicity (Caro et al., 2012). As a result, today the public prefers to consume foods which contain natural additives and are health-promoting. Thus, over the last few years, the interest of the food industry in replacing the artificial colorants with natural ones has enhanced dramatically, which is in line with the safety and regulatory issues as well as consumer concerns (Delgado-Vargas et al., 2000; Caro et al., 2012).

Generally, concerning the chemical structure of food colorants, they can be classified into six classes, namely, heme groups, carotenoids, chlorophylls, betalains, flavonoids, and miscellaneous colorants.

6.2.1 Heme Pigments

Heme is actually originated from the Greek word for blood and is responsible for the red color of two important pigments in animal species: hemoglobin, existing in the blood, and myoglobin that resides in the muscular tissues (Table 6.2). Basically, myoglobin gives the red color to meat and during processing, this pigment is subjected to chemical and color changes, for instance freshly cut meat renders a purplish color (Delgado-Vargas et al., 2000; Suman & Joseph, 2013). Upon exposure to air, a delightful red hue is observable at the surface of the meat according to the oxygenation of myoglobin (an oxygen molecule is bound to the heme compound akin to the oxygenation process of hemoglobin).

6.2.2 Chlorophylls

Several types of chlorophylls have been mentioned in the literature. Accordingly, chlorophyll *a* plus chlorophyll *b* are among two commonly used colorants in food processing because they are abundant agents in green plant tissues and are available at the relative ratio of 3:1. Tetrapyrrole derivatives are the core components of these natural colorants and in this view, they resemble heme pigments. In contrast to the heme group that contains iron as the central atom, magnesium is located at the central part in chlorophylls. Also, the pyrrole unit IV exists in the hydrogenated form within chlorophylls. Another feature of chlorophylls is their water-insoluble 20-carbon phytyl group, as shown in Table 6.2 (Delgado-Vargas et al., 2000; Delgado-Vargas & Paredes-López, 2002). Most of the leafy (lettuce, broccoli, cabbage etc.) and nonleafy (gilki, peas, asparagus, etc.) vegetables as well as unripe fruits are green owing to the chlorophyll pigment. Chlorophyll is relatively a labile molecule, but its stability can be increased by de-esterifying the chlorophyll and via the addition of copper ions. These copper complexes have a satisfactory green color and are more stable compared with its intact counterpart. In addition, the bright green chlorophylls can change to olive brown pheophytins by replacing the magnesium of the molecule with hydrogen.

6.2.3 Carotenoids

Carotenoids consist of a broad range of fat-soluble pigments that are mostly available in nature (more than 600 plant-based pigments). Carotenoids are red, yellow, and orange organic pigments, found in the cells of plants in addition to their biosynthesis in bacteria and microalgae. Carotenoids are found

TABLE 6.2 Chemical Structures and Characteristics of Common Natural Food Colorants

Natural Colorants	Main Types	Main Color	Examples of Sources	Basic Structure
Anthocyanins	Cyanidin	Reddish-orange	Apple, elderberry, blackberry, nectarine, plum, peach, red cabbage	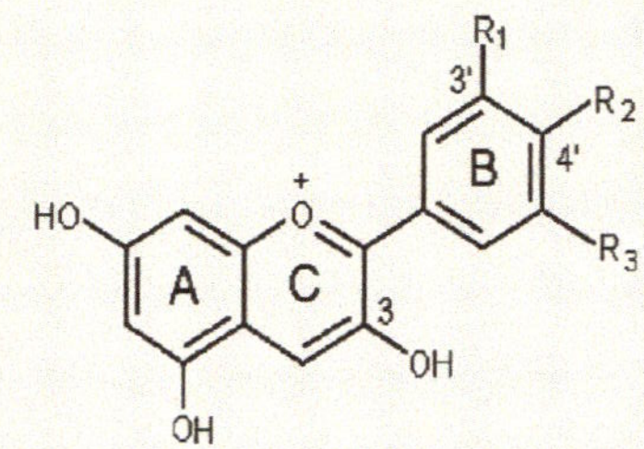
	Delphinidin	Purple, blue	Grape, beans, eggplants	
	Pelargonidin	Orange	Strawberry, red radishes, some beans	
	Malvidin	Purple	Grape	
	Peonidin	Purplish-red	Cranberries, blueberries, plums, grapes, cherries purple corn	
	Petunidin	Dark-red or purple	Grape, red berries	
Carotenoids	f-Carotene	Yellow-orange	Carrot, egg, orange, chicken fat	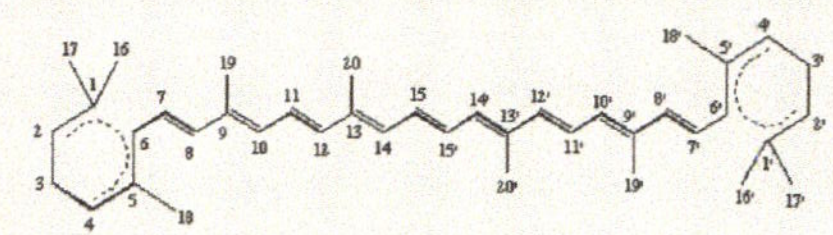
	Xanthophyll		Vegetables, egg, chicken fat	
	Zeaxanthin		Yellow corn, egg, liver	
	Cryptoxanthin		Egg, yellow corn, orange	
	Physalien		Asparagus, berries	
	Bixin		Annatto seeds	
	Lycopene		Tomato, pink grapefruit, palm oil	
	Capsanthin		Paprika	
	Astaxanthin		Lobster, shrimp, salmon	

(Continued)

TABLE 6.2 (Continued)

Natural Colorants	Main Types	Main Color	Examples of Sources	Basic Structure
Heme	Hemoglobin Myoglobin	Red- pink	Some good iron-based foods such as liver, red meat, shrimp, tofu, fortified breakfast cereals, oysters	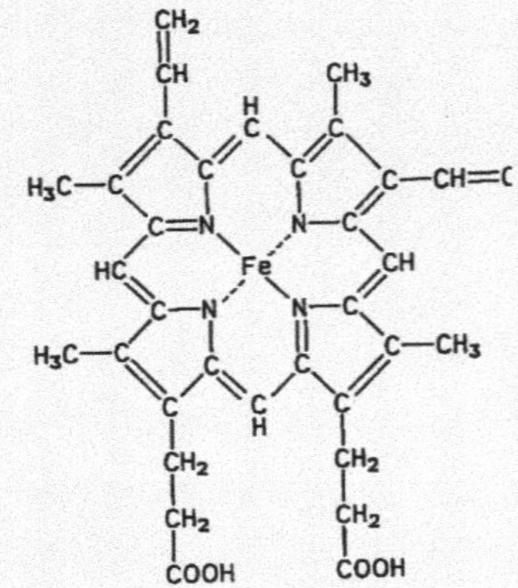
Chlorophyll	Chlorophyll a	Green	Plants	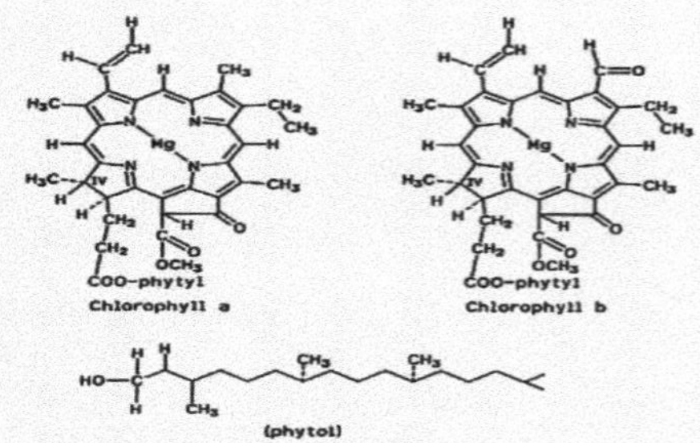
	Chlorophyll b			
	Chlorophyll c		Mostly plants	
	Chlorophyll d and f		Various algae	
			Cyanobacteria	

in several dishes including eggs, fish, shell fish along with vegetables and fruits. The basic structure of carotenoids contains eight isoprenoid units. Carotenoids are classified into two major groups, namely, carotenes composed of hydrocarbons exposing a red-orange color and xanthophylls that contain oxygen within their structure and present the color yellow. β-carotene and lycopene belong to carotenes and are water-insoluble compounds, while lutein and zeaxanthin are classified as xanthophylls (Britton, Liaaen-Jensen, & Pfander, 2009; Edge, McGarvey, & Truscott, 1997). The structures of several carotenoids, along with the foods containing these bioactive ingredients are indicated in Table 6.2. Carotenoids are normally stable ingredients against the harsh processing conditions as well as high temperature and pressure. Yet, the stability of carotenoids in foods is something that is not clear and may vary from stern loss to actual gain in carotenoid content during storage (Delgado-Vargas et al., 2000). The main factor which causes the degradation of carotenoids is oxidation. Oxygen is able to directly cleave the double bonds or to generate hydroperoxides during the lipid autoxidation process (Goodwin, 2012; Rao & Rao, 2007). Furthermore, carotenoids are proved to be efficient compounds for the prevention of a variety of lethal diseases like cancer, cardiovascular diseases, eye problems, etc. (Johnson, 2002).

6.2.4 Flavonoid Dyes

Flavonoids are water-soluble pigments, which are abundant in vascular plants. They are best known as secondary metabolites and can be found in almost all fruits and vegetables. To date, more than 5000 flavonoids have been detected and hundreds of substances resembling flavones are present in different plants. On the basis of their chemical structure, these pigments are grouped in several classes, the most important of which are listed in Table 6.2. All these compounds involve two benzene rings, linked by a heterocycle. Overall, flavonoids are categorized in accordance with their nature. Typically, flavonoids are not detected by naked eye and are visible only under ultraviolet spectrum but some display a light yellow appearance. Hence, anthocyanins can be tagged as real pigments within the flavonoids as they cover a considerable hue range of colors (Socaciu, 2007). Accordingly, anthocyanins will be discussed in the upcoming section as they have significant impact on the marketing of fruits and vegetables (Havsteen, 2002; Harborne, 2013). Considering the subsets of flavonoids other than anthocyanins, the catechins, leucoanthocyanidins, and flavonols are highly found in foodstuffs, while flavanone glycosides are normally found in citrus fruits.

6.2.4.1 Anthocyanins

Anthocyanins (being emanated from the Greek words *Anthos* and *kyanos* as flower and dark blue, respectively) are flavonoids regularly found in nature and for the most part acknowledged as the ubiquitous water-soluble pigments in the environment. They are responsible for the coloration of many fruits, flowers, and other parts of plants, and this color is an appealing factor in many food products such as beverages, jellies, marmalades, gums, smoothies, puddings, and so forth. These pigments are indeed glycosides of anthocyanidins (also named aglycones) and sugars that are composed of 3-glucosides units of the anthocyanidins. Besides, the acylation of the sugar fragment may take place via aliphatic or aromatic acids (Socaciu, 2007; Shipp & Abdel-Aal, 2010; Mahdavi, Jafari, Ghorbani, & Assadpoor, 2014). There are nearly 17 naturally occurring anthocyanidins in nature, nevertheless the majority of the entire anthocyanins isolated in nature are structurally originated from the following six anthocyanidins: pelargonidin (plg), peonidin (pnd), cyanidin (cyd), delphinidin (dpd), petunidin (ptd), and malvidin (mvd). The interest in anthocyanin pigments has dramatically increased not only for the academic sectors but also for the consumers due to their appealing color in food products along with the nutraceutical properties as natural antioxidants (Gould, Davies, & Winefield, 2008).

6.2.5 Betalains

Betalains are N-heterocyclic hydrophilic natural coloring substances that are stored in vacuoles. Betalain is almost a new term used to label a class of water-soluble plant colorants exemplified by the betacyanins that reflect a red-violet color and yellow betaxanthins. The word betalains refers to the red beet (*Beta vulgaris*), from which these pigments are extracted, moreover they are not as broadly distributed as flavonoids. Other examples of cuisines that involve betalains in their structure include pokeberries, chard, and Indian cactus fruits. Betalains are usually categorized based on their makeup and accordingly they are divided into two groups: betaxanthins and betacyanins representing yellow and red-purple colors, respectively. Betanin forms the majority of red pigments found in red beets, also the dominant yellow pigment is vulgaxanthin. Betalains preserve their structure in the pH range 3.5–7.0, which is the typical pH in common food products; however, they are susceptible to light, oxidation, and heat.

6.2.6 Miscellaneous Colorants

There are also several hundred extranatural pigments that are not common like previously described coloring items. In addition, some of the exemplified, such as anthocyanins or chlorophylls, are obviously colorants, but other compounds like quinones and xanthones (yellow pigments) are supposed to be employed in other applications. An example of a quinone is juglone, a naturally occurring component in pecans and walnuts. Mangiferin, an archetype of xanthones, is found in mangoes. Among the unusual colorants that are included in the patent lists is a yellow pigment known as gossypol considered as a harmful pigment present in cottonseeds that induces toxicity in humans and nonruminant as well (causing an abnormally high rate of hypokalemia, i.e., low levels of potassium in the blood). Its use for the animal feed is considered as a potential source of protein for human consumption (Taghvaei, Jafari, Nowrouzieh, & Alishah, 2015).

Most of the bioactive food ingredients could also be coloring agents including vitamin B_2 (riboflavin, orange-yellow), phytochrome (yellow), and vitamin B_{12} (red), although their share in the coloration of food stuff is not considerable.

6.3 STABILITY AND BIOAVAILABILITY OF FOOD COLORANTS

Despite the benefits of applying natural colorants as safe alternatives to synthetic dyes, some limitations, such as lability and lack of availability of these natural plant-based colorants, prevent the commercialization of these naturally occurring dyes. Unfortunately, all groups of natural pigments are unstable and are sensitive to degradation upon extraction. Factors that determine the stabilization or destabilization of natural colorants are (1) endogenous factors including plant enzymes like polyphenol oxidase and peroxidase, and (2) the extraction medium conditions that determine the fate of natural pigments among which water activity, temperature, oxygen content, light, and pH are considered the most effective factors (Socaciu, 2007, Vila et al., 2015).

Regarding the particular chemical structure of anthocyanins, they are likely to be degraded via external factors. Moreover, the stability of anthocyanins is influenced by pH, light, temperature, enzymes, oxygen, sugar, vitamin C and their degraded fragments, sulfur dioxide, and proteins. Also, the bioavailability of anthocyanins is highly dependent on the alterations in pH. With this in mind, anthocyanins are usually resistant at pH values of 3.5 and below; therefore, they remain relatively intact in the stomach juice. On the other hand, at higher pH values like inside the intestinal lumen ($pH = 7$) anthocyanins are disintegrated and thus their nutritional value is reduced dramatically. Thereupon, the innate instability of anthocyanins is directly attributed to the color stabilization applications so that they need to be shielded against destructive factors in order to retain the color of the product and to maintain the nutritional value of food as well (Mahdavi et al., 2014).

Carotenoids are another example of colorants, which are very sensitive to light, heat, and oxidation concerning their unsaturated chemical bonds. At the ambient temperature, they are almost insoluble in water and hardly soluble in oil (about 0.2 g/L oil), but their solubility in oil scales directly with temperature (Ribeiro, Schuchmann, Engel, Walz, & Briviba, 2010). It has been noted that a trivial part of the carotenoids that exist in raw fruits or vegetables is absorbed within the intestines, which can be attributed to their crystalline structure or in combination with protein complexes. In contrast, carotenoids dissolved in vegetable oils show a higher bioavailability (Ax, Mayer-Miebach, Link, Schuchmann, & Schubert, 2003; Ribeiro et al., 2010).

Leafy vegetables and some fruits contain high amounts of chlorophylls by which they appear green. These food products should be consumed a few days after harvesting or preserved via effective methods since they are the most perishable postharvest products. Moreover, their green color may undergo some perceptible changes because of light, temperature, and pH during storage (Socaciu, 2007).

Unfortunately, the levels of natural pigments that are effective in vitro assays are usually higher than the measurements obtained in in vivo tests. Bioactive compounds have an important role in preventing diseases but their operation depends on their bioavailability (Fang & Bhandari, 2010). This is a noteworthy issue since only a small fraction of the molecules maintain their basic structure following oral administration and the harsh food-processing operations as well as the inadequate gastric residence time, low solubility inside the intestine and the conditions within the gastrointestinal tract (pH, enzymes, reactions with nutrients). Therefore, productive formulations and protective mechanisms should be carefully engineered in order to obtain efficient delivery systems to preserve the bioactive nutrients during consumption so that the compounds can be delivered to the physiological target within living cells and tissues (Fang & Bhandari, 2010; Wildman, Wildman, & Wallace, 2016).

For this reason many studies have been conducted with the aim to increase the stability of these natural colorants. Furthermore, most of the foods derived from plants are normally cooked and processed before consumption, thus the effect of food handling and thermal processing on their physical characteristics and acceptance has been widely investigated by using kinetic models for the disintegration of natural colorants. Therefore, it is of utmost importance to maintain the original structure of the food colorants within the food matrix to keep the high quality of food products and enhance the acceptance perceived by consumers. Many techniques have been applied with the aim of increasing the stability of these natural colorants such as self-association, copigmentation, complexation and polymerization, acylation, and encapsulation (Dao, Takeoka, Edwards, & Berrios, 1998; Mahdavi et al., 2014). Based on the experiences of Cavalcanti, Santos, and Meireles (2011), among these techniques, encapsulation is a practical means to boost the stability of natural colors because chemical methods have some disadvantages, such as difficulty in controlling and orienting the interactions.

6.4 ENCAPSULATION OF NATURAL COLORS IN THE FOOD INDUSTRY

Encapsulation has recently been expanded in several sectors especially in the food industry and it is defined as the entrapment of a bioactive compound (also called core material) in a shell material (also called wall material) which can be either organic or inorganic. An important step in fabricating micro/nanocapsules is designing the right wall material to meet our needs in the encapsulation process. Indeed, encapsulation efficiency and the stability of micro/nanocapsules are strongly related to the wall material composition. This wall acts as a shield against the destructive factors that will lead to the degradation of the core material including oxygen, water, and light or could lessen their contact with the reactive materials nearby and control their diffusion. The biopolymers alone cannot provide all these conditions; therefore, they are often used in combination with other structures as well as modifiers, such as oxygen scavengers, antioxidants, chelating agents, and surfactants to obtain a high-throughput delivery system (Katouzian and Jafari, 2016). Wall material applied for encapsulation of natural colorants should have the following properties:

1. Stabilization of natural colorants;
2. Unable to react with coloring agents;
3. Programmed release in specific circumstances;
4. Palatable, tasteless, stable;
5. Do not suck in much water, low consistency, cost-effective;
6. Soluble in an aqueous medium or solvent, or melting;
7. The encapsulant can be hard, delicate, flexible, soft, etc.

The utilization of encased natural colorants instead of free (unencapsulated) bioactive compounds can be a practical solution to the instability of these ingredients, as well as masking the unpleasant tastes, odors together with improving the uptake of food colorants in vivo and in vitro.

Encapsulation brings about the preservation of heat-sensitive molecules, such as many pigments, especially anthocyanins that are highly unstable in their uncoated form. Encapsulation technology has been tested for several natural colorants, which resulted in improving their stability and better physicochemical properties. The encapsulation efficiency is an essential index for encapsulated colors and indicates the retention rate of the natural colorants inside the capsule. A pragmatic encapsulation technology must provide the highest retention rate of pigments inside the capsule and reduce a number of pigments on the surface of the wall material to the minimum dose. According to Jafari, Assadpoor, He, and Bhandari (2008), many factors influence the encapsulation efficiency among which the properties of the shell and core as well as the characteristics of emulsions and drying conditions like air flow rate, humidity along with the inlet and discharge are of utmost importance.

Based on the recently conducted investigations, the coated type of color extracts was significantly stable in comparison to their naked form. For instance, degradation kinetics and stability of entrapped anthocyanins obtained from *Berberis vulgaris* shrub was studied by Mahdavi, Jafari, Assadpoor, and Dehnad (2016a, 2016b). In this study, three types of wall materials were implemented including maltodextrin, maltodextrin–gum arabic complex, and combinations of gum arabic, maltodextrin, and gelatin. All of the prepared samples were stored for 90 days and were tested for their stability under three different storage situations. The designed wall materials were able to enhance the half-life of the colorants during the storage period compared with the nonencapsulated barberry anthocyanin extract.

Radical scavenging activity studies have also revealed that the antioxidant properties of natural colorants were preserved following the spray drying encapsulation process (Nayak & Rastogi, 2010). Moreover, in the work conducted by Jafari et al. (2008), analysis of the encapsulated particles was performed via light microscopy and particle size distribution test, which clarified that the coated particles have a spherical morphology and are dispersed homogenously.

Nowadays, various encapsulation techniques are available which can be classified as follows (Jafari et al., 2008; Mahdavi et al., 2014; Munin & Edwards-Lévy, 2011):

- Physical practices: Spray drying, fluid bed coating, extrusion, centrifugal extrusion, supercritical fluids;
- Physicochemical practices: Gelation, ionic gelation, hot melt coating, solvent evaporation extraction, coacervation;
- Chemical methods: Interfacial polymerization, interfacial polycondensation, interfacial cross-linking in situ polymerization.

The selection of an entrapment technology relies on many different parameters and it is necessary to determine in which system our engineered capsule is going to be applied. Accordingly, parameters, such as required particle size, physicochemical properties of the core–shell, release trends, formulation costs, affect the designing process of encapsulation systems. Some of the successful encapsulation methods for the retention of natural colorants are shown in Table 6.3.

6.5 NANOENCAPSULATION OF NATURAL FOOD COLORANTS

The encapsulation systems are categorized into micro (between 1.0 μm and 5000 μm), macro (larger than 5,000 μm), and nanoparticles (lower than 1.0 μm) in regard to their particle size (Huang & Given, 2009). The submicron size of nanocapsules offers a number of distinct advantages over micron size concerning their dramatically higher surface area along with enhanced solubility, uptake, controlled release, and targeted delivery that microcapsules cannot provide (Nazzaro, Fratianni, & Coppola, 2012; Neethirajan & Jayas, 2011). Nanotechnology in the food industry, according to nanoencapsulation-related review papers, has flourished considerably in recent years (Quintanilla-Carvajal, Anandharamakrishnan, 2014; Ezhilarasi, Karthik, Chhanwal, & Anandharamakrishnan, 2013).

TABLE 6.3 Overview of Common Microencapsulation Processes for Natural Food Colorants and Their Characteristics

Technology	Process Steps	Advantages	Disadvantages	Reference
Spray drying	1. Dissolving core material in aqueous wall material solution 2. Atomization 3. Dehydration	1. Low process cost 2. Wide choice of wall material 3. Good encapsulation efficiency 4. Good stability of the finished product 5. Possibility of large-scale production in continuous mode	1. Can degrade highly temperature-sensitive compounds 2. Control of particle size is difficult 3. Yields for small batches are moderate	Mahdavi et al., 2014; Mahdavi et al., 2016a; Mahdavi et al., 2016b
Freeze drying	1. Dissolving core material and wall material in water 2. Freezing the sample 3. Drying under low pressure 4. Grinding (option)	Thermosensitive substances that are unstable in aqueous solutions may be efficiently encapsulated by this technique	1. Long processing time 2. Expensive process costs 3. Expensive storage and transport of the capsules	Desobry, Netto, & Labuza, 1997; Madene, Jacquot, Scher, & Desobry, 2006
Fluid bed coating	1. Fluidization of active powder 2. Spray coating 3. Dehydration or cooling	1. Low cost process 2. It allows specific capsule size distribution and low porosities into the product	Degradation of highly temperature-sensitive compounds	DeZarn, 1995; Dewettinck & Huyghebaert 1998, 1999
Emulsification	1. Dissolving bioactive compound and emulsifiers in water or oil phase 2. Mixing oil and water phases under shear	Polar, non-polar (apolar), and amphiphilic can be incorporated	1. Limited number of emulsifiers that can be used 2. Difficult control of the capsule formation	Zuidam & Shimoni, 2010; Betz & Kulozik, 2011

(Continued)

TABLE 6.3 (Continued)

Technology	Process Steps	Advantages	Disadvantages	Reference
Melt extrusion	1. Melting the wall material 2. Dissolving active in the coating 3. Extruding with extruder 4. Cooling	1. The material is totally surrounded by the wall material 2. Any residual core is washed from outside 3. It is a relatively low-temperature entrapping method	1. The capsule must be separated from the liquid bath and dried 2. It is difficult to obtain capsules in extremely viscous carrier material melts	Zuidam & Shimoni, 2010; Nedovic, Kalusevic, Manojlovic, Levic, & Bugarski, 2011
Encapsulation by rapid expansion of supercritical fluid (RESS)	1. Creating a dispersion of core material and dissolved or swollen wall material in supercritical fluid 2. Releasing the fluid to precipitate the shell onto the active	1. Non-toxicity and easy removal of the solvent 2. Operation at low temperatures and in an inert atmosphere that allows avoiding degradation of the product	1. Both core and wall material must be very soluble in supercritical fluids 2. Low or no solubility of high molecular weight, polar compounds in CO_2 3. Poor control over the precipitated crystal morphology, size distribution	Brunner, 2005, Cocero, Martín, Mattea, & Varona, 2009; Carvallo, 2011
Ionic Gelation	1. Wall material with dissolved core material is extruded as drops within an ionic solution 2. Capsules are formed by ionic interaction	Organic solvents and extreme conditions of temperature and pH are avoided	1. Mainly used on a laboratory scale the capsules 2. In general, have high porosity which promotes intensive burst	Santos & Meireles, 2010; Zuidam & Shimoni, 2010
Thermal Gelation	It is similar to ionic gelation; no necessity of an ionic solution to form a gelled drop, the gelation is only due to thermal parameters	The same of ionic gelation	The same of ionic gelation	Ferreira et al., 2009, Santos & Meireles, 2010

According to Jafari (2017), the nanoencapsulation systems can be put together in five major groups including lipid formulation, natural nanocarriers, special equipment-based techniques, biopolymer nanoparticles, and other miscellaneous techniques. Nanoencapsulation studies on natural food colorants have been summarized in Table 6.4. The aforementioned

TABLE 6.4 Overview of Common Nanoencapsulation Processes for Natural Food Colorants

Nanoencapsulation Method	Natural Colorants	Reference
Nanoliposomes	Carotenoid	Barenholz, Diminsky, & Cohen, 2006
	Carotenoid	Socaciu, Jessel, & Diehl, 2000
	Curcumin	Takahashi et al. 2009
	Carotenoid	Tan et al., 2014
	Quercetin	Frenzel et al., 2015
	Curcumin	Shin et al., 2013
	β-carotene	de Freitas Zômpero et al., 2015
	Curcumin	Saengkrit, Saesoo, Srinuanchai, Phunpee, & Ruktanonchai, 2014
	β-carotene	Moraes et al., 2013
	β-carotene	Toniazzo et al., 2014
	Astaxanthin	Yoo et al., 2010
	Curcumin	Shaikh, Ankola, Beniwal, Singh, & Kumar, 2009
	Quercetin	Cadena et al., 2013
Nanoemulsions	β-carotene	Silva et al., 2011a; Silva et al. 2011b
	Catechin	Yi, Fan, Zhang, & Zhao, 2016
	Curcumin	Ahmed et al., 2012
	β-carotene	Salvia-Trujillo et al., 2013
	β-carotene	Liang et al., 2013
	Crocin	Esfanjani et al., 2015
	β-carotene	Qian et al., 2012
	Quercetin	Ni et al., 2017
	Crocin	Mehrnia, Jafari, Makhmal-Zadeh, & Maghsoudlou, 2016
	Lycopene	Ax et al., 2003

(Continued)

TABLE 6.4 (Continued)

Nanoencapsulation Method	Natural Colorants	Reference
Solid lipid nanoparticles	α-tocopherol	de Carvalho et al., 2013
	Curcumin	Sun et al., 2013
	β-carotene	Qian et al., 2013
	Astaxanthin	Li et al., 2015
	Anthocyanin	Ravanfar et al., 2016
Nanostructured lipid carriers	β-carotene	Hentschel, Gramdorf, Müller, & Kurz, 2008
	Lutein	Lacatusu et al., 2013
	Curcumin	Fang et al., 2012
	β-carotene	Zhang et al., 2013
	Quercetin	Liu et al., 2014
	Quercetin	Barras et al., 2009
	Astaxanthin	Tamjidi et al., 2014
	β-carotene	Hejri et al., 2013
	Quercetin	Ni et al., 2015
Cyclodextrins	Curcumin	Yadav et al., 2010
	Lycopene	Nerome et al., 2013
	Curcumin	Popat et al., 2014
	Bixin	Marcolino et al., 2011
Casein nanoparticles	β-carotene	Sáiz-Abajo et al., 2013
	Curcumin	Esmaili et al., 2011
Nanocrystals	Quercetin	Karadag et al., 2014
Electrospining	β-carotene	de Freitas Zômpero et al., 2015
Biopolymer nanoparticles	Curcumin	Rachmawati et al., 2013
	Curcumin	Xie et al., 2011
	Lutein	Arunkumar et al., 2013
	β-carotene	Yi et al., 2015
Nanogel	Curcumin	Gong et al., 2013
Supercritical fluids	Astaxanthin	Can et al., 2009
	Lutein	Heyang, Fei, Jiang, Yaping, & Lin, 2009

(*Continued*)

TABLE 6.4 (Continued)

Nanoencapsulation Method	Natural Colorants	Reference
Nanoprecipitation	Quercetin	Wu et al., 2008
	Curcumin	Anand et al., 2010
Carbon nanotubes	β-Carotene	Yanagi et al., 2006
Starch nanoparticle	β-Carotene	Kim, Seo, & Lim, 2013
Solvent displacement	β-carotene	Chu, Ichikawa, Kanafusa, & Nakajima, 2007a, 2007b
	Astaxanthin	Tachaprutinun, Udomsup, Luadthong, & Wanichwecharungruang, 2009

entrapment systems for the natural food colorants are explained in detail later. The selection of a nanoencapsulation technique requires the accurate knowledge of the physicochemical properties of the shell and core materials to pick the best materials for our encapsulation purpose and to acquire the intended results.

6.5.1 Lipid-Based Nanocarriers for Nanoencapsulation of Food Colors

Most of the natural colorants including carotenoids, tocopherols, and flavonoid pigments have a hydrophobic nature. For the accommodation of these water-insoluble pigments often emulsifiers are required to ensure the stability of these aqueous systems. Furthermore, the presence of digestible lipids is a great aid to the absorption of the bioactives via the enterocytes since it increases the content of mixed micelles responsible for the transportation of hydrophobic compounds such as natural colorants (McClements, 2012; Porter, Trevaskis, & Charman, 2007; Pouton, 2006). Thus, lipid nanovehicles are interesting candidates for the delivery of these compounds as their chemical structure fits the entrapped bioactive compounds (Tamjidi, Shahedi, Varshosaz, & Nasirpour, 2013). Various types of lipid-based nanocarriers including nanoliposomes, nanoemulsions, solid lipid nanoparticles (SLNs), and nanostructured lipid carriers (NLCs) have been reported for nanoencapsulation of natural colorants.

6.5.1.1 Nanoliposomes

Takahashi, Uechi, Takara, Asikin, and Wada (2009) incorporated curcumin as a bioactive compound in small unilamellar vesicle (SUV) nanoliposomes with the diameter of nearly 263 nm. Results indicated the enhanced rate of bioavailability as well as intensified absorption rate, compared with other tested types. This study suggested that the formulated nanoliposomes caused a higher plasma antioxidant activity of oral curcumin nanoliposomes. Therefore, nanoliposome was introduced as a promising delivery system for augmenting the absorption of nutrients. Moraes et al. (2013) analyzed the possibility of producing beta-carotene-loaded proliposomes via spray drying and to assess the encapsulation efficiency and capacity of the nanocarriers fabricated by the hydration of dry phospholipids grains. The retention of beta-carotene in the fine powders was analyzed during the storage period under ambient pressure and vacuum. The proliposome exhibited high solubility and the prepared matrix was able to retain more than 90% of the encased beta-carotene for 60 days at the refrigeration temperature and vacuum.

Also, Yoo et al. (2010) fabricated nanoliposomes as a means to increase the stability of astaxanthin. The resistance of astaxanthin accommodated within the nanoliposomes against thermal processing and UV rays was raised to twofold and threefold, respectively. Thin film evaporation is another useful technique for the production of nanoliposomes which was implemented by Tan, Xue, Abbas, et al. (2014) and Tan, Xue, Lou et al. (2014) to entrap carotenoids. Release and encapsulation properties were different for each loaded carotenoid type including lycopene, β-carotene, canthaxanthin, and lutein. Lutein and β-carotene-loaded nanoliposomes were released gradually within the GI tract media, while lycopene and canthaxanthin incorporated in nanoliposomes represented a burst and rapid liberation (Tan et al., 2014). Resultantly, the highest bioaccessibility was found for lutein-loaded nanoliposomes and the bioavailability decreased in this order: β-carotene, lycopene, and canthaxanthin incorporated within nanoliposomes. The outcome of this research suggested that the nanoliposomes are efficient candidates for improving the delivery of water-insoluble carotenoid compounds. Frenzel, Krolak, Wagner, and Steffen-Heins (2015) developed quercetin-loaded liposome and characterized their physicochemical and sensory properties after coating with whey protein isolate (WPI) and spray drying. The spray dried nanostructures were added to a functional drink composed of sweet whey permeate. In vitro gastric digestion tests revealed that the physical stability of prepared samples was dramatically improved. Also, the bitter taste of quercetin was masked following the encapsulation process. All in all, WPI-coated nanoliposomes were introduced as effective and safe food-grade nanocarriers to accommodate lipophilic bioactive compounds. In another study performed by Shin, Chung, Kim, Joung, and Park (2013), curcumin was incorporated in nanoliposomes formulated by the ethanol injection method (EIM). Later, the physicochemical properties of the prepared nanomaterials were compared to the typical dry thin film method (DTFM). Nevertheless, DTFM has some major drawbacks in order to be applied to the food products including the employment of hazardous organic solvents.

However, by using the EIM method, this limitation is removed and higher encapsulation efficiency along with better stability during the storage stage is obtained.

6.5.1.2 Nanoemulsions

Several studies have explored the encapsulation of curcumin within nanoemulsion networks and examined the impact of chain length of lipids on bioavailability levels (Ahmed, Li, McClements, & Xiao, 2012). The bioaccessibility of curcumin nanoemulsions was significantly different with chain length of the lipids. The highest bioaccessibility was observed with medium-chain triglyceride (MCT) oil. In another work, nanoparticles of saffron extract were encapsulated by spray drying of W/O/W multiple emulsions by Esfanjani, Jafari, Assadpoor, and Mohammadi (2015). The initial saffron water extract-in-oil (W/O) nanoemulsion was utilized in order to fabricate double emulsions in which a scattered mass fraction was equal to 0.25, furthermore the mixture was stabilized by implementing whey protein concentrate (WPC)/pectin blend. Also, the entrapment efficiency of crocin, picrocrocin, and saffranal were computed as well as the surface properties of the produced spray dried particles. To sum up, the formulated W/O/W multiple emulsions were able to accommodate the saffron colorant together with the other aforementioned bioactive compounds and afforded the maximum encapsulation efficiency.

Qian, Decker, Xiao, and McClements (2012) studied the impact of carrier oil type utilized in the nanoemulsions confining β-carotene as the chosen colorant. Oil-in-water nanoemulsions with diameters less than 200 nm were fabricated by employing Tween 20 as a nonionic surfactant together with long-chain triglycerides (LCT), MCT, or orange oil as efficient emulsifiers for increasing the stability of the obtained mixture. The formulated network was tested for its bioaccessibility by simulating the buccal, gastric, and small intestine media via in vitro assays. The results indicated that the utilized carrier oil character has an inevitable role in the digestion rate of lipids. The degradation rate and extent of MCT and LCT compositions were fairly alike in which almost the entire fatty acids were liberated at the final stage of digestion. Accordingly, the bioaccessibility of β-carotene was contingent upon the used lipid type and exhibited a descending trend from LCT to orange oil (LCT > MCT > orange oil). In addition, Salvia-Trujillo, Qian, Martín-Belloso, and McClements (2013) used different compositions and concentrations of oils in formulating nanoemulsions and tested the bioaccessibility of the prepared nanostructures. As a result, both samples prepared with low fat amount of LCT and high lipid level of MCT nanoemulsions proved to be highly bioaccessible. In another investigation, Liang, Shoemaker, Yang, Zhong, and Huang (2013) developed nanoemulsions loaded with β-carotene and used high starch content as a stabilizer. Indeed, the usage of modified starch has a direct impact on the droplet size, thermodynamic stability, and digestibility of the prepared nanoemulsions. The greater dispersed molecular density of modified starch leads to the formation of dense oil droplets that accelerates overall stability, increases the mean droplet size, and reduces the bioaccessibility at the time of in vitro digestion.

Ni, Hu, Sun, Zhao, and Xia, (2017) produced quercetin-loaded nanoemulsions (QT-NE) by using the high-pressure homogenization mechanism. The optimum QT-NE sample had a fair droplet size of 152 nm and entrapment efficiency of about 93.5% as well as being stable both chemically and physically. 2,2-Diphenyl-1-picrylhydrazyl (DPPH) free radical scavenging assay indicated satisfactory results of the antioxidant performance of QT-NE in comparison to bare quercetin. Moreover, the liberation style of quercetin was explored in the simulated gastrointestinal medium and modeled kinetically. Overall, the simulated digestion process depicted that the bioaccessibility of quercetin was considerably improved by means of nanoencapsulation.

6.5.1.3 Solid Lipid Nanoparticles

de Carvalho et al. (2013) developed solid lipid nanoparticles (SLNs) to entrap α-tocopherol by using the hot high homogenization technique. SLN networks were best formulated by applying response surface methodology to determine the optimized mean particle size, α-tocopherol retention value plus the zeta potential of the nanocarriers. The lecithin concentration did not alter the mean particle size, poloxamer level did not affect the restoration of α-tocopherol, and the zeta potential was not dependent on the α-tocopherol concentration. The best formulation in this experiment contained 1.1% α-tocopherol, 0.75% soy lecithin, and 1.3% poloxamer; the α-tocopherol retention rate was about 75.4. Sun et al. (2013) fabricated SLNs loaded with curcumin (Cur-SLNs) via the hot high-pressure homogenization with propylene glycol monocaprylate (Sefsol-218) as the liquid oil. Cur-SLNs presented a mean particle size of 153 nm and 90% entrapment efficiency. Cur-SLNs have the potential to boost the dispersibility and stability of curcumin and to lengthen the antitumor effect of this bioactive ingredient.

In another study, Qian, Decker, Xiao, and McClements (2013) examined the outcome of using oil as emulsifiers on the physicochemical stability of SLNs loaded with β-carotene. They noted that the liquid lipid nanoparticles (LLNs) were resistant against the degeneration of β-carotene and droplet aggregation. Furthermore, the crystals generated by cocoa butter and hydrogenated palm oil combinations were too highly organized that resulted in the liberation of β-carotene. Astaxanthin (ASTA) is another antioxidant that has been incorporated within SLNs by Li, Zahi, Yuan, Tian, and Liang (2015) via the high-pressure homogenization technique. ASTA-SLNs comprised soybean oil as the lipid phase and Tween 20 along with deionized water as the aqueous phase. The concentrations of stearic acid, glycerin monostearate, and glycerol distearates as solid lipids together with the processing mode (pressure and number of cycles) were optimized. Stearic acid (1 wt%) was chosen according to the physicochemical characteristics of ASTA-SLNs including zeta potential, mean particle size, and polydispersivity index (PDI). Moreover, the prepared nanostructure was highly stable at 4°C and 25°C and no significant alteration in the particle size was observed. Ultimately, the release assays indicated that ASTA-SLNs provide a sustained release in simulated intestinal fluids.

In a recent work, anthocyanins extracted from red cabbage (*Brassica oleracea* L. var. Capitata f. Rubra) were nanoencapsulated by SLNs (Ravanfar, Tamaddon, Niakousari, & Moein, 2016). The prepared nanoparticles were tested for particle size together with entrapment efficiency. Likewise, the formulation parameters, such as the volume of the interior aqueous state, homogenization time, amount of the entire lipid, and the whole implemented surfactant or stabilizer were also optimized. The estimated entrapment efficiency was $89.2 \pm 0.3\%$ and the average particle size was about 455 ± 2 nm. Finally, the assessment by SEM apparatus characterized the spherical shape of the particles (Ravanfar et al., 2016).

6.5.1.4 Nanostructured Lipid Carriers

Lacatusu et al. (2013) fabricated nanostructured lipid carriers (NLCs) for the nanoencapsulation of lutein by utilizing ω-3 fatty acids as liquid oil and carnauba wax and glycerol stearate as solid state lipids. They concluded that the particle sizes and entrapment efficiency of the formulation were under 200 nm and 89%, respectively (Lacatusu et al., 2013). The oxygen absorbance potential of lutein-loaded NLCs was estimated to be nearly 98%. The in vitro release assay revealed that the designed NLCs manifested a prolonged release behavior of lutein when compared with the commonly used nanoemulsions. Hence, this novel nanostructure can be employed in the production of functional food stuff in regard to its exclusive advantages mentioned before. In another study by Hentschel et al. (2008), NLCs were fabricated by using hydrophobic β-carotene in an aqueous phase. Subsequently, photon correlation spectroscopy (PCS) and laser diffraction (LD) runs were carried out to determine the physical stability and particle size of the NLC suspensions by exposure to two different storage temperatures. All the fabricated particles with adequate levels of emulsifier were smaller than 1 μm (LD diameter 100%) and the average particle size was around 0.3 μm (LD) for 9 weeks at 20°C and a minimum period of 30 weeks at 4–8°C. Differential scanning calorimetry (DSC) was also implemented to assess the required heat for the crystallization (phase conversion of the applied lipids) in both propylene glycol monostearate (PGMS) plus the NLC particles. PGMS being introduced into the NLC was actually recrystallized up to 98% during storage phase. Besides, no polymorph transitions were noticed during the 7-month storage span.

Curcumin-loaded NLCs (Cur-NLCs) were developed by Fang et al. (2012) to be directly administered into the stomach of laboratory rats. This research explored the pharmacokinetics, nanostructure diffusion in tissues, and approximate bioavailability of curcumin in rats after the rats were fed with the prepared formulation. Regarding the tissue distribution tests, curcumin was identified within heart, spleen, liver, lungs, and brain. Cur-NLCs enhanced the residence time of curcumin in the prementioned organs, particularly in the brain. They concluded that NLCs are capable of improving the oral absorption of water-insoluble bioactives, such as curcumin.

Hejri, Khosravi, Gharanjig, and Hejazi (2013) optimized β-carotene-loaded NLCs using the solvent diffusion method. The results depicted that the lipid phase concentration along with the surface-active agents level had a direct impact on the particle size of NLCs. Also, the liquid lipid contribution to the degradation of β-carotene was much higher in relation to the whole lipid ratio. Mathematical models were established and fitted to obtain the optimum formulations, subsequently they were experimentally tested and the resultant values were in agreement with the predicted measurements (least particle size of 8−15 nm and low β-carotene degradation of 0−3%). In a similar study, β-carotene was also successfully incorporated into an NLC matrix, composed of anhydrous milk fat (AMF) as the solid lipid and Tween 80 as the surface active agent through the phase inversion temperature (PIT) method (Zhang, Hayes, Chen, & Zhong, 2013). The authors noted that the advantages of PIT over the high-pressure homogenization method include the formation of transparent yield and being economical. They studied thermal processing circumstances, salinity, and the proportion of surfactant to oil in the formed NLCs and the related parameters affecting the turbidity of the emulsions. Altogether, this study supported that PIT method can be applied to fabricate stable and transparent NLCs for encapsulation of various lipophilic bioactive compounds in food and beverage systems.

NLCs have been developed for nanoencapsulation of quercetin (QT) by Ni, Sun, Zhao, and Xia (2015). The QT-NLC specimens were produced by applying the high-pressure homogenization method. The optimized QT-NLC presented a high encapsulation efficiency of about 93.50%. In addition, the formulated QT-NLCs were highly stable at the room temperature. In vitro antioxidant activity experiments demonstrated the elevated antioxidant performance of QT-NLC to uncoated QT. Furthermore, Fourier transform infrared spectroscopy report represented no chemical interaction between QT and lipid matrix as well as the successful encapsulation process. The gastrointestinal tract model approved the enhanced bioavailability obtained by encapsulation. They introduced QT-NLC as practical additives to enrich soft beverages. Similarly, Liu et al. (2014) developed a new model of quercetin-loaded cationic NLC (QT-CNLC) and analyzed its biodistribution in vivo after oral delivery. To sum up, the prepared QT-CNLC had a prolonged release and is an effective alternative to free QT as it would lead to the accumulation in several organs, such as lung, liver, and kidney following oral administration. Barras et al. (2009) implemented the phase inversion technique to prepare lipid nanocapsules (LNCs) involving two bioactive ingredients (quercetin and epigallocatechin gallate). Initially, the bioactives were mixed in the oil phase; afterwards soybean lecithin, NaCl, surfactant, and distilled water were added to the mixture and heated. The formulated combination was cooled and distilled cold water (0°C) was added under stirring to O/W nanocapsules. By using this method, it is possible to control the size of formed LNC and therefore the formulation needs to be accurately designed according to our purpose. The obtained results of this experiment exhibited the enhanced rate of nanostructure solubility by a factor of 100 as well as being stable during storage.

In another study, astaxanthin-loaded NLCs were prepared by Tamjidi, Shahedi, Varshosaz, and Nasirpour (2014). Tween 80 and lecithin were the chosen emulsifiers and oleic acid together with glyceryl behenate was the utilized lipid. Astaxanthin-NLCs were fabricated by melt emulsification sonication method and stored at 19°C during 25 days. The ideal formulation of astaxanthin-NLC (with OCL: 22.4% and LTR: 1.8) had dominant properties against the free astaxanthin. X-ray diffraction and thermal identifications suggested a novel crystalline structure presenting lower crystallinity for the ideal composition in comparison to glyceryl behenate (Tamjidi, et al. 2014).

6.5.2 Nature-Inspired Nanocarriers for Nanoencapsulation of Food Colors

Natural nanocarriers are currently popular delivery systems for incorporating food bioactive ingredients according to their safe nature and being inexpensive. Also, concerning the properties of the core material our delivery system should be accordingly chosen and engineered physically, chemically, and biochemically modifying the original substances. As an example, casein micelles are considered as beneficial nanocarriers for encasing and shielding sensitive hydrophobic nutraceuticals within other food products. Cyclodextrins are other interesting cage-like molecules that confine bioactive materials via molecular inclusion. Cyclodextrins have both water-soluble and water-insoluble parts located outside and inside their structure, respectively. This natural structure makes cyclodextrins an attractive candidate to confine apolar molecules like essential oils inside their cavity via hydrophobic interactions (Astray, Gonzalez-Barreiro, Mejuto, Rial-Otero, Simal-Gándara, 2009). Nanosuspensions or nanocrystal suspensions require the minimum surfactant for stabilization of included materials (Shin, Kim, & Park, 2015). Nanocrystals are perfect candidates for low-soluble bioactive ingredients. Reduction in the particle size leads to the increased surface area along with saturation solubility and boost the dissolution rate significantly, which are considered as particular advantages of these novel nanovehicles.

6.5.2.1 Molecular Inclusion Complexes with Cyclodextrins

Yadav et al. (2010) produced 2-hydroxypropyl-γ-cyclodextrin (HPγCD) complexes incorporating curcumin by a pH shift approach. At first, curcumin was dissolved in an alkaline solution involving HPγCD and then a modification of the pH value (at 6) resulted in the binding affinity of curcumin to the hydrophobic cavity of the CD. As a whole, the scholars reported the successful enhanced cellular uptake and antiinflammatory performance than that of curcumin. In another work, Nerome et al. (2013) made an effort to increase the dispersion of lycopene in water by applying the supercritical fluids (SEDS) approach. The parameters, such as the applied amounts of lycopene and β-CD, CO_2 volume velocity, volume velocity of the solution, and the pressure and temperature of the process were measured by running experiments. Results depicted that by applying high levels of the aforementioned parameters, particles of 40 nm were obtained in the end.

Popat, Karmakar, Jambhrunkar, Xu, and Yu (2014) applied the spray drying method to yield curcumin-loaded γ-hydroxypropyl CD (Cur-CD) hollow spheres. Then, chitosan polymer was used for encapsulation of Cur-CDs to boost their solubility and biocidal effects. Overall, CDs and chitosan were proved to be efficient nanocarriers for the delivery of hydrophobic bioactives. Another approach for accommodating bixin extracts (1:1 ratio) with curcumin in β-CDs (1:2 ratio) has been examined by Marcolino, Zanin, Durrant, Benassi, and Matioli (2011). According to the DSC and TGA tests, the complexation of curcumin and bixin yield stable networks. Curcumin and bixin complexes disintegrated at high temperatures, which indicates that the interaction between CDs and bixin-rich extracts is energetically favorable and supports the motion of water molecules inside the cavity of β-CD. The outcome of kneading process was akin to coprecipitation of β-CD, while the routine mixing procedure did not significantly change the stability of the encased carotenoids.

6.5.2.2 Caseins Nanoparticles

Casein micelles are naturally occurring nanovehicles abundant in dairy products. In a study, these nanocarriers were applied to entangle, stabilize, and preserve β-carotene during processing conditions (Sáiz-Abajo, González-Ferrero, Moreno-Ruiz, Romo-Hualde, and González-Navarro (2013)). β-Carotene was confined in the casein micelles via the self-assembly reaction. Casein micelles preserved β-carotene molecules during harsh processing conditions that are common in the food industry including baking. In another experiment, camel β-casein (B-CN) was employed in the encapsulation of curcumin (Esmaili, et al. 2011). Critical micelle value of camel B-CN was measured at 25°C, 30°C, and 37°C, furthermore the solubility of curcumin was assessed via the solvent evaporation strategy. The presence of camel B-CN increased the solubility of curcumin at least 2500-fold. Antioxidant activity of the entrapped form of curcumin was dramatically higher than that of both free B-CN and curcumin.

6.5.2.3 Nanocrystals

Quercetin is a naturally occurring flavonoid that has been exerted as nanosuspensions by simply applying Tween 80 and the high-pressure homogenization method to render it water-soluble (Karadag, Ozcelik, & Huang, 2014). Maltodextrin was used as a coating material during the spray drying method. The throughput exhibited higher antioxidant performance than the original samples. These scholars noted that taking into account both the high-pressure homogenization and spray drying is a practical means to formulate quercetin-based functional food products.

6.5.3 Nanoencapsulation of Food Colorants by Special Equipment

A variety of specialized equipment has been applied to the nanoencapsulation of several bioactive ingredients. Some of the most typical techniques for the entrapment of food bioactives are electrospinning, electrospraying, and nano-spray drying. Electrospinning was implemented to nanoencapsulate β-carotene loaded in nanoliposomes and then included within the electrospun ultrathin fibers by de Freitas Zômpero, López-Rubio, de Pinho, Lagaron, and de la Torre (2015). Hybrid encapsulation structures based on electrospinning and nanoliposomes were generated with the aim to promote the photostability of β-carotene. In this study, both the unloaded and filled nanoliposomes were immersed in polyvinyl alcohol and polyethylene oxide fluids. The modifications in the applied fluids were examined to characterize their impacts on the electrospinning processing. The mixed solutions were then loaded into the electrospinning equipment to yield nanoliposome-loaded ultrathin fibers. The availability of phospholipids in the electrospun fibers was confirmed by running FTIR analysis. The fabricated nanofibers were tested for their morphology, core material distribution along with their stability under UV rays. Moreover, following the dissolvation of electrospun fibers in water medium, liposomal liberation analysis supported the existence of liposomal entities.

6.5.4 Nanoencapsulation of Food Colorants by Biopolymer Nanoparticles

Biopolymer nanoparticles are colloidal building blocks that are gathered from several types of biopolymer molecules that meet a variety of practices in the food sector. Biopolymer nanoparticles may be particularly engineered for encasing and protecting the bioactive agents or as materials to fabricated value-added food products as well as modifying the texture, stability, and appearance of food stuff. For instance, curcumin nanosuspensions were produced by using five different stabilizers, such as polyvinyl alcohol (PVA), D-α-tocopherol polyethylene glycol 1000 succinate (TPGS), polyvinyl pyrrolidone (PVP), sodium dodecyl sulfate (SDS), and carboxy methyl cellulose sodium salt (Na-CMC) via high-pressure homogenization (Rachmawati, Shaal, Müller, & Keck, 2013). The aforestated stabilizers successfully formed curcumin nanosuspensions in the size range of 500–700 nm. The physical and momentary stability of curcumin nanosuspensions were alike for all stabilizers except PVP. Indeed, PVP was considered as the most effective compound for the stabilization of curcumin nanosuspension. As a result, the curcumin loaded in polymer nanoparticles could have great potentials of boosting the solubility and uptake of curcumin. In a similar case, Xie et al. (2011) synthetized curcumin-loaded nanoparticles by applying PLGA via the solvent evaporation technique. The buccal bioavailability of curcumin nanoparticles was promoted dramatically by improving the hydrophilicity and release speed in the intestinal fluid along with increasing of residence time in the intestinal lumen. Altogether, the PLGA polymer is a promising material to encapsulate hydrophobic bioactive molecules with a prolonged release profile and intensified bioaccessibility.

Arunkumar, Prashanth, and Baskaran (2013) prepared low molecular weight chitosan (LMWC) nanoparticles for the entanglement of lutein in order to advance its bioavailability. The bioavailability of lutein-LMWC nanostructures was detected and compared with lutein control sample by both in vitro and in vivo assays. According to the in vitro test, the bioavailability of lutein-LMWC nanoparticles was much higher (27.7%) than the control specimen. A large amount of lutein was noticeable in plasma (54.5%), liver (53.9%), and eyes (62.8%) of mice. Moreover, Yi, Lam, Yokoyama, Cheng, and Zhong (2015) fabricated protein-based nanostructures loaded with β-carotene by employing different concentrations of food-grade sodium caseinate (SC), soy protein isolate (SPI), and WPI via the homogenization evaporation mode. The generated nanovehicle was considered to be safe due to the performed cytotoxicity tests and revealed heightened cellular uptake of β-carotene, unlike the control sample. β-Carotene-loaded SC was considerably stable against oxidative stresses, which is ascribed to the different interfacial interactions of the droplet and the amino acid arrangements. All in all, the present study proposed a novel carrier system to increase the stability and absorbance of water-insoluble nutrients with the possibility of commercialization in the food sector.

6.5.5 Other Methods of Nanoencapsulation for Food Colors

Gong et al. (2013) conducted a research on both curcumin-loaded micelles (Cur-M) plus Cur-M loaded in hydrogels (Cur-M-H) susceptible to heat for yielding a biodegradable drug delivery matrix with the aim of healing wounds on skin. Cur-M bear a higher loading and encapsulation value in regard to its hydrophobic nature. In vitro experiments confirmed that Cur-M-H is able to modify the carrier into a gel within the body that causes a sustained release of curcumin because of binding to the target tissue. Also, the in vivo tests were in line with the in vitro experiments supporting the recovery effects. Altogether, the complex of curcumin and the thermosensitive hydrogel leads to the regeneration of damaged tissues and had impressive results.

Can, Johan, and Charlotta (2009) showed that RESS (rapid expansion of supercritical solutions) assisted micronization of astaxanthin-generated fine globular particles with the average particle size of 0.5 μm that is much smaller than microparticles that are produced via crystallization using n-hexane as the solvent. The application of low pressure together with high temperature simplified the formation of fine particles, which can be explained by the decreased critical nuclei size at the time of atomization.

Recently, de Paz et al. (2012a, b) developed carotenoid-incorporated delivery systems via PGSS method by utilizing soybean lecithin and poly-ε-caprolactones. The challenging factors which influenced the β-carotene encapsulation were the usage of low and high molecular weight poly-ε-caprolactones, monitoring the operation situation in the precipitation vessel (i.e., pressure, homogenization time, and temperature) and the molar proportion of β-carotene:polymeric vehicle (de Paz et al., 2012b). Low molecular weight polymers were able to significantly promote the encapsulation parameters, such as process throughput, particle average size, etc. This result can be explained by the lower viscosity (1690 vs. 9300 mPa · s) and higher distribution in supercritical CO_2. Hence, atomization is simplified and smaller droplets are formed at the end of the nozzle because of the enormous applied content of solubilized supercritical gas.

The nanoprecipitation technique was also applied to entrap curcumin in an investigation conducted by Anand et al. (2010). Poly (lactide-co-glycolide) (PLGA) and polyethylene glycol (PEG)-5000 were included in the formulation of the intended nanostructure. Nanoprecipitation has three stages: First, the bioactives and a polymer are combined with an organic solution; second, the prepared complex is added dropwise to an aqueous solution including a surfactant, and in the final step, the resultant nanoparticles are vacuum evaporated to remove the organic solvent and ultimately the particles are obtained by either centrifugation or filtration. Considering the incorporation of curcumin in nanoparticles, the entrapment efficiency was approximately 97.5% and the significant minuscule size of particles (diameter of 80.9 nm) enhanced the cellular uptake as well as the in vitro and in vivo bioaccessibility over free curcumin (Anand et al., 2010). Employing the same technique, quercetin was loaded in nanoparticles with the size less than 85 nm and the encapsulation efficiency of above 99% (Wu et al., 2008).

Finally, single-walled carbon nanotubes (SWCNTs) were implemented for nanoencapsulation of β-carotene (Yanagi, Miyata, & Kataura, 2006). The tube encompassed β-carotene and sheltered it against the radical species and inhibited the isomerization process. This protection mode resists the decomposition of π-conjugated molecules, which restricts their practice in photonic equipment.

6.6 SAFETY ASPECTS OF NANOENCAPSULATED FOOD COLORANTS

The application of food colorants is strictly controlled by the legislation at national and international levels. The legislations of each country might be different according to their particular protocols and interpretations. To date, the European Union (EU) and the United States have authorized 43 and 30 colorants as food additives, respectively. Regarding the 21 CFR 73, the FDA has clarified the regulations in regard to the safety of each colorant for food applications. In 1993, the FDA introduced the *Redbook of Colorants*, which included the toxicological effects of each food colorant and summarized the position of food additives in the United States (Socaciu, 2007). Also, the potential health effects of the nanoencapsulated colorants are not clear yet and more studies should be done on this subject. According to the FDA, the "Generally Recognized As Safe" (GRAS) materials are officially confirmed to be safe for consumption. Therefore, any novel process or material should be reported to the FDA in order to be approved before being released into the market or being introduced to mass production. Accordingly, nanoencapsulated materials belong to this novel category that must undergo an accurate safety evaluation prior to consumption by human being. Yet, the regulatory principles for nanofoods are being promoted and first it should be determined whether the nanofood does not induce immunological and carcinogenic disorders in the physiological system and when it is considered as GRAS stuff, it should be expected that in the following years the nanofood will find its way in the shelves of markets around the world. It is predicted that by implementing the nanoencapsulation technology in the food sector the public well-being is increased and this is the gift of these functional products for whole population (Gutiérrez et al., 2013).

6.7 APPLICATION OF ENCAPSULATED COLORANTS IN FOOD FORMULATIONS

Before the entry of a nanofood product in the market, it must satisfy the needs of consumers. Hence, the storage stability of the nanocarrier, the compatibility of the payload with the food matrix along with the processing condition, and release type have to be determined (Augustin & Hemar, 2009). Nanoencapsulated powders may have several applications, such as mixing with other dry ingredients or by their addition to food products during the production process or reconstituted before being introduced into potables. During the production phase of nanostructures, the encapsulated ingredients are mixed with other nutrients and undergo high temperature and shear stresses, thus the fabrication of encapsulated ingredients has to be engineered in line with the end application.

Natural colorants have a long history of application. Natural colorants together with their advantageous properties have prompted to their broad application in the food sector; preparations to incorporate them in oily or aqueous matrix including colloidal suspensions, emulsions, and complexes with proteins are fruit juices and beverages, dairy and related products, desserts, mixes, salad dressings, confectionery products, pasta, meat, canned, etc. Generally, natural colorants are preferred among people according to their origin over the synthetic dyes; however, their major limitation is ascribed to their instability. So, encapsulated colorants have been evaluated to overcome this concern. Micro/nanoencapsulation of colors is beneficial especially for water-insoluble pigments as most of the food products store a significant amount of water. Possible food products in which encapsulated colorants can be introduced are dairy, bakery, and confectionery products as shown in Table 6.5.

TABLE 6.5 Application of Encapsulated Natural Colorants in the Food Formulations

Natural Colorant	Food Model System	Reference
Anthocyanin	Jelly	Mahdavi et al., 2016a; Mahdavi et al., 2016b
	Hard candy	Selim, Khalil, Abdel-Bary, & Abdel-Azeim, 2008
	Strawberry jam	Selim et al., 2008
	Yogurt	Ersus & Yurdagel, 2007
	Soft drink	Ersus & Yurdagel, 2007
	Soft drink	Burin, Rossa, Ferreira-Lima, Hillmann, & Boirdignon-Luiz, 2011
Lycopene	Cake	Rocha, Fávaro-Trindade, & Grosso, 2012
Betacyanin	Yogurt	Azeredo, Santos, Souza, Mendes, & Andrade, 2007
	Soft drink	Azeredo et al., 2007
Betalain	Yogurt	Obón, Castellar, Alacid, & Fernández-López, 2009
β-carotene	Yogurt	Toniazzo et al., 2014

6.8 CONCLUSION AND FURTHER REMARKS

In general, natural colorants are superior to synthetic dyes. The market for the entire food colorants dedicates nearly US$1 billion to itself, whereas the natural food colorant market is only one-third of it. Already, the artificial colorants have a better position in the colorant market due to their greater stability and higher yield. Nevertheless, by the employment of nanoencapsulated natural pigments, the major drawbacks of this type of colorants are masked and their bioavailability whether it is in vivo or in vitro is improved following administration. In this chapter, the natural colorants were briefly introduced and afterwards the nanoencapsulation techniques for the entrapment and delivery of these bioactive compounds were discussed and the latest literature in this field was underlined. Among the discussed nanoencapsulation systems, entanglement by lipid-based formulations is the most common technique and carotenoids are the frequent colorants which have been studied. There are still some technologies not being applied for these natural colorants which should be considered. Another important issue is investigating the toxicological and safety of produced nanoencapsulated food colorants, which requires in vitro and in vivo studies to determine whether they are harmful for organisms or not.

REFERENCES

Ahmed, K., Li, Y., McClements, D. J., & Xiao, H. (2012). Nanoemulsion-and emulsion-based delivery systems for curcumin: encapsulation and release properties. *Food Chemistry*, *132* (2), 799–807.

Anand, P., Nair, H. B., Sung, B., Kunnumakkara, A. B., Yadav, V. R., Tekmal, R. R., & Aggarwal, B. B. (2010). RETRACTED: Design of curcumin-loaded PLGA nanoparticles formulation with enhanced cellular uptake, and increased bioactivity in vitro and superior bioavailability in vivo. *Biochemical Pharmacology*, *79*(3), 330–338.

Anandharamakrishnan, C. (2014). *Techniques for nanoencapsulation of food ingredients*. New York: Springer.

Arunkumar, R., Prashanth, K. V. H., & Baskaran, V. (2013). Promising interaction between nanoencapsulated lutein with low molecular weight chitosan: characterization and bioavailability of lutein in vitro and in vivo. *Food Chemistry*, *141*(1), 327–337.

Astray, G., Gonzalez-Barreiro, C., Mejuto, J. C., Rial-Otero, R., & Simal-Gándara, J. (2009). A review on the use of cyclodextrins in foods. *Food Hydrocolloids*, *23*(7), 1631–1640.

Augustin, M. A., & Hemar, Y. (2009). Nano- and micro-structured assemblies for encapsulation of food ingredients. *Chemical Society Reviews*, *38*(4), 902–912.

Ax, K., Mayer-Miebach, E., Link, B., Schuchmann, H., & Schubert, H. (2003). Stability of Lycopene in Oil-in-Water Emulsions. *Engineering in Life Sciences*, *3*(4), 199–201.

Azeredo, H. M., Santos, A. N., Souza, A. C., Mendes, K. C., & Andrade, M. I. R. (2007). Betacyanin stability during processing and storage of a microencapsulated red beetroot extract. *American Journal of Food Technology*, *2*(4), 307–312.

Barenholz, Y., Diminsky, D., & Cohen, R. (2006). Carotenoid-loaded liposomes: Google Patents.

Barras, A., Mezzetti, A., Richard, A., Lazzaroni, S., Roux, S., Melnyk, P., & Monfilliette-Dupont, N. (2009). Formulation and characterization of polyphenol-loaded lipid nanocapsules. *International Journal of Pharmaceutics, 379*(2), 270–277.

Betz, M., & Kulozik, U. (2011). Microencapsulation of bioactive bilberry anthocyanins by means of whey protein gels. *Procedia Food Science, 1*, 2047–2056.

Blanc, P. J. (2002). *Natural food colorants. Encyclopedia of life support systems.* Oxford: EOLSS Publishers (on line).

Britton, G., Liaaen-Jensen, S., & Pfander, H. (2009). *Carotenoids volume 5: Nutrition and health* (Vol. 5) Berlin: Springer Science & Business Media.

Brunner, G. (2005). Supercritical fluids: technology and application to food processing. *Journal of Food Engineering, 67*(1), 21–33.

Burin, V. M., Rossa, P. N., Ferreira-Lima, N. E., Hillmann, M. C., & Boirdignon-Luiz, M. T. (2011). Anthocyanins: optimisation of extraction from Cabernet Sauvignon grapes, microcapsulation and stability in soft drink. *International Journal of Food Science & Technology, 46*(1), 186–193.

Cadena, P. G., Pereira, M. A., Cordeiro, R. B., Cavalcanti, I. M., Neto, B. B., Maria do Carmo, C., & Santos-Magalhães, N. S. (2013). Nanoencapsulation of quercetin and resveratrol into elastic liposomes. *Biochimica et Biophysica Acta (BBA)-Biomembranes, 1828*(2), 309–316.

Can, Q., Johan, C., & Charlotta, T. (2009). Carotenoids particle formation by supercritical fluid technologies. *Chinese Journal of Chemical Engineering, 17*(2), 344–349.

Caro, Y., Anamale, L., Fouillaud, M., Laurent, P., Petit, T., & Dufosse, L. (2012). Natural hydroxyanthraquinoid pigments as potent food grade colorants: an overview. *Natural Products and Bioprospecting, 2*(5), 174–193.

Carvallo, R. (2011). Supercritical fluid aided microencapsulation of dry powders.

Cavalcanti, R. N., Santos, D. T., & Meireles, M. A. A. (2011). Non-thermal stabilization mechanisms of anthocyanins in model and food systems—An overview. *Food Research International, 44*(2), 499–509.

Chu, B.-S., Ichikawa, S., Kanafusa, S., & Nakajima, M. (2007a). Preparation and characterization of β-carotene nanodispersions prepared by solvent displacement technique. *Journal of Agricultural and Food Chemistry, 55*(16), 6754–6760.

Chu, B.-S., Ichikawa, S., Kanafusa, S., & Nakajima, M. (2007b). Preparation of protein-stabilized β-carotene nanodispersions by emulsification–evaporation method. *Journal of the American Oil Chemists' Society, 84*(11), 1053–1062.

Cocero, M. J., Martín, Á., Mattea, F., & Varona, S. (2009). Encapsulation and co-precipitation processes with supercritical fluids: fundamentals and applications. *The Journal of Supercritical Fluids, 47*(3), 546–555.

Dao, L. T., Takeoka, G. R., Edwards, R. H., & Berrios, J. D. J. (1998). Improved method for the stabilization of anthocyanidins. *Journal of Agricultural and Food Chemistry, 46*(9), 3564–3569.

de Carvalho, S. M., Noronha, C. M., Floriani, C. L., Lino, R. C., Rocha, G., Bellettini, I. C., & Barreto, P. L. M. (2013). Optimization of α-tocopherol loaded solid lipid nanoparticles by central composite design. *Industrial Crops and Products, 49*, 278–285.

de Freitas Zômpero, R. H., López-Rubio, A., de Pinho, S. C., Lagaron, J. M., & de la Torre, L. G. (2015). Hybrid encapsulation structures based on β-carotene-loaded nanoliposomes within electrospun fibers. *Colloids and Surfaces B: Biointerfaces, 134*, 475–482.

de Paz, E., Martín, Á., & Cocero, M. J. (2012a). Formulation of β-carotene with soybean lecithin by PGSS (particles from gas saturated solutions)-drying. *The Journal of Supercritical Fluids, 72*, 125–133.

de Paz, E., Martín, Á., Duarte, C. M. M., & Cocero, M. J. (2012b). Formulation of β-carotene with poly-(ε-caprolactones) by PGSS process. *Powder Technology*, *217*, 77–83.

Delgado-Vargas, F., Jiménez, A., & Paredes-López, O. (2000). Natural pigments: carotenoids, anthocyanins, and betalains—Characteristics, biosynthesis, processing, and stability. *Critical Reviews in Food Science and Nutrition*, *40*(3), 173–289.

Delgado-Vargas, F., & Paredes-López, O. (2002). *Natural colorants for food and nutraceutical uses*. Boca Raton, FL: CRC Press.

Desobry, S. A., Netto, F. M., & Labuza, T. P. (1997). Comparison of Spray-drying, Drum-drying and Freeze-drying for β-Carotene Encapsulation and Preservation. *Journal of Food Science*, *62*(6), 1158–1162.

Dewettinck, K., & Huyghebaert, A. (1999). Fluidized bed coating in food technology. *Trends in Food Science & Technology*, *10*(4), 163–168.

DeZarn, T. J. (1995). Food ingredient encapsulation. *Encapsulation and Controlled Release of Food Ingredients* (590, pp. 74–86).

Edge, R., McGarvey, D., & Truscott, T. (1997). The carotenoids as anti-oxidants—a review. *Journal of Photochemistry and Photobiology B: Biology*, *41*(3), 189–200.

Ersus, S., & Yurdagel, U. (2007). Microencapsulation of anthocyanin pigments of black carrot (Daucus carota L.) by spray drier. *Journal of Food Engineering*, *80*(3), 805–812.

Esfanjani, A. F., Jafari, S. M., Assadpoor, E., & Mohammadi, A. (2015). Nano-encapsulation of saffron extract through double-layered multiple emulsions of pectin and whey protein concentrate. *Journal of Food Engineering*, *165*, 149–155.

Esmaili, M., Ghaffari, S. M., Moosavi-Movahedi, Z., Atri, M. S., Sharifizadeh, A., Farhadi, M., & Moosavi-Movahedi, A. A. (2011). Beta casein-micelle as a nano vehicle for solubility enhancement of curcumin; food industry application. *LWT-Food Science and Technology*, *44*(10), 2166–2172.

Ezhilarasi, P., Karthik, P., Chhanwal, N., & Anandharamakrishnan, C. (2013). Nanoencapsulation techniques for food bioactive components: a review. *Food and Bioprocess Technology*, *6*(3), 628–647.

Fang, M., Jin, Y., Bao, W., Gao, H., Xu, M., Wang, D., & Liu, L. (2012). In vitro characterization and in vivo evaluation of nanostructured lipid curcumin carriers for intragastric administration. *International Journal of Nanomedicine*, *7*, 5395–5404.

Fang, Z., & Bhandari, B. (2010). Encapsulation of polyphenols–a review. *Trends in Food Science & Technology*, *21*(10), 510–523.

Frenzel, M., Krolak, E., Wagner, A., & Steffen-Heins, A. (2015). Physicochemical properties of WPI coated liposomes serving as stable transporters in a real food matrix. *LWT-Food Science and Technology*, *63*(1), 527–534.

Gong, C., Wu, Q., Wang, Y., Zhang, D., Luo, F., Zhao, X., & Qian, Z. (2013). A biodegradable hydrogel system containing curcumin encapsulated in micelles for cutaneous wound healing. *Biomaterials*, *34*(27), 6377–6387.

Goodwin, T. (2012). *The biochemistry of the carotenoids: Volume I plants*. New York: Springer Science & Business Media.

Gould, K., Davies, K. M., & Winefield, C. (2008). *Anthocyanins: Biosynthesis, functions, and applications*. New York: Springer Science & Business Media.

Gutiérrez, F. J., Albillos, S. M., Casas-Sanz, E., Cruz, Z., García-Estrada, C., García-Guerra, A., & González-Ferrero, C. (2013). Methods for the nanoencapsulation of β-carotene in the food sector. *Trends in Food Science & Technology*, *32*(2), 73–83.

Harborne, J. B. (2013). *The flavonoids: Advances in research since 1980*. Berlin: Springer.

Havsteen, B. H. (2002). The biochemistry and medical significance of the flavonoids. *Pharmacology & therapeutics*, *96*(2), 67–202.

Hejri, A., Khosravi, A., Gharanjig, K., & Hejazi, M. (2013). Optimisation of the formulation of β-carotene loaded nanostructured lipid carriers prepared by solvent diffusion. *Food Chemistry*, *141*(1), 117–123.

Hendry, G. A. F., & Houghton, J. (1996). *Natural food colorants*. Berlin: Springer Science & Business Media.

Hentschel, A., Gramdorf, S., Müller, R., & Kurz, T. (2008). β-Carotene-loaded nanostructured lipid carriers. *Journal of Food Science*, *73*(2), N1–N6.

Heyang, J., Fei, X., Jiang, C., Yaping, Z., & Lin, H. (2009). Nanoencapsulation of lutein with hydroxypropylmethyl cellulose phthalate by supercritical antisolvent. *Chinese Journal of Chemical Engineering*, *17*(4), 672–677.

Huang, Q., & Given, P. (2009). *Micro/nano encapsulation of active food ingredients*. New York: American Chemical Society.

Jafari, S. M. (2017). An overview of nano-encapsulation techniques and their classification. In S. M. Jafari (Ed.), *Nanoencapsulation technologies for the food and nutraceutical industries* (Chapter 1: pp. 1–34). Amsterdam: Elsevier.

Jafari, S. M., Assadpoor, E., He, Y., & Bhandari, B. (2008). Encapsulation efficiency of food flavours and oils during spray drying. *Drying Technology*, *26*(7), 816–835.

Johnson, E. J. (2002). The role of carotenoids in human health. *Nutrition in Clinical Care*, *5*(2), 56–65.

Karadag, A., Ozcelik, B., & Huang, Q. (2014). Quercetin nanosuspensions produced by high-pressure homogenization. *Journal of Agricultural and Food Chemistry*, *62*(8), 1852–1859.

Khazaei, K. M., Jafari, S., Ghorbani, M., & Kakhki, A. H. (2014). Application of maltodextrin and gum Arabic in microencapsulation of saffron petal's anthocyanins and evaluating their storage stability and color. *Carbohydrate Polymers*, *105*, 57–62.

Khazaei, K. M., Jafari, S. M., Ghorbani, M., Kakhki, A. H., & Sarfarazi, M. (2016). Optimization of Anthocyanin Extraction from Saffron Petals with Response Surface Methodology. *Food Analytical Methods*, *9*(7), 1993–2001.

Katouzian, I., & Jafari, S. M. (2016). Nano-encapsulation as a promising approach for targeted delivery and controlled release of vitamins. *Trends in Food Science & Technology*, *53*, 34–48.

Kim, J.-Y., Seo, T.-R., & Lim, S.-T. (2013). Preparation of aqueous dispersion of β-carotene nano-composites through complex formation with starch dextrin. *Food Hydrocolloids*, *33*(2), 256–263.

Lacatusu, I., Mitrea, E., Badea, N., Stan, R., Oprea, O., & Meghea, A. (2013). Lipid nanoparticles based on omega-3 fatty acids as effective carriers for lutein delivery. Preparation and in vitro characterization studies. *Journal of Functional Foods*, *5*(3), 1260–1269.

Lauro, G. J., & Francis, J. (2000). *Natural food colorants: science and technology*. Boca Raton, FL: CRC Press.

Li, M., Zahi, M. R., Yuan, Q., Tian, F., & Liang, H. (2015). Preparation and stability of astaxanthin solid lipid nanoparticles based on stearic acid. *European Journal of Lipid Science and Technology*, .

Liang, R., Shoemaker, C. F., Yang, X., Zhong, F., & Huang, Q. (2013). Stability and bioaccessibility of β-carotene in nanoemulsions stabilized by modified starches. *Journal of Agricultural and Food Chemistry*, *61*(6), 1249–1257.

Liu, L., Tang, Y., Gao, C., Li, Y., Chen, S., Xiong, T., & Chen, H. (2014). Characterization and biodistribution in vivo of quercetin-loaded cationic nanostructured lipid carriers. *Colloids and Surfaces B: Biointerfaces*, *115*, 125–131.

Madene, A., Jacquot, M., Scher, J., & Desobry, S. (2006). Flavour encapsulation and controlled release—a review. *International Journal of Food Science & Technology*, *41*(1), 1–21.

Mahdavi, S. A., Jafari, S. M., Assadpoor, E., & Dehnad, D. (2016a). Microencapsulation optimization of natural anthocyanins with maltodextrin, gum Arabic and gelatin. *International Journal of Biological Macromolecules*, *85*, 379–385.

Mahdavi, S. A., Jafari, S. M., Assadpour, E., & Ghorbani, M. (2016b). Storage stability of encapsulated barberry's anthocyanin and its application in jelly formulation. *Journal of Food Engineering*, *181*, 59–66.

Mahdavi, S. A., Jafari, S. M., Ghorbani, M., & Assadpoor, E. (2014). Spray-drying microencapsulation of anthocyanins by natural biopolymers: A review. *Drying Technology*, *32*(5), 509–518.

Marcolino, V. A., Zanin, G. M., Durrant, L. R., Benassi, M. D. T., & Matioli, G. (2011). Interaction of curcumin and bixin with β-cyclodextrin: complexation methods, stability, and applications in food. *Journal of Agricultural and Food Chemistry*, *59*(7), 3348–3357.

McClements, D. J. (2012). Crystals and crystallization in oil-in-water emulsions: Implications for emulsion-based delivery systems. *Advances in Colloid and Interface Science*, *174*, 1–30.

Mehrnia, M.-A., Jafari, S.-M., Makhmal-Zadeh, B. S., & Maghsoudlou, Y. (2016). Crocin loaded nano-emulsions: Factors affecting emulsion properties in spontaneous emulsification. *International Journal of Biological Macromolecules*, *84*, 261–267.

Moraes, M., Carvalho, J. M. P., Silva, C. R., Cho, S., Sola, M. R., & Pinho, S. C. (2013). Liposomes encapsulating beta-carotene produced by the proliposomes method: characterisation and shelf life of powders and phospholipid vesicles. *International Journal of Food Science & Technology*, *48*(2), 274–282.

Munin, A., & Edwards-Lévy, F. (2011). Encapsulation of natural polyphenolic compounds; a review. *Pharmaceutics*, *3*(4), 793–829.

Nayak, C. A., & Rastogi, N. K. (2010). Effect of selected additives on microencapsulation of anthocyanin by spray drying. *Drying Technology*, *28*(12), 1396–1404.

Nazzaro, F., Fratianni, F., & Coppola, R. (2012). *Microtechnology and nanotechnology in food science. In. Green Technologies in Food Production and Processing* (pp. 471–494). New York: Springer.

Nedovic, V., Kalusevic, A., Manojlovic, V., Levic, S., & Bugarski, B. (2011). An overview of encapsulation technologies for food applications. *Procedia Food Science*, *1*, 1806–1815.

Neethirajan, S., & Jayas, D. S. (2011). Nanotechnology for the food and bioprocessing industries. *Food and Bioprocess Technology*, *4*(1), 39–47.

Nerome, H., Machmudah, S., Fukuzato, R., Higashiura, T., Youn, Y.-S., Lee, Y.-W., & Goto, M. (2013). Nanoparticle formation of lycopene/β-cyclodextrin inclusion complex using supercritical antisolvent precipitation. *The Journal of Supercritical Fluids*, *83*, 97–103.

Ni, S., Sun, R., Zhao, G., & Xia, Q. (2015). Quercetin loaded nanostructured lipid carrier for food fortification: preparation, characterization and in vitro study. *Journal of Food Process Engineering*, *38*(1), 93–106.

Ni, S., Hu, C., Sun, R., Zhao, G., & Xia, Q. (2017). Nanoemulsions-based delivery systems for encapsulation of quercetin: Preparation, characterization, and cytotoxicity studies. *Journal of Food Process Engineering*, *40*(2), e12374.

Obón, J., Castellar, M., Alacid, M., & Fernández-López, J. (2009). Production of a red–purple food colorant from Opuntia stricta fruits by spray drying and its application in food model systems. *Journal of Food Engineering*, *90*(4), 471–479.

Popat, A., Karmakar, S., Jambhrunkar, S., Xu, C., & Yu, C. (2014). Curcumin-cyclodextrin encapsulated chitosan nanoconjugates with enhanced solubility and cell cytotoxicity. *Colloids and Surfaces B: Biointerfaces*, *117*, 520–527.

Porter, C. J., Trevaskis, N. L., & Charman, W. N. (2007). Lipids and lipid-based formulations: optimizing the oral delivery of lipophilic drugs. *Nature Reviews Drug Discovery*, *6*(3), 231–248.

Pouton, C. W. (2006). Formulation of poorly water-soluble drugs for oral administration: physicochemical and physiological issues and the lipid formulation classification system. *European Journal of Pharmaceutical Sciences*, *29*(3), 278–287.

Qian, C., Decker, E. A., Xiao, H., & McClements, D. J. (2012). Nanoemulsion delivery systems: Influence of carrier oil on β-carotene bioaccessibility. *Food Chemistry*, *135*(3), 1440–1447.

Qian, C., Decker, E. A., Xiao, H., & McClements, D. J. (2013). Impact of lipid nanoparticle physical state on particle aggregation and β-carotene degradation: Potential limitations of solid lipid nanoparticles. *Food Research International*, *52*(1), 342–349.

Rachmawati, H., Shaal, L. A., Müller, R. H., & Keck, C. M. (2013). Development of curcumin nanocrystal: physical aspects. *Journal of Pharmaceutical Sciences*, *102*(1), 204–214.

Rao, A. V., & Rao, L. G. (2007). Carotenoids and human health. *Pharmacological Research*, *55* (3), 207–216.

Ravanfar, R., Tamaddon, A. M., Niakousari, M., & Moein, M. R. (2016). Preservation of anthocyanins in solid lipid nanoparticles: Optimization of a microemulsion dilution method using the Placket–Burman and Box–Behnken designs. *Food Chemistry*, *199*, 573–580.

Ribeiro, H. S., Schuchmann, H. P., Engel, R., Walz, E., & Briviba, K. (2010). *Encapsulation of carotenoids. Encapsulation technologies for active food ingredients and food processing* (pp. 211–252). New York: Springer.

Rocha, G. A., Fávaro-Trindade, C. S., & Grosso, C. R. F. (2012). Microencapsulation of lycopene by spray drying: characterization, stability and application of microcapsules. *Food and Bioproducts Processing*, *90*(1), 37–42.

Rymbai, H., Sharma, R., & Srivastav, M. (2011). Biocolorants and its implications in health and food industry-A review. *International Journal of PharmTech Research*, *3*(4), 2228–2244.

Saengkrit, N., Saesoo, S., Srinuanchai, W., Phunpee, S., & Ruktanonchai, U. R. (2014). Influence of curcumin-loaded cationic liposome on anticancer activity for cervical cancer therapy. *Colloids and Surfaces B: Biointerfaces*, *114*, 349–356.

Sáiz-Abajo, M.-J., González-Ferrero, C., Moreno-Ruiz, A., Romo-Hualde, A., & González-Navarro, C. J. (2013). Thermal protection of β-carotene in re-assembled casein micelles during different processing technologies applied in food industry. *Food Chemistry*, *138*(2), 1581–1587.

Salvia-Trujillo, L., Qian, C., Martín-Belloso, O., & McClements, D. (2013). Influence of particle size on lipid digestion and β-carotene bioaccessibility in emulsions and nanoemulsions. *Food Chemistry*, *141*(2), 1472–1480.

Santos, D. T., & Meireles, M. A. (2010). Carotenoid pigments encapsulation: fundamentals, techniques and recent trends. *Open Chemical Engineering Journal*, *4*(1), 42–50.

Selim, K., Khalil, K., Abdel-Bary, M., & Abdel-Azeim, N. (2008). *Extraction, encapsulation and utilization of red pigments from roselle (Hibiscus sabdariffa* L.*) as natural food colourants.* Paper presented at the Alexandria Journal for Food Science and Technology. Special Volume Conference.

Shaikh, J., Ankola, D., Beniwal, V., Singh, D., & Kumar, M. R. (2009). Nanoparticle encapsulation improves oral bioavailability of curcumin by at least 9-fold when compared to curcumin administered with piperine as absorption enhancer. *European Journal of Pharmaceutical Sciences*, *37*(3), 223–230.

Shin, G. H., Chung, S. K., Kim, J. T., Joung, H. J., & Park, H. J. (2013). Preparation of chitosan-coated nanoliposomes for improving the mucoadhesive property of curcumin using the ethanol injection method. *Journal of Agricultural and Food Chemistry*, *61*(46), 11119–11126.

Shin, G. H., Kim, J. T., & Park, H. J. (2015). Recent developments in nanoformulations of lipophilic functional foods. *Trends in Food Science & Technology*, *46*(1), 144–157.

Shipp, J., & Abdel-Aal, E.-S. M. (2010). Food applications and physiological effects of anthocyanins as functional food ingredients. *The Open food Science Journal*, *4*(1), 7–22.

Silva, H. D., Cerqueira, M. A., Souza, B. W., Ribeiro, C., Avides, M. C., Quintas, M. A., & Vicente, A. A. (2011a). Nanoemulsions of β-carotene using a high-energy emulsification–evaporation technique. *Journal of Food Engineering*, *102*(2), 130–135.

Silva, H. D., Cerqueira, M. A., Souza, B. W., Ribeiro, C., Avides, M. C., Quintas, M. A., & Vicente, A. A. (2011b). Preparation and characterization of nanoemulsion encapsulating curcumin. *Journal of Food Engineering*, *102*(2), 130–135.

Socaciu, C. (2007). *Food colorants: chemical and functional properties*. Boca Raton, FL: CRC Press.

Socaciu, C., Jessel, R., & Diehl, H. A. (2000). Competitive carotenoid and cholesterol incorporation into liposomes: effects on membrane phase transition, fluidity, polarity and anisotropy. *Chemistry and Physics of Lipids*, *106*(1), 79–88.

Suman, S. P., & Joseph, P. (2013). Myoglobin chemistry and meat color. *Annual Review of Food Science and Technology*, *4*, 79–99.

Sun, J., Bi, C., Chan, H. M., Sun, S., Zhang, Q., & Zheng, Y. (2013). Curcumin-loaded solid lipid nanoparticles have prolonged in vitro antitumour activity, cellular uptake and improved in vivo bioavailability. *Colloids and Surfaces B: Biointerfaces*, *111*, 367–375.

Tachaprutinun, A., Udomsup, T., Luadthong, C., & Wanichwecharungruang, S. (2009). Preventing the thermal degradation of astaxanthin through nanoencapsulation. *International Journal of Pharmaceutics*, *374*(1), 119–124.

Taghvaei, M., Jafari, S. M., Nowrouzieh, S., & Alishah, O. (2015). The influence of cooking process on the microwave-assisted extraction of cotton oil. *Journal of Food Science and Technology*, *52*(2), 1138–1144.

Takahashi, M., Uechi, S., Takara, K., Asikin, Y., & Wada, K. (2009). Evaluation of an oral carrier system in rats: bioavailability and antioxidant properties of liposome-encapsulated curcumin. *Journal of Agricultural and Food Chemistry*, *57*(19), 9141–9146.

Tamjidi, F., Shahedi, M., Varshosaz, J., & Nasirpour, A. (2013). Nanostructured lipid carriers (NLC): A potential delivery system for bioactive food molecules. *Innovative Food Science & Emerging Technologies*, *19*, 29–43.

Tamjidi, F., Shahedi, M., Varshosaz, J., & Nasirpour, A. (2014). Design and characterization of astaxanthin-loaded nanostructured lipid carriers. *Innovative Food Science & Emerging Technologies*, *26*, 366–374. Available from http://dx.doi.org/10.1016/j.ifset.2014.06.012.

Tan, C., Xue, J., Abbas, S., Feng, B., Zhang, X., & Xia, S. (2014). Liposome as a delivery system for carotenoids: comparative antioxidant activity of carotenoids as measured by ferric reducing antioxidant power, DPPH assay and lipid peroxidation. *Journal of Agricultural and Food Chemistry*, *62*(28), 6726–6735.

Tan, C., Xue, J., Lou, X., Abbas, S., Guan, Y., Feng, B., & Xia, S. (2014). Liposomes as delivery systems for carotenoids: comparative studies of loading ability, storage stability and in vitro release. *Food & Function*, *5*(6), 1232–1240.

Tan, C., Zhang, Y., Abbas, S., Feng, B., Zhang, X., & Xia, S. (2014). Modulation of the carotenoid bioaccessibility through liposomal encapsulation. *Colloids and Surfaces B: Biointerfaces*, *123*, 692–700.

Toniazzo, T., Berbel, I. F., Cho, S., Fávaro-Trindade, C. S., Moraes, I. C., & Pinho, S. C. (2014). β-carotene-loaded liposome dispersions stabilized with xanthan and guar gums: Physico-chemical stability and feasibility of application in yogurt. *LWT-Food Science and Technology*, *59*(2), 1265–1273.

Vila, M. M. D. C., Chaud, M. V., & Balcão, V. M. (2015). Microencapsulation of natural antioxidant pigments (Chapter 19) A2. In M. C. Leonard Sagis (Ed.), *Microencapsulation and microspheres for food applications* (pp. 369–389). San Diego: Academic Press.

Wildman, R. E., Wildman, R., & Wallace, T. C. (2016). *Handbook of nutraceuticals and functional foods*. Boca Raton, FL: CRC Press.

Wu, T.-H., Yen, F.-L., Lin, L.-T., Tsai, T.-R., Lin, C.-C., & Cham, T.-M. (2008). Preparation, physicochemical characterization, and antioxidant effects of quercetin nanoparticles. *International Journal of Pharmaceutics*, *346*(1), 160–168.

Xie, X., Tao, Q., Zou, Y., Zhang, F., Guo, M., Wang, Y., & Yu, S. (2011). PLGA nanoparticles improve the oral bioavailability of curcumin in rats: characterizations and mechanisms. *Journal of Agricultural and Food Chemistry*, *59*(17), 9280–9289.

Yadav, V. R., Prasad, S., Kannappan, R., Ravindran, J., Chaturvedi, M. M., Vaahtera, L., & Aggarwal, B. B. (2010). RETRACTED: Cyclodextrin-complexed curcumin exhibits anti-inflammatory and antiproliferative activities superior to those of curcumin through higher cellular uptake. *Biochemical Pharmacology*, *80*(7), 1021–1032.

Yanagi, K., Miyata, Y., & Kataura, H. (2006). Highly Stabilized β-Carotene in Carbon Nanotubes. *Advanced Materials*, *18*(4), 437–441.

Yi, J., Fan, Y., Zhang, Y., & Zhao, L. (2016). Characterization of catechin-α-lactalbumin conjugates and the improvement in β-carotene retention in an oil-in-water nanoemulsion. *Food Chemistry*, *205*, 73–80.

Yi, J., Lam, T. I., Yokoyama, W., Cheng, L. W., & Zhong, F. (2015). Beta-carotene encapsulated in food protein nanoparticles reduces peroxyl radical oxidation in Caco-2 cells. *Food Hydrocolloids*, *43*, 31–40.

Yoo, J.-M., Kim, S.-Y., Cho, E.-A., Cho, E.-H., Choi, S.-J., Jeong, Y.-J., & Chae, H.-J. (2010). Stabilization of astaxanthin using nanoliposome. *KSBB Journal*, *25*(2), 130–136.

Zhang, L., Hayes, D. G., Chen, G., & Zhong, Q. (2013). Transparent dispersions of milk-fat-based nanostructured lipid carriers for delivery of β-carotene. *Journal of Agricultural and Food Chemistry*, *61*(39), 9435–9443.

Zuidam, N. J., & Shimoni, E. (2010). *Overview of microencapsulates for use in food products or processes and methods to make them. Encapsulation technologies for active food ingredients and food processing* (pp. 3–29). New York: Springer.

FURTHER READING

Amin, A. A., Salama, M. F., Seliem, E. I., & Mahmoud, K. F. (2015). Encapsulation of Nano Carotenoids; Evaluation of Stability and Safety. *International Journal of Current Microbiology and Applied Sciences*, *4*(2), 1017–1029.

Bernardy, N., Romio, A. P., Barcelos, E. I., Pizzol, C. D., Dora, C. L., Lemos-Senna, E., & Sayer, C. (2010). Nanoencapsulation of quercetin via miniemulsion polymerization. *Journal of Biomedical Nanotechnology*, *6*(2), 181–186.

Coradini, K., Lima, F., Oliveira, C., Chaves, P., Athayde, M., Carvalho, L., & Beck, R. (2014). Co-encapsulation of resveratrol and curcumin in lipid-core nanocapsules improves their in vitro antioxidant effects. *European Journal of Pharmaceutics and Biopharmaceutics*, *88* (1), 178–185.

Das, R. K., Kasoju, N., & Bora, U. (2010). Encapsulation of curcumin in alginate-chitosan-pluronic composite nanoparticles for delivery to cancer cells. *Nanomedicine: Nanotechnology, Biology and Medicine*, *6*(1), 153–160.

Delgado-Vargas, F. (2002). *Pigments natural colorants for food and nutraceutical uses*. Boca Raton, FL: CRC Press.

Ferrari, C. C., Germer, S. P. M., & de Aguirre, J. M. (2012). Effects of spray-drying conditions on the physicochemical properties of blackberry powder. *Drying Technology*, *30*(2), 154–163.

Ferreyra, M. L. F., Rius, S. P., & Casati, P. (2012). Flavonoids: biosynthesis, biological functions, and biotechnological applications. *Frontiers in Plant Science*, *3*, 222.

Gomes, G. Vd. L., Borrin, T. R., Cardoso, L. P., Souto, E., & Pinho, S. Cd (2013). Characterization and shelf life of β-carotene loaded solid lipid microparticles produced with stearic acid and sunflower oil. *Brazilian Archives of Biology and Technology*, *56*(4), 663–671.

Henry, B. (1996). *Natural food colours. Natural food colorants* (pp. 40–79). Berlin: Springer.

Hwang, J.-M., Kuo, H.-C., Lin, C.-T., & Kao, E.-S. (2013). Inhibitory effect of liposome-encapsulated anthocyanin on melanogenesis in human melanocytes. *Pharmaceutical Biology*, *51*(8), 941–947.

MacDougall, D. B. (2002). *Colour in food: Improving quality*. Washington DC: CRC press.

Mozafari, M. R., Flanagan, J., Matia-Merino, L., Awati, A., Omri, A., Suntres, Z. E., & Singh, H. (2006). Recent trends in the lipid-based nanoencapsulation of antioxidants and their role in foods. *Journal of the Science of Food and Agriculture*, *86*(13), 2038–2045.

Pan, K., Zhong, Q., & Baek, S. J. (2013). Enhanced dispersibility and bioactivity of curcumin by encapsulation in casein nanocapsules. *Journal of Agricultural and Food Chemistry*, *61*(25), 6036–6043.

Quintanilla-Carvajal, M. X., Camacho-Díaz, B. H., Meraz-Torres, L. S., Chanona-Pérez, J. J., Alamilla-Beltrán, L., Jimenéz-Aparicio, A., & Gutiérrez-López, G. F. (2010). Nanoencapsulation: a new trend in food engineering processing. *Food Engineering Reviews*, *2*(1), 39–50.

Rao, P. J., & Khanum, H. (2016). A green chemistry approach for nanoencapsulation of bioactive compound—Curcumin. *LWT-Food Science and Technology*, *65*, 695–702.

Sari, T., Mann, B., Kumar, R., Singh, R., Sharma, R., Bhardwaj, M., & Athira, S. (2015). Preparation and characterization of nanoemulsion encapsulating curcumin. *Food Hydrocolloids*, *43*, 540–546.

Satué-Gracia, M. T., Heinonen, M., & Frankel, E. N. (1997). Anthocyanins as antioxidants on human low-density lipoprotein and lecithin-liposome systems. *Journal of Agricultural and Food Chemistry*, *45*(9), 3362–3367.

Souguir, H., Salaün, F., Douillet, P., Vroman, I., & Chatterjee, S. (2013). Nanoencapsulation of curcumin in polyurethane and polyurea shells by an emulsion diffusion method. *Chemical Engineering Journal*, *221*, 133–145.

Soukoulis, C., & Bohn, T. (2015). A comprehensive overview on the micro-and nano-technological encapsulation advances for enhancing the chemical stability and bioavailability of carotenoids. *Critical Reviews in Food Science and Nutrition*. Available from http://dx.doi.org/10.1080/10408398.2014.971353.

Taghvaei, M., Jafari, S. M., Assadpoor, E., Nowrouzieh, S., & Alishah, O. (2014). Optimization of microwave-assisted extraction of cottonseed oil and evaluation of its oxidative stability and physicochemical properties. *Food Chemistry*, *160*, 90–97.

Xin, Y. (2002). Natural food pigment and its function. *China Food Additives*, *5*, 1–3.

Yadav, A., Lomash, V., Samim, M., & Flora, S. J. (2012). Curcumin encapsulated in chitosan nanoparticles: a novel strategy for the treatment of arsenic toxicity. *Chemico-Biological Interactions*, *199*(1), 49–61.

Yurdugul, S., & Mozafari, M. R. (2004). Recent advances in micro-and nanoencapsulation of food ingredients. *Cellular and Molecular Biology Letters*, *9*(S2), 64–65.

Chapter 7

Nanoencapsulation of Flavors

Mohsen Asghari Ghajari[1,2], Iman Katouzian[1,2], Mohammad Ganjeh[1] and Seid Mahdi Jafari[1]

[1]*Gorgan University of Agricultural Sciences and Natural Resources, Gorgan, Iran,*
[2]*Nano-encapsulation in the Food, Nutraceutical, and Pharmaceutical Industries Group (NFNPIG), Universal Scientific Education and Research Network (USERN), Tehran, Iran*

7.1 INTRODUCTION

Encapsulation is a method applied to encase one or a mixture of bioactive compounds (core material) within another ingredient (shell/wall material). The advent of microencapsulation dates back to 1950, where it was exerted in fabricating pressure-sensible coatings in non-carbon copying paper (Green, 1955). Encapsulation technique has undergone a variety of changes, and currently it is applied in food, chemical, pharmaceutical, cosmeceuticals, printing, and many other sectors (Lakkis, 2007). Until now, different food components have been micro/nanoencapsulated in order to produce functional foods, as well as preserving the volatile and delicate compounds such as enzymes, colorants, vitamins, oleoresins, aromas, and, especially, flavors (Dubey, 2009; Jafari, Assadpoor, He, & Bhandari, 2008; Jafari, He, & Bhandari, 2007b; Mahdavi, Jafari, Ghorbani, & Assadpoor, 2014; Pourashouri et al., 2014b).

Flavorings are substances in the food that are responsible for their unique taste and odor. Some of these flavors present in different food products include allylpyrazine in roasted nut, methoxypyrazines in vegetables, 2-Isobutyl-3 methoxypyrazine in green pepper, acetyl-L- pyrazines in popcorn, aldehydes in fruits, terpenoids in citrus and piney, and phenolics in smoked products (Cardinal, Cornet, Serot, & Baron, 2006; Coetzee et al., 2015; Ling, Yang, Li, & Wang, 2015; Sharma, Utreja, & Bedi, 2016; Sidhu, Lund, Kotseridis, & Saucier, 2015; Zhu, Xiao, Zhou, & Lei, 2015). Due to the instability of most flavor structures, encapsulation seems to be a logical means to preserve the properties of these compounds.

The encapsulation procedure for flavors is summarized in two steps as follows: first, emulsification is applied to the payload like a lipid-based aroma using a carrier material made from polysaccharide or protein. In the second phase, the

Nanoencapsulation of Food Bioactive Ingredients. DOI: http://dx.doi.org/10.1016/B978-0-12-809740-3.00007-6

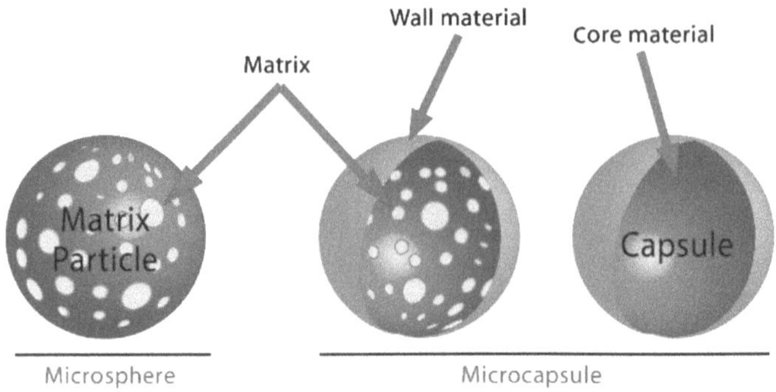

FIGURE 7.1 Simple illustration of microspheres and microcapsules.

emulsions are dried or cooled (Janda, Bernacchi, & Frieders, 1995; Mahdavi, Jafari, Assadpoor, & Dehnad, 2016). An illustration is provided for the flavor encapsulation in Fig. 7.1, which can be either in microcapsules or microspheres. Several physicochemical factors are responsible for the retention and controlled release of flavors, like molecular weight, polarity, and composition of the carrier material, etc. These parameters make possible the targeted release as well as preserving the bioactives against undesirable conditions inside and outside the body (Augustin, Sanguansri, Margetts, & Young, 2001; Gibbs, 1999; Jafari et al., 2007b; Jafari et al., 2008; Katouzian & Jafari, 2016). Regarding the implemented encapsulation technique, the encapsulants may have different shapes, such as spheres, disordered shapes plus films, and also the formed structures may be either compact or porous. All these factors together determine the diffusion pace of encapsulated flavors as well as oxygen and solvents penetration into the shell from outside. Before going through the techniques used for encapsulation of food flavors, a brief overview of food flavors is needed.

7.2 CLASSIFICATION OF FOOD BIOFLAVORS

Depending on the type and stage of bioprocessing (process pathway), flavors can be classified into two types: primary and secondary products. In this regard, the present text classify the food bioflavors based on these concepts.

7.2.1 Primary Flavors

These types of flavors are formed as a result of enzymatic reactions on raw materials. The fundamental compounds responsible for the generation of primary aromas are esters, aldehydes, terpenes, and alcohols. They are usually found in plants, fruits, and vegetables. In addition, lipids, carbohydrates, and amino acids are the main precursors in the formation of aroma compounds in fruits. Once the enzymatic hydrolysis process is initiated, most of the flavor compounds are released from their stable precursors in fruits and vegetables (Jelen, 2011).

7.2.2 Secondary Flavors

Secondary aromas are formed through three main stages: microbial activity (mainly in fermentation processes), controlled enzymatic reactions, and thermal reactions. Microbial activity triggers the release of a broad range of aroma compounds in fermented foods. In other type of food products, like cheese, and yogurt, etc., starter cultures are used for specific aroma formation (Marilley & Casey, 2004).

Heat-generated aroma compounds are another type of produced flavors that are applied to coffee and cocoa roasting, boiling, frying, grilling of meat, baking, and processes such as pasteurization of milk (Hodge, 1953).

7.3 FLAVORS FROM DIFFERENT ORIGINS

During past decades, some processed foods come with an ingredient label contained artificial flavors. A large number of chemicals, known as flavoring agents, are made from natural or artificial ingredients. Depending on the chemistry of flavors, they can be classified by their origin as listed below.

7.3.1 Flavors Derived from Saccharides

Saccharides are common components of foods, and together with lipids and proteins they constitute one of the important nutrients that provide the basic source of energy in the human diet (Jelen, 2011). Sugar, dextrin, starch, cellulose, hemicellulose, pectin, and certain gums are the most typical types of carbohydrates in foods. Besides, these basic compounds, derivatives such as deoxysugars, amino sugars, and sugar carboxylic acids have been incorporated in this class of compounds. Monosaccharides are the basic components of saccharides, and these monomers are mostly involved in the process of caramelization and Maillard reaction. These reactions are able to alter the flavor of foods.

Hydroxyl groups (−OH) are present in all saccharides as well as being available for reactions. On other hand, carbonyl groups are also available in low-molecular-weight carbohydrates. Once the carbohydrates are heated, caramelization occurs gradually creating unique flavors (Angyal, 2001). As a result of this reaction, a brown-colored product with typical caramel aroma is produced. Depending on the heating temperature and catalysts, specific color or aroma products are generated. Usually, temperatures of 150°C or greater and alkaline pH are required for the reaction to proceed (Reineccius, 2013).

Aliphatic aldehydes, ketones, and diketones are the most abundant flavor compounds formed by the Maillard reaction (Kussmann, Affolter, Nagy, Holst, & Fay, 2007). Table 7.1 represents the main flavor compounds formed by the Maillard reaction.

TABLE 7.1 Flavor Formation via the Maillard Reaction

Type of Compounds	Flavors	Reference
Carbonyl compounds	Aldehydes	Whitfield & Mottram, 1992
Nitrogen-containing Heterocyclic compounds	Pyrazine, methoxypyrazine, pyrrole, pyridine, pyrroline, pyrrolidine, pyrrolizine, and piperine	Newton, Fairbanks, Golding, Andrewes, & Gerrard, 2012
Oxygen-containing compounds	Maltol, furaneol, cyclotene, oxazole, and oxazoline	Bertrand et al., 2011; Van Boekel, 2006
Sulfur-containing Heterocyclic compounds	Thiophenes, dithioles, dithianes, dithiins, trithiolanes, trithanes, tetrathianes, thiazoles, thiazolines, and thiazolidines	Adams & De Kimpe, 2006; Yu, Tan, & Wang, 2012

7.3.2 Flavors Derived from Amino Acids

Amino acids are monomers that build up proteins, and following their degradation a large variety of volatile compounds are released. Aliphatic and branched chain alcohols, acids, carbonyls, and esters are derived from the metabolism of amino acids occurring in foods. Many plants and microorganisms can degenerate amino acids (Jelen, 2011). Since the handling of microorganisms is simple and also because of their importance in fermented foods, the knowledge of the pathways leading to the degradation of amino acids to create the flavor compounds has often obtained quicker via microorganisms (bacteria and fungi) rather than plants (Lichtenthaler, Rohmer, & Schwender, 1997). The most common pathway used by microorganisms for degrading amino acid is Erhlich–Neubauer's pathway (Reineccius, 2013). In Table 7.2, some of the flavors generated from amino acids are listed.

7.3.3 Flavors Derived from Lipids

Lipids may undergo changes during the processing of foods which will generate flavors. Precursors present in lipids are responsible for the formation of flavors. The compounds produced from lipids may be readily volatile and possess desirable odors (Jelen, 2011). Some examples of these types of compounds are low molecular weight aliphatic aldehydes, ketones, and fatty acids (Reineccius, 2013).

Deep fat fried foods, e.g., French fried potatoes, doughnuts, and snacks, are generally approved and consumed by the public. Although, their

TABLE 7.2 Examples of Flavors Produced From Amino Acids

Amino Acids	Pathway	Product	References
Valine	Converted to short chain Carbonyls	Tomato	Davidovich-Rikanati, Azulay, Sitrit, Tadmor, & Lewinsohn, 2009
Leucine			Moummou et al., 2012
Alanine,			Mathieu et al., 2009
Aspartic Acid			Tieman et al., 2012
Valine Leucine	Transformed into branched chain flavor	Banana	Torres, Pandey, & Castro, 2010
Isoleucine	Precursor of 2-methyl butyl and 2-methyl butenyl esters	Apples	Espino-Díaz, Molina-Corral, Sepulveda, González-Aguilar, & Olivas, 2016
			Maya-Meraz et al., 2014
Cinnamic acids	Esterification to methyl and ethyl cinnamate	Strawberry	Prat, Espinoza, Agosin, & Silva, 2014
		Guava	Pino & Bent, 2013
		Cranberry	Ruse et al., 2012
		Passion fruit	Macoris, Janzantti, Garruti, & Monteiro, 2011
Tyrosine	Degradation of tyrosine by *Brevibacterium linens*	Limburger	Nyberg, 2016
Cinnamic acids	decarboxylation of cinnamic acid and produce styrene (Styrene has a very strong plastic-like odor)	Cream	Pagot, Belin, Husson, & Spinnler, 2007
		Camembert	Spinnler & Leclercq-Perlat, 2007

popularity may partially lie in the physical properties imparted to the food by fats, e.g., lubricity, richness, and texture. The odor of fried flavor is unique and desirable. This flavor comes from the thermally induced changes in the food (Maillard reaction) as well as the generated flavor from the frying oil (Nawar, 1998). Lactones may be formed in foods via microbial action, extensive lipid oxidation (ambient or thermal), or heating (Maga & Katz, 1976).

Secondary reactions of the free fatty acids (e.g., oxidation) can yield a category of new flavor compounds. Acids may also be implemented from the deamination of amino acids. The products are various aliphatic (linear and branched chain) plus aromatic acids (Reineccius, 2013).

7.4 FLAVORS IN SOME FOOD PRODUCTS

In determining consumer acceptance, the flavor perception of a food is a very important factor, and flavour is consequently a significant factor in determining the commercial success of every newly introduced product. Each food has its own set of flavors, which make the product desirable for consumers. Up to 11,000 aromas in food products, constituting 18 different chemical groups, have been identified, and the data are accessible online[1]. It has been estimated that just nearly 10% of them are responsible for the generation of aromas. The action of these compounds is related to their levels plus their odor threshold. Volatile components alter the smell of a special food, when they are present in a higher dose than their threshold (Amerine, Pangborn, & Roessler, 2013). Also, it is possible to select the desirable aroma compounds among numerous volatile compounds by using the gas chromatography olfactory (GC-O) device (Berdagué, Tournayre, & Cambou, 2007). Below, some of the flavors in different types of foods being consumed daily are presented.

7.4.1 Flavors Derived from Cheeses

In cheese, lactose, citrate, proteins, and lipids catabolism is induced by microorganisms, which firstly grow in milk medium and then shift in or over the cheese curd. The catabolism process is responsible for the formation of cheese flavor. Some of the volatile compounds found in cheeses are emanated from the ones found in milk; however, their role in producing cheese flavor has not been confirmed yet (Martin et al., 2002; Cornu et al., 2005). Anaerobic flora has a higher activity in pressed cheeses because of the low surface/volume ratio in this type of cheese. Thierry and Maillard (2002) studied the volatile compounds generated by the nonstarter lactic acid bacteria (LAB), lactobacillus, or propionibacterium during the ripening of pressed cheeses.

Blue cheeses are another type of fermented dairy products in which lipolysis and proteolysis are the main reactions leading to the formation of related flavors. The amount of lipolysis in a Danish Blue cheese is much higher compared to the formation of mold on the surface of ripened cheeses. The fatty acids produced from the lipolysis are the source of a whole series of methyl ketones and secondary alcohols (Reineccius, 2013). The sequence of

1. www.vcf-online.nl

flora has been observed in Camembert or Brie cheeses. During the first phase, yeast and fungi develop and produce flavors akin to fermented apples (Galli, Martin, da Silva, Porto, & Spoto, 2016). This typical flavor formation is mainly due to the production of esters and aldehydes from amino acids during their growth phase. Yeasts such as *Kluyveromyces lactis* and *Debaryomyces hansenii* that grow initially upon draining (Reineccius, 2013) are very efficient in the breakdown of amino acids, present in the milk or produced by LAB. Their disintegration is carried out by using mainly the Ehrlich–Neubauer pathway to produce branched chain aldehydes, alcohols, and esters. Debaryomyces produce more aldehydes and is associated with malt flavors, whereas Kluyveromyces produces more esters and alcohols (Martin, Berger, Le Du, & Spinnler, 2001).

7.4.2 Flavors Derived from Red Meat

Slight changes in water-soluble compounds, lipids, cooking procedures, aging periods, packaging systems, and storage conditions influence the meat flavor. In the aging process, sugars, organic acids, peptides, free amino acids, ATP metabolites, and enzymatic activity are modified that can influence the properties of meat flavor compounds (Spanier, Flores, McMillin, & Bidner, 1997). The odor of cooked meat is basically formed due to a variety of chemical substances, presumably produced by fragmentation, such as deamination or decarboxylation of amino acids, and simultaneously with some degradation of the sulfur-bearing amino acid, e.g., cysteine, to yield H_2S and propionic acid (Sato & Hegarty, 1971).

Also, lipid autoxidation is the primary mechanism for the degradation of desirable flavors in stored meat. Lipids in muscle foods, particularly their phospholipid components, undergo degradation to produce a large number of volatile compounds (Ross & Smith, 2006). Their degradation leads to the formation of a category of secondary products, such as aldehydes, hydrocarbons, alcohols, ketones, acids, esters, furans, lactones, and epoxy compounds as well as polymers. Epoxy compounds are also known as aroma compounds, particularly aldehydes that possess low threshold values in the ppm or even ppb scales (Reineccius, 2013).

7.4.3 Flavors Derived from Bakery Products

This group comprises different products such as bread and rolls, sweet yeast dough products, biscuits, cookies and crackers, pies and pastries, cakes, and breakfast cereals (Cognat, Shepherd, Verrall, & Stewart, 2014; Tarancón, Fiszman, Salvador, & Tárrega, 2013). Proteins and starch offer substantial opportunities for flavor binding in both the baking process as well as in storage phase (staling). Loss of aromas during baking are usually determined experimentally, and an additional quantity of flavor is normally added or perhaps the flavor formulation itself is set to give the desired flavor balance and level in the final product (Brummer & Lorenz, 1991). Addition of aromas into the dough or batter prior to baking; or spraying them onto the surface of the product as it is placed in the oven; dusting onto the surface after cooking; and oiling plus introduction into the cooked product as a cream filling, glaze, or coating are four approaches for flavoring these products (Heath, 1978).

7.4.4 Flavors Derived from Fruits and Vegetables

Fruits comprise a crucial part of the human diet. In fact, the fruit quality is one of the major factors influencing the consumption of fruits (Baldwin, Plotto, & Goodner, 2007). Some features such as appearance, color, texture, aroma, and nutritional value are key factors in the quality of fruits (Song, 2007). Flavor is one of the most important quality traits for fresh fruit (Reineccius, 2016). The contribution of odor-active compounds to the fruit flavors has gained much attention, because these compounds are important for the characteristic flavors of fruits (Brückner & Wyllie, 2008). Various volatile compounds exist in fruits and determine the sensory properties of that type of fruit (Berger, 2007). Olfaction (sensorial cells located in the nasal cavity) of human are able to detect this scarce amount of volatile compounds. The diversity is partially responsible for the unique flavors found in different fruit species. Different substances are responsible for the fruit aromas that have robust odors containing low threshold values. However, knowledge of the biochemical pathways and basic regulating phases in the synthesis process of these fragrances is still incomplete and needs more research. A more comprehensive understanding of the flavor biochemistry of volatile agents present in fruits is vital to develop the flavor quality of fresh fruit that satisfy the consumer needs for better quality.

7.5 MICROENCAPSULATION OF FOOD FLAVORS

The size of microcapsules containing flavors may span from a few millimeters to less than 1 μm (Gouin, 2004; Jafari et al., 2008). Microcapsules may comprise one or a set of compounds forming the whole carrier structure. Normally, the generated microcapsules are added to larger systems so as to be adjusted to the intended system. A broad range of substances is employed to encapsulate flavors, which include lipids, carbohydrates, proteins, natural, and synthetic gums plus other biopolymers (Jafari et al., 2008; Rajabi, Ghorbani, Jafari, Mahoonak, & Rajabzadeh, 2015; Reineccius, 1988). Besides, with the rapid progress of knowledge and technology, different materials and their combinations are used as encapsulants to preserve flavors and increase the shelf life of products. In order to encapsulate flavor compounds, the micro-vehicle must not react with the payload, have a simple

TABLE 7.3 Some Wall Materials Applied to Encapsulate Flavors

Type of Wall Material	Properties	Type of Entrapped Flavor	Reference
Maltodextrin (DE < 20)	Film fabrication	Eugenol	Chatterjee & Bhattacharjee, 2013
Modified starch	Robust emulsifier	MCT and D-limonene	Paramita, Furuta, & Yoshii, 2012
Gum acacia (Arabic)	Emulsifier, film fabrication	Orange oil	Liping & Jianrong, 2013
Modified cellulose	Film fabrication	L-menthol	Ma, Tan, Dai, & Zhou, 2013
Gelatin	Emulsifier and film fabrication	linalool, limonene, and ethyl butyrate, etc.	Zafeiropoulou, Evageliou, Gardeli, Yanniotis, & Komaitis, 2012
Cyclodextrin	Molecular inclusion and emulsifier	Cinnamaldehyde and thymol	Cevallos, Buera, & Elizalde, 2010
Lecithin	Emulsifier	Isoamyl acetate	Perez et al., 2014
Whey proteins	Fine emulsifier	Beta-pinene	Koupantsis, Pavlidou, & Paraskevopoulou, 2014
Hydrogenated fat	Protection against permeation of water and oxygen	Menthol	Zhu, Lan, He, Hong, & Li, 2010

structure to be handled, i.e., can flow and does not exhibit high viscosity, protect the bioactive core against surrounding medium, and possess desirable emulsion-stabilization features plus effective dispersion performance so that flavors are released in the right place and time (Castro et al., 2016; Trubiano & Lacourse, 1988). Here, some of the typical wall materials applied for the microencapsulation of flavors are represented in Table 7.3. Finally, the properties of food components, such as polysaccharides, lipids, and proteins and their interactions with flavors should be carefully studied to design an effective delivery system (McClements, Decker, Park, & Weiss, 2009).

With the development of the encapsulation science, novel technologies are being invented for use in industrial scales. Flavors are encapsulated via different methods, and generally these techniques can be classified into chemical and physical processes (Fig. 7.2). Among these methods, spray

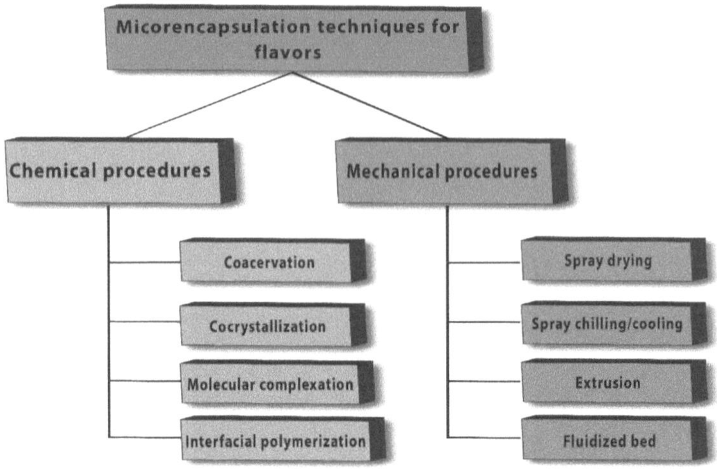

FIGURE 7.2 Different microencapsulation techniques employed for flavor molecules.

drying and extrusion are broadly employed in the food industry for food flavors as they are straightforward and cost-effective (Chew & Nyam, 2016; Jafari et al., 2008; Mahdavi et al., 2014; Rajabi et al., 2015).

7.5.1 Chemical Procedures for Encapsulation of Flavors

7.5.1.1 Coacervation

Invented by Georg (1975), it is often referred to as the original technique of encapsulation. Later, this method was exerted to microencapsulate dyes for the carbonless printing paper (Winnik, Breton, & Sanders, 1996). This process includes the isolation of colloid particles from the solution and their following aggregation, which forms a separate liquid phase called coacervate (Korus, 2001; Strzyz, 2015). Usually, the core material must be compatible with the recipient matrix plus being low-soluble in the coacervation ambient. Sengupta, Mishra, and Sweeney (2016) invented a method for preparing organoleptically-savoring films comprising plant substances. Two coating materials were provided for the encapsulation process in which the second layer was generated via coacervation and finally gelation was responsible for the formation of a gel matrix. Also, Martins, Rodrigues, Barreiro, and Rodrigues (2009) employed the coacervation method to fabricate polylactide (PLA) microcapsules entrapping thyme oil. The dissolvation of PLA in dimethylformamide (DMF) was a novel approach in their study. DMF exhibits significant solubility in water, and following dissolvation in water the uniform solution of PLA–DMF complex results in the precipitation of PLA all over the thyme flavor.

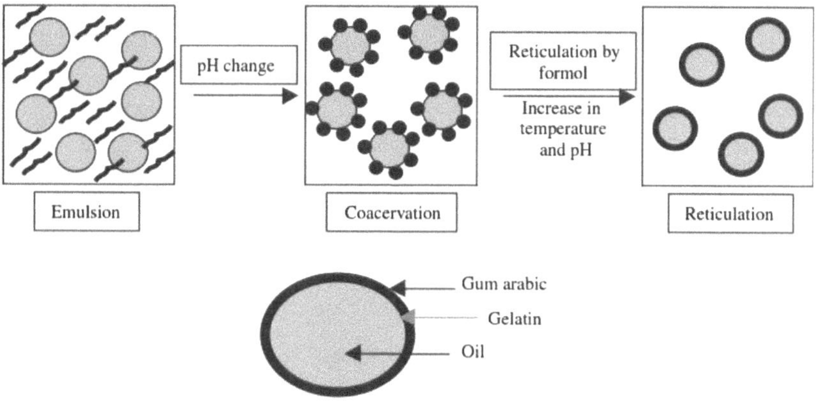

FIGURE 7.3 Description of a complex coacervation method. *Reprinted with permission from (Madene et al., 2006).*

Complex coacervation is a developed method of simple coacervation in which two or more than two sorts of polymers are utilized. The flavor can be added into the mixture during or following the phase separation; meanwhile, it is better for the flavor to be present during the coacervation process for an efficient encapsulation. Besides, it is of utmost importance to add a proper droplet stabilizer so as to avoid coagulation during the formation of microcapsules (Arshady, 1999). Until now, this technique has not been applied in a large-scale since it is intricate and not cost-effective (Schoebitz & Belchí, 2016). Yet, it is difficult to determine an optimized concentration of wall materials for the coacervation method since the desired concentration to form a fine emulsion may vary to raise the throughput of microcapsules (Nakagawa, Iwamoto, Nakajima, Shono, & Satoh, 2004). The addition of toxic materials, such as glutaraldehyde, in order to stabilize the complex coacervates and their lability are limitations for using this technology (Sanchez & Renard, 2002). Figure 7.3 summarizes the steps of complex coacervation.

7.5.1.2 Co-crystallization

Co-crystallization is a straightforward and flexible encapsulation method used for entrapping flavors (Desai & Jin Park, 2005). Studies conducted until now have applied this technique to encase some food flavors (Beristain, Vazquez, Garcia, & Vernon-Carter, 1996; Hegheş et al., 2015), essential oils (Beristain et al., 1996; Taraneh, Rahmatollah, Hassan, & Alireza, 2008), and using sugar as the outer shell (Bhandari, Datta, D'Arcy, & Rintoul, 1998; López-Córdoba, Gallo, Bucalá, Martino, & Navarro, 2016). Sucrose has a crystal morphology, by which it can entangle flavors and, thus, enhance the stability of entrapped flavors (López-Córdoba, Deladino, Agudelo-Mesa, & Martino, 2014). The final granular yield exhibits low hygroscopicity plus the ability to flow easily

(Quellet, Schudel, & Ringgenberg, 2001). Meanwhile, during encapsulation, flavors present in the liquid-state may transform into dry granular structures and degradation of sensitive compounds may occur (Bhandari et al., 1998). Furthermore, Beristain et al. (1996) used this technique to microencapsulate orange peel oil. The retained materials were as stable as spray-drying technique, and although the product was free to flow, it was important to add an anti-oxidant to decelerate the development of oxidized flavors.

7.5.2 Mechanical Procedures for Encapsulation of Flavors

7.5.2.1 Spray Drying and Spray Coating

This technique is mostly applied in the industry as it is cost-effective and has the potential to be scaled up (Mahdavi et al., 2014). According to (Jafari et al., 2007b), spray drying process has a considerable effect on the retention of flavors. Moreover, it can be employed for components that are heat-sensitive since the internal phase maintains a low temperature. In this technique, firstly the dispersion of the payload phase in the wall material is prepared. Subsequently, the mixture is atomized and sprayed into a chamber with hot air current, which yields fine and stable particles, and finally, they are moved to a cyclone separator to be recovered (Jafari et al., 2008). A schematic representation of this technique is provided in Fig. 7.4.

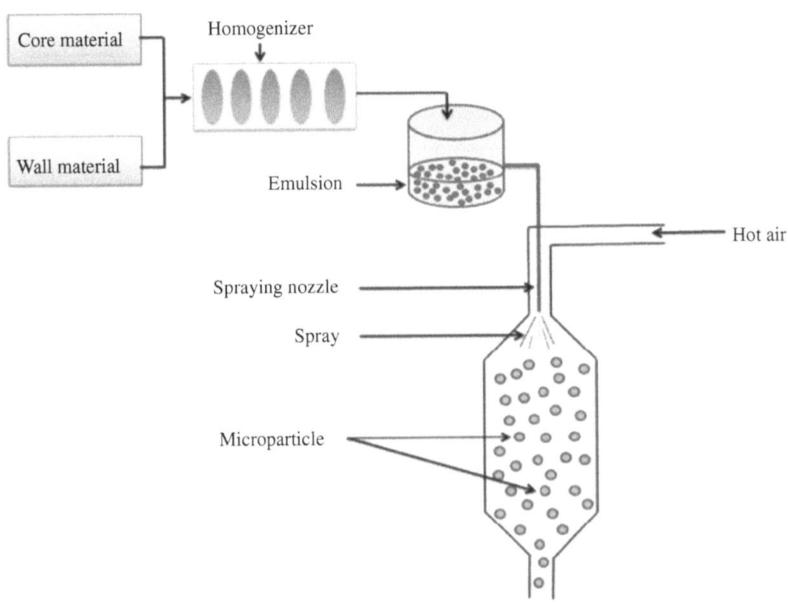

FIGURE 7.4 Fabrication of microparticles via the spray drying procedure. *Reprinted with permission from (Bakry et al., 2016).*

The physicochemical properties of the wall/core materials, encapsulant material, solid concentration of the feed, and film formation ability plus exhibiting low viscosity in high concentration medium are determining factors that guarantee the retention of flavors (Bomben, Bruin, Thijssen, & Merson, 1973; Desobry, Netto, & Labuza, 1997; Jafari et al., 2008; Leahy, Anandaraman, Bangs, & Reineccius, 1983; Pourashouri et al., 2014a). As the core materials of poor water solubility are encapsulated via spray drying technique, a matrix-type emulsion network is formed. Tiny core droplets are dispersed within this matrix network. The structure of these microcapsules are influenced by factors such as wall material properties, drying and atomization conditions, unbalanced shrinkage at the initial phase of drying, and proportion of core-to-wall plus the storage circumstances (Jafari et al., 2007b; Pourashouri et al., 2014a; Pourashouri et al., 2014b; Ré, 1998).

The prime disadvantage of this encapsulation technique is associated with the loss of heat-sensitive aromas during spray drying. Also, some extent of the core material may be present on the surface of the capsules; therefore, oxidation and flavor alterations may occur (Desobry et al., 1997; Mahdavi et al., 2014). Langrish and Fletcher (2001) explored the application of CFD[2] in transfer processes and the changes occurring during the spray drying process. As a result, the retention of volatile aromas was associated with the ultimate moisture content of the microcapsules together with the blowing air. Another issue of this method relates to the final fine powder yield (10–100 μm), which needs to be further processed to make it more soluble. Factors such as processing conditions plus carrier substances influence the agglomeration phenomena and the generated microcapsules (Buffo, Probst, Zehentbauer, Luo, & Reineccius, 2002; Mehrad, Shabanpour, Jafari, & Pourashouri, 2015).

Jafari, He, and Bhandari (2007a) fabricated submicron emulsion particles of d-limonene made by a microfluidizer and ultrasound, and then dried the mixture to obtain nanosized fine powder. Maltodextrin was applied as the wall material and incorporated in a surface active biopolymer or a small molecule surfactant. The outcome revealed that microfluidization was an effective emulsification technique leading to a powder with considerable retention (86.2%) of d-limonene, mainly thanks to its capability to produce emulsions with fairly small droplets (700–800 nm) and narrow distributions, which had a good stability throughout the process. Among totally different emulsifiers used, though Tween 20 considerably reduced the emulsion size (less than 200 nm), the resulted powder had the poorest encapsulation efficiency (Jafari et al., 2007a).

Fluidized bed spray coating is mostly applied in the pharmaceutical and cosmeceuticals industry, besides it has been employed in the food sector as a

2. Computational fluid dynamics.

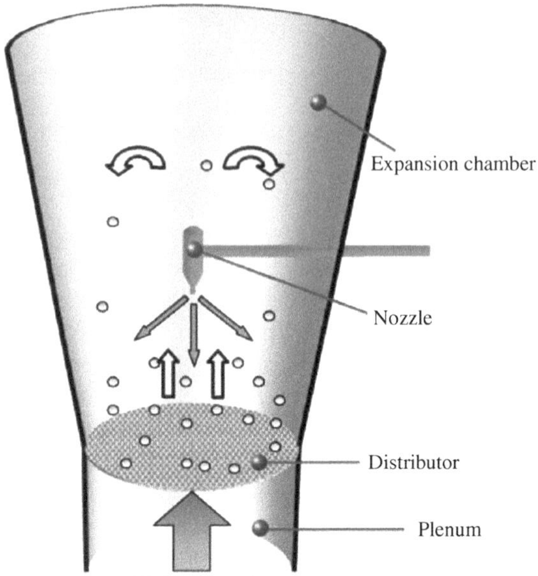

FIGURE 7.5 Spray coating system using a variety of formulations to coat the flavor molecules. *Reprinted with permission from (Depypere et al., 2009).*

means to encapsulate flavor molecules (Hampel, Bück, Peglow, & Tsotsas, 2013). This process is done in three stages as shown in Fig. 7.5. At first, particles are suspended in the coating chamber via a hot stream of air. Next, the coating agent is atomized and sprayed via a nozzle, and the film formation starts. During this time, wetting and drying phases exist together. Finally, the solvent is vaporized by the generated hot air and the coating agent is placed on the particles (Jacquot & Pernetti, 2004). The final yield ranges from 0.3 mm to 10 mm in size. It is noteworthy that the particle size distribution is specific, and low porosities are obtained through this technology. Some of the advantages of utilizing this method are as follows:

- High drying throughput because of the controlled heat and mass transfer rates.
- Limited flow area.
- Simple control of the process.
- Extreme thermal efficiency plus low maintenance costs.

7.5.2.2 Freeze Drying

It is also known as lyophilisation and is mostly exerted for the materials which are heat-sensitive; in addition, it is applied to the agents that are labile in aqueous solutions. As the freeze drying process starts, an amorphous solid is formed surrounding the solution; thus, selective diffusion is likely to occur

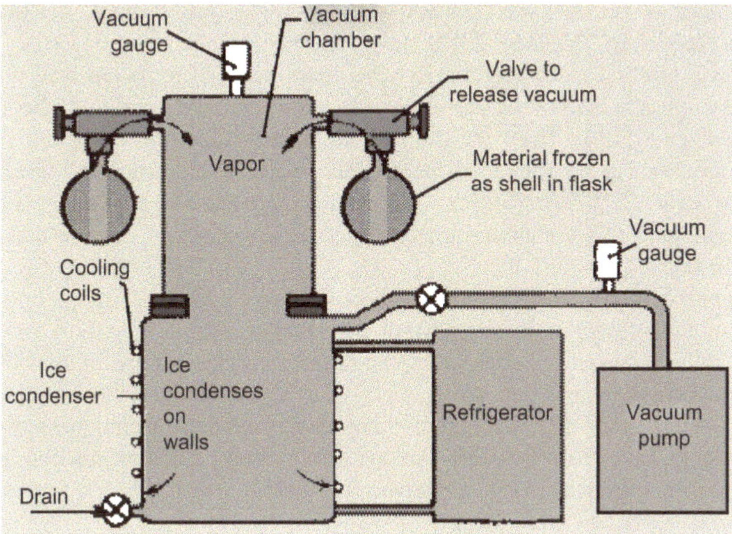

FIGURE 7.6 Main parts of a freeze drying device. *Reprinted with permission from (Menyhart, 1995).*

in this medium (Jafari, Mahdavi-Khazaei, & Hemmati-Kakhki, 2016; Rey, 2016). According to Buffo and Reineccius (2001), the retention and quality of menthyl linoleate were higher when processed with freeze drying compared with hot air drying. Parts of a freeze drying instrument are depicted in Fig. 7.6.

Argyropoulos and Müller (2014), encapsulated the lemon balm essential oils via freeze drying process. They suggested that freeze drying with high pressure condition entrapped higher concentrations of essential oil, but did not reveal a positive effect on the freeze drying duration.

In spite of the advantages of this method, it is less likely to be applied in the industry due to the high costs, as it is 50 times more expensive than freeze drying (Barbosa et al., 2015). Furthermore, the storage and handling of the products are highly expensive plus the long production time (Aalaei, Rayner, & Sjöholm, 2016).

7.5.2.3 Spray Chilling/Cooling

These techniques are considered as the cheapest encapsulation methods, and are commonly used to encase aromas and cause a sustained release in wet mediums. Moreover, they are exerted to modify liquid flavors into fine powders (Gouin, 2004). These methods are alike spray drying, where flavor substances are emulsified into the fluid wall materials. Subsequently, the mixture is atomized from the feedstock. Finally, the fine powders are generated as the droplets contact a cooling medium (Okuro, de Matos Junior, & Favaro-Trindade, 2013; Oxley, 2012).

Spray chilling method comprises the atomization of a molten wall material through a nozzle into a chamber with carbon dioxide ice bath at the temperature of −50°C. Thus, the droplets are transformed into a coating film. This process can be applied to hydrophilic materials that may be degraded during heat treatment (Sillick & Gregson, 2012). Some food products prepared with this technique are bakery products and dry soup, which are high in fat.

Spray cooling encapsulation method is akin to spray chilling, but the reactor temperature is different from the spray chilling method. A molten network, as the wall material, entrapping the core materials are spray cooled; in addition, vegetable oils are suitable wall materials due to their melting point range (45−122°C). Like other encapsulation techniques, these approaches have their own disadvantages, including particular handling plus storage circumstances (Risch & Reineccius, 1995).

7.5.2.4 *Extrusion*

Patented and advanced in 1957 by Swisher (1957), this technique has been used for the encapsulation of labile flavors via matrices of glassy carbohydrates (Castro et al., 2016; Menis, Milani, Jordano, Boscolo, & Conti-Silva, 2013; Tackenberg & Kleinebudde, 2015). Increasing the resistance of entrapped flavors against oxidation is the main benefit of this process. In fact, glassy carbohydrates act as effective barriers against the external undesirable agents and develop a productive encapsulation method (Wolf, 2010). Nevertheless, release and diffusion of flavors are increased by damage made to the capsules under processing conditions and structural imperfections like fissures, pores, and the fragile wall of capsules (Martins, Rodrigues, Barreiro, & Rodrigues, 2011). Extrusion is carried out via different approaches explained below. Furthermore, the extrusion methods used for microencapsulation of flavors are depicted in Fig. 7.7.

7.5.2.4.1 Simple Extrusion

First, the aroma is dissolved in the polymer network at the temperature of 110°C. Then, the mixture is pushed through a dye, afterwards the yields (filaments) are immersed in a desiccant liquid (solutions with high affinity for water vapor), which result in hardening of the extruded mixture along with entrapping the bioactive compounds (Tackenberg, Krauss, Schuchmann, & Kleinebudde, 2015). Isopropyl alcohol is mostly utilized for the purpose of dehydration in this approach. The dehydrated hard filaments are next shredded, separated, and dried (Tackenberg et al., 2015).

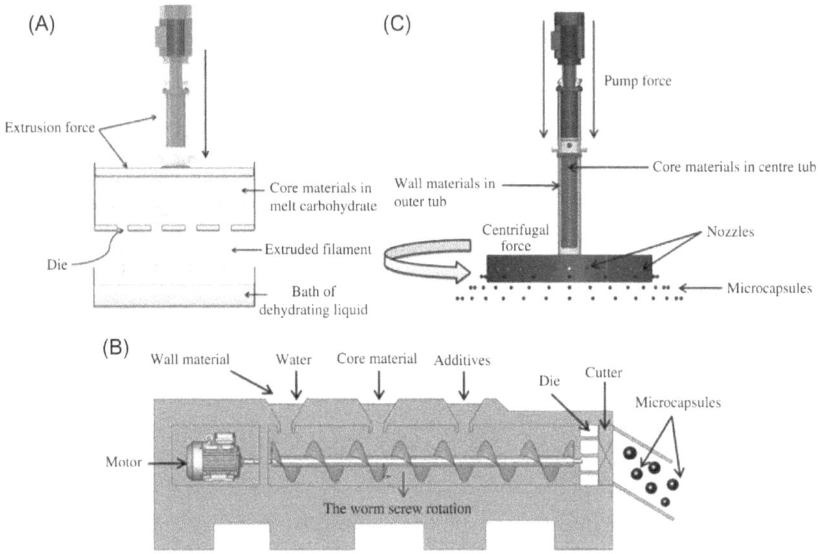

FIGURE 7.7 Microencapsulation of flavors by (A) simple extrusion, (B) melt extrusion, and (C) centrifugal extrusion (dual capillary) processes. *Reprinted with permission from (Bakry et al., 2016).*

7.5.2.5 Double Capillarity Extrusion Instruments

7.5.2.5.1 Dual Capillary (coaxial)

The payload and micro-vehicle are inserted through the internal and external opening of the device consecutively. The payload is normally a liquid, and the carrier may be a typical solution or its molten state. It is of utmost importance that the vehicle material and payload have to be immiscible. A uniform jet flow is formed at the end of the coaxial nozzle, which degrades and generates the final microdroplets (Zuidam & Shimoni, 2010).

7.5.2.5.2 Centrifugal Extrusion

The nozzles of this device are located on the external surface area of the rolling cylinder. The intended flavor plus the carrier agent are pumped via the inner and outer orifices, respectively; thus, the generated rod-shaped flavor compounds are encased by the shell material. During the rotation of the device, the formed rods degrade into droplets, resulting in the formation of microcapsules (Tackenberg, 2014).

7.5.2.5.3 Recycling Centrifugal Extrusion

As it is obvious from its name, the additional fluid used for coating the core material is recycled in the typical approach of spinning disc extrusion. In general, the payload is dispersed within the coating material and subsequently the mixture is extruded along the spinning disc. As noted, the excess fluid used for coating is atomized and isolated from the final products being coated. The generated yield is then hardened and fixed via solvent extraction or lowering the temperature. Taking into account the most proper process temperature, pressure, emulsifier concentration, and maintenance time, etc., microcapsules are fabricated with high levels of flavors. Accordingly, fissures and cracks in the wall material enhance the release process during or following production procedure (Barbosa-Cánovas, Ortega-Rivas, Juliano, & Yan, 2005; Zuidam & Shimoni, 2010). The main drawback of this procedure is associated with emulsion stability. In high viscous carbohydrate-based molten networks, it is rather complicated to acquire a stable emulsion (Donnelly et al., 2009).

7.6 NANOENCAPSULATION TECHNOLOGIES FOR FOOD FLAVORS

In recent years, the eating habits of the public have oriented towards healthy and safe food products, as well as being tasty at the same time. The perception of food taste is generally affected by the applied flavors and aromas during consumption (Sozer & Kokini, 2009). Because of the natural sensitivity of flavors and aromas, the encapsulation procedure is an effective approach to preserve their functional properties. As mentioned before, the advantages ascribed to encapsulation are reflected in easier handling of liquid flavors by its conversion into a dry state, improved stability once exposed to oxygen, light, and/or high temperatures, prolonged shelf-life, controlled and even targeted release of food flavors, masking of off-flavors, and the ability to alter the textural properties of the yield (Kohane, Yeo, Given, & Langer, 2015). Herein, the nanoencapsulation techniques employed in the entrapment of flavors are discussed below; also these methods plus the related studies are summarized in Table 7.4.

7.6.1 Nanoemulsification of Food Flavors

Nanoemulsions are functional and efficient nanocarriers due to their clarity, stability, and bioavailability. Structure–function relationships are usually applied to determine an optimized formulation of food-grade nanoemulsions in the food and beverage sector (Jafari, Fathi, & Mandala, 2015).

Liang et al. (2012) blended the mixture of a flavor (peppermints oil) with medium-chain triacylglycerol in water and stabilized it with modified starch in order to fabricate nanoemulsions. The impact of emulsifying conditions, including homogenization pressure, process cycles, and oil loading, on the mean diameters and viscosities of nanoemulsions were assessed by rheological measurements, dynamic light scattering (DLS), and optical microscopy. The developed flavor nanoemulsions with mean diameters normally less than

TABLE 7.4 Different Nanoencapsulation Techniques Implemented for Food Flavors

Nanoencapsulation Technique/Carrier	Wall Material	Core Material	Aim	References
Nanoemulsification	Medium-chain triacylglycerol (MCT)-starch	Peppermint oil	Increasing stability/antibacterial activities	Liang et al., 2012
	MCT-buffer solution (citric acid/sodium hydroxide/sodium chloride)	Citral	Stopping develop of off-flavor	Zhao et al., 2013
	Lemon oil/water	β-carotene	Increasing in physical stability and bioaccessibility	Rao, 2013
Electrospinning	Polymethyl methacrylate	Menthol/cyclodextrin	Increasing thermal stability	Uyar et al, 2009
	Polyvinyl alcohol	Vanillin/cyclodextrin	Increasing the shelf-life and thermal stability	Kayaci & Uyar, 2012
	Polyvinyl alcohol	Ethyl vanillin	Thermoresistance film	Lević et al., 2014
	Polyvinyl alcohol	Geraniol/cyclodextrin	Increasing the shelf-life and thermal stability	Kayaci et al., 2014
Nanoemulsification	Maltodextrin	D-limonene	Increasing of stability during process	Jafari et al., 2007
Biopolymer nanoparticles	Starch	Menthone, menthol and limonene	Optimizing the release of core material/increasing of encapsulation efficiency	Ades et al., 2012
	Zein	Oregano, red thyme, and cassia	Controlling the release	Pariss et al., 2005

(Continued)

TABLE 7.4 (Continued)

Nanoencapsulation Technique/Carrier	Wall Material	Core Material	Aim	References
	β-cyclodextrin polymer	Hypericin	Increasing the water-solubility of the core material	Zhang et al., 2013
	β-cyclodextrin	Lemon flavoring	Impose a regular-fat taste for fat-free yogurt	Kant et al.,2004
	β-cyclodextrin and 2-hydroxypropyl-β-cyclodextrin	benzyl acetate and linalool	Yield a controlled release along with the stabilization of the selected aromas	Numanoğlu et al., 2007
Granules	Globular protein	Capsaicin	Increasing of encapsulation efficiency	Hong et al., 2012
Milk proteins	Whey protein	Ethyl hexanoate	Controlling the release as well as enhancing the stability during process	Giroux & Britten, 2011

200 nm showed high stability over at least 30 days of storage time. Furthermore, antimicrobial properties of flavors were determined by the minimum inhibitory concentration and time-kill dynamic processes, against two gram-positive bacteria. Compared with bulk flavor, the flavor loaded nanoemulsions showed prolonged antibacterial activities. Altogether, these results recommend that the nanoemulsion systems will offer novel applications of essential oils in extending the shelf life of liquid food products.

Capsaicin is a key polyphenolic compound of chili peppers, and it is responsible for the pungency of chili pepper (the capsicum fruits) and their products (Sricharoen et al., 2016). Pressurization at different pressures and concentrations of proteins was done and analyzed subsequently. Moreover, the morphology, encapsulation efficiency (EE) and release profile were studied to determine the effects of high hydrostatic pressurization upon the stability of the emulsion. As a result, the size of droplets was reduced due to the changes in the protein structure (denaturation) and increased rate of protein adsorption on the oil droplet surface. It can be suggested that the structure of the globular protein was altered into filamentous structure under high hydrostatic pressure circumstances. EE (%) of the WPI, SPI, and CSP emulsions was enhanced by the increase in pressure level. The release rate of the core material in the pressurized protein emulsions was rather slow compared to non-pressurized conditions. The remaining volume of encapsulated capsicum oleoresin of non-pressured emulsions of WPI and CSP was less than emulsions treated under high pressurization. (Hong, Surassmo, Chun, Min, & Choi, 2012). Therefore, these types of proteins can be considered as applicable encapsulating materials.

In another study, the citral-loaded oil-in-water (O/W) nanoemulsions and effects of various concentrations of ubiquinol-10 (Q10H2) on the stability of citral was investigated (Zhao, Ho, & Huang, 2013). Solid phase microextraction gas chromotography (SPME-GC) was used to monitor the degradation of citral and the formation of off-flavor compounds at 25°C and 45°C. The optimum concentration of Q10H2 to protect citral from chemical degradation and oxidation within the current formulation was approximately 0.10 wt% in the medium (Q10H2/citral quantitative relation 1:1). However, results suggested that a low concentration of Q10H2 might induce oxidoreduction transition concerning the majority of the ubisemiquinone/ubiquinone (Q10) that probably rendered Q10H2 with pro-oxidant properties. Additional increase in Q10H2 concentration beyond a particular level also inhibited its effect due to the complex properties of radicals plus the overall environment. With proper concentrations of Q10H2 given within the system, major citral oxidation off-flavor compounds (p-cresol; α,p-dimethylstyrene; and p-methylacetophenone) and some products resulted from the lipid degradation can be hindered to lower levels. However, ubiquinone-10 (Q10) had an insignificant impact on the chemical stability of citral and generation of off-flavor.

In a study conducted by Rao (2013), nanoemulsions were prepared by the phase inversion temperature (PIT) technique, carried out by heating a surfactant, oil, and water (SOW) systems close to the PIT and then cooled quickly by stirring the slurry. Preliminary experiments were undertaken by a model system consisting of hydrocarbon oil, water, and a non-ionic surfactant. Mixtures close to nanoemulsion's PIT were formed by: (1) holding SOW (38.5°C) and (2) cooling them quickly at 10°C. The PIT was measured via electrical conductivity and turbidiness strategies. The optimum storage temperature for the tested nanoemulsions was about 27°C under the PIT. Also, the stability of nanoemulsions at ambient temperatures may be improved by adding either Tween 80 (T80; 0.2 wt%) or SDS (0.1 wt%) to displace the surfactant from the nanodroplet surfaces. Experiments were then applied to determine if stable nanoemulsions might be formed through the PIT technique from food-grade ingredients. Nanoemulsions were fabricated from a T80 (non-ionic surfactant) and flavor oil (lemon oil) under heat treatment. Depending on the ratio of surfactant-to-oil (SOR), different types of colloidal dispersion could be developed by simple heat treatment (90°C, 30 min). The results suggested that there was a mechanical energy barrier in the SOW system at ambient temperature that prevented it from moving from an extremely labile system toward a nanoemulsion system. The conditions in which stable nanoemulsions could be generated were also established as sucrose monopalmitate (SMP) together with lemon oil were utilized as the surfactant and oil phase. Nanoemulsions (size less than 100 nm) were set at low surfactant to oil ratios (SOR $<$ 1) depending on the conditions of homogenization, while microemulsions (size less than 10 nm) were formed at higher ratios (SOR $>$ 1). Relatively, stable nanoemulsions could be formed at pH = 6 and 7; however, extensive particle growth/aggregation was found at higher and lower pH values. Besides, flavor-oil nanoemulsions were produced by a technique called "emulsion titration," which includes the titration of emulsion droplets in surfactant-based micelle solutions.

Later, the effectiveness of nanoemulsion formation using nonionic surfactants SMP and/or T80 was investigated by Rao (2013). Flavor was transferred from emulsion droplets into the micelle phase till a critical flavor concentration was reached. The solubilization process was relatively fast (less than few minutes), furthermore the solubilization rate enhanced as the concentration of surfactant increased. The value of critical flavor concentration increased with increase in surfactant concentration, and the value was considered to be higher for SMP than T80.

Rao (2013) investigated the effect of flavor composition on the formation and properties of oil-in-water nanoemulsions. They concluded that the lower fold oils were extremely unstable to droplet growth throughout storage, with the growth rate increasing with extending storage temperature and reducing oil level. Moreover, the oil level affected the solubilization and stability of flavor-loaded nanoemulsions titrated into a T80 solution. Finally, it was observed that the movement of oil molecules toward surfactant micelles from nanoemulsion droplets, was enhanced by the increase in flavor level.

7.6.2 Electrospraying and Electrospinning for Nanoencapsulation of Flavors

Till date, electrospinning has drawn a great attention, since it is a simple and cost-effective technique for fabricating functional nanofibers (NF) that have a nanoporous structure as well as wide surface area to volume ratio and high encapsulation efficiency (Ghorani & Tucker, 2015; Esfanjani & Jafari, 2016).

Torres-Giner, Martinez-Abad, Ocio, and Lagaron (2010) encapsulated omega-3 fatty acid (DHA) in zein capsules using the electrospraying method. The nanoencapsulated DHA reduced off-flavor extent and increased the chemical stability against degradation under harsh environmental conditions, such as high temperature and relative humidity.

Electrospinning of NF with cyclodextrins (CDs) is an interesting nanoencapsulation method used by food scientists. Uyar, Nur, Hacaloglu, and Besenbacher (2009b) utilized menthol as a model fragrance/flavor material. They run electrospinning of polymethyl methacrylate NF containing inclusion complexes of cyclodextrin and menthol so as to fabricate functional NF that contain fragrances/flavors with high thermal stability. The above-mentioned NF were generated by three forms of CD: α-CD, β-CD, and γ-CD. The data collected from direct pyrolysis mass spectrometry revealed high heat stability of nanostructures. Also, the thermal evaporation of flavor shifted to a high and broad temperature range (100–355°C). Moreover, they suggested that the strength of interaction between menthol and the CD raised together with the heat stability in the order of the cavity size (gamma-CD was suggested to be the highest) (Uyar, Hacaloglu, & Besenbacher, 2009a).

Nanoencapsulation of vanillin with polyvinyl alcohol (PVA) webs incorporating cyclodextrin inclusion (CD-IC) was also studied (Kayaci & Uyar, 2012) using the electrospinning technique to yield prolonged shelf-life and high temperature stability. The vanillin/CD-IC was prepared with three kinds of CDs (α-CD, β-CD, and γ-CD) to find out the most favorable CD type for the stabilization of vanillin. PVA/vanillin/CD-IC nanofibres were successfully electrospun from a liquid mixture of PVA and vanillin/CD-IC with the diameter of about 200 nm. Results indicated that vanillin with increased durability and high temperature stability was achieved from PVA/vanillin/CD-IC nanowebs due to the complexation of vanillin with CD, whereas the PVA nanofibers could not effectively preserve the vanillin per se. Accordingly, PVA/vanillin/γ-CD-IC nanoweb was more effective for the stabilization similar to the investigation of (Uyar et al., 2009b). In essence, the slow release of vanillin implies that the power of interaction between flavor and the γ-CD cavity is stronger when put next to α-CD and β-CD.

Ethyl-vanillin (3-ethoxy-4-hydroxybenzaldehyde: EV) is used as a flavor in food products and cosmetics. This artificial flavor has additional intensive flavoring power compared to vanillin (3-methoxy 4-hydroxybenzaldehyde). On the contrary, vanillin is more soluble than EV. An example of that is lower solubility of EV in water compared with vanillin, whereas it is soluble in organic solvents (i.e., ethanol) (Burdock, 2009). PVA nanofibers with EV as an active compound were fabricated by electrospinning technique (Lević et al., 2014). The diameter of PVA/EV nanofibers was about 100–1700 nm. The thermal properties of PVA/EV nanofibers were examined by differential scanning calorimetry (DSC) and subsequently a significant influence on the immobilization process was observed for EV. Moreover, it was noticed that the melting point of immobilized EV was relatively low (about 55°C) compared with that of the free flavor (about 77°C). Lević et al. (2014) suggested that the films which are based on PVA/EV nanofibers are mechanically stable.

Geraniol is a natural fragrance/flavor having rose-like odor and taste, usually used in the food industry. In a study, (Kayaci, Sen, Durgun, & Uyar, 2014) fabricated nanofibers of CD and geraniol incorporated in PVA via electrospinning. They noticed higher thermal resistance of geraniol within the electrospun PVA/geraniol/γ-CD-IC nanofiber; however, as geraniol is a volatile compound it cannot be preserved without CD-IC throughout electrospinning and following storage. On the contrary, the loss of geraniol was insignificant (less than 10%) for PVA/geraniol/γ-CD-IC nanofibers even after storage for two years. According to their investigation, this phenomenon is related to inclusion complexation; thus, electrospun nanofiber incorporating CD-IC may be quite applicable in the food industry because of its nanoporous structure and the vast surface area, thermal resistance, and developed durability of flavors (Kayaci et al., 2014).

7.6.3 Nanoencapsulation of Flavors by Biopolymeric Nanocarriers

Starch is an abundant, basic food component which is also widely used as a texturing agent. Owing to its strong ability to interact with aroma compounds, many studies have focused on using starches for flavor retention. A key feature is the formation of complexes of amylose molecules with volatile molecules. It is recognized that starch is able to form inclusion complexes with volatile compounds, especially small nonpolar molecules (Putseys, Lamberts, & Delcour, 2010). The addition of volatile compounds induces the formation of a single left-handed α-helix, known as V-type amylose, resulting in the compounds being encased in the helical cavities (Tusch, Krüger, & Fels, 2011). This complex formation, i.e., the ordered helical structure, promotes the formation of a starch microstructure interconnected by an amylose network. These structures of the microstructures are dependent on the type of included substances plus the rate and extent of the inclusion complex formation. Starch can build a wide variety of nano and microstructures based on crystalline, glassy, and gel types (Delcour et al., 2010). The characteristics of the physical adsorption of aroma compounds onto the starchy matrix are also influenced by these structural properties. Moreover, it was reported that volatiles which interacted weakly with the starch were released from the matrix at a rate controlled by the nano and microstructure (Lafarge, Bard, Breuvart, Doublier, & Cayot, 2008).

Ades, Kesselman, Ungar, and Shimoni (2012) ran an experiment to study the release of starch complex systems with aroma materials in the buccal cavity. Starches with various amylose contents were employed to formulate complexes with menthone, menthol, and limonene (as model flavor compounds). Unlike menthone and menthol V-amylose complexes, limonene does not form such complexes efficiently. Considering the ultimate results, as the amylose content increased, less free core content and more included aroma complexation occurred (i.e., higher complexation yield). The results of oral digestion revealed that the complexes can release the core materials in a saliva medium, especially menthone and menthol. Therefore, these complexes could be useful platforms to control the release of aroma. According to the desired long-lasting release of flavors, applying release kinetic models could be helpful to predict the concentration of complexed aroma in the products.

Cyclodextrins are widely employed to encapsulate bioactives including flavors (Reineccius, 1988; Sanchez & Renard, 2002). They are obtained via applying cyclodextrin glucosyltransferase on starch (Hedges & McBride, 1999). The inclusion complexation can be defined as the interactions made between materials, leading to the fitting of a guest molecule into the lattice of the host compound (Kenar, Compton, Felker, & Fanta, 2015). The internal cavity of the β-cyclodextrin molecules possesses hydrophobic properties, while the outside surface represents hydrophilic characteristics. This exclusive structure as shown in Fig. 7.8 begets the physicochemical features of cyclodextrins (Viswalingam, Prabu, Sivakumar, & Rajamohan, 2016).

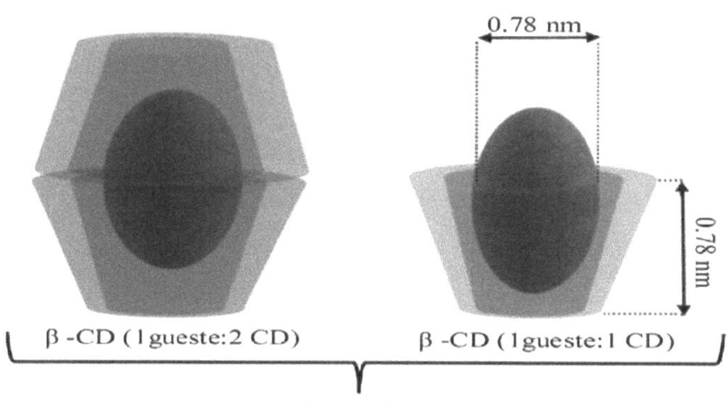

FIGURE 7.8 Schematic representation of encapsulation process by β-cyclodextrin. *Reprinted with permission from (Esfanjani and Jafari, 2016).*

As Goubet, Le Quere, and Voilley (1998) explained, retention of flavors is significantly influenced by the shape, chemical structure, steric obstruction and polarity of the guest molecule. As an example, if the flavor is small in relation to the cavity, only some parts of the compound are in contact with the cavity walls. Therefore, the reaction potential of the guest molecule with the cyclodextrin is unpredictable (Hedges, Shieh, & Sikorski, 1995). Also, high temperature or reaction with water are two factors required for the release of core materials (Reineccius, Reineccius, & Peppard, 2002).

According to Bhandari, D'Arc, and Padukka (1999), the mostly utilized methods for the complexation of guest molecules by β-cyclodextrins are:

1. Preparing an aqueous solution of cyclodextrin plus flavors and then have them shaked or stirred, and finally the precipitate is removed;
2. Blending the cyclodextrin with the core material using a robust mixer and then bubbling the core material (flavor) via a solution of cyclodextrin;
3. The flavor is kneaded with the mixture of cyclodextrin-water.

These scholars proved that the encapsulation of lemon oil in β-cyclodextrin was done more rapidly and efficiently compared to the vacuum oven. In spite of the advantages of applying cyclodextrins, they are relatively expensive as the latest cost for 50 g, in the powder form, is 10.15$[3].

Zhang et al. (2013), applied β-cyclodextrin polymer (CDP) to entrap the hypericin (HY) molecules. Results revealed that the water-solubility was greatly raised due to the hydrophilic nature of the host molecule. UV–Vis spectrophotometry assays showed that the dissociated level of HY-CDP was about 1.47×10^{-7} mol/L^{-1} at the temperature of 25°C. In essence, it was suggested that CDP is a robust solubilizer for various applications, particularly in the pharmaceutical sector.

Also, Kant, Linforth, Hort, and Taylor (2004) employed lemon-flavored-loaded β-cyclodextrin in a fat-free yogurt system and noticed that the release of the commercial fragrance altered the release of the product to a regular fat yogurt. Thus, it was suggested that β-cyclodextrin is able to change the flavor delivery in food systems.

In another investigation, Numanoğlu et al. (2007) utilized β-cyclodextrin (βCD) plus 2-hydroxypropyl-β-cyclodextrin (2-HPβCD) to impose controlled release and stabilize the benzyl acetate and linalool flavor compounds. Results indicated that the 2-HPβCD enhanced the solubility of linalool 5.9 fold and benzyl acetate to 4.2 fold. Finally, the controlled release of aromas was made possible by employing the prepared β-cyclodextrins.

Zein is another natural biopolymer commonly used in the field of encapsulation, especially nanoencapsulation. As an example, nanoencapsulation of three different essential oils (oregano, red thyme, and cassia) by phase

3. http://www.cyclodex.com/cyclodextrins/trappsol-cyclo.html

separation into zein nanospheres has been carried out by (Parris, Cooke, & Hicks, 2005). Topographic images indicated that the powders were made from irregularly formed particles (~50 μm) containing compact nanospheres. Around 31% of the oregano encapsulated particles had mean diameters greater than 100 nm compared with 19% for only the zein particles. Finally, the different rates of release were attributed to the different locations of the closely packed nanospheres and their different sizes.

7.6.4 Nanoencapsulation of Flavors by Milk Proteins

Emulsification, stabilization, coacervation, and desolvation methods are normally used to prepare nanoparticles from proteins like casein, albumin and β-lactoglobulin (Arnedo, Espuelas, & Irache, 2002; Bhattacharjee & Das, 2001; Chen & Subirade, 2005; Ko & Gunasekaran, 2006).

Hong et al. (2012) studied the results of high hydrostatic pressure on the physical properties of capsicum oleoresin encapsulated with globular proteins [whey protein isolates (WPI), soybean protein isolates (SPI), and casein protein (CSP)].

Giroux and Britten (2011), prepared ethyl hexanoate[4]-loaded nanoparticles (size less than 300 nm) by cross-linking the denatured whey proteins through pH-cycling. The effect of nanoparticulation conditions and ethyl hexanoate concentration on the physicochemical characteristics of nanoparticles and aroma retention was studied. It should be noted that pH value might be effective on aroma retention. According to this article, the highest aroma retention was obtained from nanoparticles made at pH = 5.0 and 5.5 without calcium addition. These nanoparticles were characterized by a poor compact and extra porous internal structure providing higher loading of aroma. Increasing aroma concentration enhanced the diameter and volume of the ethyl hexanoate-loaded nanoparticles. The percentage of aroma retention displayed an increase over denatured whey protein. As the encapsulation of ethyl hexanoate in whey protein nanoparticles decreased the mass transfer of aroma at the surface of the matrix (Giroux & Britten, 2011), milk proteins can be considered as a reliable source for the nanoencapsulation of food flavors.

7.7 CONCLUSION

Today, nano and microstructures have provided the public with the opportunity to incorporate different natural and modified flavors in various food systems. The main goal of engineering delivery systems for flavors is to induce a targeted delivery and to retain their properties until they are transferred to the right place of the body and then they can be released. In recent years,

4. Apple-like aroma.

nanoencapsulation has attracted the attention of many scientists according to their benefits over the micro-scale delivery systems. Some of these advantages are faster dissociation, higher surface area, which renders them more reactive, reinforced physical stability against aggregation, and gravitational separation, etc. As an example for flavor retention, proteins by stabilizing oil droplets in nanoemulsion media are able to retain aromas to a greater extent, as well as enhancing the retention rate of flavors. In addition, amylose nanostructures have been applied to encase flavors by their unique starch-based structures enabling this process. Overall, it seems that implementing naturally-occurring compounds like different types of cyclodextrins, casein molecules, and other natural molecules, which are considered as GRAS (generally considered as safe), is a green and safe method to entrap and deliver the food flavors. Furthermore, novel encapsulation techniques as well as effective analytical methods and simulations should be employed to ensure an efficient and targeted delivery.

REFERENCES

Aalaei, K., Rayner, M., & Sjöholm, I. (2016). Storage stability of freeze-dried, spray-dried and drum-dried skim milk powders evaluated by available lysine. *LWT-Food Science and Technology*, *73*, 675–682.

Adams, A., & De Kimpe, N. (2006). Chemistry of 2-acetyl-1-pyrroline, 6-acetyl-1, 2, 3, 4-tetrahydropyridine, 2-acetyl-2-thiazoline, and 5-acetyl-2, 3-dihydro-4 H-thiazine: Extraordinary Maillard flavor compounds. *Chemical Reviews*, *106*(6), 2299–2319.

Ades, H., Kesselman, E., Ungar, Y., & Shimoni, E. (2012). Complexation with starch for encapsulation and controlled release of menthone and menthol. *LWT-Food Science and Technology*, *45*(2), 277–288.

Amerine, M. A., Pangborn, R. M., & Roessler, E. B. (2013). *Principles of sensory evaluation of food*. Elsevier.

Angyal, S. J. (2001). *The Lobry de Bruyn-Alberda van Ekenstein transformation and related reactions*. *Glycoscience* (pp. 1–14). Springer.

Argyropoulos, D., & Müller, J. (2014). Effect of convective-, vacuum-and freeze drying on sorption behaviour and bioactive compounds of lemon balm (Melissa officinalis L.). *Journal of Applied Research on Medicinal and Aromatic Plants*, *1*(2), 59–69.

Arnedo, A., Espuelas, S., & Irache, J. (2002). Albumin nanoparticles as carriers for a phosphodiester oligonucleotide. *International Journal of Pharmaceutics*, *244*(1), 59–72.

Arshady, R. (1999). Manufacturing methodology of microcapsules. *MML Series*, *1*, 279.

Augustin, M., Sanguansri, L., Margetts, C., & Young, B. (2001). Microencapsulation of food ingredients. *Food Australia*, *53*(6), 220–223.

Baldwin, E. A., Plotto, A., & Goodner, K. (2007). Shelf-life versus flavour-life for fruits and vegetables: How to evaluate this complex trait. *Stewart Postharvest Review*, *3*(1), 1–10.

Barbosa-Cánovas, G. V., Ortega-Rivas, E., Juliano, P., & Yan, H. (2005). Encapsulation processes. *Food Powders: Physical Properties, Processing, and Functionality*, 199–219.

Barbosa, J., Borges, S., Amorim, M., Pereira, M., Oliveira, A., Pintado, M., & Teixeira, P. (2015). Comparison of spray drying, freeze drying and convective hot air drying for the production of a probiotic orange powder. *Journal of Functional Foods*, *17*, 340–351.

Berdagué, J., Tournayre, P., & Cambou, S. (2007). Novel multi-gas chromatography–olfactometry device and software for the identification of odour-active compounds. *Journal of Chromatography A*, *1146*(1), 85–92.

Berger, R. G. (2007). Flavours and fragrances: Chemistry, *bioprocessing and sustainability*. Springer Science and Business Media.

Beristain, C. I., Vazquez, A., Garcia, H. S., & Vernon-Carter, E. J. (1996). Encapsulation of orange peel oil by co-crystallization. *LWT-Food Science and Technology*, *29*(7), 645–647.

Bertrand, E., Machado-Maturana, E., Chevarin, C., Portanguen, S., Mercier, F., Tournayre, P., ... Berdagué, J. L. (2011). Heat-induced volatiles and odour-active compounds in a model cheese. *International Dairy Journal*, *21*(10), 806–814.

Bhandari, B. R., D'Arc, B. R., & Padukka, I. (1999). Encapsulation of lemon oil by paste method using β-cyclodextrin: Encapsulation efficiency and profile of oil volatiles. *Journal of Agricultural and Food Chemistry*, *47*(12), 5194–5197.

Bhandari, B. R., Datta, N., D'Arcy, B. R., & Rintoul, G. B. (1998). Co-crystallization of honey with sucrose. *LWT-Food Science and Technology*, *31*(2), 138–142.

Bhattacharjee, C., & Das, K. (2001). Characterization of microcapsules of β-lactoglobulin formed by chemical cross linking and heat setting. *Journal of Dispersion Science and Technology*, *22*(1), 71–78.

Bomben, J. L., Bruin, S., Thijssen, H. A., & Merson, R. L. (1973). Aroma recovery and retention in concentration and drying of foods. *Advances in Food Research*, *20*, 1–111.

Brückner, B., & Wyllie, S. G. (2008). *Fruit and vegetable flavour: Recent advances and future prospects*. Elsevier.

Brummer, J. M., & Lorenz, K. (1991). European developments in wheat sourdoughs. *Cereal Foods World*, *36*, 310–314.

Buffo, R., & Reineccius, G. (2001). Comparison among assorted drying processes for the encapsulation of flavors. *Perfumer & Flavorist*, *26*, 58–67.

Buffo, R., Probst, K., Zehentbauer, G., Luo, Z., & Reineccius, G. (2002). Effects of agglomeration on the properties of spray-dried encapsulated flavours. *Flavour and Fragrance Journal*, *17*(4), 292–299.

Cardinal, M., Cornet, J., Serot, T., & Baron, R. (2006). Effects of the smoking process on odour characteristics of smoked herring (Clupea harengus) and relationships with phenolic compound content. *Food Chemistry*, *96*(1), 137–146.

Castro, N., Durrieu, V., Raynaud, C., Rouilly, A., Rigal, L., & Quellet, C. (2016). Melt Extrusion Encapsulation of Flavors: A Review. *Polymer Reviews*, *56*(1), 137–186.

Cevallos, P. A. P., Buera, M. P., & Elizalde, B. E. (2010). Encapsulation of cinnamon and thyme essential oils components (cinnamaldehyde and thymol) in β-cyclodextrin: Effect of interactions with water on complex stability. *Journal of Food Engineering*, *99*(1), 70–75.

Chatterjee, D., & Bhattacharjee, P. (2013). Comparative evaluation of the antioxidant efficacy of encapsulated and un-encapsulated eugenol-rich clove extracts in soybean oil: Shelf-life and frying stability of soybean oil. *Journal of Food Engineering*, *117*(4), 545–550.

Chen, L., & Subirade, M. (2005). Chitosan/β-lactoglobulin core–shell nanoparticles as nutraceutical carriers. *Biomaterials*, *26*(30), 6041–6053.

Chew, S. C., & Nyam, K. L. (2016). Microencapsulation of kenaf seed oil by co-extrusion technology. *Journal of Food Engineering*, *175*, 43–50.

Coetzee, C., Brand, J., Emerton, G., Jacobson, D., Silva Ferreira, A. C., & Toit, W. (2015). Sensory interaction between 3-mercaptohexan-1-ol, 3-isobutyl-2-methoxypyrazine and oxidation-related compounds. *Australian Journal of Grape and Wine Research*, *21*(2), 179–188.

Cognat, C., Shepherd, T., Verrall, S. R., & Stewart, D. (2014). Relationship between volatile profile and sensory development of an oat-based biscuit. *Food Chemistry*, *160*, 72–81.

Cornu, A., Kondjoyan, N., Martin, B., Verdier-Metz, I., Pradel, P., Berdagué, J. L., & Coulon, J. B. (2005). Terpene profiles in Cantal and Saint-Nectaire-type cheese made from raw or pasteurised milk. *Journal of the Science of Food and Agriculture*, *85*(12), 2040–2046.

Davidovich-Rikanati, R., Azulay, Y., Sitrit, Y., Tadmor, Y., & Lewinsohn, E. (2009). Tomato aroma: Biochemistry and biotechnology. In D. Havkin-Frenkel, & F. C. Belanger (Eds.), Biotechnology in flavour production (pp. 118–129). London:: Blackwell Publishing Ltd.

Delcour, J. A., Bruneel, C., Derde, L. J., Gomand, S. V., Pareyt, B., Putseys, J. A., ... Lamberts, L. (2010). Fate of starch in food processing: From raw materials to final food products. *Annual Review of Food Science and Technology*, *1*, 87–111.

Desai, K. G. H., & Jin Park, H. (2005). Recent developments in microencapsulation of food ingredients. *Drying Technology*, *23*(7), 1361–1394.

Desobry, S. A., Netto, F. M., & Labuza, T. P. (1997). Comparison of Spray-drying, Drum-drying and Freeze-drying for β-Carotene Encapsulation and Preservation. *Journal of Food Science*, *62*(6), 1158–1162.

Donnelly, R. F., Morrow, D. I., Singh, T. R., Migalska, K., McCarron, P. A., O'Mahony, C., & Woolfson, A. D. (2009). Processing difficulties and instability of carbohydrate microneedle arrays. *Drug Development and Industrial Pharmacy*, *35*(10), 1242–1254.

Dubey, R. (2009). Microencapsulation technology and applications. *Defence Science Journal*, *59*(1), 82.

Espino-Díaz, M., Molina-Corral, F., Sepulveda, D., González-Aguilar, G., & Olivas, G. (2016). Alginate coatings containing high levels of isoleucine improve aromatic and standard quality in fresh-cut apple. *European Journal of Horticultural Science*, *81*(3), 175–184.

Faridi Esfanjani, A., & Jafari, S. M. (2016). Biopolymer nano-particles and natural nano-carriers for nano-encapsulation of phenolic compounds. *Colloids and Surfaces B: Biointerfaces*, *146*, 532–543.

Gibbs, F. B., Kermasha, S., Alli, I., & Mulligan, C. N. (1999). Encapsulation in the food industry: a review. *International Journal of Food Sciences and Nutrition*, *50*(3), 213–224.

Galli, B. D., Martin, J. G. P., da Silva, P. P. M., Porto, E., & Spoto, M. H. F. (2016). Sensory quality of Camembert-type cheese: Relationship between starter cultures and ripening molds. *International Journal of Food Microbiology*, *234*, 71–75.

Georg, H. (1975). Encapsulation process by simple coacervation using inorganic polymers. *Google Patents*.

Ghorani, B., & Tucker, N. (2015). Fundamentals of electrospinning as a novel delivery vehicle for bioactive compounds in food nanotechnology. *Food Hydrocolloids*, *51*, 227–240.

Giroux, H. J., & Britten, M. (2011). Encapsulation of hydrophobic aroma in whey protein nanoparticles. *Journal of Microencapsulation*, *28*(5), 337–343.

Goubet, I., Le Quere, J. L., & Voilley, A. (1998). Retention of aroma compounds by carbohydrates: Influence of their physicochemical characteristics and of their physical state. A review. *Journal of Agricultural and Food Chemistry*, *46*(5), 1981–1990.

Gouin, S. (2004). Microencapsulation: Industrial appraisal of existing technologies and trends. *Trends in Food Science and Technology*, *15*(7), 330–347.

Green, B.K. (1955). Pressure-Sensitive Record Material. U.S. Patent no. 2,712,507.

Hampel, N., Bück, A., Peglow, M., & Tsotsas, E. (2013). Continuous pellet coating in a Wurster fluidized bed process. *Chemical Engineering Science*, *86*, 87–98.

Heath, H. B. (1978). Flavor technology, profiles, products, applications. *Avi Pub. Co*, .

Hegheş, A., Hădărugă, N. G., Fuliaş, A. V., Bandur, G. N., Hădărugă, D. I., & Dehelean, C. A. (2015). Capsicum annuum extracts/β-cyclodextrin complexes. *Journal of Thermal Analysis and Calorimetry*, *120*(1), 603–615.

Hedges, A., & McBride, C. (1999). Utilization of β-cyclodextrin in food. *Cereal Foods World, 44*(10), 700–704.

Hedges, A. R., Shieh, W. J., & Sikorski, C. T. (1995). Use of cyclodextrins for encapsulation in the use and treatment of foods. *ACS Symposium Series, 59*, 60–71.

Hodge, J. E. (1953). Dehydrated foods, chemistry of browning reactions in model systems. *Journal of Agricultural and Food Chemistry, 1*(15), 928–943.

Hong, G. P., Surassmo, S., Chun, J. Y., Min, S. G., & Choi, M. J. (2012). Influence of high hydrostatic pressure on the capsicum oleoresin encapsulated by globular protein. *International Journal of Food Engineering, 8*, 2.

Re, I. M., (1998). Microencapsulation by spray drying. *Drying Technology, 16*(6), 1195–1236.

Jacquot, M., & Pernetti, M. (2004). *Spray coating and drying processes. Fundamentals of cell immobilisation biotechnology* (pp. 343–356). Springer.

Jafari, S. M., Fathi, M., & Mandala, I. (2015). *Emerging product formation. Food Waste Recovery: Processing Technologies and Industrial Techniques* (pp. 293–317). Elsevier Inc.

Jafari, S. M., Assadpoor, E., He, Y., & Bhandari, B. (2008). Encapsulation efficiency of food flavours and oils during spray drying. *Drying Technology, 26*(7), 816–835.

Jafari, S. M., He, Y., & Bhandari, B. (2007a). Encapsulation of nanoparticles of d-limonene by spray drying: Role of emulsifiers and emulsifying techniques. *Drying Technology, 25*(6), 1069–1079.

Jafari, S. M., He, Y., & Bhandari, B. (2007b). Role of powder particle size on the encapsulation efficiency of oils during spray drying. *Drying Technology, 25*(6), 1081–1089.

Jafari, S. M., Mahdavi-Khazaei, K., & Hemmati-Kakhki, A. (2016). Microencapsulation of saffron petal anthocyanins with cress seed gum compared with Arabic gum through freeze drying. *Carbohydrate Polymers, 140*, 20–25.

Janda, J., Bernacchi, D., & Frieders, S. (1995). Microencapsulation process. US Patent 5418010.

Jelen, H. (2011). *Food flavors: Chemical, sensory and technological properties*. CRC Press.

Kant, A., Linforth, R. S., Hort, J., & Taylor, A. J. (2004). Effect of β-cyclodextrin on aroma release and flavor perception. *Journal of Agricultural and Food Chemistry, 52*(7), 2028–2035.

Katouzian, I., & Jafari, S. M. (2016). Nano-encapsulation as a promising approach for targeted delivery and controlled release of vitamins. *Trends in Food Science and Technology, 53*, 34–48.

Kayaci, F., Sen, H. S., Durgun, E., & Uyar, T. (2014). Functional electrospun polymeric nanofibers incorporating geraniol–cyclodextrin inclusion complexes: High thermal stability and enhanced durability of geraniol. *Food Research International, 62*, 424–431.

Kayaci, F., & Uyar, T. (2012). Encapsulation of vanillin/cyclodextrin inclusion complex in electrospun polyvinyl alcohol (PVA) nanowebs: Prolonged shelf-life and high temperature stability of vanillin. *Food Chemistry, 133*(3), 641–649.

Kenar, J., Compton, D., Felker, F., & Fanta, G. (2015). Amylose inclusion complexation of ferulic acid via lipophilization. *Meeting Abstract*, .

Ko, S., & Gunasekaran, S. (2006). Preparation of sub-100-nm β-lactoglobulin (BLG) nanoparticles. *Journal of Microencapsulation, 23*(8), 887–898.

Kohane, D.S., Yeo, Y., Given, P., & Langer, R.S., (2015). Delivery and controlled release of encapsulated lipophilic nutrients. *Google Patents*.

Korus, J. (2001). Microencapsulation of flavours in starch matrix by coacervation method. *Polish Journal of Food and Nutrition Sciences, 10*(1), 17–23.

Koupantsis, T., Pavlidou, E., & Paraskevopoulou, A. (2014). Flavour encapsulation in milk proteins–CMC coacervate-type complexes. *Food Hydrocolloids, 37*, 134–142.

Kussmann, M., Affolter, M., Nagy, K., Holst, B., & Fay, L. B. (2007). Mass spectrometry in nutrition: Understanding dietary health effects at the molecular level. *Mass Spectrometry Reviews, 26*(6), 727–750.

Lafarge, C., Bard, M. H., Breuvart, A., Doublier, J. L., & Cayot, N. (2008). Influence of the structure of cornstarch dispersions on kinetics of aroma release. *Journal of Food Science, 73* (2), S104–S109.

Lakkis, J. M. (2007). Introduction. In J. M. Lakkis (Ed.), *Encapsulation and controlled release technologies in food systems* (pp. 1–11). Ames, IA: Blackwell Publishing.

Langrish, T., & Fletcher, D. (2001). Spray drying of food ingredients and applications of CFD in spray drying. *Chemical Engineering and Processing: Process Intensification, 40*(4), 345–354.

Leahy, M., Anandaraman, S., Bangs, W., & Reineccius, G. (1983). Spray drying of food flavors II. A comparison of encapsulating agents for the drying of artificial flavors. *Perfumer and Flavorist, 8*, 49–58.

Lević, S., Obradović, N., Pavlović, V., Isailović, B., Kostić, I., Mitrić, M., ... Nedović, V. (2014). Thermal, morphological, and mechanical properties of ethyl vanillin immobilized in polyvinyl alcohol by electrospinning process. *Journal of Thermal Analysis and Calorimetry, 118*(2), 661–668.

Liang, R., Xu, S., Shoemaker, C. F., Li, Y., Zhong, F., & Huang, Q. (2012). Physical and antimicrobial properties of peppermint oil nanoemulsions. *Journal of Agricultural and Food Chemistry, 60*(30), 7548–7555.

Lichtenthaler, H. K., Rohmer, M., & Schwender, J. (1997). Two independent biochemical pathways for isopentenyl diphosphate and isoprenoid biosynthesis in higher plants. *Physiologia Plantarum, 101*(3), 643–652.

Ling, B., Yang, X., Li, R., & Wang, S., (2015). Physicochemical properties, volatile compounds, and oxidative stability of cold pressed kernel oils from raw and roasted pistachio (Pistacia vera L. Var Kerman). *European Journal of Lipid Science and Technology.*

Liping, D. Z. Z. S. S., & Jianrong, L. (2013). Preparation of CMC/Gum Arabic/Gelatin Microcapsules Encapsulating Orange Oil by Complex Coacervation [J]. *Journal of Chinese Institute of Food Science and Technology, 6*, 019.

López-Córdoba, A., Deladino, L., Agudelo-Mesa, L., & Martino, M. (2014). Yerba mate antioxidant powders obtained by co-crystallization: Stability during storage. *Journal of Food Engineering, 124*, 158–165.

López-Córdoba, A., Gallo, L., Bucalá, V., Martino, M., & Navarro, A. (2016). Co-crystallization of zinc sulfate with sucrose: A promissory strategy to render zinc solid dosage forms more palatable. *Journal of Food Engineering, 170*, 100–107.

Ma, M., Tan, L., Dai, Y., & Zhou, J. (2013). An investigation of flavor encapsulation comprising of regenerated cellulose as core and carboxymethyl cellulose as wall. *Iranian Polymer Journal, 22*(9), 689–695.

Macoris, M. S., Janzantti, N. S., Garruti, Dd. S., & Monteiro, M. (2011). Volatile compounds from organic and conventional passion fruit (Passiflora edulis F. Flavicarpa) pulp. *Food Science and Technology (Campinas), 31*(2), 430–435.

Maga, J. A., & Katz, I. (1976). Lactones in foods. *Critical Reviews in Food Science and Nutrition, 8*(1), 1–56.

Mahdavi, S. A., Jafari, S. M., Assadpoor, E., & Dehnad, D. (2016). Microencapsulation optimization of natural anthocyanins with maltodextrin, gum Arabic and gelatin. *International Journal of Biological Macromolecules, 85*, 379–385.

Mahdavi, S. A., Jafari, S. M., Ghorbani, M., & Assadpoor, E. (2014). Spray-drying microencapsulation of anthocyanins by natural biopolymers: A review. *Drying Technology, 32*(5), 509–518.

Marilley, L., & Casey, M. (2004). Flavours of cheese products: Metabolic pathways, analytical tools and identification of producing strains. *International Journal of Food Microbiology, 90* (2), 139–159.

Martin, B., Verdier-Metz, I., Cornu, A., Pradel, P., Hulin, S., Buchin, S., & Berdagué, J. (2002). Do terpenes influence the flavour of cheeses? II. *Cantal Cheese, Caseus Intern, 3*, 25–27.

Martin, N., Berger, C., Le Du, C., & Spinnler, H. (2001). Aroma compound production in cheese curd by coculturing with selected yeast and bacteria. *Journal of Dairy Science, 84*(10), 2125–2135.

Martins, I. M., Rodrigues, S. N., Barreiro, F., & Rodrigues, A. E. (2009). Microencapsulation of thyme oil by coacervation. *Journal of Microencapsulation, 26*, 667–675.

Martins, I. M., Rodrigues, S. N., Barreiro, M. F., & Rodrigues, A. E. (2011). Release of thyme oil from polylactide microcapsules. *Industrial and Engineering Chemistry Research, 50*(24), 13752–13761.

Mathieu, S., Dal Cin, V., Fei, Z., Li, H., Bliss, P., Taylor, M. G., ... Tieman, D. M. (2009). Flavour compounds in tomato fruits: Identification of loci and potential pathways affecting volatile composition. *Journal of Experimental Botany, 60*(1), 325–337.

Maya-Meraz, I. O., Espino-Díaz, M., Molina-Corral, F. J., González-Aguilar, G. A., Jacobo-Cuellar, J. L., Sepulveda, D. R., & Olivas, G. I. (2014). Production of Volatiles in Fresh-Cut Apple: Effect of Applying Alginate Coatings Containing Linoleic Acid or Isoleucine. *Journal of Food Science, 79*(11), C2185–C2191.

McClements, D. J., Decker, E. A., Park, Y., & Weiss, J. (2009). Structural design principles for delivery of bioactive components in nutraceuticals and functional foods. *Critical Reviews in Food Science and Nutrition, 49*(6), 577–606.

Mehrad, B., Shabanpour, B., Jafari, S., & Pourashouri, P. (2015). Characterization of dried fish oil from Menhaden encapsulated by spray drying. *AACL Bioflux, 8*(1), 57–69.

Menis, M. E. C., Milani, T. M. G., Jordano, A., Boscolo, M., & Conti-Silva, A. C. (2013). Extrusion of flavored corn grits: Structural characteristics, volatile compounds retention and sensory acceptability. *LWT-Food Science and Technology, 54*(2), 434–439.

Moummou, H., Tonfack, L. B., Chervin, C., Benichou, M., Youmbi, E., Ginies, C., & Van Der Rest, B. (2012). Functional characterization of SlscADH1, a fruit-ripening-associated short-chain alcohol dehydrogenase of tomato. *Journal of Plant Physiology, 169*(15), 1435–1444.

Nakagawa, K., Iwamoto, S., Nakajima, M., Shono, A., & Satoh, K. (2004). Microchannel emulsification using gelatin and surfactant-free coacervate microencapsulation. *Journal of Colloid and Interface Science, 278*(1), 198–205.

Nawar, W. (1998). Volatile components of the frying process. *Grasas y Aceites, 49*(3–4), 271–274.

Newton, A. E., Fairbanks, A. J., Golding, M., Andrewes, P., & Gerrard, J. A. (2012). The role of the Maillard reaction in the formation of flavour compounds in dairy products—not only a deleterious reaction but also a rich source of flavour compounds. *Food and Function, 3*(12), 1231–1241.

Numanoğlu, U., Şen, T., Tarimci, N., Kartal, M., Koo, O. M., & Önyüksel, H. (2007). Use of cyclodextrins as a cosmetic delivery system for fragrance materials: Linalool and benzyl acetate. *AAPS PharmSciTech, 8*(4), 34–42.

Nyberg, J., (2016). Microorganisms influence on quality and flavor of cheese.

Okuro, P. K., de Matos Junior, F. E., & Favaro-Trindade, C. S. (2013). Technological challenges for spray chilling encapsulation of functional food ingredients. *Food Technology and Biotechnology, 51*(2), 171–182.

Oxley, J. D. (2012). Spray cooling and spray chilling for food ingredient and 55 nutraceutical encapsulation. In N. Garti, & D. J. McClements (Eds.), *Encapsulation technologies and delivery systems for food ingredients and nutraceuticals* (pp. 110–130). Oxford, U.K: Woodhead Publishing.

Pagot, Y., Belin, J. M., Husson, F., & Spinnler, H. E. (2007). Metabolism of phenylalanine and biosynthesis of styrene in Penicillium camemberti. *Journal of Dairy Research, 74*(02), 180–185.

Paramita, V., Furuta, T., & Yoshii, H. (2012). High-Oil-Load Encapsulation of Medium-Chain Triglycerides and d-Limonene Mixture in Modified Starch by Spray Drying. *Journal of Food Science, 77*(2), E38–E44.

Parris, N., Cooke, P. H., & Hicks, K. B. (2005). Encapsulation of essential oils in zein nanospherical particles. *Journal of Agricultural and Food Chemistry, 53*(12), 4788–4792.

Perez, R., Gaonkar, A.G., Akashe, A., Coates, R., Anastasiou, T., & Bheemreddy, R., (2014). Chewing gum composition comprising a micro-encapsulated flavour in a matrix comprising protein. *Google Patents.*

Pino, J. A., & Bent, L. (2013). Odour-active compounds in guava (Psidium guajava L. cv. Red Suprema). *Journal of the Science of Food and Agriculture, 93*(12), 3114–3120.

Pourashouri, P., Shabanpour, B., Razavi, S. H., Jafari, S. M., Shabani, A., & Aubourg, S. P. (2014a). Impact of wall materials on physicochemical properties of microencapsulated fish oil by spray drying. *Food and Bioprocess Technology, 7*(8), 2354–2365.

Pourashouri, P., Shabanpour, B., Razavi, S. H., Jafari, S. M., Shabani, A., & Aubourg, S. P. (2014b). Oxidative stability of spray-dried microencapsulated fish oils with different wall materials. *Journal of Aquatic Food Product Technology, 23*(6), 567–578.

Prat, L., Espinoza, M. I., Agosin, E., & Silva, H. (2014). Identification of volatile compounds associated with the aroma of white strawberries (Fragaria chiloensis). *Journal of the Science of Food and Agriculture, 94*(4), 752–759.

Putseys, J., Lamberts, L., & Delcour, J. (2010). Amylose-inclusion complexes: Formation, identity and physico-chemical properties. *Journal of Cereal Science, 51*(3), 238–247.

Quellet, C., Schudel, M., & Ringgenberg, R. (2001). Flavors & fragrance delivery systems. *CHIMIA International Journal for Chemistry, 55*(5), 421–428.

Rajabi, H., Ghorbani, M., Jafari, S. M., Mahoonak, A. S., & Rajabzadeh, G. (2015). Retention of saffron bioactive components by spray drying encapsulation using maltodextrin, gum Arabic and gelatin as wall materials. *Food Hydrocolloids, 51*, 327–337.

Rao, J., (2013). Rationalizing lipid nanoemulsion formation for utilization in the food and beverage industry.

Reineccius, G. A. (1988). Spray drying of food flavors. In S. J. Risch, & G. A. Reineccius (Eds.), *Flavor encapsulation* (pp. 55–66). Washington, DC: Amer Chem Soc.

Reineccius, G. (2013). *Source book of flavors.* Springer Science and Business Media.

Reineccius, G. (2016). *Flavor chemistry and technology.* CRC press.

Reineccius, T., Reineccius, G., & Peppard, T. (2002). Encapsulation of flavors using cyclodextrins: Comparison of flavor retention in alpha, beta, and gamma types. *Journal of Food Science, 67*(9), 3271–3279.

Rey, L. (2016). *Freeze-drying/lyophilization of pharmaceutical and biological products.* CRC Press.

Risch, S. J., & Reineccius, G. (1995). *Encapsulation and controlled release of food ingredients,* ACS symposium series. United States: American chemical society.

Ross, C. F., & Smith, D. M. (2006). Use of volatiles as indicators of lipid oxidation in muscle foods. *Comprehensive Reviews in Food Science and Food Safety, 5*(1), 18–25.

Ruse, K., Sabovics, M., Rakcejeva, T., Dukalska, L., Galoburda, R., & Berzina, L. (2012). The effect of drying conditions on the presence of volatile compounds in cranberries. *Proceedings of World Acadamey of Science, Engineering and Technology, 64*, 854–860.

Sanchez, C., & Renard, D. (2002). Stability and structure of protein–polysaccharide coacervates in the presence of protein aggregates. *International journal of pharmaceutics, 242*(1), 319–324.

Sato, K., & Hegarty, G. R. (1971). Warmed-over flavor in cooked meats. *Journal of Food Science*, *36*(7), 1098–1102.

Schoebitz, M., & Belchí, M. D. L. (2016). *Encapsulation Techniques for Plant Growth-Promoting Rhizobacteria. Bioformulations: For Sustainable Agriculture* (pp. 251–265). Springer.

Sengupta, T., Mishra, M.K., & Sweeney, W.R., (2016). Dissolvable films impregnated with encapsulated tobacco, tea, coffee, botanicals, and flavors for oral products. *US Patent*, 20, 160, 044, 942.

Sharma, P., Utreja, D., & Bedi, S. (2016). Chemical Transformations and Biological Studies of Terpenoids Isolated from Essential Oil of Cyperus scariosus. *Asian Journal of Chemistry*, *28* (10), 2153.

Sidhu, D., Lund, J., Kotseridis, Y., & Saucier, C. (2015). Methoxypyrazine analysis and influence of viticultural and enological procedures on their levels in grapes, musts, and wines. *Critical Reviews in Food Science and Nutrition*, *55*(4), 485–502.

Sillick, M., & Gregson, C. M. (2012). Spray chill encapsulation of flavors within anhydrous erythritol crystals. *LWT-Food Science and Technology*, *48*(1), 107–113.

Song, J. (2007). Flavour volatile production and regulation in apple fruit. *Stewart Postharvest Review*, *3*(2), 1–8.

Sozer, N., & Kokini, J. L. (2009). Nanotechnology and its applications in the food sector. *Trends in Biotechnology*, *27*(2), 82–89.

Spanier, A., Flores, M., McMillin, K., & Bidner, T. (1997). The effect of post-mortem aging on meat flavor quality in Brangus beef. Correlation of treatments, sensory, instrumental and chemical descriptors. *Food Chemistry*, *59*(4), 531–538.

Spinnler, H., & Leclercq-Perlat, M. (2007). 136 What causes bitterness and other flavour defects in Camembert? *Cheese problems solved*, 282.

Sricharoen, P., Lamaiphan, N., Patthawaro, P., Limchoowong, N., Techawongstien, S., & Chanthai, S. (2016). Phytochemicals in Capsicum oleoresin from different varieties of hot chilli peppers with their antidiabetic and antioxidant activities due to some phenolic compounds, Ultrason. *Sonochem*, http://dx.doi.org/10.1016/j.ultsonch.2016.08.018.

Strzyz, P. (2015). Molecular networks: Protein droplets in the spotlight. *Nature Reviews Molecular Cell Biology*, *16*(11), 639–639

Tackenberg, M.W., (2014). Investigations in the mechanisms of encapsulating liquid active ingredients via extrusion processing. Shaker.

Tackenberg, W. M., & Kleinebudde, P. (2015). Encapsulation of liquids via extrusion-A review. *Current Pharmaceutical Design*, *21*(40), 5815–5828.

Tackenberg, M. W., Krauss, R., Schuchmann, H. P., & Kleinebudde, P. (2015). Encapsulation of orange terpenes investigating a plasticisation extrusion process. *Journal of Microencapsulation*, *32*(4), 408–417.

Tarancón, P., Fiszman, S., Salvador, A., & Tárrega, A. (2013). Formulating biscuits with healthier fats. Consumer profiling of textural and flavour sensations during consumption. *Food Research International*, *53*(1), 134–140.

Taraneh, J. B., Rahmatollah, G., Hassan, A., & Alireza, D. (2008). Effect of wax inhibitors on pour point and rheological properties of Iranian waxy crude oil. *Fuel Processing Technology*, *89*(10), 973–977.

Thierry, A., & Maillard, M. B. (2002). Production of cheese flavour compounds derived from amino acid catabolism by Propionibacterium freudenreichii. *Le Lait*, *82*(1), 17–32.

Tieman, D., Bliss, P., McIntyre, L. M., Blandon-Ubeda, A., Bies, D., Odabasi, A. Z., ... Goulet, C. (2012). The chemical interactions underlying tomato flavor preferences. *Current Biology*, *22*(11), 1035–1039.

Torres-Giner, S., Martinez-Abad, A., Ocio, M. J., & Lagaron, J. M. (2010). Stabilization of a nutraceutical omega-3 fatty acid by encapsulation in ultrathin electrosprayed zein prolamine. *Journal of Food Science*, *75*(6), N69–N79.

Torres, S., Pandey, A., & Castro, G. R. (2010). Banana flavor: Insights into isoamyl acetate production. *Cell*, *549*(155.778), 776.

Trubiano, P., & Lacourse, N., (1988). Emulsion-stabilizing starches: Use in flavor encapsulation, *ACS symposium series*. United States: American chemical society.

Tusch, M., Krüger, J., & Fels, G. (2011). Structural stability of V-amylose helices in water-DMSO mixtures analyzed by molecular dynamics. *Journal of Chemical Theory and Computation*, *7*(9), 2919–2928.

Uyar, T., Hacaloglu, J., & Besenbacher, F. (2009a). Electrospun polystyrene fibers containing high temperature stable volatile fragrance/flavor facilitated by cyclodextrin inclusion complexes. *Reactive and Functional Polymers*, *69*(3), 145–150.

Uyar, T., Nur, Y., Hacaloglu, J., & Besenbacher, F. (2009b). Electrospinning of functional poly (methyl methacrylate) nanofibers containing cyclodextrin-menthol inclusion complexes. *Nanotechnology*, *20*(12), 125703.

Van Boekel, M. (2006). Formation of flavour compounds in the Maillard reaction. *Biotechnology Advances*, *24*(2), 230–233.

Viswalingam, M., Prabu, S., Sivakumar, K., & Rajamohan, R. (2016). Preparation and characterization of a imipramine-ß-cyclodextrin inclusion complex. *Instrumentation Science and Technology*, 1–21.

Whitfield, F. B., & Mottram, D. S. (1992). Volatiles from interactions of Maillard reactions and lipids. *Critical Reviews in Food Science and Nutrition*, *31*(1–2), 1–58.

Winnik, F.M., Breton, M.P., & Sanders, D.J., (1996). Ink jet printing of concealed images on carbonless paper. *Google Patents*.

Wolf, B. (2010). Polysaccharide functionality through extrusion processing. *Current Opinion in Colloid and Interface Science*, *15*(1), 50–54.

Yu, A. N., Tan, Z. W., & Wang, F. S. (2012). Mechanism of formation of sulphur aroma compounds from L-ascorbic acid and L-cysteine during the Maillard reaction. *Food Chemistry*, *132*(3), 1316–1323.

Zafeiropoulou, T., Evageliou, V., Gardeli, C., Yanniotis, S., & Komaitis, M. (2012). Retention of selected aroma compounds by gelatine matrices. *Food Hydrocolloids*, *28*(1), 105–109.

Zhang, W., Gong, X., Cai, Y., Zhang, C., Yu, X., Fan, J., & Diao, G. (2013). Investigation of water-soluble inclusion complex of hypericin with β-cyclodextrin polymer. *Carbohydrate Polymers*, *95*(1), 366–370.

Zhao, Q., Ho, C. T., & Huang, Q. (2013). Effect of ubiquinol-10 on citral stability and off-flavor formation in oil-in-water (O/W) nanoemulsions. *Journal of Agricultural and Food Chemistry*, *61*(31), 7462–7469.

Zhu, G., Xiao, Z., Zhou, R., & Lei, D. (2015). Preparation and simulation of a taro flavor. *Chinese Journal of Chemical Engineering*, *23*(10), 1733–1735.

Zhu, L., Lan, H., He, B., Hong, W., & Li, J. (2010). Encapsulation of menthol in beeswax by a supercritical fluid technique. *International Journal of Chemical Engineering*, 1–7. Available from http://dx.doi.org/10.1155/2010/608680.

Zuidam, N. J., & Shimoni, E. (2010). Overview of microencapsulates for use in food products or processes and methods to make them. In N. J. Zuidam, & V. A. Nedovic (Eds.), *Encapsulation technologies for active food ingredients and food processing* (pp. 3–29). New York: Springer.

Chapter 8

Nanoencapsulation of Enzymes, Bioactive Peptides, and Biological Molecules

Muhammed Yusuf Çağlar[1], Mehmet Demirci[1], Kadir Bayrambaş[2], Bilal Çakır[1] and İbrahim Gülseren[1]
[1]*İstanbul Sabahattin Zaim University (İZÜ), İstanbul, Turkey,*
[2]*Tunceli University, Tunceli, Turkey*

8.1 INTRODUCTION

Throughout history, humanity has utilized natural resources in the pursuit for food and bioactive components for curing poisoning and also treatment of various diseases. The utilization of various bioactive metabolites, especially secondary metabolites are being directly attributed to the promotion of human health which further outlines the necessity to utilize such compounds in novel formulations such as functional foods. In the manufacture of industrial products, there is a clear need to deliver functional/bioactive components, including the food, pharmaceutical, cosmetic, and biomedical industries. These components represent a diverse range of substances, including proteins, enzymes, bioactive lipids, antimicrobials, antioxidants, nucleic acids, and medical drugs (Chen, Remondetto, Subirade, 2006; Chen, Weiss, & Shahidi, 2006; Shefer & Shefer, 2003; Ubbink, 2002; Ubbink & Kruger, 2006). In this chapter, we will primarily focus on the bioactive components and utilization of a variety of nanoencapsulation techniques relevant to these components with different methods.

Due to the advances in modern biological techniques, clinical symptoms of deficiency may be understood as a result of a long-term deficiency of micronutrients. Unfortunately, prior to such symptoms, the inadequate delivery of micronutrients may trigger the development of chronic diseases. At the same time, despite an optimum intake of micronutrients, some individuals still may exhibit a high risk for chronic diseases (Boushey, Beresford, Omenn, & Motulsky, 1995). All these findings highlight the importance of balanced diets and consumption of functional food products.

Nanoencapsulation of Food Bioactive Ingredients. DOI: http://dx.doi.org/10.1016/B978-0-12-809740-3.00008-8

Bioactive compounds consist of essential and nonessential natural compounds (e.g., vitamins or polyphenols), and they represent a small fraction of components in the food chain, but the critical point is that they most certainly were shown to demonstrate biological effect(s) on human health. In some cases, the presence of bioactive substances in food products provide health benefits that is well beyond the caloric value of the corresponding product.

The basis for the increasing interest in bioactive compounds is that knowledge from epidemiologic studies, such as the correlation between the consumption of certain compounds and occurrence risk of chronic diseases, has been increased. While until recently, vitamins and other micronutrients have been recommended only to avoid the classic symptoms of deficiency, current claims and lists of bioactive substances expanded significantly.

Bioactive compounds are expected to generate beneficial effects on human health even at considerably small amounts. The potential bioactivity and effectiveness of a certain bioactive material may be determined based on the monitoring of biomarker molecules *in vivo*. Incorporation of such compounds into our daily diet possibly decreases the risk of chronic diseases and improves our overall health conditions. For example, antioxidants found in plants are secondary metabolites serving several protective functions to the plants such as repelling insects (i.e., anthocyanins) and shielding against UV radiation, or regulating osmotic pressure, while exhibiting an astringent effect (Sepúlveda, Ascacio, Rodriguez-Herrera, Aguilera-Carbó, & Aguilar, 2011; Zarei, Azizi, & Bashiri-Sadr, 2010). Upon their consumption, bioactive effects in the human body proves advantageous such as inhibition of enzymatic and nonenzymatic oxidation reactions, binding of pro-oxidative substances, or scavenging free radicals.

For instance, antioxidants tend to stabilize and protect polyunsaturated fatty acids (PUFA) in food products due to their contribution to a few different stabilization mechanisms including reactions with free radicals, chelation of metals, and inhibiting the propagation of lipid oxidation. In the food products, antioxidant may form complexes with different juice components; they may induce a cloudy appearance or formation of a precipitate, especially, when complexed with proteins (Oziyci, Karhan, Tetik, & Turhan, 2013). Based on these findings, generation of bioactive bearing foods have to be carefully executed in order to optimize both sensory and biological quality. Due to the above mentioned advantages, antioxidants are among the most commonly investigated plant or fruit components (Bhandari, 2012; Fischer, Carle, & Kammerer, 2011; Gil et al., 2000; Jurenka, 2008; Viladomiu, Hontecillas, Lu, & Basseganya-Riera, 2013).

In this section, some of the most important types of bioactive compounds that need to be present in a balanced diet will be briefly discussed, along with a discussion of their physicochemical characteristics and the current challenges to their application in functional food formulations.

Among the priorities in the nanoencapsulation technologies, it is possible to mention the prevention of molecular degradation of bioactive compounds based on oxidation, hydrolysis, etc. Protection against the stresses that take place in processing, storage, transportation, and also during digestion are regarded as important. Consequently, potential exploitation of nanoencapsulation system in biomolecules is reviewed.

8.2 ENZYMES

Enzymes are a group of protein based molecules that catalyze biochemical reactions (Wiseman, 1987). In their molecular structure, some enyzmes also contain nonprotein components such as metal ions (i.e., copper, iron, magnesium, etc.) and vitamins. In our daily diets, we also consume a low concentration of enzymes both in processed and raw food products, whereas there have not been any clear findings on any damage that these enzymes caused to human up to date. Industrial utilization of enyzmes has been carried out since 1874, when Christian Hansen extracted rennin from calf stomach to manufacture cheese products (Nielsen et al., 1994). Since environmental conditions have a bearing on enzyme activity, their influence has to be accounted for.

8.2.1 Stability of Enzymes

There, generally, is a temperature range where each enzyme demonstrates its optimal performance. Unless enzymes were originally extracted from thermophilic organisms, this range mostly tends to lie between 30–50°C. In certain processes, such as the preparation of corn sirups, utilization of thermostable enzymes is necessary. Similarly, both for digestive and technological processes, there is an optimum pH range where enzymes demonstrate maximum stability and performance. While for pepsin the optimum pH range is between 1.5–3, trypsin is known to function most properly between pH 7–9. The changes in medium pH clearly affects the rate of enzymatic reactions. In that sense, encapsulation systems can stabilize the structural characteristics of enzymes and prolong their period of activity. In the ripening of cheese products, the two sets of major enzymes that are responsible tend to be endopeptidases and exopeptidases. When these enzymes were encapsulated in combination, ripening period and storage costs were found to be clearly reduced (e.g., Kınık, Kavas, & Yılmaz, 2003). Cheese products prepared with encapsulated enzymes were found to demonstrate enhanced taste, aroma, and textural approval compared to the control group samples (Peker & Arslan, 2011).

8.3 PEPTIDES

Peptides that are generated from food peptides under *in vitro* or *in vivo* conditions and that demonstrate biological functions and/or physiological effects are classified as bioactive peptides (Smacchi & Gobbetti, 2000). They are generally composed of short amino acid chains (i.e., 3–20 amino acids). Eggs, beans, fish, and corn proteins possibly constitute the most important sources of bioactive peptides along with dairy proteins (Ahn, Park, Atwal, Gibbs, & Lee, 2009; Majumder & Wu, 2009; Narva, 2004).

Although some sequences in proteins remain as inactive constituents, once hydrolyzed with digestive enzymes or during fermentative or other technological processes, they tend to demonstrate their bioactive properties (Korhonen & Pihlanto, 2006; Möller, Scholz-Ahrens, Roos, & Schrezenmeir, 2008; Roufik, Gauthier, & Turgeon, 2006; Schanbacher, Talhouk, & Murray, 1997; Schanbacher, Talhouk, Murray, Gherman, & Willett, 1998; Smacchi & Gobbetti, 2000). Based on their bioactivities, dietary intake of bioactive peptides may reduce the risk of chronic diseases and boost the immune system of the consumers. Consequently, a variety of peptides that demonstrate antimicrobial, antioxidative, antithrombotic, immune system regulating, and blood pressure lowering activities were found, and their bioactivities are possibly related to their amino acid composition and sequence (Korhonen & Pihlanto, 2006). Based on such bioactivities, functional food products can be manufactured (Park, Morimae, Matsumura, Nakamura, & Sato, 2008) and encapsulation systems may be utilized accordingly. Especially at reasonable cost levels, due to their safety, reliability, and high nutritional values, peptides and proteins may be utilized as functional food ingredients (Park et al., 2008).

8.4 PHYTOSTEROLS

Phytosterols are a group of phytochemicals that include compounds such as stigmasterol, β-sitosterol, and campesterol. Plant stanols, which are found naturally at lower concentrations than sterols, can be produced by the hydrogenation of phytosterols. Phytosterol concentrations in vegetable oils range from 0.1% to 1.0% (Chaiyasit, Elias, McClements, & Decker, 2007) and typical phytosterol consumption is in the range of 200 to 400 mg/day. The production of phytosterol fortified foods has become popular due to the ability of phytosterols to decrease total and low-density lipoprotein cholesterol in humans by inhibiting the absorption of dietary cholesterol (Wong, 2001; Ostlund, 2004).

Daily intake of 1.6 g phytosterols was found to lead to an approximately 10% reduction in LDL cholesterol (Hallikainen, Sarkkinen, & Uusitupa, 2000). The intestinal absorption of phytosterols is very low so dietary phytosterols do not have adverse effects on health. Phytosterol incorporation into food formulations is quite challenging due to their high melting point and tendency to form insoluble crystals. However, PUFA esterified phytosterols demonstrate a higher solubility. Upon ingestion of phytosterols esters, lipases hydrolyze the fatty acid to produce free phytosterols. Phytosterols have mostly been added to high fat foods (e.g., margarine) where solublization and dispersion were relatively simple. Phytosterol introduction into water-based foods require them either being suspended or emulsified. Phytosterol oxidation products have been observed in model systems, oils, and food products (Bortolomeazzi, Cordaro, Pizzale, & Conte, 2003; Cercaci, Rodriguez-Estrada, Lercker, & Decker, 2007; Dutta, 1997; Lambelet et al., 2003; Soupas, Juntunen, Lampi, & Piironen, 2004). It is not clear whether oxidized phytosterols tend to lose their bioactive properties or demonstrate any toxic effects *in vivo* in a manner similar to oxidized cholesterol. As with other bioactive lipids that are susceptible to oxidative reactions, nanoencapsulation of phytosterols could potentially increase their oxidative stability and consequently render enhance their bioavailability.

8.5 NUCLEIC ACIDS

DNA is a critical biomolecule used in gene therapy (Putnam, 2006), diagnostics (Dharmadi & Gonzalez, 2004), nanorobotics (Yurke, Turberfield, Mills, Simmel, & Neumann, 2000), and molecular evolution (Chakrabarti, Breaker, Joyce, & Deamer, 1994; Tawfik & Griffiths, 1998; Ghadessy, Ong, & Holliger, 2001). A general problem in the effective and efficient utilization of DNA in these areas is to prevent its degradation, which can occur due to mechanical shear forces (Murphy, Cano, Fox, & Willson, 2006), chemical degradation by nucleases (Putnam, 2006), or other processing related variables (Zelikin et al., 2007). It is logical to concentrate and/or preserve DNA by a physical barrier (Zelikin et al., 2007) possibly based on nanoencapsulation techniques. There are a number of methods utilized in gene therapy to restrict DNA degradation, including complexation of DNA with polycations (Putnam, 2006), blockcopolymer micelles (Kataoka et al., 2001), cationic lipids, or liposomes (de Lima, Simoes, Pires, Faneca, & Düzgüneş, 2001). Furthermore, DNA molecules can be entrapped within gels (Goh et al., 2004), micellar (Csaba, Caamaño, Sánchez, Domínguez, & Alonso, 2005), or polymeric microparticles (Ando, Putnam, Pack, & Langer, 1999). In order to prepare transcriptionally active nucleic acid formulations, DNA may be encapsulated within liposomes (Edwards & Baeumner, 2007; Tsumoto, Nomura, Nakatani, & Yoshikawa, 2001), water-in-oil emulsions and polyelectrolyte capsules, fibers, or nanoparticles.

Among the priorities in the nanoencapsulation technologies, it is possible to mention the prevention of molecular degradation of bioactive compounds based on oxidation, hydrolysis, etc. Protection against the stresses that take place in processing, storage, transportation, and also during digestion are regarded as important. In this chapter, potential exploitation of nanoencapsulation systems in biomolecules will be reviewed.

8.6 LIPID FORMULATION TECHNOLOGIES FOR NANOENCAPSULATION OF BIOLOGICAL MOLECULES

8.6.1 Nanoemulsions

Nanoemulsions (also known as miniemulsions or submicron emulsions) are nanoscale droplets of multiphase colloidal dispersions formed by dispersing at least one liquid phase in other immiscible liquid(s) by physical shear-induced rupturing mechanisms (Liu, Sun, Li, Liu, & Xu, 2006; Mason, Wilking, Meleson, Chang, & Graves, 2006; Meleson, Graves, & Mason, 2004; Russel, Saville, & Schowalter, 1989). Different size ranges have been reported in the literature for nanoemulsions, i.e., less than 100 nm (Guo et al., 2007; Porras, Solans, Gonzalez, & Gutierrez, 2008; Shakeel & Ramadan, 2010), 10–100 nm (Talegaonkar, Mustafa, Akhter, & Iqbal, 2010), 100–500 nm (Anton, Benoit, & Saulnier, 2008; Constantinides, Chaubal, & Shorr, 2008; Rossi & Leroux, 2007; Tadros, Izquierdo, Esquena, & Solans, 2004), and 100–600 nm (Solans et al., 2005). Physicochemical characteristics and biological fate of emulsion systems highly depend on the mean particle size characteristics; therefore, the definition of size range is not a trivial point. From a nanotechnological perspective, having size ranges of less than 100 nm and possessing different properties than ordinary emulsions could be considered, since nanoemulsions have some interesting physical properties that can be applied to distinguish them from conventional emulsion systems.

Nanoemulsions are appropriate systems for the delivery of poorly water-soluble food ingredients, such as fish oil and lipophilic vitamins, because of their ability to enhance the solubilization and the possibility of absorption in the gastrointestinal (GI) tract (Jafari, Fathi, & Mandala, 2015). After ingestion in the human body, droplets readily disperse in stomach to small droplets of nanoemulsion, which promotes the rapid release of the encapsulated bioactive throughout the GI tract (Talegaonkar et al., 2010).

Couvreur, Blanco-Prieto, Puisieux, Roques, and Fattal (1997) reviewed the possibilities of peptide and olipeptide delivery via multiple emulsion systems. Since the inner aqueous phase has the capacity to stabilize aqueous peptides, W/O/W (water-in-oil-in-water) emulsions represent an appropriate means of peptide delivery. Similarly, W/O/W nanoemulsions that were decorated with alginate and/or chitosan were successfully utilized in the delivery of insulin to rats orally (Li et al., 2013), and pronounced hypoglycemic activity was observed. Balcão et al. (2013) demonstrated that lactoferrin bearing W/O/W nanoemulsions demonstrated antimicrobial activity.

8.6.2 Nanoliposomes

Hydrophobic/hydrophilic interactions among lipid/lipid and lipid/water interfaces are responsible for the formation of liposomes. Liposomes are formed in single and bilayer arrangements. Lipo-soluble and water-soluble vitamins can be entrapped in these nanocarriers for maintaining their stability in different media.

Nanoliposomal systems have been commonly utilized in antimicrobial peptide encapsulation and also in the stabilization of enzymes in cheese manufacture. Liposomal encapsulation of antimicrobial peptides, including nisin, other bacteriocins, or bacteriocin-like compounds (Laridi et al., 2003; Teixeira, dos Santos, Silveira, & Brandelli, 2008; Were, Bruce, Davidson, & Weiss, 2003), have been investigated in detail in the literature. Since antimicrobial resistance of microorganisms is a major concern, enhancing the efficacy of antimicrobials through means of encapsulation could be a critical contribution to industrial food safety issues. Nisin or nisin plus EDTA bearing liposomes demonstrated pronounced inhibition towards *Listeria monocytogenes*. Coencapsulation of nisin and EDTA demonstrated maximum inhibition of *E. coli* O157:H7 (Taylor, Bruce, Weiss, & Davidson, 2008).

Liposomal encapsulation enhanced the antimicrobial potential of pediocin AcH against *L. monocytogenes* (Degnan, Buyong, & Luchansky, 1993). In addition to commonly used dairy, soy, or egg phospholipids, Imran et al. (2015) utilized marine phospholipids in the preparation of nisin bearing liposomes. Liposomal encapsulation could be also instrumental in the protection of proteins or peptides from thermal and/or pressure induced denaturation or degradation (da Silva Malheiros, Sant'Anna, Micheletto, da Silveira, & Brandelli, 2011, da Silva Malheiros, Sant'Anna, de Souza Barbosa, Brandelli, & de Melo Franco, 2012).

Liposomal bacteriocins from *Lactobacillus sakei* subsp. sakei 2a inhibited *L. monocytogenes* growth in both microbial media and intentionally contaminated goat milk (Malheiros, Cuccovia, & Franco, 2016). Not in all cases though the encapsulated peptides were more effective than free peptides. For example, *L. monocytogenes* was inhibited more significantly by free bacteriocin than liposomal P34 (da Silva Malheiros et al., 2011).

Liposomal encapsulation of enzymes in cheese manufacture has also been utilized for a considerably long period (Kirby, Brooker, & Law, 1987). Liposomal enzymes may enhance product texture, reduce storage time necessary for flavor formation and reduce the enzyme concentrations used in the process (Mozafari, 2006); thus, increasing the profitability in cheese manufacture.

8.6.3 Nanostructured Lipid Carriers

Since an increasing extent of commercial applications are anticipated in the delivery of proteins, oligonucleotides, and DNA, the importance of encapsulation systems including emulsions and liposomes increase. In addition, other lipid based carriers have been reviewed. Solid lipid nanoparticles (SLNs) and nanostructured lipid carriers (NLCs) could also be instrumental based on their biocompatible features (Martins, Sarmento, Ferreira, & Souto, 2007). Especially, hydrophobic proteins and peptides can be immediately stabilized in these systems without requiring an additional aqueous phase unlike W/O/W emulsions. Especially since 40% of new molecules manufactured as drugs have hydrophobic characteristics, the importance of SLN and NLC systems will increase over time, while toxicity concerns related to emulsifiers have to be keenly investigated (Martins et al., 2007).

8.7 NATURAL NANOCARRIERS FOR NANOENCAPSULATION OF BIOLOGICAL MOLECULES

8.7.1 Casein Nanocapsules

Cow's milk contains approximately 30–35 g of protein per liter. Casein accounts for about 80% of all bovine milk proteins and is coordinated in casein micelles. Casein micelles are successful nanovehicles that concentrate, stabilize, and transport essential nutrients, mainly calcium, phosphate, and protein, for the neonate (DeKruif & Holt, 2003). Therefore, it is a natural nanodelivery system that is common to a variety of mammals, although the composition and physicochemical characteristics vary between species.

The micelles are spherical colloids, 50–500 nm in diameter (150 nm in average) (Fox, 2003), formed by the presence of four major casein phosphoproteins: α_{s1}-casein (α_{s1}-CN), α_{s2}-CN, β-CN, and κ-CN (molar ratio 4:1:4:1, respectively) (DeKruif & Holt, 2003; Fox, 2003).

The molecular size range of casein proteins are approximately 19–25 kDa (Eigel et al., 1984), whereas more than half of their amino acid residues have polar side chains (Morimoto, 1970), which in turn promote intra- and intermolecular hydrogen bonding (Khadka & Haynie, 2012). All these critical properties render caseins a suitable nanodelivery system for food and pharmaceutical applications in addition to being an important source of minerals and essential amino acids (Brugman et al., 2004).

A number of studies have been carried out in order to encapsulate bioactive compounds in casein micelles, and a few brief examples are reviewed in this section.

The β-casein can self-associate to compose micelles that are able to dissolve lipophilic compounds but is not economical for food applications at this time. (Anema & De Kruif, 2013; Haratifar, Meckling, & Corredig, 2014; Pan, Zhong, & Baek, 2013; Yazdi et al., 2013). Casein micelles can be dissociated into nanoclusters by high-pressure homogenization (Orlien, Boserup, & Olsen, 2010) which was used for the incorporation of vitamin D2 in reassembled nanoparticles of sodium caseinate composed upon depressurization (Semo, Kesselman, Danino, & Livney, 2007), which possibly can be further utilized for the encapsulation of peptides, proteins, and DNA molecules. Similarly, casein micelles can also be dissociated by calcium chelating agents such as citrate and ethylenediaminetetraacetate (Udabage, Mckinnon, & Augustin, 2000) or at alkaline conditions (McMahon, Du, McManus, & Larsen, 2009). These agents could represent an alternative strategy to reform casein micelles and casein based fractions upon which bioactive materials may be encapsulated. Ghasemi and Abbasi (2014) investigated the effects of alkaline pH and ultrasound treatment on the formation, structural attributes, entrapment efficiency, and protective properties of natural casein micelles. When medium pH was raised, size of casein micelles increased and the hydrophobic material encapsulation capabilities. Power intensity and duration of ultrasonic treatment clearly affected the entrapment efficiency of casein micelles as well.

Crosslinking of casein micelles with the enzyme transglutaminase cause the formation of stable nanoparticles that compose of a covalently linked casein network. Transglutaminase caused crosslinking of casein micelles formed intramicellar bonds, but the substructure of crosslinked casein micelles stayed identical to that of native casein micelles. The intramicellar crosslinking raised the stability of casein micelles (Huppertz & de Kruif, 2008).

Pan, Yu, Yao, and Shao (2007) formed polyelectrolyte complex micelles by mixed β-casein with lysozyme. The micelles were heated, which induced the gelation of lysozyme and the entrapment of β-casein in the gel, generating narrowly dispersed spherical nanoparticles. The synthesis of β-casein-lysozyme nanoparticles was primarily a combination of polyelectrolyte complex formation and heat caused gelation. These nanoparticles showed pH dependent hydrophobicity, which may be funtional in the encapsulation and release of relatively hydrophobic compounds (Pan et al., 2007).

In conclusion, casein micelles are potential nanovehicles for biomolecules. Casein micelles can be utilized for nanoencapsulation of hydrophobic biomolecule substances for potential enrichment of low-fat or nonfat food products.

8.7.2 Nanocrystal Nanocapsules

Nanocapsules are defined in the literature as mostly an oily or hydrophobic cavity surrounded by a thin polymer wall. A broad variety of polymer materials, such as poly(lactide-co-glycolide), poly-e caprolactone, and polyalkylcyanoacrylate, can be used for the preparation of nanocapsules (Skiba, Nemati, Puisieux, Duchêne, & Wouessidjewe, 1996). Nanocapsules are promising applications, since they are ideal for the encapsulation of lipophilic bioactives (Vauthier-Holtzscherer, Benabbou, Spenlehauer, Veillard, & Couvreur, 1991). On the other hand, nanocrystals are crystalline clusters of a few hundred to a few thousand atoms with the corresponding mean sizes of a few nanometers (Parak et al., 2003).

Nanocrystals are particularly utilized for poorly water soluble molecules (Caban, Aytekin, Sahin, & Capan, 2014). Their contribution to bioavailability occurs by enhancing both solubility and bioadhesion to the intestinal wall (Caban et al., 2014). In their utilization in pharmaceutical applications, nanocrystals enhance dissolution velocity and saturation solubility, reproducibility and repeatability of oral absorption, the amount of bioavailable dose, and patient compliance by the reduction of number of oral units that need to be used (Müller, Jacobs, & Kayser, 2001; Rabinow, 2005). Nanocrystals are especially favorable formulations for poorly soluble nutraceuticals like coenzyme Q10, rutin, hesperidin, apigenin, etc. (Shegokar & Müller, 2010). The potential findings on nanocrystals render them promising for food applications (Gülseren & Corredig, 2013; Tzoumaki, Moschakis, & Biliaderis, 2011a, Tzoumaki, Moschakis, Kiosseoglou, & Biliaderis, 2011b; Tzoumaki, Moschakis, Scholten, & Biliaderis, 2013).

8.7.3 Cyclodextrin Nanocapsules

Cyclodextrins (CDs) are cyclic oligosaccharides, which are produced from starch via cyclodextrin glucanotransferase (CGTase) enzyme (Szetjli, 1998). These starch derivatives are known as cycloamyloses, cyclomaltoses, and Schardinger dextrins (Villiers, 1891; Eastburn & Tao, 1994). CDs are comprised of six (α-CD), seven (β-CD), eight (γ-CD), or more glucopyranose units linked by α-(1,4) bonds (Del Valle, 2004). They are nontoxic ingredients, are not absorbed in the upper gastrointestinal tract, and are completely metabolized by the colon microflora. β-CD has been on the GRAS list since 1998, as a flavor carrier and protectant, at a level of 2% in numerous food products (Szente & Szejtli, 2004).

The preparation of CDs is simpler and cheaper than most other methods of nanoencapsulation (Marques, 2010) and has the potential to provide the following benefits:

- Stabilization of light- or oxygen-sensitive substances.
- Modification of the chemical reactivity of guest molecules.
- Fixation of very volatile substances.
- Improvement of solubility of substances.
- Modification of liquid substances to powders.
- Protection against degradation of substances by microorganisms.
- Masking the bitter taste of certain compounds.
- Masking pigments or the color of substances.
- Catalytic activity of cyclodextrins with guest molecules.
- Controling the release of materials.
- Reducing material toxicity.

All of these advantages of cyclodextrins make them favorable for applications in analytical chemistry, agriculture, pharmaceutical, and food and cosmetic manufacturing areas (Astray, Gonzalez-Barreiro, Mejuto, Rial-Otero, & Simal-Gándara, 2009; Cabral Marques, 1994; Hedges, Shieh, & Sikorski, 1995; Loftsson & Duchêne, 2007; Jackson & Lee, 1991; Marques, 2010; Singh, Sharma, & Banerjee, 2002). Cyclodextrins have sweet taste; so this reason make them suitable for masking the unwanted taste and odor (Marques, 2010). They are resistant to many rigorous food-processing methods such as freezing, thawing, and microwaving (Astray et al., 2009).

CDs can be utilized to solubilize and stabilize several biomedically-significant peptides and proteins, including growth hormones (Brewster, Hora, Simpkins, & Bodor, 1991; Charman, Mason, & Charman, 1993), interleukin-2 (Brewster et al., 1991), monoclonal antibody MN12 (Ressing et al., 1992), aspartame (Prankerd, Stone, Sloan, & Perrin, 1992), tumor necrosis factor (Hora, Rana, & Smith, 1992), albumin (Katakam & Banga, 1995), g-globulin (Katakam & Banga, 1995), lactate dehydrogenase (Izutsu, Yoshioka, & Kojima, 1995), etc. For example, α-CD raised the solubility of cyclosporin A, an immunosuppressive agent, in eyedrop form. α-CD provided the drug to penetrate into the cornea with the least local toxicity (Ichikawa, Kanai, & Yamazaki, 1995; Kanai et al., 1989).

Peptides and proteins cannot always be fully encapsulated into inclusion complexes of cyclodextrins (Prokai et al., 1996). Appropriate complex forming sidechains can bind cyclodextrins, resulting in modified solubility, stability, and/or membrane transport characteristics (Szejtli, 1994). Prokai et al. (1996) characterized interactions of amino acids with cyclodextrins by electrospray ionization mass spectrometry. They binded α-cyclodextrin with protonated tryptophan and β-cyclodextrin with human insulin. They concluded that electrospray ionization mass spectrometry is a powerful technique for the analysis of binding CDs with amino acids and peptides.

In conclusion, cyclodextrins are very useful agents for nanoencapsulation of biomolecules, and their further exploitation is anticipated in biomolecule delivery in functional foods.

8.8 EQUIPMENT BASED TECHNOLOGIES FOR NANOENCAPSULATION OF BIOLOGICAL MOLECULES

Electrospinning and electrospraying are electrohydrodynamic processes. In these processes, a polymer solution can be spun or sprayed by the application of a high potential electric field to obtain fibers or particles, respectively (Faridi Esfanjani & Jafari, 2016; Katouzian & Jafari 2016). Electrospinning and electrospraying are used to produce new materials including encapsulation systems. Electrically charged jet of polymer solutions are used in electrohydynamic processes because of the producing fibers or particles at micron, submicron, or nanoscales. When compared to the other encapsulation methods, these processes are relatively easier, cheaper, and flexible and present many advantages, while their utilization in the field of food processing is novel.

Electrohydrodynamic encapsulation methods comprise of four main components: (i) A high voltage source (1–30 kV) usually activated in direct current mode, though alternating current mode is also possible (Kessick, Fenn, & Tepper, 2004), (ii) a blunt ended stainless steel needle or capillary, (iii) a syringe pump, and (iv) a grounded collector either flat plate or rotating drum (Bhushani & Anandharamakrishnan, 2014). Electrospinning and electrospraying are accepted as "similar" technologies; however, there are several discrepancies which distinguish the two electrohydrodynamic processes.

In electrospinning, when electrical forces suppress the forces of surface tension in the charged polymer liquid, a charged jet excluded from the tip of a capillary tube elongates, which is known as Taylor cone, and goes towards a grounded surface. The solvent in the jet is vaporized during the flight, inducing mat of nanofibers deposited on the surface. Electrospun fibers are continuous and can vary in diameter of several nanometers to micrometers. These fibers can be produced by changing the collector or the electrical field (Chakraborty, Liao, Adler, & Leong, 2009). Polymer types, molecular weight, and concentration are critical factors for electrospinning because the feasibility of polymer electrospinning is determined by these factors (Chakraborty et al., 2009). For example, elastin-like polypeptides have been used as polymer materials in electrospinning methods. Electrospining is a simple and rapid encapsulation technique for the stabilization of bioactive compounds (Bhushani & Anandharamakrishnan, 2014; Brandenberger, Nüssli, Piech, & Widmer, 1999).

On the other hand, electrospraying is a process of liquid atomization by electrical forces. The difference between electrospinning and electrospraying techniques is the concentration of the polymer solution. When the solution concentration is high, the jet from Taylor cone is stabilized, and elongation takes place by whipping instability mechanism. If the solution concentration is low, the jet is destabilized due to varicose waves on the surface of the jet, leading to the formation of small and highly charged liquid droplets and, hence, fine droplets are formed. These highly charged droplets are self-dispersing in space, thereby preventing droplet agglomeration and coagulation (Brandenberger et al., 1999; Jaworek & Sobczyk, 2008).

Consequently, electrospinning and electrospraying techniques are commonly used for encapsulation of bioactive molecules; since they provide efficient means of bioactive encapsulation and enzyme immobilization, these techniques are also suitable for various food coating and development of novel filtration materials and active food packaging applications.

8.8.1 Enzyme Immobilization Based on Electrohydrodynamic Processes

Enzyme immobilization is a widely utilized method for most large-scale (i.e., industrial) applications since catalyst recycling, continuous operation, and product purification become more controllable and reproducible with this approach (Li, Chen, & Wu, 2007). Electrohydrodynamic processes are quite instrumental in the encapsulation of enzymes since they are nonthermal process, which in turn protect the native structure of enzymes. When enzymes were immobilized onto several insoluble or solid materials, their thermal and operational stabilities were found to increase significantly (Mozhaev, Melik-Nubarov, Sergeeva, Šikšnis, & Martinek, 1990). Due to the reduced biocatalytic efficiency in many immobilized enzyme investigations, industrial applicability is generally low. Biocatalytic efficiency of enzymes can be enhanced though by the tailoring of the structures formed by carrier materials (Shinkai, Honda, & Kobayashi, 1991). Although nonporous materials severely limit the diffusion of enzymes, their capabilities of enzyme loading are quite limited as well. While, porous materials can provide high enzyme loading, their corresponding release rater are significantly faster (Hayashi et al., 1993). Based on these findings, nanofibers were utilized as the supports for enzyme immobilization with respect to both enzyme loading and enzymatic activity.

For example, Li et al. (2007) produced *Candida rugosa* lipase bearing polyacrylonitrile nanofibers by electrospinning. The immobilization was based on the amidination of polyacrylonitrile nanofibers. Enzyme molecules were determined to be covalently attached to the support and produced aggregates on the fiber surface, which also became more hydrophilic and resistant after enzyme immobilization. In another application, an amperometric biosensor was developed by electrospining poly(vinyl alcohol) (PVA) in the presence of glucose oxidase (Ren et al., 2006). The immobilized glucose oxidase was placed on the surface of Au electrode. The immobilized enzyme remained active inside the electrospun PVA fibrous membranes. The enzyme immobilization within PVA membranes demonstrated by the infrared (IR) and UV–Vis spectra and the scanning electronic microscope (SEM) micrographs while enzyme was shown to remain active. The biosensors displayed a rapid response time (1 s) and a higher output current (μA level) to glucose in the normal and diabetic level (Ren et al., 2006). PVA fibrous membranes were used in enzyme immobilization due to their high specific surface area and porous structure.

8.8.2 Nucleic Acid Encapsulation Based on Electrohydrodynamic Processes

As an alternative and promising technology, electrohydrodynamic processes are used for DNA encapsulation. DNA molecules were observed to lose their bioactivity and degrade within a few days following their conjugation to electrospun fiber surfaces and slowly being released into the nearby tissue would significantly preserve their functionality (Zamani, Prabhakaran, & Ramakrishna, 2013). While the particle core encapsulated bioactive molecules, the shell can be incorporated with bioactive molecules like nonviral gene-delivery vectors, which can act as a nonviral delivery vector for encapsulated DNA upon release. In another study, improved a coaxial electrospray process to produce oligodeoxynucleotide encapsulated lipoplex nanoparticles for gene delivery. Incorporation of oligodeoxynucleotide and lipoplex nanoparticles is used in inhalation therapy. Compared to emulsion technologies, this process facilitated better controllability.

8.8.3 Protein Encapsulation Based on Electrohydrodynamic Processes

Native structure of proteins is formed by a one-dimensional linear chain of amino acids that has folded into a complex three-dimensional structure. In most cases, a "folded" protein will show various levels of structural organization. The structural and functional properties of peptides and proteins, the biocompatibility of these molecules, the nutritional necessity for protein in animals, and the tissue toxicity of several synthetic organic biopolymers, including those confirmed by the US Food and Drug Administration (FDA), have together encouraged interest in developing biomaterials made of proteins and polypeptides (Khadka & Haynie, 2012). Proteins demonstrate a wide range of functions, including the processes of enzymatic catalysis, cellular signaling, immune responses, cellular adhesion, and cellular cycle. Some other proteins are organized as structural fibrous materials. Protein fibers can be considered as building blocks of organisms, enabling scaffolding, stabilization, protection, elasticity, and motility at length scales ranging from nanometers to meters. Structural and functional attributes of fibrous proteins are increasingly being exploited to enhance the performance of synthetic biomaterials.

Electrospinning was studied to structure proteins on a nanometer length scale in a specific way (Nieuwland et al., 2013). Proteins are very difficult to electrospin, principally, because of their complex secondary and tertiary structures. Firstly, in order to achieve a spinnable system, protein molecules should be well dispersed in a random spiral conformation. Globular proteins have little interaction with each other that could facilitate entanglements during the spinning process. Caseinates, as a group of protein with a random spiral structure in water (i.e., no well-defined tertiary structure), may appear spinnable at the first glance. However, spinning pure caseinate from an aqueous dispersion was not possible. The inability to spin caseinate was most probably provoked by the aggregation of the caseinate molecules (Pitkowski, Durand, & Nicolai, 2008).

Proteins can be electrospun under conditions where they dissolve well (Bhardwaj & Kundu, 2010) in a random spiral conformation. When proteins are dissolved in solvents with high solvent qualities, it could be possible to distinguish between both intra- and intermolecular interactions between proteins and solubilize the resulting unstructured protein. This strategy was used for collagen and gelatin, where they were spun from hexafluoro-2-propanol (HFP), trifluoroethanol (TFE), or aqueous acid (with at least 40% acetic acid or formic acid) (Boland et al., 2004; Chen, Li, Li, & Guo, 2007; Chen,

Mo, & Qing, 2007; Huang, Gao, et al., 2004; Huang, Zhang, Ramakrishna, & Lim, 2004; Ki et al., 2005; Li et al., 2005; Matthews, Wnek, Simpson, & Bowlin, 2002; Songchotikunpan, Tattiyakul, & Supaphol, 2008). Silk and fibrinogen were spun from HFP or 98% formic acid, (Kawahara, Nakayama, Matsumura, Yoshioka, & Tsuji, 2008; Min et al., 2004; Wnek, Carr, Simpson, & Bowlin, 2003) and various globular proteins (among others hemoglobin and BSA) were spun from TFE (Barnes et al., 2006; Dror et al., 2008). Although these harsh solvents are usable for biomedical applications, their food applications are not permitted (Nieuwland et al., 2013).

Stephansen, Chronakis, and Jessen (2014) electrospun cod sarcoplasmic proteins. They determined that fish sarcoplasmic protein (FSP) fibers were insoluble in water and FSP concentration influenced the morphology of the electrospun FSP fibers. Aceituno-Medina, Lopez-Rubio, Mendoza, and Lagaron (2013) electrospun amaranth protein isolate in the presence of formic acid solution. They determined the effect of pH, type of solvent, and surfactant addition on the spinnability, morphology, and molecular organization of electrospun fibers. In their study, hexafluoro-2-propanol was used as a solvent for electrospinning process. In another study, whey protein isolate (WPI) and beta-lactoglobulin molecules were electrospun with poly(ethylene oxide) (PEO). It was determined that WPI/PEO composite nanofibers maintained their fibrous morphology at temperatures as high as 100°C, which is well-above the melting point of PEO (i.e., 60°C).

In conclusion, electrospinning and electrospraying methods can accelerate the expansion of food processing sector in encapsulation applications, food packaging, and edible coating. Electrospun nanofibers and electrosprayed nanoparticles present structural and functional advantages. The encapsulation of food bioactive compounds in electrospun fibers and electrosprayed particles contributes significantly to their stability and controlled release characteristics.

8.9 BIOPOLYMER BASED TECHNOLOGIES FOR NANOENCAPSULATION OF BIOLOGICAL MOLECULES

8.9.1 Protein Nanogels

Ovalbumin and lysozyme, which are hen egg proteins, can be utilized in order to manufacture nanogels. Once these proteins are thermally denatured, they are bonded by hydrophobic interactions, hydrogen bonds, and disulfide linkages. To stabilize surface structure of nanogels, electrostatic repulsive forces can be effectively tailored. Alkali reagents have been used in nanogel formation process for rendering protein nanogels edible and nutritional (Yu et al., 2006). It is possible to manufacture protein nanogels from food proteins (e.g., Chen, Lin, Sun, & Zhao, 2014), which in turn can be utilized in the encapsulation of bioactive compounds. Polypeptide nanogels could also be utilized in the delivery of bioactives (Kim, Han, Kweon, Park, & Lim, 2013; Kim, Park, Kim, & Lim, 2013; Kim et al., 2013).

8.9.2 Chitosan-Based Nanogels

Chitosan is a widely utilized biopolymer in the delivery of bioactive compounds, especially macromolecules, due to its both physicochemical and biological properties (Kato, Onishi, & Machida, 2003). Because chitosan is water-soluble (i.e., in dilute acetic acid solutions, etc.) and positively charged, it can easily interact with negatively charged polymers or polyanions in aqueous environments. Furthermore, chitosan has been utilized as an absorption enhancer through the mucosal barrier in a variety of studies. In addition, chitosan based nanogels display low toxicity and good biocompatibility (Shutava & Lvov, 2006; Knapczyk et al., 1989). Consequently, they are also appropriate systems for the delivery of bioactive molecules such as proteins, peptides, and DNA. For example, peptide conjugation to chitosan nanogels was shown to successfully facilitate peptide delivery, while the gel structure remained unaffected by the inclusion of a bioactive (i.e., cardiac) peptide (Reis et al., 2012).

8.9.3 Alginate-Based Nanogels

Alginate-based nanogels represents nanogels formed by nano particles within a range of 250–850 nm with sodium alginate solution, calcium chloride, and poly-lysine. The interest for these nanogels is increasing in many fields such as antitubercular and antifungal drugs, therapeutic agents, and even gene delivery. Pescosolido et al. (2010) showed that, in combination with other polysaccharides, alginate is an appropriate raw material in protein delivery and release.

Alternative technologies based on poly(vinyl alcohol), poly(ethylene oxide), poly (ethyleneimine), poly(vinylpyrrolidone), poly-N-isopropylacrylamide, and hyaluronic acid nanogels are also utilized in the delivery of bioactive compounds and medicinal drugs.

8.9.4 Nanotubes

Nanotubes were discovered by Sumio Iijima, who is a Japanese electron microscopist, in 1991. Nanotubes demonstrate a characteristic hexagonal shape similar to the structure of an empty carbon tube, with the slight difference of having additional atom groups on two sides (Scott, 2005). They display thermal resistance at elevated temperatures and have a solid and flexible structure that might be used in many areas, including food industries, nanomedicine, and medical devices, etc. As an example relevant to food applications, it is noteworthy that when milk protein α-lactalbumin (α-La) was hydrolyzed partially, it transformed into self-assembled nanotubes under suitable conditions (Graveland-Bikker & De Kruif, 2006, Graveland-Bikker, Fritz, Glatter, & de Kruif, 2006).

Recently, researchers indicated that in addition to α-La, some globular whey proteins, such as β-lactoglobulin (β-lg) and bovine serum albumin (BSA), also self-assembled into fibrillar structures at elevated temperatures and low pH values. These nanofibrils had a characteristic diameter of approximately 5 nm and range up to 15 μm in length. Protease enzymes from *Bacillus licheniformis* were found to partially hydrolyze α-lactalbumin molecules, and immediately afterwards hydrolyzed protein was exposed to calcium ions. Consequently, the formation of linear nanotubes was observed (Graveland-Bikker & de Kruif, 2006).

All of the major physicochemical characteristics of α-lactalbumin nanotubes render this component a promising encapsulation agent (Gouin, 2004). Firstly, since α-lactalbumin is a milk protein, its utilization in food industry is already quite common (Rajagopal & Schneider, 2004). Much similar to the casein micelle, which is a promising natural nanovehicle for delivering hydrophobic bioactives including probiotics, self-assembled α-lactalbumin nanotubes could play similar roles (Augustin & Hemar, 2009). Similar to casein micelles, in addition to the potential delivery of bioactives, protein nanotubes, as a compact system, concentrate protein molecules and, thus, facilitate their delivery.

There are also a variety of investigations in the literature where the encapsulation of proteins, peptides, and DNA molecules were carried out by carbon, lipid, and protein nanotubes (Davis et al., 1998; de la Escosura, Janssen, Schenning, Nolte & Cornelissen, 2010; Kameta et al., 2007; Kang et al., 2009).

8.9.5 Starch Nanoparticles

Starch, which is a storage carbohydrate in plants, is mostly presented in vegetables (30–50 g/100 g), cereal grains (40–90 g/100 g), immature fruits (40–70 g/100 g), and tubers (65–85 g/100 g). Starch is structurally made of amylose and amylopectin chains. While amylose is a linear chain that contains D-glucose units gathered by glycosidic α-1,4 bonds; amylopectin, which is a branched structure, consists of D-glucose units joined by α-1,4 and α-1,6 bonds (Gonçalves, Noreña, da Silveira & Brandelli, 2014). In addition to its critical importance in baked goods, native and modified starch products have been widely used in many areas such as fat replacers, excipients for tableting, drug delivery matrix formers, and food emulsifiers (Mahkam, 2010). Due to its biodegradable and biocompatible properties, starch is a promising, versatile, and inexpensive polysaccharide for drug delivery applications (El-Hag & Al Arifi, 2011; Chen, Li, et al., 2007; Chen, Mo, et al., 2007). In a variety of investigations, starch nanoparticles (SNPs) were utilized and their size range is approximately between 10–1000 nm (Chin, MohdYazid & Pang, 2014). Utilization of modified starch in the manufacture of nanosized starch particles is practical, since the crystallinity, hydrophobicity and stability characteristics against enzymatic and thermal degradation need to be improved (Xu et al., 2010). SNPs have distinct advantages such as lower viscosity and small particle size, while considerable amounts of active ingredients can be loaded in these systems (Chen, Remondetto, et al., 2006; Chen, Weiss, et al., 2006).

When starch is used as an encapsulation agent, its crystalline arrangement is vital due to the possibility to produce more linkages among the polymer chains and generation of a wall material with improved characteristics such as reduced diffusion rate is necessary (Palma-Rodriguez et al., 2012). Crosslinking method is also widely used in order to enhance the functional properties of starch products. When the number of crosslinks in starch granules increase, the viscosity, water absorption capacity, and textural characteristics of starch suspensions will be improved. Based on these advantages, SNPs display a desirable elevated viscosity and optimum extent of water holding (Kramer, 2009).

Winarti, Sunarti, Mangunwidjaja, and Richana (2014) discussed the potential of SNPs in the encapsulation of bioactives. Both in the research related to SNPs and starch spherulites (Kong & Ziegler, 2014; Ma, 2010), there is an increasing amount of information in the area of bioactive molecule encapsulation, especially that of small molecules. In the following years, we anticipate a significant extent of progress in this area.

8.10 RELEASE OF PEPTIDES FROM NANOSCALE DELIVERY SYSTEMS

In order to establish peptide nanoencapsulation systems with the capabilities of slow-release of peptides, mostly a particle size range of 50–100 nm is utilized due to the prolonged stability of peptides (Couvreur & Puisieux, 1993). Environmental factors deeply affect the stability of encapsulated peptides, and an appropriate encapsulation system has to be used to prevent inactivation.

Also, since the major characteristics of proteins may be altered during metabolic activities, bioactive material (i.e., peptide) release needs to be tailored via the design of successful delivery systems. Immunoglobulins, serum albumin, lactoferrin, lactoperoxidase, and proteose-peptone are among the physiologically active milk serum proteins. During their enzymatic degradation, a variety of bioactive peptides, such as lactoferricin, alfa-lactophorin, serophorin, lactokinin, beta-lactophorin, beta-lactotensin, and albutensin, were shown to be formed (Cabuka et al., 2014). The correlation between the breakdown of nanoparticles and the extent of enyzme released from these systems was investigated (Martins et al., 1996). In addition to the importance of the extent of release, the interactions with the nanoparticle matrices have a bearing on the activity of incorporated enzymes (Martins et al., 1996).

Rapid (i.e., burst) release was related to the loading of proteins on the particle surface (i.e., surface adsorption) rather than the incorporation within particle interior (Giovagnoli et al., 2005). Consequently, the design of delivery systems are quite critical. For example, lipid matrices are clearly good encapsulation systems for lipophilic bioactives rather than hydrophilic proteins (Reithmeier et al., 2001). However, for lipophilic peptides as well, it is possible to utilize SLN systems. In the case of leuprolide in prostate cancer treatment, encapsulation in SLNs and *in vitro* release characteristics were clarified by Gallarate et al. (2011). Relatively large peptides (13–43 amino acids) or recombinant human growth hormones were rapidly released (up to 70%) within a day, whereas the extent of insulin release from PLGA nanoparticles accounted for about 30–40% within 24 h (Yang et al., 2012).

BSA and immunoglobulin (IgG) bearing PEO–PLGA particles offered high encapsulation efficiency (EE = 58.9%) and slow *in vitro* release rate (Santander et al., 2010), which pointed out the potential viability of such commercial products.

High affinity of the core and wall materials is paramount to the formation of stable encapsulated peptide products that can withstand food processing and storage conditions with limited diffusion losses of the core materials. Contrary to EE, a recent study demonstrated that the release kinetics of peptides encapsulated in protein microbeads in aqueous environment was inversely proportional to the peptide hydrophobicity with average release rate constants of 0.1 and 0.014 min1 for Phe–Trp and Leu–Trp–Met–Arg–Phe, respectively, after 1 h. Conversely, the modification of rapeseed protein by acylation and high pressure treatment that resulted in higher EE was found to increase the % release of the encapsulated peptide compared with the native protein after 24 h using the dynamic dialysis method. This indicates weaker interaction of the peptides with the modified protein carriers. Although theoretically promising, the dearth of experimental information on the biostability of encapsulated protein hydrolysate and peptides makes it difficult to evaluate the prospects of encapsulation in oral delivery of bioactive peptides. A myriad of bioactive peptides derived from various food proteins have been reported, and it is becoming increasingly apparent that the focus needs to be shifted to the translation of the peptides into commercial functional food products.

Studies focused on characterizing the digestion and release of encapsulated peptides during gastrointestinal processing are crucial in understanding the effect of encapsulation on biostability. One study evaluated the biostability of bioactive peptides encapsulated with a carboxymethylated gum and sodium alginate, and found minimal (up to 10%) and maximal (up to 60%) release of protein materials after simulated gastric and intestinal digestion phases, respectively. The released peptides at the intestinal phase can then be presented for absorption into the enterocytes, and subsequently into circulation where they are still susceptible to further peptidolytic modification. Therefore, it is imperative to assess the digestion kinetics and biostability of encapsulated peptides and their bioavailability in different physiological sites to ensure the release of intact bioactives at appropriate time and target location. Finally, the potential success of encapsulation applications depends on the fundamental characteristics of nanocarrier systems and the bioactive compound that they bear. There are rapid, robust, and reliable methods that can be used to characterize the basic characteristics of encapsulation systems such as particle size, morphology, wall thickness, chemical composition, and structure of the particle and the encapsulated material.

8.11 SAFETY OF NANOENCAPSULATION SYSTEMS LOADED WITH BIOLOGICAL MOLECULES

Nanotechnology has been gaining popularity in food industry in the recent years. However, consumers are suspicious about food products and packaging materials that contain nanocomponents and expect that either nanosystems should be excluded from food systems totally or not affect the human body in any negative way possible (Dowling, 2004).

Although there is a lack of long-term risks and regulatory inefficiencies, considerable variety of food products are in the market that contain nanomaterials, such as nutritional drinks, oils rich in vitamins and minerals on its micelles, etc. Toxicity issue is one of the major problems that must be investigated. There is strong possibility that certain nanocomponents can cause an increase of oxidative stress, generate free radicals, cancer, and DNA mutation (Moraru et al., 2003). Nanotechnological applications in food products require precautions related strongly to their impact on human health and environment (Chau, Wu, & Yen, 2007). Firstly, in order to minimize the possible side effects caused by nanosystem bearing food products, strong and reliable tests must be carried out that demonstrate conformity to food safety standards (Bowman & Hodge, 2007). In European (EU) regulations, food products containing nanomaterials are required to satisfy specific safety standards and testing procedures. In the USA, these products are regulated by the United States Food and Drug Administration (FDA). Also in Australia and New Zealand, Food Standards Australia and New Zealand (FSANZ) is responsible of regulating nanotechnological applications (Afroz et al., 2012).

8.12 CONCLUSION AND FURTHER REMARKS

Recently, there has been a considerable interest in the development of nanoscale delivery systems for biological molecules in order to improve the bioavailability and stability of such compounds. Consequently, studies on nanoencapsulation and delivery of bioactives by polymeric and natural carrier based systems have been reviewed in the present work, including the preparation methods for efficient delivery of bioactives. Critical aspects of food compatible nanoencapsulation systems should pertain the analysis on toxicity characteristics, removal of any residual solvents, and the investigations on the biological fate of bioactives during digestion, absorption, and excretion of these systems.

Many beneficial food bioactives, such as proteins, peptides, enzymes and phytosterols, are known to be limited in terms of aqueous solubility, and consequently their inclusion in food systems is generally difficult. Especially regarding their unstable characteristics and tendency of degradation, delivery systems that are utilized in their implementation is of critical importance. The challenge for the food industry is the development of appropriate methods and techniques that can be used to manufacture these systems. Their utilization in foods has significant potential in the generation of novel functional foods.

Nanoassemblies could potentially alter the lipid digestibility characteristics of food products, supply the producers with novel tools to prepare functional foods with desirable sensory attributes, enhance the pH, heat, saline, oxidative, etc. stability of food components, tailor and control the release kinetics of bioactives from food matrices among other potential benefits. Although in most value added applications, some extra costs might be necessary to generate novel foods with enhanced features, industrial utilization seems to be steadily increasing which enhances the chances of novel applications in the marketplace. However, as of now, although there is an extended history of safe usage for natural encapsulation systems such as casein micelles, we cannot conclusively claim that the potential risks of nanomaterials to the health of humans are completely well-known and characterized.

Several technological and bioactive properties of peptides are characterized with an emphasis on functional food production. Especially their manufacture from protein-rich waste or by-products of industrial processing is currently being investigated in detail. Utilization of by-products can alter the economic difficulties in manufacture of bioactive added food products. In these areas, considerable extent of studies is still needed in order to figure out the specifics of cost-effective solutions that can generate a myriad of novel products and consequently, promote public health globally.

ACKNOWLEDGMENTS

This study was partially funded by a grant from TAGEM R&D Support Program, Republic of Turkey Ministry of Food, Agriculture and Livestock (Grant No. TAGEM-14/AR-GE/27).

REFERENCES

Aceituno-Medina, M., Lopez-Rubio, A., Mendoza, S., & Lagaron, J. M. (2013). Development of novel ultrathin structures based in amaranth (*Amaranthus hypochondriacus*) protein isolate through electrospinning. *Food Hydrocolloids*, *31*(2), 289–298.

Afroz, Q. M., Swaminathan, K., Karthikeyan, P., Ahmed, K. P., Sudhir, U., & Kumar, M. U. (2012). Application of nanotechnology in food and dairy processing: An overview. *The Pakistan Journal of Food Science and Technologies*, *22*(1), 23–31.

Ahn, J. E., Park, S. Y., Atwal, A., Gibbs, B. F., & Lee, B. H. (2009). Angiotensin Iconvertingenzyme (ACE) inhibitory peptides from whey fermented by lactobacillus species. *Journal of Food Biochemistry*, *33*, 587–602.

Ando, S., Putnam, D., Pack, D. W., & Langer, R. (1999). PLGA microspheres containing plasmid DNA: Preservation of supercoiled DNA via cryopreparation and carbohydrate stabilization. *Journal of Pharmaceutical Sciences*, *88*(1), 126–130.

Anema, S. G., & De Kruif, C. G. (2013). Protein composition of different sized casein micelles in milk after the binding of lactoferrin or lysozyme. *Journal of Agricultural and Food Chemistry*, *61*(29), 7142–7149.

Anton, N., Benoit, J. P., & Saulnier, P. (2008). Design and production of nanoparticles formulated from nano-emulsion templates—A review. *Journal of Controlled Release*, *128*(3), 185–199.

Astray, G., Gonzalez-Barreiro, C., Mejuto, J. C., Rial-Otero, R., & Simal-Gándara, J. (2009). A review on the use of cyclodextrins in foods. *Food Hydrocolloids*, *23*(7), 1631–1640.

Augustin, M. A., & Hemar, Y. (2009). Nano- and micro-structured assemblies for encapsulation of food ingredients. *Chemical Society Reviews*, *38*(4), 902–912.

Balcão, V. M., Costa, C. I., Matos, C. M., Moutinho, C. G., Amorim, M., Pintado, M. E., & Teixeira, J. A. (2013). Nanoencapsulation of bovine lactoferrin for food and biopharmaceutical applications. *Food Hydrocolloids*, *32*(2), 425–431.

Barnes, C. P., Smith, M. J., Bowlin, G. L., Sell, S. A., Tang, T., & Matthews, J. A. (2006). Feasibility of electrospinning the globular proteins hemoglobin and myoglobin. *Journal of Engineered Fibers and Fabrics*, *1*(2), 16–29.

Bhandari, P. R. (2012). Pomegranate (Punica granatum L). Ancient seeds for modern cure? Review of potential therapeutic applications. *International Journal of Nutrition, Pharmacology, Neurological Diseases*, *2*(3), 171–184.

Bhardwaj, N., & Kundu, S. C. (2010). Electrospinning: A fascinating fiber fabrication technique. *Biotechnology Advances*, *28*(3), 325–347.

Bhushani, J. A., & Anandharamakrishnan, C. (2014). Electrospinning and electrospraying techniques: Potential food based applications. *Trends in Food Science and Technology*, *38*(1), 21–33.

Boland, E. D., Matthews, J. A., Pawlowski, K. J., Simpson, D. G., Wnek, G. E., & Bowlin, G. L. (2004). Electrospinning collagen and elastin: Preliminary vascular tissue engineering. *Frontiers in Bioscience*, *9*, 1422–1432.

Bortolomeazzi, R., Cordaro, F., Pizzale, L., & Conte, L. S. (2003). Presence of phytosterol oxides in crude vegetable oils and their fate during refining. *Journal of Agricultural and Food Chemistry*, *51*(8), 2394–2401.

Boushey, C. J., Beresford, S. A., Omenn, G. S., & Motulsky, A. G. (1995). A quantitative assessment of plasma homocysteine as a risk factor for vascular disease. Probable benefits of increasing folic acid intakes. *Journal of the American Medical Association*, *274*, 1049–1057.

Bowman, D., & Hodge, G. (2007). A small matter of regulation: An international review of nanotechnology regulation. *Columbia Science Technology Law Review*, *8*, 1–32.

Brandenberger, H., Nüssli, D., Piech, V., & Widmer, F. (1999). Monodisperse particle production: A method to prevent drop coalescence using electrostatic forces. *Journal of Electrostatics*, *45*(3), 227–238.

Brewster, M. E., Hora, M. S., Simpkins, J. W., & Bodor, N. (1991). Use of 2-hydroxypropyl-β-cyclodextrin as a solubilizing and stabilizing excipient for protein drugs. *Pharmaceutical Research*, *8*(6), 792–795.

Brugman, S., Klatter, F. A., Visser, J., Bos, N. A., Elias, D., & Rozing, J. (2004). Neonatal oral administration of DiaPep277, combined with hydrolysed casein diet, protects against Type 1 diabetes in BB-DP rats. An experimental study. *Diabetologia*, *47*(7), 1331–1333.

Caban, S., Aytekin, E., Sahin, A., & Capan, Y. (2014). Nanosystems for drug delivery. *OA Drug Design & Delivery*, *2*(1), 2.

Cabral Marques, H. (1994). Applications of cyclodextrins: Thermodynamic aspects of cyclodextrin complexes. *Revista Portuguesa De Farmacia*, *44*(2), 85–96.

Cabuka, B., Okuklu, B., Stanciuc, N., & Harsa, S. T. (2014). Nanoencapsulation of biologically active peptides from whey proteins. *Journal of Nutrition & Food Sciences*, *2*, 1–4.

Cercaci, L., Rodriguez-Estrada, M. T., Lercker, G., & Decker, E. A. (2007). Phytosterol oxidation in oil-in-water emulsions and bulk oil. *Food Chemistry*, *102*(1), 161–167.

Chaiyasit, W., Elias, R. J., McClements, D. J., & Decker, E. A. (2007). Role of physical structures in bulk oils on lipid oxidation. *Critical Reviews in Food Science and Nutrition*, *47*, 299–317.

Chakrabarti, A. C., Breaker, R. R., Joyce, G. F., & Deamer, D. W. (1994). Production of RNA by a polymerase protein encapsulated within phospholipid vesicles. *Journal of Molecular Evolution*, *39*(6), 555–559.

Chakraborty, S., Liao, I. C., Adler, A., & Leong, K. W. (2009). Electrohydrodynamics: A facile technique to fabricate drug delivery systems. *Advanced Drug Delivery Reviews*, *61*(12), 1043–1054.

Charman, S. A., Mason, K. L., & Charman, W. N. (1993). Techniques for assessing the effects of pharmaceutical excipients on the aggregation of porcine growth hormone. *Pharmaceutical Research*, *10*(7), 954–962.

Chau, C., Wu, S., & Yen, G. (2007). The development of regulations for food nanotechnology. *Trends in Food Science and Technology*, *18*(5), 269–280.

Chen, H., Weiss, J., & Shahidi, F. (2006). Nanotechnology in nutraceuticals and functional foods. *Food Technology*, *60*, 30–36.

Chen, L., Li, X., Li, L., & Guo, S. (2007). Acetylated starch-based biodegradable materials with potential biomedical applications as drug delivery systems. *Current Applied Physics*, *7*, e90–e93.

Chen, L. Y., Remondetto, G. E., & Subirade, M. (2006). Food protein-based materials as nutraceutical delivery systems. *Trends in Food Science and Technology*, *17*(5), 272–283.

Chen, N., Lin, L., Sun, W., & Zhao, M. (2014). Stable and pH-sensitive protein nanogels made by self-assembly of heat denatured soy protein. *Journal of Agricultural and Food Chemistry*, *62*(39), 9553–9561.

Chen, Z. G., Mo, X. M., & Qing, F. L. (2007). Electrospinning of collagen–chitosan complex. *Materials Letters*, *61*(16), 3490–3494.

Chin, S. F., MohdYazid, S. N. A., & Pang, S. C. (2014). Preparation and characterization of starch nanoparticles for controlled release of curcumin. *International Journal of Polymer Science*, *2014*(1a), 1–8.

Constantinides, P. P., Chaubal, M. V., & Shorr, R. (2008). Advances in lipid nanodispersions for parenteral drug delivery and targeting. *Advanced Drug Delivery Reviews*, *60*(6), 757–767.

Couvreur, P., & Puisieux, F. (1993). Nano-and microparticles for the delivery of polypeptides and proteins. *Advanced Drug Delivery Reviews*, *10*(2–3), 141–162.

Couvreur, P., Blanco-Prieto, M. J., Puisieux, F., Roques, B., & Fattal, E. (1997). Multiple emulsion technology for the design of microspheres containing peptides and oligopeptides. *Advanced Drug Delivery Reviews*, *28*(1), 85–96.

Csaba, N., Caamaño, P., Sánchez, A., Domínguez, F., & Alonso, M. J. (2005). PLGA: Poloxamer and PLGA: Poloxamine blend nanoparticles: New carriers for gene delivery. *Biomacromolecules*, *6*(1), 271–278.

da Silva Malheiros, P., Sant'Anna, V., de Souza Barbosa, M., Brandelli, A., & de Melo Franco, B. D. G. (2012). Effect of liposome-encapsulated nisin and bacteriocin-like substance P34 on Listeria monocytogenes growth in Minas frescal cheese. *International Journal of Food Microbiology*, *156*(3), 272–277.

da Silva Malheiros, P., Sant'Anna, V., Micheletto, Y. M. S., da Silveira, N. P., & Brandelli, A. (2011). Nanovesicle encapsulation of antimicrobial peptide P34: Physicochemical characterization and mode of action on Listeria monocytogenes. *Journal of Nanoparticle Research*, *13*(8), 3545–3552.

Davis, J. J., Green, M. L., Hill, H. A. O., Leung, Y. C., Sadler, P. J., Sloan, J., & Tsang, S. C. (1998). *The immobilisation of proteins in carbon nanotubes. Inorganica Chimica Acta*, *272*(1), 261–266.

Degnan, A. J., Buyong, N., & Luchansky, J. B. (1993). Antilisterial activity of pediocin AcH in model food systems in the presence of an emulsifier or encapsulated within liposomes. *International Journal of Food Microbiology*, *18*(2), 127–138.

De Kruif, C. G., & Holt, C. (2003). *Casein micelle structure, functions and interactions. Advanced Dairy Chemistry-1 Proteins* (pp. 233–276). New York: Springer.

de la Escosura, A., Janssen, P. G., Schenning, A. P., Nolte, R. J., & Cornelissen, J. J. (2010). Encapsulation of DNA-templated chromophore assemblies within virus protein nanotubes. *Angewandte Chemie International Edition*, *49*(31), 5335–5338.

de Lima, M. C. P., Simoes, S., Pires, P., Faneca, H., & Düzgüneş, N. (2001). Cationic lipid–DNA complexes in gene delivery: From biophysics to biological applications. *Advanced Drug Delivery Reviews*, *47*(2), 277–294.

Del Valle, E. M. (2004). Cyclodextrins and their uses: A review. *Process Biochemistry*, *39*(9), 1033–1046.

Dharmadi, Y., & Gonzalez, R. (2004). DNA microarrays: Experimental issues, data analysis, and application to bacterial systems. *Biotechnology Progress*, *20*(5), 1309–1324.

Dowling, A. P. (2004). Development of nanotechnologies. *Materials Today*, *7*, 30–35.

Dror, Y., Ziv, T., Makarov, V., Wolf, H., Admon, A., & Zussman, E. (2008). Nanofibers made of globular proteins. *Biomacromolecules*, *9*(10), 2749–2754.

Dutta, P. C. (1997). Studies on phytosterol oxides. 2. Content in some vegetable oils and in french fries prepared in these oils. *Journal of the American Oil Chemists' Society*, *74*(6), 659–666.

Eastburn, S. D., & Tao, B. Y. (1994). Applications of modified cyclodextrins. *Biotechnology Advances*, *12*(2), 325–339.

Edwards, K. A., & Baeumner, A. J. (2007). DNA-oligonucleotide encapsulating liposomes as a secondary signal amplification means. *Analytical Chemistry*, *79*(5), 1806–1815.

Eigel, W. N., Butler, J. E., Ernstrom, C. A., Farrell, H. M., Harwalkar, V. R., Jenness, R., & Whitney, R. M. (1984). Nomenclature of proteins of cow's milk: Fifth revision. *Journal of Dairy Science*, *67*(8), 1599–1631.

El-Hag Ali, A., & Al-Arifi, A. S. (2011). Swelling and drug release profile of poly (2-ethyl-2-oxazoline)-based hydrogels prepared by gamma radiation-induced copolymerization. *Journal of Applied Polymer Science*, *120*(5), 3071–3077.

Faridi Esfanjani, A., & Jafari, S. M. (2016). Biopolymer nano-particles and natural nano-carriers for nano-encapsulation of phenolic compounds. *Colloids and Surfaces B: Biointerfaces*, *146*, 532–543.

Fischer, U. A., Carle, R., & Kammerer, D. R. (2011). Identification and qualification of phenolic compounds from pomegranate (Punica granatum L.) peel, mesocarp, aril and differently produced juices by HPLC-DAD-ESI/MS. *Food Chemistry*, *127*, 807–821.

Fox, P. F. (2003). Milk proteins: General and historical aspects. In P. F. Fox, & P. L. H. McSweeney (Eds.), *Advanced Dairy Chemistry-1 Proteins Part A* (pp. 1–48). New York: Kluwer Academic/Plenum Publishers.

Gallarate, M., Battaglia, L., Peira, E., & Trotta, M. (2011). Peptide-loaded solid lipid nanoparticles prepared through coacervation technique. *International Journal of Chemical Engineering*, 2011.

Ghadessy, F.J., Ong, J.L., & Holliger, P. (2001). Directed evolution of polymerase function by compartmentalized self-replication. *Proceedings of the National Academy of Sciences*, *98*(8), 4552–4557.

Ghasemi, S., & Abbasi, S. (2014). Formation of natural casein micelle nanocapsule by means of pH changes and ultrasound. *Food Hydrocolloids*, *42*, 42–47.

Gil, M. J., Tomás-Barberán, F. A., Hees-Pierce, B., Holcroft, D. M., & Kader, A. A. (2000). Antioxidant activity of pomegranate juice and its relationship with phenolic composition and processing. *Journal of Agricultural and Food Chemistry*, *48*(10), 4581–4589.

Giovagnoli, S., Luca, G., Casaburi, I., Blasi, P., Macchiarulo, G., Ricci, M., … Rossi, C. (2005). Long-term delivery of superoxide dismutase and catalase entrapped in poly (lactide-co-glycolide) microspheres: In vitro effects on isolated neonatal porcine pancreatic cell clusters. *Journal of Controlled Release*, *107*(1), 65–77.

Goh, S. L., Murthy, N., Xu, M., & Fréchet, J. M. (2004). Cross-linked microparticles as carriers for the delivery of plasmid DNA for vaccine development. *Bioconjugate Chemistry*, *15*(3), 467–474.

Gonçalves, P. M., Noreña, C. P. Z., da Silveira, N. P., & Brandelli, A. (2014). Characterization of starch nanoparticles obtained from Araucaria angustifolia seeds by acid hydrolysis and ultrasound. *LWT-Food Science and Technology*, *58*(1), 21–27.

Gouin, S. (2004). Microencapsulation: Industrial appraisal of existing technologies and trends. *Trends in Food Science and Technology*, *15*(7), 330–347.

Graveland-Bikker, J. F., & De Kruif, C. G. (2006). Unique milk protein based nanotubes: Food and nanotechnology meet. *Trends in Food Science and Technology*, *17*(5), 196–203.

Graveland-Bikker, J. F., Fritz, G., Glatter, O., & de Kruif, C. G. (2006). Growth and structure of α-lactalbumin nanotubes. *Journal of Applied Crystallography*, *39*(2), 180–184.

Gülseren, İ., & Corredig, M. (2013). Interactions of chitin nanocrystals with β-lactoglobulin at the oil–water interface, studied by drop shape tensiometry. *Colloids and Surfaces B: Biointerfaces*, *111*, 672–679.

Guo, H., Wilking, J. N., Liang, D., Mason, T. G., Harden, J. L., & Leheny, R. L. (2007). Slow, nondiffusive dynamics in concentrated nanoemulsions. *Physical Review E*, *75*, 1–8.

Hallikainen, M. A., Sarkkinen, E. S., & Uusitupa, M. I. J. (2000). Plant stanol eaters affect serum cholesterol concentrations of hypercholesterolemic men and women in a dosedependent manner. *The Journal of Nutrition*, *130*(4), 767–776.

Haratifar, S., Meckling, K. A., & Corredig, M. (2014). Bioefficacy of tea catechins encapsulated in casein micelles tested on a normal mouse cell line (4D/WT) and its cancerous counterpart (D/v-src) before and after in vitro digestion. *Food and Function*, *5*(6), 1160–1166.

Hayashi, T., Hyon, S. H., Cha, W. I., & Ikada, Y. (1993). Lipoprotein lipase immobilization onto porous polyvinyl alcohol beads. *Journal of Applied Polymer Science*, *49*(12), 2121–2127.

Hedges, A.R., Shieh, W.J., & Sikorski, C.T. (1995). Use of cyclodextrins for encapsulation in the use and treatment of food products. *ACS Symposium Series*, *59*, 60–71.

Hora, M. S., Rana, R. K., & Smith, F. W. (1992). Lyophilized formulations of recombinant tumor necrosis factor. *Pharmaceutical Research*, *9*(1), 33–36.

Huang, G., Gao, J., Hu, Z., John, J. V. S., Ponder, B. C., & Moro, D. (2004). Controlled drug release from hydrogel nanoparticle networks. *Journal of Controlled Release*, *94*(2), 303–311.

Huang, Z. M., Zhang, Y. Z., Ramakrishna, S., & Lim, C. T. (2004). Electrospinning and mechanical characterization of gelatin nanofibers. *Polymer*, *45*(15), 5361–5368.

Huppertz, T., & de Kruif, C. G. (2008). Structure and stability of nanogel particles prepared by internal cross-linking of casein micelles. *International Dairy Journal*, *18*(5), 556–565.

Ichikawa, T., Kanai, A., & Yamazaki, Y. (1995). Tear secretion-stimulating effect of cyclosporin A eyedrops in rabbits. *Atarashii Ganka (J Eye)*, *12*, 983–987.

Imran, M., Revol-Junelles, A. M., Paris, C., Guedon, E., Linder, M., & Desobry, S. (2015). Liposomal nanodelivery systems using soy and marine lecithin to encapsulate food biopreservative nisin. *LWT-Food Science and Technology*, *62*(1), 341–349.

Izutsu, K. I., Yoshioka, S., & Kojima, S. (1995). Increased stabilizing effects of amphiphilic excipients on freeze-drying of lactate dehydrogenase (LDH) by dispersion into sugar matrices. *Pharmaceutical Research*, *12*(6), 838–843.

Jackson, L. S., & Lee, K. (1991). Microencapsulation and the food industry. *Lebensmittel-Wissenschaft und Technologie*, *24*(4), 289–297.

Jafari, S. M., Fathi, M., & Mandala, I. (2015). *Emerging product formation. Food Waste Recovery: Processing Technologies and Industrial Techniques* (pp. 293–317). Elsevier Inc.

Jaworek, A. T. S. A., & Sobczyk, A. T. (2008). Electrospraying route to nanotechnology: An overview. *Journal of Electrostatics*, *66*(3), 197–219.

Jurenka, J. S. (2008). Therapeutic applications of pomegranate (*Punica granatum* L.): A review. *Alternative Medicine Review*, *13*(2), 128–144.

Kameta, N., Masuda, M., Minamikawa, H., Mishima, Y., Yamashita, I., & Shimizu, T. (2007). Functionalizable organic nanochannels based on lipid nanotubes: Encapsulation and nanofluidic behavior of biomacromolecules. *Chemistry of Materials*, *19*(14), 3553–3560.

Kanai, A., Alba, R.M., Takano, T., Kobayashi, C., Nakajima, A., Kurihara, K., ... Fukami, M. (1989). The effect on the cornea of alpha cyclodextrin vehicle for cyclosporin eye drops. *Transplantation Proceedings* (Vol. 21, No. 1 Pt 3, pp. 3150–3152).

Kang, Y., Liu, Y. C., Wang, Q., Shen, J. W., Wu, T., & Guan, W. J. (2009). On the spontaneous encapsulation of proteins in carbon nanotubes. *Biomaterials*, *30*(14), 2807–2815.

Katakam, M., & Banga, A. K. (1995). Aggregation of proteins and its prevention by carbohydrate excipients: Albumins and γ-globulin. *Journal of Pharmacy and Pharmacology*, *47*(2), 103–107.

Kataoka, K., Harada, A., & Nagasaki, Y. (2001). Block copolymer micelles for drug delivery: Design, characterization and biological significance. *Advanced Drug Delivery Reviews*, *47*(1), 113–131.

Kato, Y., Onishi, H., & Machida, Y. (2003). Application of chitin and chitosan derivatives in the pharmaceutical field. *Current Pharmaceutical Biotechnology*, *4*(5), 303–309.

Katouzian, I., & Jafari, S. M. (2016). Nano-encapsulation as a promising approach for targeted delivery and controlled release of vitamins. *Trends in Food Science & Technology*, *53*, 34–48.

Kawahara, Y., Nakayama, A., Matsumura, N., Yoshioka, T., & Tsuji, M. (2008). Structure for electrospun silk fibroin nanofibers. *Journal of Applied Polymer Science*, *107*(6), 3681–3684.

Kessick, R., Fenn, J., & Tepper, G. (2004). The use of AC potentials in electrospraying and electrospinning processes. *Polymer*, *45*(9), 2981–2984.

Khadka, D. B., & Haynie, D. T. (2012). Protein-and peptide-based electrospun nanofibers in medical biomaterials. *Nanomedicine: Nanotechnology, Biology and Medicine*, *8*(8), 1242–1262.

Ki, C. S., Baek, D. H., Gang, K. D., Lee, K. H., Um, I. C., & Park, Y. H. (2005). Characterization of gelatin nanofiber prepared from gelatin-formic acid solution. *Polymer*, *46*(14), 5094–5102.

Kim, H. Y., Han, J. A., Kweon, D. K., Park, J. D., & Lim, S. T. (2013). Effect of ultrasonic treatments on nanoparticle preparation of acid-hydrolyzed waxy maize starch. *Carbohydrate Polymers*, *93*, 582–588.

Kim, H. Y., Park, D. J., Kim, J. Y., & Lim, S. T. (2013). Preparation of crystalline starch nanoparticles using cold acid hydrolysis and ultrasonication. *Carbohydrate Polymers*, *98*(1), 295–301.

Kim, K., Bae, B., Kang, Y. J., Nam, J. M., Kang, S., & Ryu, J. H. (2013). Natural polypeptide-based supramolecular nanogels for stable noncovalent encapsulation. *Biomacromolecules*, *14* (10), 3515–3522.

Kınık, Ö., Kavas, G., & Yılmaz, E. (2003). Mikroenkapsülasyon Tekniği ve Süt Teknolojisindeki Kullanım Olanakları. *Gıda*, *28*(4), 401–407.

Kirby, C. J., Brooker, B. E., & Law, B. A. (1987). Accelerated ripening of cheese using liposome-encapsulated enzyme. *International Journal of Food Science and Technology*, *22*(4), 355–375.

Knapczyk, J., Krowczynski, L., Krzck, J., Brzeski, M., Nirnberg, E., Schenk, D., & Struszcyk, H. (1989). Requirements of chitosan for pharmaceutical and biomedical applications. *Chitin and Chitosan: Sources, Chemistry, Biochemistry, Physical Properties and Applications* (pp. 657–663). London: Elsevier.

Kong, L., & Ziegler, G. R. (2014). Formation of starch-guest inclusion complexes in electrospun starch fibers. *Food Hydrocolloids*, *38*, 211–219.

Korhonen, H., & Pihlanto, A. (2006). Bioactive peptides: Production and funtionality. *International Dairy Journal*, *16*, 945–960.

Kramer, M. E. (2009). Structure and function of starch-based edible films and coatings. In K. C. Huber, & M. E. Embuscado (Eds.), *Edible films and coatings for food applications* (pp. 113–134). New York: Springer.

Lambelet, P., Grandgirard, A., Gregoire, S., Juaneda, P., Sebedio, J. L., & Bertoli, C. (2003). Formation of modified fatty acids and oxyphytosterols during refining of low erucic acid rapeseed oil. *Journal of Agricultural and Food Chemistry*, *51*(15), 4284–4290.

Laridi, R., Kheadr, E. E., Benech, R. O., Vuillemard, J. C., Lacroix, C., & Fliss, I. (2003). Liposome encapsulated nisin Z: Optimization, stability and release during milk fermentation. *International Dairy Journal*, *13*(4), 325–336.

Li, M. Y., Mondrinos, M. J., Gandhi, M. R., Ko, F. K., Weiss, A. S., & Lelkes, P. I. (2005). Electrospun protein fibers as matrices for tissue engineering. *Biomaterials*, *26*(30), 5999–6008.

Li, S. F., Chen, J. P., & Wu, W. T. (2007). Electrospun polyacrylonitrile nanofibrous membranes for lipase immobilization. *Journal of Molecular Catalysis B: Enzymatic*, *47*(3), 117–124.

Li, X., Qi, J., Xie, Y., Zhang, X., Hu, S., Xu, Y., & Wu, W. (2013). Nanoemulsions coated with alginate/chitosan as oral insulin delivery systems: Preparation, characterization, and hypoglycemic effect in rats. *International Journal of Nanomedicine*, *8*, 23.

Liu, W., Sun, D., Li, C., Liu, Q., & Xu, J. (2006). Formation and stability of paraffin oil-in-water nano-emulsions prepared by the emulsion inversion point method. *Journal of Colloid and Interface Science*, *303*, 557–563.

Loftsson, T., & Duchêne, D. (2007). Cyclodextrins and their pharmaceutical applications. *International Journal of Pharmaceutics*, *329*(1), 1–11.

Ma, U.V.L. (2010). Encapsulating fatty acid esters of bioactive compounds in starch (Doctoral dissertation, The Pennsylvania State University).

Mahkam, M. (2010). Starch-based polymeric carriers for oral-insulin delivery. *Journal of Biomedical Materials Research Part A*, *92*(4), 1392–1397.

Majumder, K., & Wu, J. (2009). Angiotensin I converting enzyme inhibitory peptides from simulated in vitro gastrointestinal digestion of cooked eggs. *Journal of Agricultural and Food Chemistry*, *57*(2), 471–477.

Malheiros, P. S., Cuccovia, I. M., & Franco, B. D. (2016). Inhibition of Listeria monocytogenes in vitro and in goat milk by liposomal nanovesicles containing bacteriocins produced by Lactobacillus sakei subsp. sakei 2a. *Food Control*, *63*, 158–164.

Marques, H. M. C. (2010). A review on cyclodextrin encapsulation of essential oils and volatiles. *Flavour and Fragrance Journal*, *25*(5), 313–326.

Martins, M. B. F., Simoes, S. I. D., Cruz, M. E. M., & Gaspar, R. (1996). Development of enzyme-loaded nanoparticles: Effect of pH. *Journal of Materials Science: Materials in Medicine*, *7*(7), 413–414.

Martins, S., Sarmento, B., Ferreira, D. C., & Souto, E. B. (2007). Lipid-based colloidal carriers for peptide and protein delivery-liposomes versus lipid nanoparticles. *International Journal of Nanomedicine*, 2(4), 595.

Mason, T. G., Wilking, J. N., Meleson, K., Chang, C. B., & Graves, S. M. (2006). Nanoemulsions: Formation, structure, and physical properties. *Journal of Physics: Condensed Matter*, *18*, R635–R666.

Matthews, J. A., Wnek, G. E., Simpson, D. G., & Bowlin, G. L. (2002). Electrospinning of collagen nanofibers. *Biomacromolecules*, *3*(2), 232–238.

McMahon, D. J., Du, H., McManus, W. R., & Larsen, K. M. (2009). Microstructural changes in casein supramolecules during acidification of skim milk. *Journal of Dairy Science*, *92*(12), 5854–5867.

Meleson, K., Graves, S., & Mason, T. G. (2004). Formation of concentrated nanoemulsions by extreme shear. *Soft Materials*, *2*, 109–123.

Min, B. M., Lee, G., Kim, S. H., Nam, Y. S., Lee, T. S., & Park, W. H. (2004). Electrospinning of silk fibroin nanofibers and its effect on the adhesion and spreading of normal human keratinocytes and fibroblasts in vitro. *Biomaterials*, *25*(7–8), 1289–1297.

Möller, N. P., Scholz-Ahrens, K. E., Roos, N., & Schrezenmeir, J. (2008). Bioactive peptides and proteins from foods: Indication for health effects. *European Journal of Nutrition*, *47*, 171–182.

Moraru, C. I., Panchapakesan, C., Huang, Q., Takhistov, P., Liu, S., & Kokini, J. L. (2003). Nanotechnology: A new frontier in food science. *Food Technology*, *7*, 24–29.

Morimoto, S. (1970). Silk-lice fiber K-6 (chinon). *Symp Industrial and Engineering Chemistry*, *62*, 23–32.

Mozafari, M. R. (2006). Bioactive entrapment and targeting using nanocarrier technologies: An introduction. In *Nanocarrier Technologies* (pp. 1–16). Netherlands: Springer.

Mozhaev, V. V., Melik-Nubarov, N. S., Sergeeva, M. V., Šikšnis, V., & Martinek, K. (1990). Strategy for stabilizing enzymes part one: Increasing stability of enzymes via their multipoint interaction with a support. *Biocatalysis*, *3*(3), 179–187.

Müller, R. H., Jacobs, C., & Kayser, O. (2001). Nanosuspensions as particulate drug formulations in therapy: Rationale for development and what we can expect for the future. *Advanced Drug Delivery Reviews*, *47*, 3–19.

Murphy, J. C., Cano, T., Fox, G. E., & Willson, R. C. (2006). Compaction agent protection of nucleic acids during mechanical lysis. *Biotechnology Progress*, *22*(2), 519–522.

Narva, M. (2004). *Effects of Lactobacillus helveticus fermented milk and milkderived bioactive peptides (CPP, IPP and VPP) on calcium and bone metabolism*, Ph.D. thesis, Institute of Biomedicine, Pharmacology, University of Helsinki, Helsinki (p. 83).

Nielsen, P. H., Malmos, H., Damhus, T., Diderichsen, B., Nielsen, H. K., Simonsen, M., . . . Nielsen, T. K. (1994). Enzyme applications (industrial). *Kirk-Othmer Encyclopedia of Chemical Technology* (Vol. 9, 4th ed., pp. 567–620). New York, NY: John Wiley & Sons, Inc.

Nieuwland, M., Geerdink, P., Brier, P., Van Den Eijnden, P., Henket, J. T., Langelaan, M. L., . . . Martin, A. H. (2013). Food-grade electrospinning of proteins. *Innovative Food Science and Emerging Technologies*, *20*, 269–275.

Orlien, V., Boserup, L., & Olsen, K. (2010). Casein micelle dissociation in skim milk during high-pressure treatment: Effects of pressure, pH, and temperature. *Journal of Dairy Science*, *93*(1), 12–18.

Ostlund, R. E. (2004). Phytosterols and cholesterol metabolism. *Current Opinion in Lipidology*, *15*(1), 37–41.

Oziyci, H. R., Karhan, M., Tetik, N., & Turhan, I. (2013). Effects of processing method and storage temperature on clear pomegranate juice turbidity and color. *Journal of Food Processing Preservation*, *37*(5), 899–906.

Palma-Rodriguez, H. M., Agama-Acevedo, E., Mendez-Montealvo, G., Gonzalez-Soto, R. A., Vernon-Carter, E., & Bello-Perez, L. A. (2012). Effect of acid treatment on the physicochemical and structural characteristics of starches from different botanical sources. *Starch/Staerke*, *64*, 115–125.

Pan, K., Zhong, Q., & Baek, S. J. (2013). Enhanced dispersibility and bioactivity of curcumin by encapsulation in casein nanocapsules. *Journal of Agricultural and Food Chemistry*, *61*(25), 6036–6043.

Pan, X., Yu, S., Yao, P., & Shao, Z. (2007). Self-assembly of β-casein and lysozyme. *Journal of Colloid and Interface Science*, *316*(2), 405–412.

Parak, W. J., Gerion, D., Pellegrino, T., Zanchet, D., Micheel, C., Williams, S. C., & Alivisatos, A. P. (2003). Biological applications of colloidal nanocrystals. *Nanotechnology*, *14*(7), R15.

Park, E. Y., Morimae, M., Matsumura, Y., Nakamura, Y., & Sato, K. (2008). Antioxidant activity of some protein hydrolysates and their fractions with different isoelectric points. *Journal of Agricultural and Food Chemistry*, *56*(19), 9246–9251.

Peker, H., & Arslan, S. (2011). Mikroenkapsülasyon ve süt teknolojisinde kullanım alanları. *Akademik Gıda*, *9*(6), 70–80.

Pescosolido, L., Miatto, S., Di Meo, C., Cencetti, C., Coviello, T., Alhaique, F., & Matricardi, P. (2010). Injectable and in situ gelling hydrogels for modified protein release. *European Biophysics Journal*, *39*(6), 903–909.

Pitkowski, A., Durand, D., & Nicolai, T. (2008). Structure and dynamical mechanical properties of suspensions of sodium caseinate. *Journal of Colloid and Interface Science*, *326*(1), 96–102.

Porras, M., Solans, C., Gonzalez, C., & Gutierrez, J. M. (2008). Properties of water-in-oil (W/O) nano-emulsions prepared by a low-energy emulsification method. *Colloids and Surfaces A: Physicochemical and Engineering Aspects*, *324*, 181–188.

Prankerd, R. J., Stone, H. W., Sloan, K. B., & Perrin, J. H. (1992). Degradation of aspartame in acidic aqueous media and its stabilization by complexation with cyclodextrins or modified cyclodextrins. *International Journal of Pharmaceutics*, *88*(1), 189–199.

Prokai, L., Ramanathan, R., Nawrocki, J., & Eyler, J. (1996). Electrospray ionization mass spectrometry of cyclodextrin complexes of amino acids and peptides. *Journal of Inclusion Phenomena and Molecular Recognition in Chemistry*, *25*(1–3), 117–120.

Putnam, D. (2006). Polymers for gene delivery across length scales. *Nature Materials*, *5*(6), 439–451.

Rabinow, B. (2005). Pharmacokinetics of drugs administered in nanosuspension. *Discovery Medicine*, *5*, 74–79.

Rajagopal, K., & Schneider, J. P. (2004). Selfassembling peptides and proteins for nanotechnological applications. *Current Opinion in Structural Biology*, *14*(4), 480–486.

Reis, L. A., Chiu, L. L., Liang, Y., Hyunh, K., Momen, A., & Radisic, M. (2012). A peptide-modified chitosan–collagen hydrogel for cardiac cell culture and delivery. *Acta Biomaterialia*, *8*(3), 1022–1036.

Reithmeier, H., Herrmann, J., & Göpferich, A. (2001). Development and characterization of lipid microparticles as a drug carrier for somatostatin. *International Journal of Pharmaceutics*, *218*(1), 133–143.

Ren, G., Xu, X., Liu, Q., Cheng, J., Yuan, X., Wu, L., & Wan, Y. (2006). Electrospun poly(vinyl alcohol)/glucose oxidase biocomposite membranes for biosensor applications. *Reactive and Functional Polymers*, *66*, 1559–1564.

Ressing, M. E., Jiskoot, W., Talsma, H., van Ingen, C. W., Beuvery, E. C., & Crommelin, D. J. (1992). The influence of sucrose, dextran, and hydroxypropyl-β-cyclodextrin as lyoprotectants for a freeze-dried mouse IgG2a monoclonal antibody (MN12). *Pharmaceutical Research*, *9*(2), 266–270.

Rossi, J., & Leroux, J. C. (2007). *Principles in the development of intravenous lipid emulsions. Role of Lipid Excipients in Modifying Oral and Parenteral Drug Delivery* (pp. 88–123). Hoboken, New Jersey: Wiley-Interscience.

Roufik, S., Gauthier, S. F., & Turgeon, S. L. (2006). In vitro digestibility of bioactive peptides derived from bovine β-lactoglobulin. *International Dairy Journal*, *16*, 294–302.

Russel, W. B., Saville, D. A., & Schowalter, W. R. (1989). *Colloidal dispersions*. Cambridge: Cambridge University Press.

Santander-Ortega, M. J., Stauner, T., Loretz, B., Ortega-Vinuesa, J. L., Bastos-González, D., Wenz, G., ... Lehr, C. M. (2010). Nanoparticles made from novel starch derivatives for transdermal drug delivery. *Journal of Controlled Release*, *141*(1), 85–92.

Schanbacher, F. L., Talhouk, R. S., & Murray, F. A. (1997). Biology and origin of bioactive peptides in milk. *Livestock Production Science*, *50*, 105–123.

Schanbacher, F. L., Talhouk, R. S., Murray, F. A., Gherman, L. I., & Willett, L. B. (1998). Milk-borne bioactive peptides. *International Dairy Journal*, *8*, 393–403.

Scott, N. R. (2005). Nanotechnology and animal health. *Revue Scientifique Et Technique (International Office of Epizootics)*, *24*(1), 425–432.

Semo, E., Kesselman, E., Danino, D., & Livney, Y. D. (2007). Casein micelle as a natural nano-capsular vehicle for nutraceuticals. *Food Hydrocolloids*, *21*(5), 936–942.

Sepúlveda, L., Ascacio, A., Rodriguez-Herrera, R., Aguilera-Carbó, A., & Aguilar, C. N. (2011). Ellagic acid: Biological properties and biotechnological development for production processes. *African Journal of Biotechnology*, *10*(22), 4518–4523.

Shakeel, F., & Ramadan, W. (2010). Transdermal delivery of anticancer drug caffeine from water-in-oil nanoemulsions. *Colloids and Surfaces B: Biointerfaces*, *75*, 356–362.

Shefer, A., & Shefer, S. (2003). Novel encapsulation system provides controlled release of ingredients. *Food Technology*, *57*, 40–43.

Shegokar, R., & Müller, R. H. (2010). Nanocrystals: industrially feasible multifunctional formulation technology for poorly soluble actives. *International Journal of Pharmaceutics, 399*(1), 129–139.

Shinkai, M., Honda, H., & Kobayashi, T. (1991). Preparation of fine magnetic particles and application for enzyme immobilization. *Biocatalysis, 5*(1), 61–69.

Shutava, T. G., & Lvov, Y. M. (2006). Nano-engineered microcapsules of tannic acid and chitosan for protein encapsulation. *Journal of Nanoscience and Nanotechnology, 6*(6), 1655–1661.

Singh, M., Sharma, R., & Banerjee, U. C. (2002). Biotechnological applications of cyclodextrins. *Biotechnology Advances, 20*(5), 341–359.

Skiba, M., Nemati, F., Puisieux, F., Duchêne, D., & Wouessidjewe, D. (1996). Spontaneous formation of drug-containing amphiphilic β-cyclodextrin nanocapsules. *International Journal of Pharmaceutics, 145*(1), 241–245.

Smacchi, E., & Gobbetti, M. (2000). Bioactive peptides in dairy products: Synthesis and interaction with proteolytic enzymes. *Food Microbiology, 17*, 129–141.

Solans, C., Izquierdo, P., Nolla, J., Azemar, N., & Garcia-Celma, M. J. (2005). Nano-emulsions. *Current Opinion in Colloid & Interface Science, 10*(3), 102–110.

Songchotikunpan, P., Tattiyakul, J., & Supaphol, P. (2008). Extraction and electrospinning of gelatin from fish skin. *International Journal of Biological Macromolecules, 42*(3), 247–255.

Soupas, L., Juntunen, L., Lampi, A. M., & Piironen, V. (2004). Effects of sterol structure, temperature, and lipid medium on phytosterol oxidation. *Journal of Agriculture and Food Chemistry, 52*(21), 6485–6491.

Stephansen, K., Chronakis, I. S., & Jessen, F. (2014). Bioactive electrospun fish sarcoplasmic proteins as a drug delivery system. *Colloids and Surfaces B: Biointerfaces, 122*, 158–165.

Szente, L., & Szejtli, J. (2004). Cyclodextrins as food ingredients. *Trends in Food Science and Technology, 15*(3), 137–142.

Szetjli, J. (1998). Introduction and general overview of cyclodextrin chemistry. *Chemical Reviews, 98*, 1743–1753.

Szejtli, L. (1994). Medicinal applications of cyclodextrins. *Medical Research Reviews, 14*, 353–386.

Tadros, T., Izquierdo, P., Esquena, J., & Solans, C. (2004). Formation and stability of nano-emulsions. *Advances in Colloid and Interface Science, 108*, 303–318.

Talegaonkar, S., Mustafa, G., Akhter, S., & Iqbal, Z. I. (2010). Design and development of oral oil-in-water nanoemulsion formulation bearing atorvastatin: In vitro assessment. *Journal of Dispersion Science and Technology, 31*, 690–701.

Tawfik, D. S., & Griffiths, A. D. (1998). Man-made cell-like compartments for molecular evolution. *Nature Biotechnology, 16*(7), 652–656.

Taylor, T., Bruce, B. D., Weiss, J., & Davidson, P. M. (2008). Listeria monocytogenes and Escherichia coli O157:H7 inhibition in vitro by liposome-encapsulated nisin and ethylene diaminetetraacetic acid. *Journal of Food Safety, 28*(2), 183–197.

Teixeira, M. L., dos Santos, J., Silveira, N. P., & Brandelli, A. (2008). Phospholipid nanovesicles containing a bacteriocin-like substance for control of Listeria monocytogenes. *Innovative Food Science and Emerging Technologies, 9*(1), 49–53.

Tsumoto, K., Nomura, S. I. M., Nakatani, Y., & Yoshikawa, K. (2001). Giant liposome as a biochemical reactor: Transcription of DNA and transportation by laser tweezers. *Langmuir, 17*(23), 7225–7228.

Tzoumaki, M. V., Moschakis, T., & Biliaderis, C. G. (2011a). Mixed aqueous chitin nanocrystal–whey protein dispersions: Microstructure and rheological behaviour. *Food Hydrocolloids, 25*(5), 935–942.

Tzoumaki, M. V., Moschakis, T., Kiosseoglou, V., & Biliaderis, C. G. (2011b). Oil-in-water emulsions stabilized by chitin nanocrystal particles. *Food Hydrocolloids*, *25*(6), 1521–1529.

Tzoumaki, M. V., Moschakis, T., Scholten, E., & Biliaderis, C. G. (2013). In vitro lipid digestion of chitin nanocrystal stabilized o/w emulsions. *Food and Function*, *4*(1), 121–129.

Ubbink J. 2002. Flavor delivery systems: Trends, technologies and applications. In *Abstracts of Papers of the American Chemical Society* (Vol. 223, pp. U34–U34). 1155 16th ST, NW, Washington DC, USA: American Chemical Society.

Ubbink, J., & Kruger, J. (2006). Physical approaches for the delivery of active ingredients in foods. *Trends in Food Science and Technology*, *17*(5), 244–254.

Udabage, P., Mckinnon, I. R., & Augustin, M. A. (2000). Mineral and casein equilibria in milk: Effects of added salts and calcium-chelating agents. *Journal of Dairy Research*, *67*(03), 361–370.

Vauthier-Holtzscherer, C., Benabbou, S., Spenlehauer, G., Veillard, M., & Couvreur, P. (1991). Methodology for the preparation of ultra-dispersed polymer systems. *STP Pharma Sciences*, *1*(2), 109–116.

Viladomiu, M., Hontecillas, R., Lu, P., & Basseganya-Riera, J. (2013). Preventive and prophylactic mechanisms of action of pomegranate bioactive compounds. *Evidance-Based Complementary Alternative Medicine*, 1–18.

Villiers, A. (1891). Sur la fermentation de la fécule par l'action du ferment butyrique. *Compte Rendu de l Académie De Sciences*, *112*, 536–538.

Were, L. M., Bruce, B. D., Davidson, P. M., & Weiss, J. (2003). Size, stability, and entrapment efficiency of phospholipid nanocapsules containing polypeptide antimicrobials. *Journal of Agricultural and Food Chemistry*, *51*(27), 8073–8079.

Winarti, C., Sunarti, T. C., Mangunwidjaja, D., & Richana, N. (2014). Preparation of arrowroot starch nanoparticles by butanol-complex precipitation, and its application as bioactive encapsulation matrix. *International Food Research Journal*, *21*, 6.

Wiseman, A., 1987. Handbook of Enzymes Biotechnology. *The Application of Enzymes in Industry* (2nd, Chapter 3, pp. 274–373).

Wnek, G. E., Carr, M. E., Simpson, D. G., & Bowlin, G. L. (2003). Electrospinning of nanofiber fibrinogen structures. *Nano Letters*, *3*(2), 213–216.

Wong, N. C. W. (2001). The beneficial effects of plant sterols on serum cholesterol. *Canadian Journal of Cardiology*, *17*(6), 715–721.

Xu, Y., Ding, W., Liu, J., Kennedy, J. F., Gu, Q., & Shao, S. (2010). Preparation and characterization of organic soluble acetylated starch nanocrystal. *Carbohydrate Polymers*, *80*, 1078–1084.

Yang, J., Sun, H., & Song, C. (2012). Preparation, characterization and in vivo evaluation of pH-sensitive oral insulin-loaded poly (lactic-co-glycolicacid) nanoparticles. *Diabetes, Obesity and Metabolism*, *14*(4), 358–364.

Yazdi, S. R., Bonomi, F., Iametti, S., Miriani, M., Brutti, A., & Corredig, M. (2013). Binding of curcumin to milk proteins increases after static high pressure treatment of skim milk. *Journal of Dairy Research*, *80*(02), 152–158.

Yu, S., Yao, P., Jiang, M., & Zhang, G. (2006). Nanogels prepared by self-assembly of oppositely charged globular proteins. *Biopolymers*, *83*(2), 148–158.

Yurke, B., Turberfield, A. J., Mills, A. P., Simmel, F. C., & Neumann, J. L. (2000). A DNA-fuelled molecular machine made of DNA. *Nature*, *406*(6796), 605–608.

Zamani, M., Prabhakaran, M. P., & Ramakrishna, S. (2013). Advances in drug delivery via electrospun and electrosprayed nanomaterials. *International Journal of Nanomedicine*, *8*(1), 2997–3017.

Zarei, M., Azizi, M., & Bashiri-Sadr, Z. (2010). Studies on physicochemical properties and bioactive compounds of six pomegranate cultivars grown in Iran. *Journal of Food Technology*, *8*(3), 112–117.

Zelikin, A. N., Becker, A. L., Johnston, A. P., Wark, K. L., Turatti, F., & Caruso, F. (2007). A general approach for DNA encapsulation in degradable polymer microcapsules. *ACS Nano*, *1*(1), 63–69.

FURTHER READINGS

Ahn, C. H., Choi, J. W., & Cho, H. J. (2004). Nanomagnetics for biomedical applicationsIn H. S. Nalwa (Ed.), *Encyclopedia of Nanoscience and Nanotechnology* (Vol. 6, pp. 815–821). California, CA: American Scientific Publishers, Stevenson Ranch.

Bharali, D. J., Sahoo, S. K., Mozumdar, S., & Maitra, A. (2003). Cross-linked polyvinylpyrrolidone nanoparticles: a potential carrier for hydrophilic drugs. *Journal of Colloid and Interface Science*, *258*(2), 415–423.

Bock, N., Dargaville, T. R., & Woodruff, M. A. (2012). Electrospraying of polymers with therapeutic molecules: State of the art. *Progress in Polymer Science*, *37*(11), 1510–1551.

Brown, L. O., & Hutchison, J. E. (1999). Controlled growth of gold nanoparticles during ligand exchange. *Journal of the American Chemical Society*, *121*(4), 882–883.

Brust, M., Fink, J., Bethell, D., Schiffrin, D. J., & Kiely, C. (1995). Synthesis and reactions of functionalised gold nanoparticles. *Journal of the Chemical Society, Chemical Communications*, *16*, 1655–1656.

Chen, H., Zhang, Y., & Zhong, Q. (2015). Physical and antimicrobial properties of spray-dried zein–casein nanocapsules with co-encapsulated eugenol and thymol. *Journal of Food Engineering*, *144*, 93–102.

Chen, L. Y., & Subirade, M. (2009). Elaboration and characterization of soy/zein protein microspheres for controlled nutraceutical delivery. *Biomacromolecules*, *10*(12), 3327–3334.

Colvin, V. L., & Alivisatos, A. P. (1992). CdSe nanocrystals with a dipole moment in the first excited state. *Journal of Chemical Physics*, *97*, 730–733.

Couvreur, P. C., & Dubernet, F. (1995). Puisieux, Controlled drug delivery withnanoparticles: Current possibilities and future trends. *Europen Journal of Pharmaceutics and Biopharmaceutics*, *41*, 2–13.

Diehl, M. R., Yu, J. Y., Heath, J. R., Held, G. A., Doyle, H., Sun, S., & Murray, C. B. (2001). Crystalline, shape, and surface anisotropy in two crystal morphologies of superparamagnetic cobalt nanoparticles by ferromagnetic resonance. *The Journal of Physical Chemistry B*, *105*(33), 7913–7919.

Ershov, B. G., Sukhov, N. L., & Janata, E. (2000). Formation, absorption spectrum, and chemical reactions of nanosized colloidal cobalt in aqueous solution. *The Journal of Physical Chemistry B*, *104*(26), 6138–6142.

Eychmüller, A., & Rogach, A. L. (2000). Chemistry and photophysics of thiol-stabilized II-VI semiconductor nanocrystals. *Pure and Applied Chemistry*, *72*(1–2), 179–188.

Gonçalves, C., Pereira, P., Schellenberg, P., Coutinho, P. J., & Gama, F. M. (2012). Self-assembled dextrin nanogel as curcumin delivery system. *Journal of Biomaterials and Nanobiotechnology*, *3*, 178–184.

Guzelian, A. A., Katari, J. B., Kadavanich, A. V., Banin, U., Hamad, K., Juban, E., & Heath, J. R. (1996). Synthesis of size-selected, surface-passivated InP nanocrystals. *The Journal of Physical Chemistry*, *100*(17), 7212–7219.

Handley, D. A. (1989). In M. A. Hayat (Ed.), *Colloidal Gold: Principles, Methods, and Applications* (Vol 1, pp. 13–32). New York, NY: Academic.

Hyeon, T., Lee, S. S., Park, J., Chung, Y., & Na, H. B. (2001). Synthesis of highly crystalline and monodisperse maghemite nanocrystallites without a size-selection process. *Journal of the American Chemical Society, 123*(51), 12798–12801.

Jouquand, C., Ducruet, V., & Giampaoli, P. (2004). Partition coefficients of aroma compounds in polysaccharide solutions by the phase ratio variation method. *Food Chemistry, 85*, 467–474.

Lee, H., Mok, H., Lee, S., Oh, Y. K., & Park, T. G. (2007). Target-specific intracellular delivery of siRNA using degradable hyaluronic acid nanogels. *Journal of Controlled Release, 119*(2), 245–252.

Li, D., & Xia, Y. (2004). Electrospinning of nanofibers: Reinventing the wheel? *Advanced Materials, 16*(14), 1151–1170.

Li, J. K., Wang, N., & Wu, X. S. (1998). Poly (vinyl alcohol) nanoparticles prepared by freezing–thawing process for protein/peptide drug delivery. *Journal of Controlled Release, 56* (1), 117–126.

Li, M., Witt, T., Xie, F., Warren, F. J., Halley, P. J., & Gilbert, R. G. (2015). Biodegradation of starch films: The roles of molecular and crystalline structure. *Carbohydrate Polymers, 122*, 115–122.

Li, W., Dobraszczyk, B. J., & Wilde, P. J. (2004). Surface properties and locations of gluten proteins and lipids revealed using confocal scanning laser microscopy in bread dough. *Journal of Cereal Science, 39*(3), 403–411.

Li, Y., Du, W., Sun, G., & Wooley, K. L. (2008). pH-responsive shell cross-linked nanoparticles with hydrolytically labile cross-links. *Macromolecules, 41*(18), 6605–6607.

Li, Y., Lokitz, B. S., Armes, S. P., & McCormick, C. L. (2006). Synthesis of reversible shell cross-linked micelles for controlled release of bioactive agents. *Macromolecules, 8*(39), 2726–2728.

Li, Z., Percival, S. S., Bonard, S., & Gu, L. (2011). Fabrication of nanoparticles using partially purified pomegranate ellagitannins and gelatin and their apoptoticeffects. *Molecular Nutrition and Food Research, 55*, 1096–1103.

Liu, J., Wang, B., Lin, L., Zhang, J., Liu, W., Xie, J., & Ding, Y. (2014). Functional, physicochemical properties and structure of cross-linked oxidized maize starch. *Food Hydrocolloids, 36*, 45–52.

Marcus, M. A., Flood, W., Stiegerwald, M., Brus, L., & Bawendi, M. (1991). Structure of capped cadmium selenide clusters by exafs. *The Journal of Physical Chemistry, 95*(4), 1572–1576.

Micic, O. I., Cheong, H. M., Fu, H., Zunger, A., Sprague, J. R., Mascarenhas, A., & Nozik, A. J. (1997). Size-dependent spectroscopy of InP quantum dots. *The Journal of Physical Chemistry B, 101*(25), 4904–4912.

Murray, C., Norris, D. J., & Bawendi, M. G. (1993). Synthesis and characterization of nearly monodisperse CdE (E = sulfur, selenium, tellurium) semiconductor nanocrystallites. *Journal of the American Chemical Society, 115*(19), 8706–8715.

Olshavsky, M. A., Goldstein, A. N., & Alivisatos, A. P. (1990). Organometallic synthesis of gallium-arsenide crystallites, exhibiting quantum confinement. *Journal of the American chemical society, 112*(25), 9438–9439.

Park, Y. W., Juárez, M., Ramos, M., & Haenlein, G. F. W. (2007). Physico-chemical characteristics of goat and sheep milk. *Small Ruminant Research, 68*, 88–113.

Peng, X., Manna, L., Yang, W., Wickham, J., Scher, E., Kadavanich, A., & Alivisatos, A. P. (2000). Shape control of CdSe nanocrystals. *Nature, 404*(6773), 59–61.

Peng, Z. A., & Peng, X. (2001). Formation of high-quality CdTe, CdSe, and CdS nanocrystals using CdO as precursor. *Journal of the American Chemical Society*, *123*(1), 183–184.

Price, R., BP Gaber, Y., & Lvov, R. (2001). In-vitro release characteristics of tetracycline HCl, khellin and nicotinamideadeninedineculeotide from halloysite; a cylindrical mineral. *Journal of Microencapsulation*, *18*(6), 713–722.

Puntes, V. F., & Krishnan, K. M. (2001). Science 291 2115 Manna L, Scher EC and Alivisatos AP 2000. *Journal of American Chemical Society*, *122*, 12700.

Puntes, V. F., Krishnan, K. M., & Alivisatos, P. (2001). Synthesis, self-assembly, and magnetic behavior of a two-dimensional superlattice of single-crystal ε-Co nanoparticles. *Applied Physics Letters*, *78*(15), 2187–2189.

Quaroni, L., & Chumanov, G. (1999). Preparation of polymer-coated functionalized silver nanoparticles. *Journal of the American Chemical Society*, *121*(45), 10642–10643.

Reis, C. P., Neufeld, R. J., Vilela, S., Ribeiro, A. J., & Veiga, F. (2006). Review and current status of emulsion/dispersion technology using an internal gelation process for the design of alginate particles. *Journal of Microencapsulation*, *23*(3), 245–257.

Reis, C. P., Neufeld, R. J., Ribeiro, A. J., & Veiga, F. (2006). Nanoencapsulation I. Methods for preparation of drug-loaded polymeric nanoparticles. *Nanomedicine*, *2*, 8–21.

Rivas, L., Sanchez-Cortes, S., Garcia-Ramos, J. V., & Morcillo, G. (2001). Growth of silver colloidal particles obtained by citrate reduction to increase the Raman enhancement factor. *Langmuir*, *17*(3), 574–577.

Rockenberger, J., Scher, E. C., & Alivisatos, A. P. (1999). A new nonhydrolytic single-precursor approach to surfactant-capped nanocrystals of transition metal oxides. *Journal of the American Chemical Society*, *121*(49), 11595–11596.

Rogach, A. L., Katsikas, L., Kornowski, A., Su, D., Eychmüller, A., & Weller, H. (1996). Synthesis and characterization of thiol-stabilized CdTe nanocrystals. *Berichte der Bunsengesellschaft für physikalische Chemie*, *100*(11), 1772–1778.

Ryu, J. H., Chacko, R. T., Jiwpanich, S., Bickerton, S., Babu, R. P., & Thayumanavan, S. (2010a). Self-cross-linked polymer nanogels: A versatile nanoscopic drug delivery platform. *Journal of the American Chemical Society*, *132*(48), 17227–17235.

Ryu, J. H., Jiwpanich, S., Chacko, R., Bickerton, S., & Thayumanavan, S. (2010b). Surface-functionalizable polymer nanogels with facile hydrophobic guest encapsulation capabilities. *Journal of the American Chemical Society*, *132*(24), 8246–8247.

Santra, S., Tapec, R., Theodoropoulou, N., Dobson, J., Hebard, A., & Tan, W. (2001). Synthesis and characterization of silica-coated iron oxide nanoparticles in microemulsion: The effect of nonionic surfactants. *Langmuir*, *17*(10), 2900–2906.

Schmid, G., & Lehnert, A. (1989). The complexation of gold colloids. *Angewandte Chemie International Edition in English*, *28*(6), 780–781.

Sozer, N., & Kokini, J. L. (2009). Nanotechnology and its applications in the food sector. *Trends in Biotechnolology*, *27*(2), 82–89.

Steigerwald, M. L., & Brus, L. E. (1990). Semiconductor crystallites: a class of large molecules. *Accounts of Chemical Research*, *23*(6), 183–188.

Sun, M., Gao, Y., Guo, C., Cao, F., Song, Z., Xi, Y., ... Zhai, G. (2010). Enhancementof transport of curcumin to brain in mice by poly (n-butylcyanoacrylate) nanoparticle. *Journal of Nanoparticle Research*, *12*, 3111–3122.

Talapin, D. V., Haubold, S., Rogach, A. L., Kornowski, A., Haase, M., & Weller, H. (2001a). A novel organometallic synthesis of highly luminescent CdTe nanocrystals. *The Journal of Physical Chemistry B*, *105*(12), 2260–2263.

Talapin, D. V., Rogach, A. L., Kornowski, A., Haase, M., & Weller, H. (2001b). Highly luminescent monodisperse CdSe and CdSe/ZnS nanocrystals synthesized in a hexadecylamine-trioctylphosphine oxide-trioctylphospine mixture. *Nano Letters*, *1*(4), 207–211.

Tolbert, S. H., & Alivisatos, A. P. (1994). Size dependence of a first order solid-solid phase transition: The wurtzite to rock salt transformation in CdSe nanocrystals. *Science*, *265*(5170), 373–376.

Tomalia, D. A., & Frechet, J. M. J. (2002). Discovery of dendrimers and dendritic polymers: A brief historical perspective. *Journal of Polymer Science Part A: Polymer Chemistry*, *40*, 2719–2728.

Trentler, T. J., Denler, T. E., Bertone, J. F., Agrawal, A., & Colvin, V. L. (1999). Synthesis of TiO2 nanocrystals by nonhydrolytic solution-based reactions. *Journal of the American Chemical Society*, *121*(7), 1613–1614.

Wang, Y., Dave, R. N., & Pfeffer, R. (2004). Polymer coating/encapsulation of nanoparticles using a supercritical anti-solvent process. *Journal of Supercritical Fluids*, *28*, 85–99.

Wu, C., Böttcher, C., & Haag, R. (2015). Enzymatically crosslinked dendritic polyglycerol nanogels for encapsulation of catalytically active proteins. *Soft Matter*, *11*(5), 972–980.

Xia, Y., & Whitesides, G. M. (1998). Soft lithography. *Annual Review of Materials Science*, *28*(1), 153.

Chapter 9

Nanoencapsulation of Minerals

Seyed Mohammad Taghi Gharibzahedi[1] and Seid Mahdi Jafari[2]
[1]*Young Researchers and Elites Club, Science and Research Branch, Islamic Azad University, Tehran, Iran,* [2]*Gorgan University of Agricultural Sciences and Natural Resources, Gorgan, Iran*

9.1 INTRODUCTION

Minerals are one of the most significant components involved in the cellular metabolism and biological functions. Plants and animals are unable to synthesize these micronutrients (GharibZahedi and Jafari, 2017a). Nonetheless, these nutrient ions present in the soil can be mostly absorbed via plant roots. Main dietary sources of minerals are vegetables and fruits (Singh, Rathi, Angal, Parida, & Rautaray, 2016). In some cases, they may also be present in drinking water (Watzke, 1998).

In general, essential minerals are broadly divided into major minerals (macrominerals) and trace minerals (microminerals). Main minerals are including calcium (Ca), magnesium (Mg), sodium (Na), potassium (K), phosphorus (P), chloride (Cl), and sulfur (S), whereas iron (Fe), manganese (Mn), copper (Cu), iodine (I), zinc (Zn), cobalt (Co), molybdenum (Mo), fluoride (F), selenium (Se), chromium (Cr), and boron (B) are classified as trace minerals. Although the amounts needed of minerals in the body are not a sign of their significance, the body requires trace minerals in smaller levels than major minerals. Since, there is a high risk in mineral deficiency for people with increased mineral demands (pregnancy, adolescence) and dietary restrictions (geriatric patients, vegetarians, low-calorie diets), existence of a balanced diet can commonly support all of the essential minerals for body (Berginc, 2015; Lukaski, 2004; Singh et al., 2016).

Fortification strategy of foods is considered a feasible and cost-effective food-based process for the control of micronutrient deficiencies. The objective in food fortification with functional ingredients is to guarantee that these nutraceuticals are delivered and released at the proper time and location in the body. Nevertheless, there are a lot of solubility and stability restrictions in relation to their direct addition to food formulations. Some of these micronutrients can also interact with many other food components and/or can

Nanoencapsulation of Food Bioactive Ingredients. DOI: http://dx.doi.org/10.1016/B978-0-12-809740-3.00009-X

highly reduce sensory evaluation scores by creating an unfavorable flavor, taste, or color. Furthermore, many minerals are prone to oxidative degradation, and accordingly it is necessary to protect them in fortified/supplemented food products during their shelf-life period (Akhtar, Anjum, Rehman, & Munir, 2010).

Among the various strategies, micro/nanoencapsulation can be successfully used to overcome these challenges with the protection and delivery of minerals and other bioactive ingredients in food uses (Faridi Esfanjani & Jafari, 2016; Katouzian & Jafari, 2016). This process can highly provide stable conditions to control the minerals release from two perspectives of the release-rate and the release-start, and also to separate unstable or reactive ingredients in final food formulations (Mozafari et al., 2008). Since there is a robust relationship between the delivery rate of any bioactive constituent and its particle size (Hughes, 2005), the incorporation, absorption, or dispersion of minerals can be facilitated by the forming capsules with nanometer or submicron diameters (Oehlke et al., 2014). Nanocapsules having minerals and trace elements can notably increase gastrointestinal maintenance time and subsequently bioadhesiveness of poorly soluble mineral compounds in the mucus covering the intestinal epithelium. Apart from those qualities, nanoencapsulation of minerals is a valuable opportunity to have better stability and protection against oxidation, pleasant taste masking and/or long-lasting sensory perception, and ideal controlled release (Hurrell, 2002a,b).

Therefore, the purpose of this chapter is to review the current techniques for designing delivery systems of minerals in fortifying food products in terms of micro- and nanocapsules, their importance in bioavailability, safety and release viewpoints, and also to present some future prospects.

9.2 ENCAPSULATION TECHNIQUES OF MINERALS

A coating agent in encapsulation processes is commonly used to cover small-size ingredients with the solid, liquid, and gaseous nature. The developed small micro/nanocapsules have diverse morphology forms (e.g., spherical and irregular). They are composed of a core area surrounded by a continuous wall which covers numerous fine particles or droplets. Biopolymers, fats, and waxes generally are examples of the wall materials for capsules (Jafari, Fathi, & Mandala 2015).

Encapsulation has numerous benefits for minerals in terms of nutritional and industrial perspectives including: (1) mineral protection from various environmental parameters, such as moisture, oxygen, heat, acids, etc.; (2) bio-engineering of functional components, minerals not only can be protected from processing losses but also can be favorably released under exposure to specific conditions; (3) minimizing unsatisfying flavors and odors related to a certain mineral can significantly improve consumer preference and willingness using this technology; (4) providing an easy handling for dry and free-flowing constituents; (5) potential of measuring and delivery of exact levels of the favorite mineral due to the stability obtained by encapsulation; and (6) high efficiency of this process to produce functional foods with excellent stability, bioavailability, and delivery aspects (Ghosh, 2006; Gouin, 2004).

Nowadays, a number of techniques to microencapsulate minerals especially Fe, I, and Zn has been used. These methods can be generally categorized into two groups including: (1) physical/mechanical techniques, such as freeze-dying, fluidized bed coating (FBC), spray-drying, spray-chilling (cooling), extrusion, and co-extrusion; and (2) chemical techniques, such as liposome/niosome entrapment, fatty acid esters (FAEs), gelation, emulsification, and solvent evaporation. Table 9.1 summarizes some microencapsulation techniques for minerals which have been used to fortify dairy products and salts. It also represents coating materials in various encapsulation methods, microcapsules size and geometry, and encapsulation efficiency (EE) values.

9.2.1 Physicomechanical Methods

9.2.1.1 Freeze-Drying

An appropriate and multistage method for microencapsulation of many heat-sensitive components is freeze-drying (lyophilization), which can be applied to dehydrate and prepare powders. Although this technique compared with other drying methods (e.g., spray drying) has advantages, such as intact form of microcapsules with a more porous structure and low oxidation rate, it usually is a time-consuming process which can last up to 20 h with regard to the used components and the loading rate (Desai & Park, 2005). A recent study in the field of mineral microencapsulation using freeze-drying method was conducted by Gupta, Chawla, and Arora (2015a). For this work, Na–alginate and modified starch–Na–alginate were separately dissolved into distilled water during a stirring rate of 500 rpm for 0.5 h in order to encapsulate Fe salt. A sonicator was used to degas the solution for 15 min and then Fe salt and vitamin C (15:1) were added to the prepared solutions under the abovementioned mixing conditions. To solidify the formed microencapsules, the mixture was sprayed into chilled calcium chloride ($CaCl_2$) solution and kept for 3–4 h. In the next step, $CaCl_2$ was removed by filtering under vacuum conditions. The final solution was frozen ($-20°C$) and lyophilized using a freeze-drying system to produce the powders. Results showed that EE of Fe microencapsules increased with decreasing $CaCl_2$ level from 1.0 to 0.1 M. The EE of the microcapsules prepared based on Na–alginate-modified starch blend at 0.1 M $CaCl_2$ was much more than ones only developed by Na–alginate. Therefore, it is obvious that modified starch had a key role in the EE increase.

TABLE 9.1 Different Encapsulation Technologies to Prepare Encapsulated Minerals for Fortification of Some Food Supplements/Foods

Product	Mineral	Wall Material[a]	Technology/ Process	Size (μm)	Shape	EE (%)	Reference
Milk	Fe	PGMS	Liposome	–	Irregular spherical	81.3	Abbasi and Azari (2011)
Milk	Fe	Phosphatidylcholine	Liposome	–	–	62.9–74.8	Gupta et al. (2015a)
Soymilk	Ca	Lecithin	Liposome	–	–	–	Hirotsuka et al. (1984)
Yogurt	Fe	Peceol, Plurol Oleique, Span 80	Niosome	0.35–0.82	Spherical	72–84	Gutiérrez et al. (2016)
Yogurt	Fe	Tween 61, Cholesterol	Niosome (supercritical CO_2)	1.44, 7.21	Spherical	25.1	Wagner et al. (2016)
Milk	Fe	PGMS	Fatty acid esters (FAE)	–	Irregular spherical	85.0	Abbasi and Azari (2011)
Milk	Fe	PGMS	FAE	–	–	62.9–74.8	Gupta et al. (2015a)
Milk	Fe	PGMS	FAE	2–5	Spherical	75.0	Kwak et al. (2003b)
Drink yogurt	Fe	PGMS	FAE	2–5	Spherical	75.0	Kim et al. (2003)
Cheddar cheese	Fe	PGMS	FAE	2–5	Spherical	72.0	Kwak et al. (2003a)
Supplement additive	Fe	Na–alginate	Coacervation	400	Spherical	–	Khosroyar and Arastehnodeh (2007)
Hydrated/ dehydrated food	Fe	Na–alginate	Coacervation	400–600	Spherical	98	Khosroyar et al. (2013)
Supplement additive	Fe	Gelatin–acacia gum	Coacervation	>200	Spherical	13–40	Al-Gawhari (2016)

Supplement additive	Zn	Ethylcellulose	Coacervation	250,400	Spherical	–	Oner et al. (1988)
Milk	Fe	AG, MD, MS	Modified solvent evaporation	15.54	Spherical	91.58	Gupta et al. (2015b)
Milk	Fe	Na–alginate	Emulsification	–	–	62.9–74.8	Gupta et al. (2015a)
Yogurt	Fe	Ca–alginate beads	Emulsification	–	Spherical	–	Subash and Elango (2015)
Supplement additive	Fe	Na–alginate	Emulsification	1–2	Spherical	90–95.8	Khosroyar et al. (2012)
Milk	Fe	Medium-chain triglyceride, WPI	Emulsification[b]	2–12	Spherical	93.63	Chang et al. (2016)
Soymilk	Ca	Gelatin–agar	Emulsification[b]	–	Irregular spherical	91.18	Saeidy et al. (2014b)
Dairy products	Mg	Na–caseinate	Emulsification[b]	~9–10	Regular spherical	>99	Bonnet et al. (2009)
Supplement additive	Fe	LMP, AG, WPC, MG	Emulsification[b]	2.05	–	22.2–66.3	Jiménez-Alvarado et al. (2009)
Milk	Fe	Na–alginate, Ms, Pectin	Freeze-drying	–	–	62.9–74.8	Gupta et al. (2015a)
Yogurt	Fe	WPI gel	Salt-induced cold gelation	1	–	82.0	Bagci and Gunasekaran (2016a)
Supplement additive	Fe	WPI gel	Salt-induced cold gelation	–	Spherical	>90	Martin and de Jong (2012)

(Continued)

TABLE 9.1 (Continued)

Product	Mineral	Wall Material[a]	Technology/Process	Size (μm)	Shape	EE (%)	Reference
Dual fortified salt	Fe, I	HPMC	Fluidized bed coating	50–100	Cylindrical	–	Li et al. (2011)
Dual fortified salt	Fe, I	HPMC	Fluidized bed coating	~50	Spherical	–	Yadava et al. (2012)
Dual fortified salt	Fe, I	MD (DE = 7), HPMC	Spray-drying	<20	Spherical	85.0	Romita et al. (2011)
Additive powder[c]	Zn	Arabic gum, MD, MS	Spray-drying	16.13	Spherical	13.12[b]	Porrarud and Pranee (2010)
Dual fortified salt	Fe, I	Eudragit EPO, chitosan	Spray-drying	<20	Irregular spherical	94.0	Dueik and Diosady (2016)
Triple fortified salt[d]	Fe, I	Hydrogenated palm fat	Spray-cooling	2.5	Spherical	–	Wegmüller et al. (2006)
Dual fortified salt	Fe, I	HPMC(10% w/w)	Extrusion agglomeration	~50	Irregularly porous	87.0–94.0	Li et al. (2010)
Dual fortified salt	Fe, I	HPMC	Extrusion agglomeration	50–100	Cylindrical	–	Li et al. (2011)

[a] *PGMS, polyglycerol monostearate; LMP, low methoxyl pectin; Ms, modified starch; WPC, whey protein concentrate; WPI, whey protein isolate; HPMC, hydroxypropyl methylcellulose; MG, mesquite gum; AG, Arabic gum; MD, maltodextrin.*
[b] *As W/O/W double-emulsion.*
[c] *Additive powder containing chlorophyll pigments extracted from pandan leaf; EE is based on mg/kg.*
[d] *Vitamin A was fortified as the third component.*

9.2.1.2 Fluidized Bed Coating

In this encapsulation technique, a bed/column of particles suspends into a stirring air stream and consequently the particles are uniformly covered by spraying a coating solution containing one or more biomacromolecules, such as proteins, polysaccharides, lipids, and even emulsifiers. Finally, drying process after cycling the newly coated solid particles into a specific zone is conducted using solvent evaporation or cooling. Depending on the difference in the nozzle's arrangement/setting to spray the coating liquid, there are three main kinds of accessible fluidized beds including bottom-spray, tangential-spray, and top-spray. Although this process can effectively be implemented by providing suitable rates of mass and heat transfer, it has some serious restrictions because this method can only produce microcapsules ($>10\ \mu m$) enveloping solid particles (Gouin, 2004). Frankly speaking, FBC technique can be applied to produce premixes and supplements containing different types of vitamins and minerals, such as vitamins B group and C, ferrous fumarate ($C_4H_2FeO_4$), ferrous sulfate ($FeSO_4$), potassium iodide (KI), Na–ascorbate, etc. (Turchiuli et al., 2005).

Diosady, Alberti, and Venkatesh Mannar (2002) reported a hard control for FBC system so that the particles tend to clump together in the system by adding slight quantities of solution containing wall materials. Re-suspension of large-size particles aggregated into the bed using the water evaporation and their abrasion action of deposit materials on the bed rest is very time-consuming. Although they reported that the best output can be obtained for KI particles (2–300 μm) with high iodine-loading rate, unfortunately these researchers could not redevelop microcapsules with smaller sizes. Moreover, high amounts of iodine in KI microcapsules prepared by FBC technique were lost after 60-day storage.

In another study, Yadava, Li, Diosady, and Wesley (2012) optimized the operating factors involved in the FBC and found a temperature of 65–75°C for fluidizing air, a flow rate of coating solution of 1.5 mL/min, a fluidizing air flow rate of 35%–45% of the full flap opening, a nozzle air pressure of 1.8–2.2 bar, and finally a vertical position for nozzle in 30 cm above the chamber bottom which can provide an improved encapsulation process for preparing grain-sized Fe-premix to fortify salt. To determine the defined processing parameters, they earlier selected the most suitable coating polymers along with their optimal concentrations including 2.5% Methocel E6 [hydroxypropyl methylcellulose (HPMC), high resistance against abrasion], 4.0% Opadry white [HPMC-based with unique brightness due to the presence of titanium dioxide (TiO_2)], 6.0% Opadry AMB [polyvinyl alcohol (PVA)-based with strong ability in film formation], 3.0% Sepifilm LP770 (HPMC-based), 3.0% Eudragit EPO (reverse enteric, a uniform and glossy coating with high resistant to abrasion), 5.0% Aquacoat ECD (enteric coating with moderate uniformity), and 5.0% soy stearine (lipid material as a uniform and

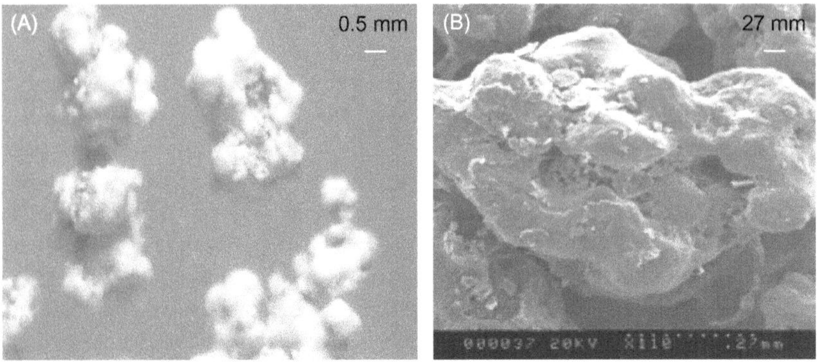

FIGURE 9.1 Photographs obtained from optical microscopy (A) and SEM (B) for the premix microcapsules of $FeSO_4$ prepared using fluidized-bed agglomeration.

smooth coating) (Yadava et al., 2012). Also, Li, Yadava, Lo, Diosady, and Wesley (2011) using fluidized-bed agglomeration produced the low-density and high-porosity double fortified salt (DFS) particles. Fig. 9.1 illustrates the microencapsulated $FeSO_4$ premix obtained by the fluidized-bed agglomeration process with lipid coating of soy stearine.

9.2.1.3 *Spray-Drying*

This technique is the most common technique used to microencapsulate functional food ingredients (1−150 μm) because it is a relatively low-cost and easy scale-up process. Although the developed microparticles in this method can rapidly release the core components because of high-water solubility of the applied carriers, there is a main limitation in choice of ideal wall materials. The limited application of organic solvents due to their toxicity and flammability, and the low payload rates (<40%) are other disadvantages of spray-drying (Madene, Jacquot, Scher, & Desobry, 2006). Romita, Cheng, and Diosady (2011) microencapsulated Fe (<20 μm) in both forms of aqueous and suspended $C_4H_2FeO_4$ (9% w/v) into HPMC (6% w/v) with 63% Na−fumarate, and 22% TiO_2 using a mini-spray drier. The produced spherical particles with Fe-loading (up to 20%) added into iodized salt for developing stable DFS under harsh storage conditions for 6 months. Hence, spray-drying technique as a simply scaled and single-step process can be used to produce the $C_4H_2FeO_4$ premix. Dueik and Diosady (2016) also used a minispray dryer to develop reverse enteric coated Fe microparticles using two wall materials of chitosan (CS) and Eudragit EPO. They sprayed the prepared mixture with CS at the inlet temperature of 120°C, air flow rate of 667 L/h, feed flow rate of 1.8 mL/min, atomizer pressure of 618 kPa and aspirator rate at 24.5 Pa. However, the inlet temperature and feed flow rate for solutions containing Eudragit EPO were 105°C and 2.7 mL/min, respectively. They generally concluded that the polysaccharide of CS was a better candidate than Eudragit EPO to produce Fe-premix for preparing stable DFS fortified with iodine and Fe (Dueik & Diosady, 2016). Porrarud and Pranee (2010) also produced microparticles of Zn−chlorophyll obtained from pandan leaf in a spherical shape with a high surface/volume ratio using this technique. The applied operating parameters for microencapsulation were feed rate of 300 mL/h, inlet temperature of 150°C, outlet temperature of 90°C, air flow rate of 0.70 m^3/min, and atomizer pressure of 50 kPa.

9.2.1.4 Spray-Chilling/Cooling

In spray-cooling, similar to spray-drying, a solution containing a bioactive constituent and a melted wall material is sprayed/atomized into a chamber chilled by liquid nitrogen or cold air. The used carrier in created cooling medium is solidified and forms fine spherical particle powders with the improved stability (Fig. 9.2). However, solubility of these powders can be variable depending on the lipophilic/hydrophilic nature of the applied wall materials. Okuro, Junior, and Favaro-Trindade (2013) mentioned that a significant drawback present in this process is a considerable percentage of

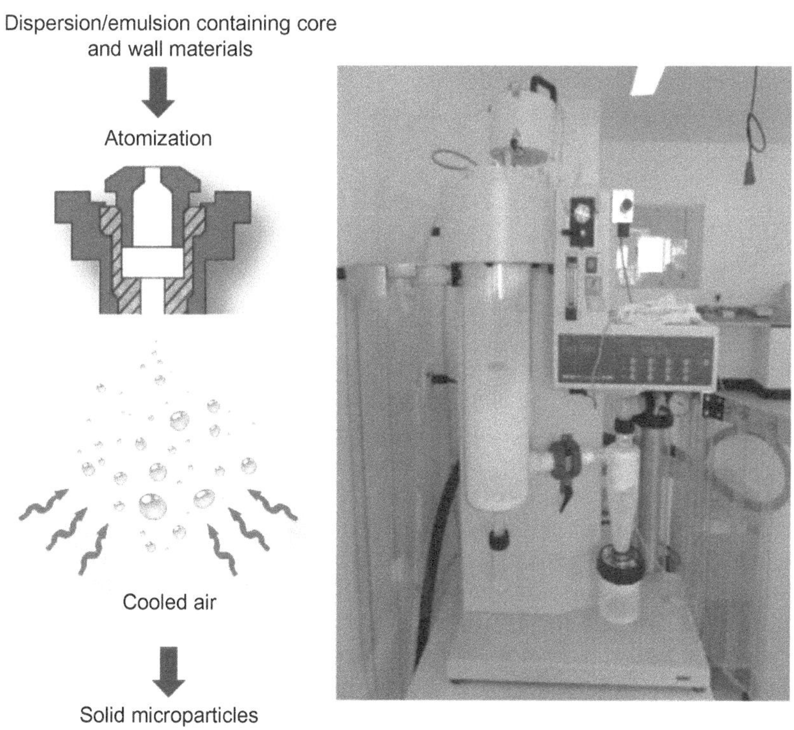

FIGURE 9.2 A view of process and apparatus of spray chilling used to produce mineral microcapsules. *Modified from Alvim et al. (2013).*

unprotected core components on the particle surface or sticking out of the carrier which can significantly reduce the EE.

Microcapsules containing Fe (40% w/w ferric pyrophosphate ($Fe_4O_{21}P_6$)), iodine, and vitamin A with the help of 1% w/w lecithin and wall material of molten hydrogenated palm fat (HPF, 80–90°C) using spray-chilling method have been prepared (Wegmüller, Zimmermann, Buhr, Windhab, & Hurrell, 2006). A system with two liner tubes to transfer the suspension of HPF into stainless-steel spraying tower was considered, whereas warm water (~90°C) flowed in the vicinity of the external tube to prevent obstruction because of lipid solidification at ambient temperature. The formulated suspension via the two-fluid nozzle and air at 90°C and 2.5 bar was directly sprayed into the pre-cooled tower with liquid nitrogen. After the operation, the collected powders at the tower bottom showed a small particle size of about 2.5 μm to produce triple fortified salt (TFS). To achieve a similar target, Zimmermann et al. (2004) produced 100 μm microcapsules comprising potassium iodate (KIO_3), $Fe_4O_{21}P_6$, and retinyl palmitate using spray chilling method. A spray chilling system equipped with a two-fluid nozzle and a de-humidifier used to encapsulate complex of flavor compound of 2-acetylpyridine (2-APri)–zinc chloride ($ZnCl_2$) by a molten (melting point of 60–63°C) vegetable stearin (Moran, 2013). Therefore, a heating temperature of 94°C was set to prevent blockage in the system for attaining a 2-APri–$ZnCl_2$ loading of 5%. Moran (2013) pointed out that application of spray-chilling encapsulation method can be very useful to protect less-stable nutritional complexes in food and pharma industries.

9.2.1.5 Extrusion

In extrusion technique, the plasticized composite matrices like combination of starch with fat or polyethylene glycol (PEG) are used to encapsulate the core components. After adding water (20% w/w) to the obtained dry mixture, the prepared wet paste is extruded using an extruder machine to attain 500–1000 μm portions and finally air-dried. In contrast with spray-cooling method, a nearly full protection for the core components can be provided in this process. Although this method gives a valuable opportunity to use glassy polymers for reducing the oxygen permeability rate and also increasing the shelf life of final food products, production of very large microparticles can be a serious obstacle for the practical usages (Li, 2009). The design of a new extrusion-based process for preparing $C_4H_2FeO_4$ microencapsules to fortify salt was done by Li, Diosady, and Wesley (2010). A stable DFS at high temperatures and relative humidities (RHs) was obtained by the incorporating/blending $C_4H_2FeO_4$ premix developed through extrusion agglomeration, color masking by 25 wt% TiO_2, and polymer coating containing 10 wt% HPMC (Fig. 9.3). The three main steps, respectively, led to a significant decrease in the unprotected surface area of $C_4H_2FeO_4$, the formation of

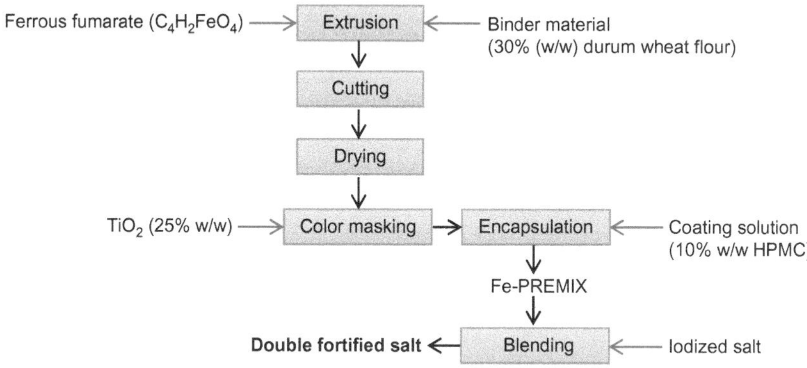

FIGURE 9.3 A schematic view of production process flow of $C_4H_2FeO_4$ microcapsules. *Modified from Li et al. (2010).*

stable water-insoluble layer for filling pores on the extrudates surface, and the creation of coating layer to reduce/avoid the diffusion of iodate and water molecules into the core environment. Therefore, these processing steps by decreasing interaction between the various ingredients can notably provide an improved physicochemical stability for the DFS fortified with $C_4H_2FeO_4$ and KIO_3 (Li et al., 2010, 2011). Moreover, this process can significantly diminish the costs of capital and operating expenses (Li et al., 2011). In another study, Moretti, Lee, Zimmermann, Nuessli, and Hurrell (2005) earlier used two extrusion premix tactics: hot (50 and 70°C) and cold (30 and 40°C) extrusion processes using a Brabender single-screw extruder to fortify rice flour with $Fe_4O_{21}P_6$ without any discoloration.

9.2.2 Chemical Methods

9.2.2.1 *Liposome Entrapment*

At first, the pharmaceutical industries were used liposomal encapsulation techniques to design drug-based delivery systems. But food and cosmetic industries were gradually applied it to produce particles with a relatively fine size range by dispersing a bilayer-forming polar lipid (e.g., lecithin) into an aqueous phase comprising dissolved bioactive components. The liposomal particles can have amphiphilic polymolecular films as mono-layer or multilayer similar with cell membrane of many plant materials (Taylor & Davidson, 2005). Since this structure like emulsions can be kinetically unstable during storage time, many researchers have focused on the liposome development using fundamentals and preparing methods of emulsion fabrication (Gharibzahedi, Razavi, & Mousavi, 2015a,b). Li (2009) stated that although the liposomal structures can cause a high stability for water-soluble ingredients due to their intrinsic aqueous solubility, there is a serious restriction so long as the coated ingredients need to be under dry environment.

To the best of our knowledge, the first study on the liposomal mineral encapsulation was performed by Hirotsuka, Taniguchi, Narita, and Kito (1984), who sonicated soy lecithin in aqueous solutions of calcium lactate ($C_6H_{10}CaO_6$) with different concentrations to form a stable liposomal structure containing Ca^{2+}. They did not observe any precipitation and/or coagulation phenomenon by adding the liposomal structure covering 60 mM Ca^{2+} to solution of soy protein. It was pointed out that soymilk can be efficiently fortified with up to 120 mg/100 g liposome system developed based on Ca–lecithin (Hirotsuka et al., 1984). Boccio et al. (1997) worked on the production of fortified fluid milk and other dairy products with high Fe bioavailability using $FeSO_4$ encapsulated in lecithin without any unwanted interactions with milk constituents. Lecithin was also used to develop liposomes encapsulating a nutritious Fe composition for the beverages fortification with the lowest off-flavor attributes (Mehansho & Mannar, 1999).

Xia and Xu (2005) applied four specific methods including freeze–thawing, thin-film, reverse-phase evaporation (RPE), and thin film and sonication to prepare $FeSO_4$-loaded liposomes in the hydrating medium of deionized water. First, they found that the optimal concentrations of cholesterol and Tween 80 can significantly increase physical stability (electrostatic and steric) of empty liposomes to encapsulate Fe^{3+}. Moreover, a considerable impact on the quantity of encapsulated Fe^{3+} was observed by investigating the amounts of $FeSO_4$ and deionized water and also the sonication magnitude. The liposome formation using RPE method led to the highest EE (67%) of $FeSO_4$. It was also concluded that Fe concentration can be increased up to 15 mg/L in milk fortified with $FeSO_4$ loaded liposomes. Under these conditions, the product had a high stability against heat sterilization processing (100°C for 0.5 h) and storage period (4°C for 7 days). Similarly, RPE technique was used to prepare liposomes of ferrous glycinate as a novel Fe fortifier (Ding, Xia, Hayat, & Zhang, 2009). A mean diameter of ~0.56 μm was recorded in the hydrating medium of pH 7.0. Moreover, a high EE (84.8%) along with an ideal Fe-bioavailability rate were acquired at the optimal conditions. The encapsulation of ferrous glycinate into liposomes could considerably enhance its stability or protection against the extra-capsular disruption by lipid bilayer under strong acidic conditions (Ding et al., 2009). Liposomes have also been used to encapsulate Fe-forms of $Fe_4O_{21}P_6$ and ferrous gluconate ($C_{12}H_{24}FeO_{14}$). A similar and favorable bioavailability rate was obtained during an in-vivo study (Navas-Carretero, Perez-Granados, Sarria, Schoppen, & Vaquero, 2007). It was concluded that the mentioned Fe sources can be applied to fortify meat products and their derivatives.

Liposome encapsulation method has been recently used to involve Fe (ammonium iron(III) sulfate ($NH_4Fe(SO_4)_2$) and $FeSO_4$) to fortify pasteurized milk. Subsequently, the impacts of Fe/lipid ratio, Tween 80 emulsifier and ratio of polyglycerol monostearate (PGMS) to Fe on the EE were evaluated (Abbasi & Azari, 2011). A 5% Tween 80 and a ratio of Fe/lipid of 0.04 led to a high EE of about 86%. Gupta et al. (2015a) also prepared and characterized Fe microencapsules prepared by different methods in order to fortify milk. Fig. 9.4 shows a simple flow chart for the liposome encapsulating Fe salt. EE of Fe liposomal microcapsules developed using egg lecithin (64%) was meaningfully higher compared to soy lecithin (<40%). The EE of 64% was in a good agreement with results obtained by Xia and Xu (2005). However, the sensory evaluation of the fortified milk with liposomal microcapsules showed the lowest scores of color, appearance, odor, taste, mouthfeel, and total acceptability in comparison with the formed microcapsules by other methods (Gupta et al., 2015a).

Xu et al. (2014) have recently evaluated the anemia reduction in rats with high-intensity running exercises using supplements containing ammonium ferric citrate ($(NH_4)_5Fe(C_6H_4O_7)_2$) liposomes or heme Fe–liposomes for 28 days. They found that the use of Fe liposomes as an ideal supplement can considerably decrease Fe-deficiency with the lowest side effects. Yuan et al. (2013) using rotary-evaporated film–ultrasonication technique developed the $(NH_4)_5Fe(C_6H_4O_7)_2$ liposomes and heme liposomes as a Fe-supplement to fortify food products. They demonstrated that the liposomes entrapping Fe compared with free Fe-supplements had a high ability to intake Fe in the body.

Developing a liposome technique to microencapsulate Ca^{2+} using egg phosphatidyl choline (EPC) and injection of the produced Ca^{2+}-microparticulates into rabbit before slaughter were carried out by Kim, Kim, Jeong,

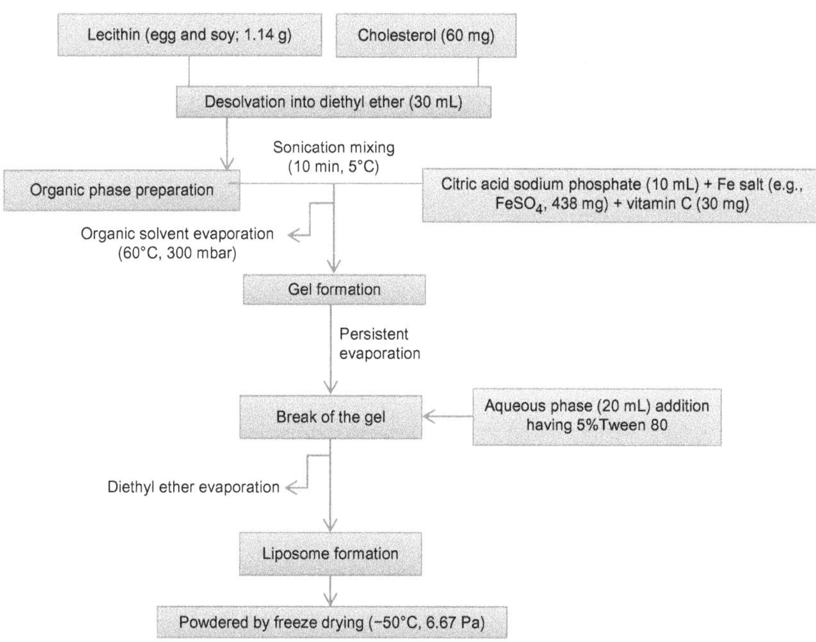

FIGURE 9.4 A scheme representing how the formation of liposomal microcapsules containing Fe salt are prepared by the RPE method. *Retrieved from Gupta et al. (2015a).*

and Kim (2006). They concluded that the Ca level injected into the muscle not only can change amount of Ca^{2+} concentration but also can efficiently reduce the aging time of meat. The highest rate of Ca encapsulated in liposome was obtained at Ca concentrations of 1.0 M (63.8%) and 2.0 M (61.7%), while the higher levels (3.0 and 4.0 M) had lower Ca encapsulation rate (54.3–54.5%). This fact was affirmed by the analysis of microscopy images of microcapsules entrapping the various levels of Ca^{2+} solution (1.0–0.4 M). Many uniform microcapsules (0.25–1.8 μm) at 1.0–2.0 M Ca^{2+} were found in comparison with the higher concentrations, as shown in Fig. 9.5.

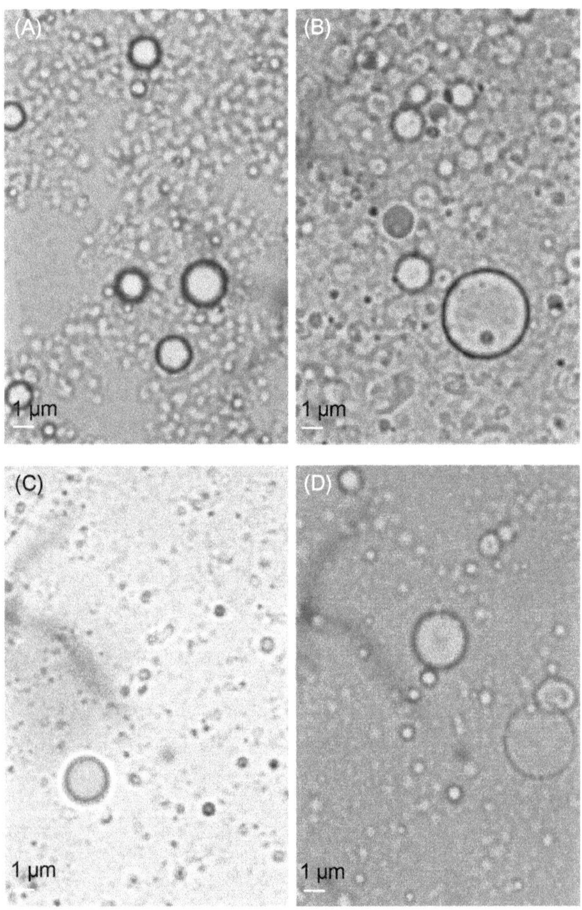

FIGURE 9.5 Phase-contrast microscopy images of EPC–liposomes as a function of Ca^{2+} concentration (A–D are 1.0, 2.0, 3.0 and 4.0 M Ca^{2+}, respectively). *Reproduced from Kim et al. (2006) with permission of Taylor and Francis Group.*

9.2.2.2 Niosome Entrapment

Niosomes as nonionic surfactant-based vesicles are capable to entrap both hydrophilic and lipophilic compounds in the aqueous phases between the bilayers and inside the surfactant bilayer, respectively (Devaraj et al., 2002). This unique character along with low-cost and high storage stability has led to their wide usages in cosmetic, pharma, bio, and food industries (Marianecci et al., 2014). Conventional ethanol injection (CEI) technique is one of the most important methods to prepare niosomes because it is a simple and easily scaled approach without use of any possibly dangerous material (Pham, Jaafar-Maalej, Charcosset, & Fessi, 2012). The production of Fe microcapsules using niosomes prepared by CEI to enrich yogurt has been recently reported by Gutiérrez et al. (2016). These researchers distinctly produced two aqueous and organic phases. The suitable levels of three common surfactants (Span 80, Peceol, and Plurol Oleique), and a membrane stabilizer agent (1-dodecanol) in absolute ethanol were dissolved to form the organic phase. Components of $FeSO_4$, vitamin C, di-Na phosphate, and citric acid in distilled water were dissolved to create the aqueous phase. In the next step, the developed organic phase with a flow rate of 130 mL/h at 40°C was injected into the water phase using a syringe pump. Prior to the contact of two produced phases, a number of spontaneous niosomes were formed. However, there were high quantities of niosomes with very fine particle size and narrow distribution after blending and stirring at 5000 rpm. Finally, the spherical microcapsules (0.35–0.82 μm, EE = 72–84%) by removing ethanol in a rotary evaporator were stored to formulate the yogurts. Evaluating the Fe-bioavailability by determining its oxidation rate and EE under the simulated gastrointestinal conditions showed that the best formulation for niosomal systems was use of Peceol surfactant and 1-dodecanol which could significantly improve the quality attributes of yogurt compared with the control (Gutiérrez et al., 2016). Wagner, Spoth, Kourkoutis, and Rizvi (2016) have also studied concurrent encapsulation of hydrophilic vitamin D_3 and lipophilic $FeSO_4$ supplements into niosomes using a new supercritical CO_2 technique. They found that this encapsulation type can lead to the capsules with 1.44 and 7.21 μm with a $FeSO_4$-EE of ~25% and a vitamin D_3-EE of ~96%. A better storage stability for the encapsulating niosomes at 20°C was monitored than those stored at 4°C for 21 days.

9.2.2.3 Fatty Acid Esters

A multistep method is usually applied to develop water-soluble mineral-containing microcapsules using FAE. In this technique, a mixture of a coating agent, PGMS as a FAE, a mineral salt type and distilled water is prepared after heating and stirring processes under certain conditions. In the next step, the resulted mixture is nebulized into the surfactant (e.g., Tween 60) dispersion in distilled water with a sprayer. Having centrifuged, the separated phases are removed. Finally, the microcapsules are collected and stored at a low temperature till the use time. Fig. 9.6 illustrates a schematic view of FAE microencapsulation steps for a Fe salt.

Many scientists have used FAE technique to microencapsulate various Fe salts including $FeSO_4$, ferric chloride ($FeCl_3$) and $NH_4Fe(SO_4)_2$ (Jackson & Lee, 1992a,b; Kwak, Ihm, & Ahn, 2001; Kim, Ahn, & Kwak, 2003; Kwak, Ju, Ahn, Ahn, & Lee, 2003a,b). Abbasi and Azari (2011) also have recently investigated the efficiency of FAE method to encapsulate Fe salt to fortify milk. A significant EE (~82%) using FAE technique could be obtained in optimal ratio of PGMS/Fe salt of 15:1. However, Gupta et al. (2015a) found that the microcapsules prepared with FAE method had the lowest EE among

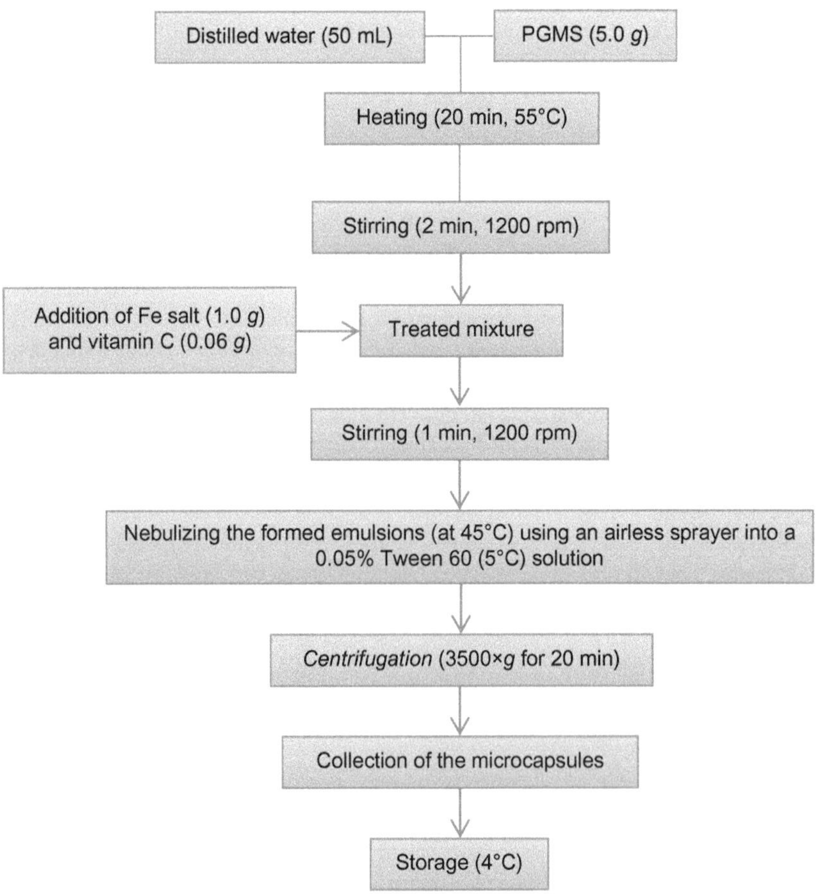

FIGURE 9.6 A flow chart illustration for Fe encapsulation using FAE method. *Retrieved from Kwak et al. (2001, 2003a,b) and Abbasi and Azari (2011).*

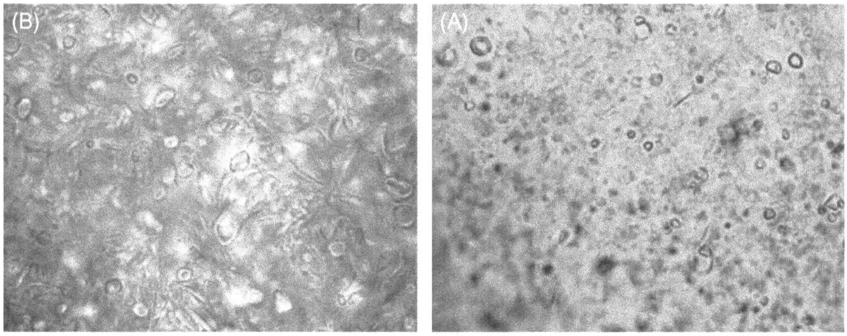

FIGURE 9.7 Morphology of $FeSO_4$ microcapsules prepared by liposome entrapment (A) and FAEs (B) techniques with a magnification level of 1000×. *Reproduced from Abbasi and Azari (2011) with permission of Wiley and Sons.*

samples developed by other encapsulation techniques. Abbasi and Azari (2011) reported that the developed microcapsules by FAE and liposome entrapment method had an irregularly spherical structure with smooth surface morphology (Fig. 9.7). The best sensory attributes of pasteurized milk was for samples of control and fortified with FAE-microencapsulated Fe at a low level of 7 mg/L. These researchers emphasized that this encapsulation technique can be considered as an appropriate, simple, low-cost, fast, and efficient approach to encapsulate different Fe salts and subsequently preparation of Fe-fortified pasteurized milk (Abbasi & Azari, 2011).

9.2.2.4 Coacervation

Coacervation technique as a polymer aggregation process is based on phase separation (concentrated and diluted colloidal phases) owing to the partial desolvation of wholly solvated polymers. Changing the macromolecule solubility by environmental variations like adding a salt or a reverse-charged polymer can meaningfully begin this chemical process (de Kruif, Weinbreck, & de Vries, 2004). Accordingly, this method can be classified into two main groups including simple and complex. The first difference between these groups is the number of polymers involving in the process so that one and two macromolecules are participated in the simple and complex coacervation, respectively. The second difference is phase separation mechanism. The phase separation in simple coacervation is conducted by the addition of a salt (e.g., Ca salt to alginate polysaccharide) and/or pH and temperature alterations, but coacervation in complex type is happened by anion–cation interactions, particularly in gelatin and gum Arabic (GA) biopolymers (Shewan & Stokes, 2013).

A coacervation technique was earlier applied to produce zinc sulfate ($ZnSO_4$) microcapsules using ethylcellulose as wall material in two wall:core ratios of 1:1 and 1:2; obtained microcapsule groups had a particle size of 250 and 500 μm, respectively (Oner, Kas, & Hincal, 1988). It was inferred that the microparticle size and ratio of wall:core had an influential effect on the in-vitro Fe-bioavailability. Khosroyar and Arastehnodeh (2007) using the coacervation method produced microcapsules containing ferric saccharate with coating material of alginate at a core/wall ratio of 30:70. 400 μm-particles formed in the presence of $CaCl_2$ had a low Fe-release rate ($<0.04\%$) at mild and harsh temperatures and RHs. Nevertheless, a high Fe-release rate was obtained by increasing the RH amount. Ferric saccharate was also microencapsulated into wall material of Na–alginate by Khosroyar et al. (2013) using three techniques of coacervation (by needle), emulsification, and spray-drying. Stirring speed had a key role in producing Fe microparticles so that an increase in quantity of this parameter led to the development of spherical and smooth microparticles. A complex coacervation approach has been recently utilized to incorporate $FeSO_4$ into coating mixture of gelatin–Arabic gum (Al-Gawhari, 2016). It was found that this technique cannot be useful to microencapsulate this Fe source probably because of its leaking from the coacervation phase due to high $FeSO_4$ solubility in water.

9.2.2.5 Modified Solvent Evaporation

Gupta, Chawla, Arora, Tomar, and Singh (2015b) designed the present technique to prepare $FeSO_4$-microcapsules using three coating agents of GA, maltodextrin (MD), and Ms in a ratio of 4:1:1. After dissolving 6.0 g of this carbohydrate blend in distilled water (10 mL, 60°C), the mixture for further hydration (12 h) was maintained at 4–7°C. $FeSO_4$ and antioxidant agent of vitamin C (15:1) after dissolving in water (10 mL) were added to the rehydrated solution and mixed thoroughly and sonicated for 15 min in a waterbath at 5°C. It seems that the carboxylic groups present in GA structure can effectively interact with Fe^{2+} through strong electrostatic (or ionic) bonds. The obtained mixture was sprayed in chilled alcohol using an airless paint sprayer and filtered after a 5-min rest using a Whatman filter paper (No. 1) under vacuum conditions. The residual alcohol in the microcapsules (6.8–33.4 μm; EE = 91.58%) collected on the filter paper was removed by the evaporation process at low temperatures and stored at 5°C for 13 h. The modified solvent evaporation is an efficient method for Fe microencapsulation compared with other techniques because the used solvent can be easily recycled via distillation process of filtered alcohol. After preparing the Fe microcapsules, the fortified milk based on these microcapsules was developed and its sensory attributes and oxidative stability was compared with milk samples containing free Fe-salt during storage. More sensory scores, bioavailability rates, and thiobarbituric acid (TBA) values were recorded for the milk fortified with Fe microcapsules from the modified solvent evaporation (Gupta et al., 2015b).

9.2.2.6 *Emulsification*

Jiménez-Alvarado, Beristain, Medina-Torres, Román-Guerrero, and Vernon-Carter (2009) encapsulated ferrous bisglycinate ($C_4H_8FeN_2O_4$, 15 wt%) into the inner phase (W_1) of W_1/mineral oil/W_2 multiple emulsions stabilized with a emulsifier combination (5 wt%) of Panodan SDK (esters of mono- and di-glycerides of di-acetyl tartaric acid) and polyglycerol polyricinoleate (PGPR) in a ratio of 4:6. Having prepared the preliminary emulsions, they were re-emulsified for producing the last multiple emulsions and stabilized with one of polysaccharides of mesquite gum (MG), GA and low methoxyl pectin and whey protein concentrate (WPC). Results exposed that the protein–polysaccharide complex between WPC and MG led to the favorable Fe-EE, Fe-bioavailability, and emulsion physicochemical stability in terms of the lowest coalescence rate, droplet size, and oxidation rate and release of $C_4H_8FeN_2O_4$.

In another study, Fe capsules (2–12 μm) were produced during two steps by W/O/W emulsion method to formulate into milk (Chang, Lee, & Kwak, 2016). $FeSO_4$ was initially dissolved in distilled water and then added into the medium-chain triglyceride (MCT) obtained from palm oil as a lipid phase which was containing hydrophobic emulsifier of PGPR too. Iron solution (core agent) was first blended with MCT (preliminary wall agent) in a ratio of 4:6. The obtained emulsion was later mixed with an outer coating material [30% whey protein isolate (WPI) solution] with a ratio of 2.5:7.5 and also hydrophilic emulsifier of polyoxyethylene sorbitan monolaurate and once more homogenized to create a W/O/W emulsion with fine droplets. A high EE (93.63%) and an outstanding stability in simulated gastric conditions (1.2%–1.9% Fe release during 0.5–2 h of incubation) were obtained for the formed spherical microcapsules having a smooth and serrated surface. An analogous process was designed to fabricate Ca microcapsules to fortify soy-milk (Saeidy, Keramat, & Nasirpour, 2014b). It was revealed that the fortified products during pasteurization had a high physical stability. These researchers in another study optimized EE and payload of Ca microcapsules by response-surface methodology (RSM). Saeidy, Keramat, and Nasirpour (2014a) revealed an increase in concentration of protective oil and wall components can lead to a decrease in Ca-release rate. The maximum values of EE and payload of Ca were predicted in an emulsion system formulated with 3% gelatin, 0.5% agar, 48.58% W_1/O, and 47.92% W_2.

Bonnet et al. (2009) encapsulated Mg (as $MgCl_2$) in the internal aqueous phase (W_1, 1 μm) of multiple W_1/O/W_2 systems emulsified with Na–caseinate and PGPR and studied the emulsion-storage stability and Mg-release rate. The used oil kind had a significant influence on the Mg leakage from thermo-stable multiple emulsions so that the oils with more content of saturated fatty acids and lesser viscosity showed higher levels of Mg leakage. A high bioavailability in the GT was achieved using pancreatic lipase which could hydrolyze triglycerides present in the emulsions representing Mg-loaded emulsions. Zhang et al. (2016) developed $FeSO_4$ microparticles with wall materials of alginate–agarose. They utilized a low-energy emulsification method namely "premix membrane emulsification" based on four phase-emulsion structures ($W_1/O_1/W_2/O_2$) mixed with ionic solidification technique. A remarkable rise in mechanical strength level was obtained by adding agarose to alginate-based microcapsules. Moreover, they inferred that an increase in the volume proportion of W_1/O_1, density and viscosity of O_2 phase and content of the emulsifier can considerably enhance the emulsions stability with fine droplet sizes.

Fig. 9.8 depicts the preparing steps of mineral (Fe) microcapsules using alginate emulsification method. To collect the microcapsules, having

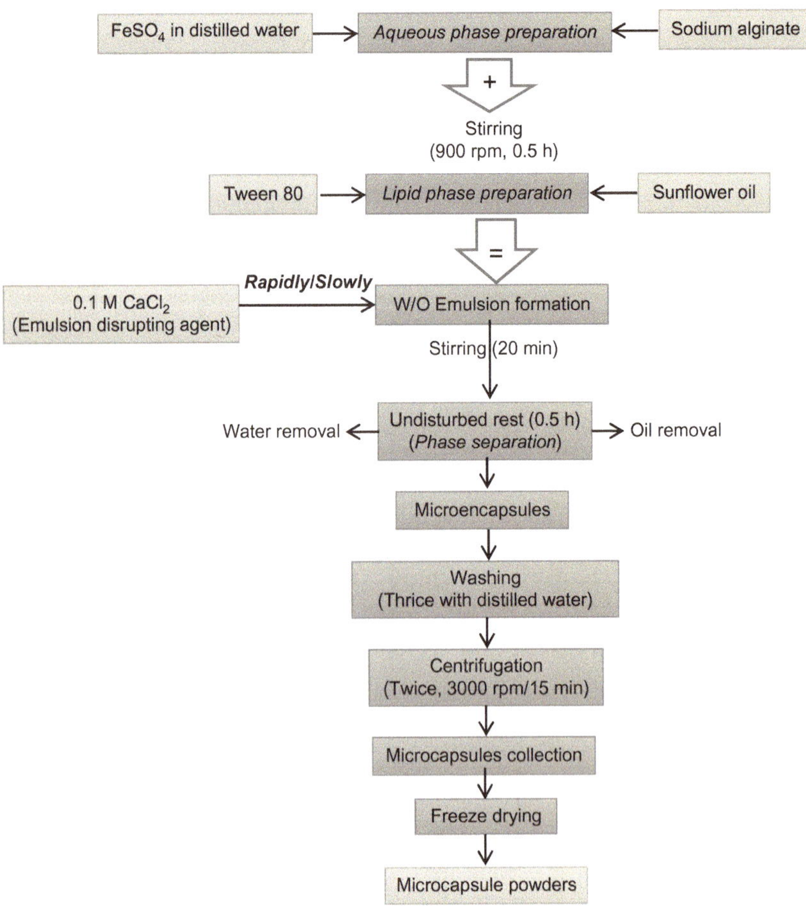

FIGURE 9.8 The production steps of Fe microcapsules using emulsification method. *Retrieved from Gupta et al. (2015a).*

formed the emulsion system, $CaCl_2$ as an emulsion disrupting agent is added to separate oil and water phases. For instance, Gupta et al. (2015a) added cooled solution of 0.1 M $CaCl_2$ into W/O emulsions rapidly or slowly to investigate its effect on the EE. Results showed that the fast addition of this component to create Fe microcapsules can considerably result in a higher EE than those developed by slow addition of $CaCl_2$. This fact demonstrated that a high EE can be obtained by fast deemulsification. A research group had earlier explained that rapid addition of $CaCl_2$ during the encapsulation process of ferric saccharate could provide an instant disruption of the emulsion and a very short time for coalescence which thus led to the formation of smaller size microcapsules with higher EE (Khosroyar, Akbarzade, Arjoman, Safekordi, & Mortazavi, 2012). Khosroyar et al. (2012) using coating material of Ca–alginate and emulsification approach microencapsulated ferric saccharate. A substantial increase in Fe-bioavailability rate and a notable improvement in sensory attributes were obtained for the Fe microcapsules. They also demonstrated that the best conditions to produce the spherical microcapsules with smoother surfaces were combination of ~0.8 g Fe, 1.5% Ca–alginate, and 0.1 M $CaCl_2$ under stirring speed of 100 rpm.

9.2.2.7 *Salt-Induced Cold Gelation*

Gelation techniques based on cross-linking using Ca–alginate beads, cooling processes, and chemical reactions can produce the gelled microcapsules containing functional components. Gelation mechanism to design microcapsules is an expensive and difficult-scaled process which leads to a uniform porous structures with controlled diffusion ability of liquids like water (Li, 2009). Bagci and Gunasekaran (2016a) have recently optimized production of Fe-encapsulated cold-set WPI gels using salt-induced cold gelation method (Fig. 9.9) in order to the achieve the highest Fe-EE and the lowest color variations. In-vitro gastrointestinal experiments revealed that the WPI–Fe particles (1 μm) generated under optimum conditions can be considered as an outstanding system to site-specific delivery of Fe by their addition into dairy products like yogurt (Bagci & Gunasekaran, 2016b). In another study, Fe microparticles were developed by entrapping $FeSO_4$ into WPI matrix using cold-set gelation (Martin & de Jong, 2012). Moreover, the effect of various heating intensities (slight, middle, and vigorous) and Fe-loading amount on the effectiveness of encapsulation process was also evaluated in order to optimize the strength of protein gel. They inferred found an excellent Fe-EE and Fe-bonding ability by WPI because the high and low Fe-release rates were monitored at neutral and acidic pH values, respectively. Since an increase in heating intensity lead to an increase in the release level of Fe, this fact can be considered in utilizing particles of Fe–WPI to fortify the food products pasteurized/sterilized at high temperatures.

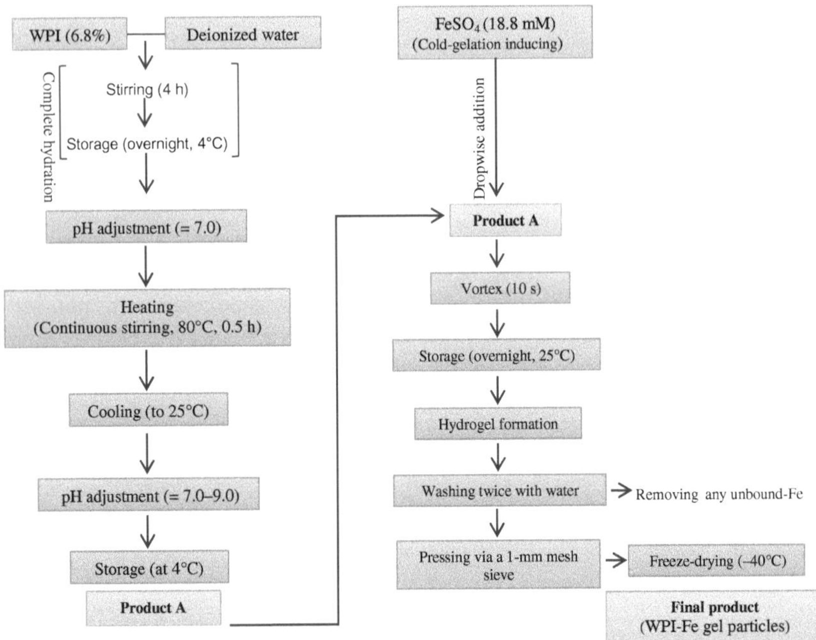

FIGURE 9.9 Synthesis of WPI–Fe gel particles using salt-induced cold gelation method. *Retrieved from Bagci and Gunasekaran (2016a).*

9.3 NANOENCAPSULATION OF MINERALS

Nanoencapsulation is considered as a novel technology to cover bioactive substances into a matrix at a size lower than 1000 nm. This method can possibly present new delivery systems for minerals and other functional ingredients with enhanced physicochemical stability, water solubility, bioaccessibility, and bioavailability. Since use of mineral nanocapsules can considerably increase the production of novel functional food formulations, a high focus on the design of applicable approaches has been recently done to nanoencapsulate minerals for fortifying food products. In our appraisal, we realized the chemical processes much more practical (>95%) than the physicomechanical ones to produce the mineral nanoparticles (NPs). In the following, we clarify some new methods for nanoencapsulating minerals particularly Fe. Table 9.2 summarizes the nanoencapsulation techniques of mineral ions and salts using different wall materials and represents size, shape, and EE properties of the developed nanocapsules.

9.3.1 Nanoliposomes

Kosaraju, Tran, and Lawrence (2006) developed $FeSO_4$ nanocapsules using the techniques of pro-liposome (PLS) and microfluidization (MF) with the

TABLE 9.2 Production Processes/Technologies of the Mineral Ions/Salts-Nanocapsules Using Different Wall Materials and their Size, Geometrical, and EE Characteristics

Technology	Process	Mineral Form	Wall Material[c]	Size (nm)	Shape	EE (%)	References
Nanoliposome	Pro-liposome entrapment	$FeSO_4 \cdot 7H_2O$	SPC	~5	Spherical	58	Kosaraju et al. (2006)
Nanoliposome	Microfluidization	$FeSO_4 \cdot 7H_2O$	Emultop	~150–200	Spherical	11	Kosaraju et al. (2006)
Nanoliposome	Reverse-phase evaporation	Ferrous glycinate	Egg-PC	<100	Spherical	69.6–76.2	Ding et al. (2011b)
Nanoliposome	Microfluidization	$FeSO_4$	SPC, HSPC, CHOL, CS, CPL	400–750	Spherical	–	Hermida et al. (2011)
Nanoemulsification	Liquid whistle, sonication, HPH[a]	$C_4H_2FeO_4$	Gelatin	600	Spherical	–	Tang and Sivakumar (2013)
Nanoemulsification	High-speed homogenization	Fe ion	CHOL, PC	59	Spherical	–	Naveen and Kanum (2014)
Cyclodextrin inclusion	Co-precipitation	$C_4H_2FeO_4$	β-CD, HP-β-CD	–	Toroidal	–	Kapor et al. (2012)
Cyclodextrin inclusion	Not reported	$C_6H_{10}FeO_6$	α-CD, β-CD, γ-CD	–	Toroidal	–	Leite et al. (2003)

(*Continued*)

TABLE 9.2 (Continued)

Technology	Process	Mineral Form	Wall Material[c]	Size (nm)	Shape	EE (%)	References
Solid lipid NPs (SLNs)	Hot homogenization/ Ultrasonication	$FeSO_4$	Compritol 888 ATO, lecithin	24.2–1015	–	65.1–92.5	Hosny et al. (2015)
Solid lipid NPs (SLNs)	Double emulsion solvent evaporation	$FeSO_4$	CS–HCl, stearic acid, PVA	300–495	Spherical	53.9–87.7	Zariwala et al. (2013)
Biopolymeric NPs	Controlled ionic gelation	$FeSO_4$	Na–alginate	15–30	Spherical	75	Katuwavila et al. (2016)
Biopolymeric NPs	Nanoprecipitation	Na_2SeO_3	Guar gum	41–173	Spherical	–	Soumya et al. (2013, 2014)
Biopolymeric NPs	Mild stirring with NPs produced by ionotropic gelation	Na_2SeO_3	CS/TPP	–	Spherical	–	Zhang et al. (2011)
Biopolymeric NPs	Mild stirring with NPs produced by ionotropic gelation	Na_2SeO_3	CS/TPP[d]	100–400	Spherical	60–95	Luo et al. (2010)
Biopolymeric NPs	Ethanol desolvation	$ZnCl_2$	WPI	~100	–	>95	Gülseren et al. (2012)
Biopolymeric NPs	Ethanol desolvation	$ZnCl_2/FeCl_3$	Alginate	90–135	Spherical	70–85	Sharifi et al. (2013)

Ionotropic gelation	Ionic gelation with TPP[a] anions	$FeSO_4$	WPC, CS	44.4	Spherical	99.7	El-Sayed et al. (2015)
Ionotropic gelation	Ionic gelation with Na–TPP	Cu^{2+}, Zn^{2+}, Mn^{2+}, Fe^{2+}	CS	95.8–210.9	–	–	Du et al. (2009)
Ionotropic gelation	Ionic gelation with Na–TPP (Hydrothermal procedure)	$FeCl_3$, $MnCl_2$, $ZnCl_2$	CS	14	Large agglomerations	34	Zahraei et al. (2015)
Ionotropic gelation	Ionic gelation with Na–TPP	$C_4H_6O_4Zn$	β-Cs	84.55	Spherical	97.33	Zhang and Zhao (2015)
Ionotropic gelation	Ionic gelation with Na–TPP	$C_4H_6O_4Zn$	β-Cs	208–591	–	53–89	Zhang et al. (2016)
Ionotropic gelation	Ionic gelation in 0.05 M NaCl	$ZnCl_2$	Na–alginate	200–230	–	–	Pistone et al. (2015)
Coacervation	Complex coacervation	ICS[b]	CS (casein)	830–1070	Spherical	–	Min et al. (2016)

[a]*High-pressure homogenization; TPP, tripolyphosphate.*
[b]*ICS, iron casein succinylate.*
[c]*PC, phosphatidylcholine; SPC, soybean phosphatidylcholine; HSPC, hydrogenated phosphatidylcholine; CHOL, cholesterol; CS, chitosan; CPL, cationic phospholipids; PVA, polyvinyl alcohol; WPC, whey protein concentrate.*
[d]*With or without zein coating.*

size of ~5 and ~150–200 nm, respectively. Also, the EE values of $FeSO_4$ for liposomes produced with MF (6% lipid and 7.5% $FeSO_4$) and PLS (1.5% lipid and 15% $FeSO_4$) were 58 and 11%, respectively. Though Fe–liposomes fabricated by PLS method had a higher oxidative stability, the oxidative stability of $FeSO_4$ in PLS and MF methods was the same after 77-day storage. According to the inclusive findings, the nanoliposomal delivery systems produced based on MF technique had a favorable possibility to nanoencapsulate Fe for the industrial applications.

Fe-deficiency anemia in rats by feeding free ferrous glycinate and $FeSO_4$, and ferrous glycinate nanoliposomes were evaluated by Ding, Zhang, and Xia (2011a). It was concluded that use of the nanometric forms of liposomes entrapping ferrous glycinate which were prepared by RPE technique from egg phosphatidylcholine had a more significant effect in decreasing Fe-deficiency in rats than the two free Fe-forms (Ding et al., 2011a). In another study by Ding et al. (2011b), these researchers assessed the EE, particle size and distribution, structure and stability of spherical nanoliposomes of ferrous glycinate prepared by RPE method. They produced stable nanoliposomes (<100 nm) without any agglomeration with a unique polydispersity index (0.361). These delivery systems showed an acceptable EE (69.6%–76.2%) and zeta-potential (6.3 mV) at pH 7.0. The nanoliposomes showed a reduction in the retention ratio of Fe source by prolonging storage so that this ratio after 3 months dropped to 91.9%, probably due to degradation/hydrolyzation of the bilayer membranes or merger/combination of nanoliposomes (Flaten, Bunjes, Luthman, & Brandl, 2006). It was claimed use of these nanovesicles to fortify different food products can be outstanding with regard to the obtained data and also a 5-h stability under the simulated conditions of gastrointestinal environment at 37°C (Ding et al., 2011b).

Hermida, Roig, Bregni, Sabés-Xamaní, and Barnadas-Rodríguez (2011) using MF process fabricated several $FeSO_4$ nanoliposomes which were formulated with CS, soy phosphatidylcholine, hydrogenated phosphatidylcholine (HSPC), HSPC-cholesterol, and 3β [*N*-(*N*′,*N*′-dimethylaminoethane)-carbamoyl] cholesterol (DC-cholesterol)–HSPC. The maximum incorporation of Fe into the nanoliposomal matrix was monitored for cholesterol component. The CS and DC–cholesterol-based nondigested nanoliposomes showed a low stability; nevertheless, the digested nanoliposomes made of HSPC–cholesterol represented the maximum Fe-uptake and bioavailability according to the CaCO-2 cells. The fabricated nanoliposomes can be applied in the final food products because of having a strong barrier against oxidative reactions and high absorption of Fe in the simulated GT (Hermida et al., 2011).

9.3.2 Nanoemulsification

Benefits of nanoemulsions (NEs) in overcoming the solubility and stability problems of functional bioactive additives in aqueous solutions have caused an impressive growth in scientific and industrial researches. In general, preparation techniques of NEs with droplet sizes of 50–1000 nm are categorized into two main groups including low-energy and high-energy methods (McClements & Rao, 2011). Sometimes, a significant level of mechanical energy is required to form NEs using high-pressure homogenization (HPH), MF, and sonication. Besides, a number of different low-energy approaches have been used to develop NEs, such as spontaneous emulsification, phase inversion temperature, phase inversion composition, and emulsion inversion point (Gharibzahedi & Mohammadnabi, 2016). Jafari, He, and Bhandari (2007) showed that the nanoscale emulsions can be usually applied as liquid or spray-dried powders to enhance the stability and EE of bioactive ingredients.

A stable multiple W/O/W NE with mean diameter of 600 nm and polydispersity index of 0.35–0.40 was designed to entrap $C_4H_2FeO_4$ using the liquid whistle, HPH, and ultrasound approaches (Tang & Sivakumar, 2013). Although these NEs were stable against phase separation after storage at $-4°C$ for 10 days, milder color development along with a sluggish creaming percentage (1.6%) were obtained over the same time at 45°C. Naveen and Kanum (2014) also using high-speed homogenization technique produced a stable phospholipid/O/W Fe-NE with *z*-average of 59 nm for fortifying milk. Their results proved that two emulsifiers of cholesterol and phosphatidylcholine could lead to the lowest droplet sizes and polydispersity index. Furthermore, a blend of the used emulsifiers resulted in a considerable stability for the NE. Increasing storage temperature not only decreased the physical stability but also a substantial chemical degradation of Fe (22% loss) was happened at 25°C. A more in-vivo bioavailability enhancement in rats fed with the fortified milk was observed compared with those fed with milk enriched by direct addition of Fe. No toxicity in twice-daily dosing regimens (10 mL/kg) of milks fortified with NE encapsulating Fe during 15-day feeding was observed. Therefore, Naveen and Kanum (2014) concluded that the developed NEs can be used as a functional supplement alone or in combination with other food products.

9.3.3 Cyclodextrin Inclusion

Unique "molecular nanoencapsulation" of minerals can be provided by complexes of molecular inclusion. Empty cyclodextrin (CD) nanocapsules with a specified molecular size are able to nanoencapsulate many different bioactive components through the entry of appropriate "guest" ingredients into their cavity. The CD cavity has a less polarity than water linked by the atoms of glycosidic oxygen and hydrogen; these cyclic oligosaccharides with α-(1$\rightarrow$4)-linked glucopyranose units in a cylindrical structure have a water-insoluble internal portion and a water-soluble outer (Gharibzahedi & Jafari, 2017).

Kapor, Nikolić, Nikolić, and Stanković (2012) developed inclusion complexes of $C_4H_2FeO_4$ in a molar ratio 1:1 with CDs [β-CD and 2-hydroxypropyl-β-CD (HP-β-CD)] using coprecipitation technique. Having mixed into a solution for 3 days at room temperature, the volume was reduced by evaporation process and then placed in a desiccator containing concentrated H_2SO_4 in order to dry or maintain the gradient of RH. Solubility determined by UV/vis method revealed that the formed complexes between CDs and $C_4H_2FeO_4$ had an enhanced solubility. Yet, nanocapsules prepared with HP-β-CD showed a better solubility than β-CD and thus it is possible to fortify the pharma and food products (Kapor et al., 2012). Leite, Lino, and Takahata (2003) had theoretically and experimentally studied the inclusion complex of ferrous lactate ($C_6H_{10}FeO_6$) with three main CDs (α-, β-, and γ-CD). The highest and lowest solubility rates were for β- and α-CDs, respectively. Moreover, the Fe-nanocapsules produced based on the CD inclusion complexes could considerably protect $C_6H_{10}FeO_6$ against oxidation process and improve its delivery.

9.3.4 Solid Lipid Nanoparticles

Solid lipid nanoparticles (SLNs) developed via congealing process is considered as an ideal superseded for nanodispersions (Fig. 9.10). Fabrication of $FeSO_4$–SLNs (Fig. 9.10) using combination of two techniques of ultrasonication (1–7 min) and hot homogenization (1–10 min) was done by Hosny, Banjar, Hariri, and Hassan (2015). Bioavailable Fe–SLNs with particles size

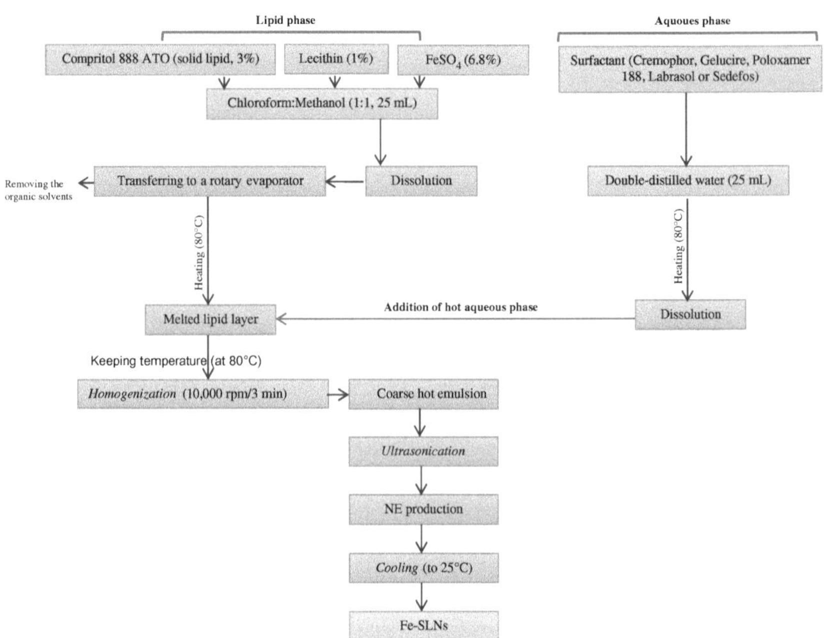

FIGURE 9.10 The $FeSO_4$-loaded SLNs fabrication using hot homogenization–ultrasonication. *Retrieved from Hosny et al. (2015).*

of 25 nm and EE of 92.3% were obtained by preparing an optimal formulation containing 3% Compritol 888 ATO, 1% lecithin, 3% Poloxamer 188 surfactant, and 0.2% dicetylphosphate. The produced Fe-SLNs could resolve the problems related to the consumption of similar commercial samples, for example, constipation disorder, blood existence in stool, and extensive dissimilarities in the Fe-uptake and bioaccessibility level (Hosny et al., 2015). A new technique of double emulsion solvent evaporation to produce $FeSO_4$–SLNs based on stearic acid (SA) was before developed (Zariwala et al., 2013). An oily phase by dissolving SA into 1:1 mixture of dichloromethane and methanol at 60°C was first made. The aqueous phase was prepared at temperature of 60°C by blending 10% PVA, 1%–3% $FeSO_4$ and 0.1%–0.4% CS–HCl. Homogenization (21,000 rpm, 4 min) was done by mixing two immiscible phases at 60°C to reach a coarse microemulsion. Then, this emulsion was mixed with 1% PVA and finally rehomogenized for 7 min at the same temperature and stirring speed. In a hierarchy process, the resulted emulsion was evaporated, cooled, washed, freeze-dried (for 25 h at −40°C), and again dried (for 8 h at 20°C) in order to obtain Fe-loaded SLNs (300.3–495.1 nm). A rise in CS–HCl content in the studied range led to a significant improvement in Fe-EE of SLNs. Studying the *in-vitro* Fe-uptake based on the Caco-2 cell model showed that this parameter in Fe–SLN formulations was considerably more than the reference of $FeSO_4$. Production of these exciting delivery systems containing Fe can improve bioavailability of this bioactive ingredient in the body.

9.3.5 Biopolymer Nanoparticles

Incorporation of many functional bioactive ingredients in pharma and food products can be done using the delivery systems fabricated by biopolymer nanoparticles. Na–alginate as a wall material for $FeSO_4$ in the NPs preparation has recently exerted by a controlled ionic gelation technique (Katuwavila et al., 2016). The alginate NPs (15–30 nm) loaded with Fe (0.06% w/v) revealed an extended in vitro release kinetic for 4 days. The Fe-release rate was highly depended on pH so that Fe-release at pH values of 2.0 (Fickian diffusion), and 6.0 and 7.4 (non-Fickian diffusion) were 20% and 65%–70%, respectively. The new developed delivery systems for Fe-loaded Na–alginate NPs can be a fantastic option for simple remedy of oral Fe. In another study, Gülseren, Fang, and Corredig (2012) produced WPI-NPs by means of ethanol desolvation technique to encapsulate $ZnCl_2$ under acidic conditions. An increase in ratio of ethanol to water and loaded-$ZnCl_2$ level increased the size of WPI-NPs. Moreover, a controlled dilution after desolvation process provided a high EE for Zn. As WPI-NPs represented a significant capability to encapsulate $ZnCl_2$ with good storage stability for 30 days at 22°C, the prepared delivery systems can be effectively applied to incorporate minerals and unstable water-soluble ingredients in the development of acidic food formulations.

In another study, a novel carrier based on alginate-NPs to fortify ice cream with $FeCl_3$ and $ZnCl_2$ was used (Sharifi, Golestan, & Sharifzadeh Baei, 2013). The alginate-NPs loaded with Fe/Zn using ethanol desolvation procedure had a regular spherical shape with a mean size of 90–135 nm. The NPs revealed a favorable EE (70%–85%) and a steady-state release pattern. Assessing the organoleptic and rheological characteristics of the final ice cream formulations also demonstrated that the addition of these NPs did not have any negative effect on the product quality. Soumya, Vineetha, Reshma, and Raghu (2013) scrutinized the effect of spherical NPs of sodium selenite (Na_2SeO_3)–guar gum (41–173 nm) prepared by nanoprecipitation method on cardiac H9c2 cells. The cell viability analysis revealed that use of NPs equal to 25 nM (10.3%) was safe, however, additional rise caused to a cytotoxicity initiation. A more amount in cell absorption and permeability for the produced Se-loaded NPs than the normal Se was also monitored. Moreover, the developed NPs provided a strong barrier against free radicals of hydroxyl brought DNA damage (Soumya et al., 2013). Soumya, Vineetha, Salin Raj, and Raghu (2014) reported the promising Se-loaded NPs also can significantly defend cardiomyoblast cells against ischemic toxicity. NPs of Na_2SeO_3-loaded CS/tripolyphosphate (TPP) were prepared by mild mixing Na_2SeO_3 and the NPs developed based on ionic gelation. Zhang et al. (2011) explained that application of these spherical NPs could improve Se maintenance in cells and reduce response of cellular sensitivity and DNA breakage when exposed to Se. Luo, Zhang, Cheng, and Wang (2010) found that creation of a zein coating on the NPs could lead to an increase in size (from 100–300 to 200–400 nm) and EE (from 60% to 95%), and a reduction in selenite-release rate (from 85% to 30% for 4 h in phosphate butter saline). Also, a higher antioxidant activity was observed for the fabricated NPs than selenite.

9.3.6 Ionotropic Gelation

In ionotropic gelation method, polyelectrolytes in the presence of ions can create crosslinks to develop hydrogel beads "gelispheres." Nature of these spherical crosslinked water-soluble polymeric materials provides the considerable gel-forming and swelling abilities into simulated bio-fluids with a high control in release of bioactive ingredients, such as minerals. A hydrogel bead can be produced by adding/dropping the biocomponent-loaded polymeric solution to the water solution containing polyvalent cations. A 3-dimensional framework under this condition is developed by incorporating/diffusing biocomponent into the gelispheres matrix (Patil, Kamalapur, Marapur, & Kadam, 2010).

El-Sayed, Hassan, Mervat, Awad, and Salama (2015) via ionic gelation with TPP anions formed complexes of CS–whey protein to produce $FeSO_4$-loaded NPs with the improved Fe-bioaccessibility. The surface charge and association and loading efficiencies of the NPs were significantly affected by pH. A positive-charge for Fe after its incorporation at low level of whey protein was resulted, while a negative charge at whey protein concentrations of 8 and 12% (pH = 5.5) was found. An exceptional Fe-bioavailability amount and an ideal stability regarding acidic and enzymatic degradation were also reported for the fabricated NPs (0–12 mg/g protein) after 6 h under the simulated GT conditions. Du, Niu, Xu, Xu, and Fan (2009) designed novel antibacterial NPs according to the ionic gelation between CS and Na–TPP (Fig. 9.11) and incorporated mineral ions of Cu^{2+} (121.9 nm), Mn^{2+} (102.3 nm), Zn^{2+} (210.9 nm), and Fe^{2+} (95.81 nm). Evaluating of in-vitro antimicrobial activity of the fabricated NPs showed that apart from Fe^{2+}-loaded NPs, the other NPs had an important impact on the activity reduction of *Salmonella choleraesuis*, *Escherichia coli*, and *Staphylococcus aureus* bacteria. This fact had a positive correlation with zeta-potential value so that Fe^{2+}-loaded CS-NPs had lower zeta-potential (+71.42 mV) than the CS-NPs entrapping the other ions (+75.74 to +88.69 mV). Overall, the Gram-positive *S. aureus* bacterium represented more resistance than the Gram-negative *S. choleraesuis* and *E. coli* bacteria against the produced CS-NPs loaded mineral ions.

Zhang et al. (2016) using ionic gelation procedure and β-CS NPs encapsulated catechins or complex of catechins-Zn and then evaluated their antimicrobial activities on *Listeria innocua* and *E. coli*. The prepared β-CS NPs

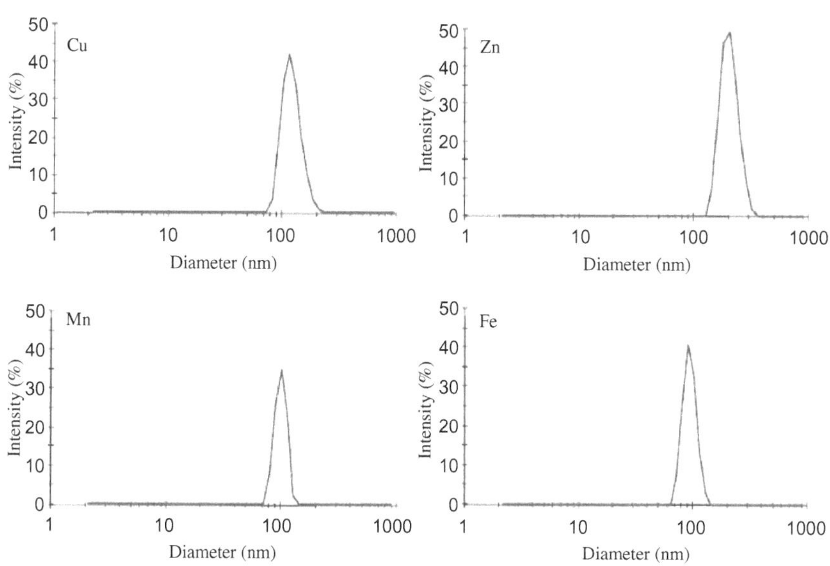

FIGURE 9.11 Particle size distribution of CS-NPs loaded Cu^{2+}, Zn^{2+}, Mn^{2+}, and Fe^{2+} prepared with ionic gelation mechanism. *Reproduced from Du et al. (2009) with permission of Elsevier.*

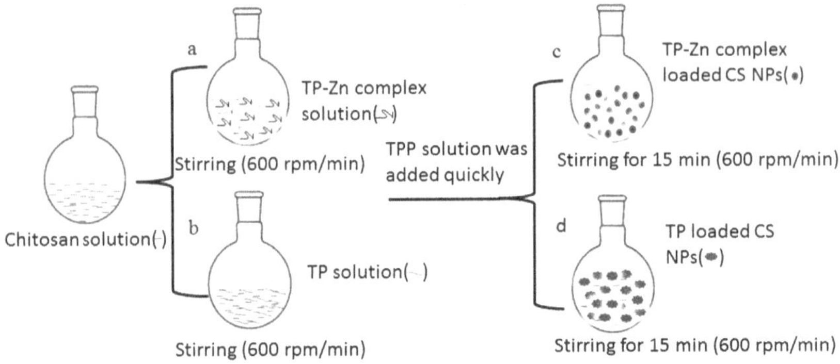

FIGURE 9.12 A pattern for the fabrication of TP–Zn complex and TP loaded CS-NPs. *Reproduced from Zhang and Zhao (2015) with permission of Elsevier.*

encapsulating catechins-Zn with lower particle size (208–591 nm) had a more potential to inactive the mentioned bacterial strains compared to ones with greater size. These NPs represented an excellent stability at acidic conditions with a pH range of 2.0–4.5. Thus, the developed new NPs can be used to incorporate into the food formulations or active packaging materials to improve the antibacterial capacity. Zhang and Zhao (2015) earlier using the same method prepared β-CS NPs (84.55 nm) to load tea polyphenol (TP)–$C_4H_6O_4Zn$ complex with an outstanding EE (97.33%) (Fig. 9.12). Antiradical scavenging capacities of TP–Zn complex-loaded β-CS NPs were much higher than those of β-CS NPs loading TP. The in-vitro release examinations in simulated GT conditions also showed that these NPs could tolerate pH values of 4.5 and 7.4 for 5.5 h. Therefore, presence of antioxidant nanocapsules in different drug and food supplement formulations can be applied to improve bioavailability and bioaccessibility of Zn and even tea polyphenols in the body.

The production feasibility of alginate NPs cross-linked with Zn^{2+} ($ZnCl_2$) was also effectively studied by the scrutinizing impact of solvent's ionic strength (0.05 M NaCl) and Zn content (Pistone, Qoragllu, Smistad, & Hiorth, 2015). A rise in ionic strength of 0.05 M NaCl and a stepwise increase in Zn amount could provide fine NPs with narrow monomodal distribution. Since an outstanding storage stability (70 days) was obtained for the prepared alginate NPs, an opportunity can arise to eliminate the additional processing steps (e.g., freeze-drying) during the NPs preparation.

9.3.7 Coacervation

So far, few studies have conducted on the mineral NPs fabricated using the coacervation approach. A promising stable delivery system for Fe (830–1070 nm) with ideal bioavailability level has been lastly designed based on the iron casein succinylate (ICS)–CS coacervate using complex coacervation process (Min, Cho, Song, & Kim, 2016). It was pointed out that the addition of 2% PEG 400 can significantly increase the physicochemical stability of the developed coacervates. According to the Caco-2 cell permeability tests, CS–ICS NPs also increased the Fe transport in comparison with other Fe complexes, such as $C_{12}H_{24}FeO_{14}$, Na–ferric gluconate, and ferric-hydroxide polymaltose.

9.4 APPLICATION OF ENCAPSULATED MINERALS IN THE FOOD INDUSTRY

With regard to the current disadvantages in fortifying food products with mineral salts through direct addition/mixing, application and incorporation of minerals encapsulated with a variety of coatings into food and drug formulations can provide unique benefits in developing novel functional products with improved physicochemical and sensorial characteristics. Majeed, Jamshaid Qazi, Safdar, and Fang (2013) pointed out that encapsulated mineral salts can have advantages, such as inhibition of interactions with substances present in the created matrix, discoloring avoidance, off-flavor reduction by masking of taste and smells, controlled release of the mineral components, perfect preservation in the production and storage processes, and improvement of the product's physical properties.

During the recent two decades, most studies of researchers have been concentrated on the mineral fortification of dairy products and table salt with encapsulated minerals especially Fe, Ca, Zn, and I. Fe encapsulated forms used as core materials in fortifying dairy products are electrolytic-Fe, $FeSO_4$, $FeSO_4 \cdot 7H_2O$, $C_4H_8FeN_2O_4$, $NH_4Fe(SO_4)_2$, and $C_6H_{10}FeO_6$ (Table 9.3). Encapsulated forms of two Ca-salts of tricalcium phosphate ($Ca_3(PO_4)_2$) and calcium citrate ($Ca_3(C_6H_5O_7)_2$) have also been applied to fortify soy-yogurt and soymilk (Table 9.3). The most popular encapsulated formulas of iodine and Fe for fortifying edible salts are KI and KIO_3, and, $C_4H_2FeO_4$ and $FeSO_4$, respectively. Nevertheless, some attempts have been made to develop fortified bakery products based on flours encapsulated with minerals as suitable vehicles. Hence, current industrial progresses and new opportunities in applying encapsulated Fe, Ca, Zn, and I in key food products will be discussed henceforward with more details.

9.4.1 Fortification of Dairy Products with Encapsulated Minerals

9.4.1.1 Milk

Although milk is one of the most important nutritious foods in the world, it has a very low content of Fe (Kwak, Yang, & Ahn, 2003b). Thus, use of encapsulated-Fe in fortifying milk would be an advantageous means of

TABLE 9.3 Fe- and Ca-Encapsulation for Fortifying Dairy Products and their Effect on the Quality and Bioavailability

Dairy Products	Mineral Type (Core) Wall Materials (WM)	Encapsulation Method	Special Mention of Determined Aspects	References
Milk	*Core*: $NH_4Fe(SO_4)_2$, $FeSO_4$ *WM*: Tween 80, Polyglycerol monostearate (PGMS)	Liposome Fatty acid esters (FAE)	An unfavorable sensory score for metallic taste and smell was obtained. FAE method was an applicable method for microencapsulating different Fe-salts into pasteurized milk	Abbasi and Azari (2011)
Milk	*Core*: $NH_4Fe(SO_4)_2$ *WM*: PGMS	Airless paint sprayer	A notable difference between capsulated and un-capsulated groups in sensory scores of astringency, metallic, color, and overall acceptability	Kwak et al. (2003b)
Milk	*Core*: $FeSO_4 \cdot 7H_2O$ *WM*: egg phosphatidylcholine liposomes, Na–alginate and Modified starch (Ms)	Liposome FAE Freeze-drying emulsificationEmulsification	Milk fortified with liposomes had an oily and unacceptable odor and taste (the lowest sensory scores) due to the inherent flavor of phosphatidylcholine	Gupta et al. (2015a)
Milk	*Core*: $FeSO_4$ *WM*: Arabic gum (AG) + Maltodextrin (MD) + MS	Modified solvent evaporation	Negative effects have been not reportedFe microcapsules fortified milk (63.78%) showed significantly higher in-vitro bioavailability of Fe as compared to control (unfortified, 19.86%) and Fe-salt-fortified milk (54.31%)	Gupta et al. (2015b)
Milk	*Core*: $FeSO_4$ *WM*: Medium-chain triglyceride from palm oil + whey protein isolate	W/O/W double-emulsion	The 0.1% (w/v) Fe microcapsules can be used for the production of the Fe-microcapsule-fortified milk without the deterioration of sensory characteristics	Chang et al. (2016)
Powdered milk	*Core*: $FeSO_4$ *WM*: Not reported	Spray-drying	Fe-fortified powdered milk can be produced from fluid milk fortified with microencapsulated $FeSO_4$ (SFE-171). The bioavailability of SFE-171 in the rat model was not changed by the manufacturing process	Lysionek et al. (2002)

Soymilk	*Core*: $C_6H_{10}CaO_6$ *WM*: Lecithin	Liposome	Not reported any negative quality effect	Hirotsuka et al. (1984)
Soymilk	*Core*: $Ca_3(PO_4)_2$ *WM*: Gelatin–agar	W/O/W double-emulsion	Encapsulation was an ideal Ca-fortification method in soymilk regarding higher stability of the product over pasteurization shelf life	Saeidy et al. (2014b)
Yogurt and Pasteurized milk	*Core*: $C_4H_8FeN_2O_4$; $C_6H_{10}FeO_6$; $FeSO_4$ *WM*: 50% vegetable fats	Spray-drying	The highest TBA and peroxide (PV) values were in ones fortified with $FeSO_4$ microencapsulate	Gilliard Nkhata (2013)
Yogurt	*Core*: $FeSO_4$ *WM:* Ca–alginate beads	Emulsification	Encapsulated whey protein chelated Fe with a high bioavailability can be added (up to 80 mg/L) without altering the accepted appearance and sensorial attributes	Subash et al. (2015) Subash and Elango (2015)
Yogurt	*Core*: $FeSO_4$ *WM:* Whey protein isolate gel	Gelation	Synthesis optimization of WPI-Fe particles to fortify yogurt. Releasing 95% of the particles in the intestinal condition (pH = 7.5)	Bagci and Gunasekaran (2016a)
Yogurt	*Core:* $FeSO_4$ *WM:* Whey protein isolate gel	Gelation	High maintenance of the physicochemical and sensorial traits during storageThe similar quality of fortified yogurt (60 mg Fe/kg) with the control	Bagci and Gunasekaran (2016b)
Yogurt	*Core*: $FeSO_4$, electrolytic-Fe *WM:* Na–alginate	Polymer complex	TBA and PV remained unchanged when yogurt was fortified with microencapsules containing Fe–whey protein complexThe yogurt fortified with un-encapsulated $FeSO_4$ had metallic taste	Azzam (2009)
Probiotic yogurt (*L. acidophilus*)	*Core*: $FeSO_4$ *WM*: not reported	Polymer complex	Oxidized flavor of Fe was suppressed and TBA absorption was low in the sample fortified with microencapsulated Fe-whey protein complex	Jayalalitha et al. (2012)

(*Continued*)

TABLE 9.3 (Continued)

Dairy Products	Mineral Type (Core) Wall Materials (WM)	Encapsulation Method	Special Mention of Determined Aspects	References
Drink yogurt	*Core*: $NH_4Fe(SO_4)_2$ *WM*: PGMS	Airless paint sprayer	TBA values remained unaffected in encapsulated Fe-fortified yogurt. However, sensory attributes (astringency/bitterness) of the fortified yogurt had significant difference compared with un-encapsulated one	Kim et al. (2003)
Soy yogurt	*Core*: $FeSO_4 \cdot 7H_2O$ + $Ca_3(C_6H_5O_7)_2$ *WM*: Partially hydrogenated lecithin	Spray-drying	Negative effects have been not reported	Cavallini and Rossi (2009)
Cheddar cheese	*Core*: $NH_4Fe(SO_4)_2$ *WM*: PGMS	Airless paint sprayer	Lower TBA in microencapsulated treatments during ripeningSensory aspects (bitterness/astringency/ sourness) were higher in Cheddar cheese fortified with microencapsulated Fe	Kwak et al. (2003a)
Feta cheese	*Core*: $FeSO_4$ *WM*: ~50% vegetable fats	Spray-drying	Fortification of cheese with 80 mg/kg microencapsulated Fe and 150 mg/kg L-ascorbic acid is technically feasibleA small increase in lipid oxidation was found by measuring TBA valueNo off-flavor was detected by trained sensory panelistsAscorbic acid had a hopeful impact on decreasing negative effects of Fe	Jalili (2016)

attaining more Fe intake. For instance, a lecithin liposome system was applied to enrich milk with microencapsulated $FeSO_4$ and the Fe bioavailability has been investigated. There was no significant decrease in Fe bioavailability of the fortified milk after the heat treatment and 6 month-storage. A similar Fe-bioavailability rate compared with its absorption from high-bioavailable $FeSO_4$ was also reported (Boccio et al., 1997; Uicich et al., 1999).

In another study by Kwak et al. (2003b), a system microencapsulating Fe based on PGMS as a coating material was designed to fortify milk. Albeit just low quantities of Fe (3%–5%) was in vitro released in simulated gastric fluid ($pH < 6.0$), a considerable enhancement in Fe release rate was occurred during 1 h-incubation in simulated intestinal fluid by rising pH from 5.0 (12.3%) to 8.0 (95.7%). The sensory analysis at 3-day storage revealed that there were no significant differences in most sensory traits except for metallic taste and color between samples of control and fortified with microencapsulated-Fe. Moreover, TBA value in the sample fortified with un-encapsulated Fe was higher than the encapsulated Fe (Kwak et al., 2003b). Abbasi and Azari (2011) also found the TBA value can be meaningfully increased in milk fortified with free Fe. No notable alteration in TBA of milk fortified with high loadings of microencapsulated-Fe was observed. Microencapsulation not only can decrease the rate of lipid oxidation by 60% but also can significantly mask the metallic taste of Fe in milk. However, sensory attributes (e.g., astringency or bitterness) of milk containing microencapsulated Fe were comparable to those of the control (Abbasi & Azari, 2011).

The fortifying feasibility of Fe microcapsules into milk and the effects on the physicochemical and sensory properties of the final products during storage time have been studied by Chang et al. (2016). Milk fortification with the Fe microcapsules at low levels (0.1–0.3% w/v) did not significantly vary TBA levels. The optimum content of Fe-microcapsule powder for the production of fortified milk was 0.1% (w/v), according to the obtained data from the pH, TBA, color, and sensory analysis during 16-day storage at 4°C (Chang et al., 2016). Lysionek et al. (2002) by investigating the Fe-bioavailability of microencapsulated $FeSO_4$ in a diet based on powdered milk in rats found that the Fe-bioavailability values were significantly higher than that of the control diet.

Gupta et al. (2015a) prepared Fe microcapsules using four different techniques and then selected three microcapsules for fortifying milk on the basis of better EE (62.97%–74.85%). The organoleptic scores of Fe-fortified milk containing Na–alginate and modified starch microcapsules (10 mg/L Fe) were highly similar with the control milk. In another investigation by Gupta et al. (2015b), Fe microcapsules with average size of 15.54 mm were produced by mixing GA, MD, and modified starch using a modified solvent evaporation method. Panelists gave lower sensory scores to the Fe salt-fortified milk compared with fortified milk with Fe microcapsules during 5-day cold storage. Also, the fortified milk with Fe microcapsules had significantly higher in-vitro Fe-bioavailability as compared to the control and Fe salt fortified milk (Gupta et al., 2015b). Also, Gilliard Nkhata (2013) after fortifying pasteurized liquid milk with Fe microcapsules obtained from three Fe salts investigated their sensory properties. It was proved that there were no substantial differences in appearance and flavor in all treatments. Nevertheless, the control milk and fortified one with $C_6H_{10}FeO_6$ showed no significant variations in taste. Overall, the best option to fortify pasteurized liquid milk was $FeSO_4$ microcapsulate.

Ca content of soya milk (12 mg/100 g soya milk) is a lot lower than cow's milk (120 mg/100 g). As a result, a lot of investigators have tried to overcome this nutritious defect through Ca-fortification of soymilk. Weingartner, Nelson, and Erdman (1983) by fortifying soy beverage with Ca salts ($Ca_3(PO_4)_2$ and $Ca_3(C_6H_5O_7)_2$) found that this process was ineffective because it led to an undesirable Ca–protein interaction as well as protein coagulation and precipitation. One year later, Hirotsuka et al. (1984) could fortify soymilk by entrapping $C_6H_{10}CaO_6$ into a lecithin liposome structure. They fortified soya milk with Ca (110 mg Ca/100 g soya milk) to obtain an equivalent Ca level with usual cow's milk. The fortified soymilk showed a high stability along with a high Ca-bioavailability for at least 1 week at 4°C (Hirotsuka et al., 1984). Saeidy et al. (2014b) using direct addition and microencapsulation technique fortified soymilk with Ca ($Ca_3(PO_4)_2$; 2000 mg/L). They added potassium citrate ($C_6H_5K_3O_7$; <30 g/L) as a metal-chelating agent to soymilk samples to inhibit Ca–protein interaction and improve soymilk stability. Nonetheless, mixture of $Ca_3(PO_4)_2$ and $C_6H_5K_3O_7$ led to a less stability in soymilk. They concluded that addition of encapsulated Ca can be an ideal alternative for the addition of $C_6H_5K_3O_7$ in order to attain a more stable soymilk enriched with high amounts of Ca.

9.4.1.2 Yogurt

Yogurt among different dairy products is an interesting medium for mineral fortification because most age groups, such as children, menstruating, pregnant or lactating women, and adolescents are main consumers with a high risk in Fe deficiency (Bagci & Gunasekaran, 2016a,b). Azzam (2009) fortified yogurt with $FeSO_4$ and, Fe–whey protein complex and its microencapsulated form. This researcher evaluated lipid oxidation and sensory characteristics of samples during 7-day cold storage. No considerable discrepancy in oxidation rate (TBA and peroxide) between the control and those fortified with Fe–whey protein complex, and microencapsulated Fe–whey protein complex was observed. Yet, a highly oxidized and metallic taste was distinguished for yogurt enriched by $FeSO_4$. Generally speaking, the finest yogurt from the perspective of sensory panelists was yogurt fortified with microencapsulated Fe–whey protein complex because this sample had a similar flavor and overall quality compared with the control.

Jayalalitha, Balasundaram, Palanidorai, and Naresh (2012) have recently fortified probiotic yogurt containing *Lactobacillus acidophilus* with Fe microcapsules based on Fe–whey protein complex. The fortified yogurts represented a satisfactory sensory quality with a notable decrease in the oxidized flavor intensity of Fe. High TBA values in samples fortified with un-encapsulated Fe can be attributed to the interaction of Fe with casein proteins of milk, the pro-oxidant activity of oxygen species, and activated lipid oxidation. TBA level under the oxidative conditions was increased by accumulating free fatty acids. Gilliard Nkhata (2013) also evaluated the organoleptic quality of yogurts enriched with Fe microcapsules made of three commercial Fe-salts. The sensory test was carried out after a 7-day storage and showed that there were no significant differences in scores of appearance, mouthfeel, flavor, and overall acceptability among the samples of control and fortified with $FeSO_4$ microcapsules. However, an obvious difference was identified between control yogurt and yogurts fortified with $C_4H_8FeN_2O_4$ and $C_6H_{10}FeO_6$. Yogurt fortified with $FeSO_4$ had the highest values of TBA and peroxide. The samples of control and enriched by $C_6H_{10}FeO_6$ also had the lowest TBA and peroxide values, respectively. He concluded that $FeSO_4$ microcapsules can be the best choice to fortify yogurt (Gilliard Nkhata, 2013).

Subash, Elango, Pandiyan, Karthikeyan, and Kumaresan (2015) and Subash and Elango (2015) developed novel yogurts fortified with microencapsulated whey protein-chelated Fe (Fe–WP) and compared their starter survival rate, oxidation speed, and sensory properties with control and Fe-free ones. There was no noteworthy variation in starter bacteria count in the control and various Fe-fortification treatments on 0, 1, 2, and 3 weeks. But a falling trend by extending the storage was observed in number of these bacteria in control and Fe-fortified yogurt. As a result, the bacteria viability in yogurt was not affected by the fortified Fe. The TBA values of control and encapsulated Fe-fortified yogurts were significantly lower than the fortified yogurts with free Fe. Significant variances in scores of astringent and oxidized flavor and overall acceptability were observed between control and different fortification treatments of yogurt during 21-day storage. Finally, they recommended that addition of 80 mg Fe-WP/L to yogurt can meaningfully increase Fe-bioavailability without any change in the accepted appearance and organoleptic aspects (Subash & Elango, 2015; Subash et al., 2015).

Kim et al. (2003) studied the chemical and sensory features of yogurt drink during 20-day storage as affected by the fortification with Fe microencapsulated. It was revealed that Fe fortification did not affect the fermentation time required for the yogurt drink to attain pH 4.2, while the incorporation of un-encapsulated Fe led to a decrease in pH and an increase in titratable acidity during storage. Microbial analysis demonstrated that fortification with encapsulated and un-encapsulated Fe did not affect the microbial count of yogurt drink. A lower kinetic rate for TBA in encapsulated treatments compared to un-encapsulated ones was also recorded. Though a substantial change in astringency and bitterness was found in yogurt drink fortified with unencapsulated Fe, encapsulated-Fe successfully masked off the taste and flavor of Fe. Stability assessment of soy yogurts enriched with Fe ($FeSO_4 \cdot 7H_2O$, 12 mg of Fe/L) and Ca ($Ca_3(C_6H_5O_7)_2$, 600 mg of Ca/L) during 28-day storage at 10°C has also been investigated (Cavallini & Rossi, 2009). Although a low viscosity for the fortified sample was measured, this parameter did not have any considerable change by prolonging the storage period. Moreover, there were no significant differences related to the all sensory properties and acidity amounts during the investigated storage time.

The physicochemical and sensory quality characteristics of Fe-fortified yogurts during 14-day storage have been recently investigated by Bagci and Gunasekaran (2016b). The yogurts were fortified by direct addition of $FeSO_4$ solution (20–60 mg Fe/kg) and by gel powder of Fe-encapsulated cold-set WPI (WPI-Fe). Although yogurts enriched with direct addition of $FeSO_4$ exhibited obvious negative impacts even at the lowest fortification amounts, the quality properties (in particular, color and flavor) of yogurt fortified with WPI-Fe particles (up to 60 mg Fe/kg) were similar to those of unfortified control sample. Results acquired from the physicochemical and sensory evaluations showed that the quality parameters of samples fortified with WPI-Fe can be well maintained throughout 2 weeks of storage independently of the used Fe concentration. It was also claimed that a Fe-enriched typical yogurt (~225 g) can provide women's daily Fe requirement (up to 60%), without any obvious deviations in color and flavor in comparison with the control yogurt. This great attainment would be impossible with direct addition of Fe to yogurt even at much lower levels of fortification (Bagci & Gunasekaran, 2016b). Bagci and Gunasekaran (2016a) previously optimized the effects of three independent variables involved in the Fe-WPI synthesis on EE and HunterLab color characteristics using a RSM-central composite design with a second-order polynomial model. The certified optimum parameters for preparing WPI-Fe particles were pH 7.0, WPI content of 6.8%, and Fe concentration of 18.8 mM. The in-vitro GI studies revealed that WPI-Fe particles prepared under the optimum conditions can be an excellent network for site-specific delivery of Fe because there was a high release rate (95%) for the particles in the intestinal condition (provided with pancreatin, pH 7.5). Nevertheless, findings of Bagci and Gunasekaran (2016a) indicated that only approximately 28% of the Fe-encapsulated powders in the gastric condition (created with pepsin, pH 1.2) could be released.

9.4.1.3 Cheese

Cheese has a very low content of Fe similar to milk. Most of the performed studies to fortify cheeses with encapsulated-Fe have been focused on Cheddar cheese. According to the data reported by IDFA (2015), 28.5% of the total cheese consumption in the United States alone belonged to consumption of Cheddar one.

The earliest explorations on cheese enrichment were conducted by Zhang and Mahoney between 1988 and 1999. These investigators fortified Cheddar cheese with four Fe sources including $FeCl_3$ or Fe–casein, ferripolyphosphate–whey protein (FIP–WP), and Fe–WP complexes during the early steps of the manufacturing process. In this research, $FeSO_4$ was used as the reference Fe source. A slight increase in TBA level of Fe-fortified cheeses was determined, but this amount was comparable with that of the unfortified cheeses. There were low correlation coefficients between Fe level and oxidized off-flavor, cheese flavor, or TBA value. Besides, a weak correlation was found between TBA value and oxidized off-flavor or flavor in cheese. The aging process of cheese up to 3 months did not affect TBA values or oxidized off-flavor and cheese flavor scores. Their results also showed that the maximal Fe bioavailabilities for $FeCl_3$ or complexes of Fe-casein, FIP-WP, and Fe-WP were 85%, 71%, 73%, and 72%, respectively ($P > 0.05$). Basal Fe-bioavailabilities for 10-day- and 14-day feeding of the corresponding fortification of Fe sources were 5% and 4%, 8% and 4%, 6% and 3%, and 7% and 3%, respectively ($P > 0.05$). Nevertheless, maximal and basal bioavailabilities for $FeSO_4$ were 85% and 5%, respectively. The best Fe-recovery percent reported for $FeCl_3$ and Fe microencapsulated into WP-complex, but these sources accelerated lipid oxidation reactions (Zhang & Mahoney, 1988, 1989). Zhang and Mahoney (1990) repeated their past fortification method in Cheddar cheese, but added WP–$FeCl_3$ salt to their list of exanimated Fe-salts. $FeSO_4$-fortification induced oxidized- and off-flavors for Cheddar cheeses aged for 5 months. However, the most acceptable result for sensory perception was in relation to the samples fortified with Fe–WP complexes.

Kwak et al. (2003a) designed a study to determine the addition effect of encapsulated-Fe on the critical chemical and sensory attributes of Cheddar cheese. The facts obtained from this research indicated that TBA value was significantly lower in encapsulated chesses than those un-encapsulated cheeses during the ripening period. Yet, short-chain FFAs and neutral volatile compounds in the diverse experiments were insignificantly produced during the ripening. The enriched samples with encapsulated Fe had the highest scores for negative sensory traits of bitterness, astringency, and sourness. These results demonstrated that Fe-fortification did not cause any defect in the quality of Cheddar cheese. Hence, it is possible to enrich Cheddar cheese with Fe for achieving a high bioavailability.

Arce (2016) fortified Cheddar cheese with LMFS (large microencapsulated $FeSO_4$, 700–1000 μm) or SMFS (small micro-encapsulated $FeSO_4$, 220–422 μm). A successful Fe recovery for LMFS (66%) and SMFS (91%) after 90-day aging was acquired. No considerable variances in the lipid oxidation rate and content of chemical composition (e.g., dry matter, fat, protein, ash, Mg, Zn, and Ca) were observed. Since microencapsulation of $FeSO_4$ could not mask distinct taste, odor, and color of Fe, this process negatively affected the sensory properties of Cheddar cheese. In general, Fe-retention and sensory traits of Cheddar cheeses fortified with SMFS were a little better than those of LMFS.

Rice and McMahon (1998) fortified Mozzarella cheese using $FeCl_3$, and WP–Fe and casein–Fe salts (25 and 50 mg Fe/kg). Even though fortification operation with 50 mg Fe/kg did not affect its apparent viscosity, color, and lipid oxidation values of Mozzarella cheeses, the trained panelists assessed them with off-odor, metallic/oxidized flavor, and other unwanted flavors. Therefore, a low consumer preference was resulted for all the Mozzarella formulations owing to the tangible metallic and off-flavors. Overall, some interesting properties in Cheddar cheese were arisen with incorporation of $FeCl_3$ and Fe–protein compounds, but not in Mozzarella cheese, representing that study on Fe-fortification cannot be generalized to all kinds of cheeses.

Jalili (2016) have recently fortified Feta cheese with 80 mg/kg Fe compounds ($FeSO_4$, $FeCl_3$, and microencapsulated $FeSO_4$) with or without vitamin C (150 mg/kg). Chemical composition and FFAs of control samples and fortified ones with Fe or Fe–vitamin C did not have any significant difference. However, a remarkable variation in Fe content was found between Fe or Fe–vitamin C fortified cheeses and the control ones. The lowest TBA value and the best organoleptic scores were for Feta cheeses fortified with 80 mg/kg microencapsulated Fe and 150 mg/kg of vitamin C.

9.4.2 Salt Fortification With Encapsulated Iron and Iodine

9.4.2.1 Dual Fortified Salt

Since Fe compounds are rather more stable than iodine compounds, fortifying salt with Fe and iodine was primarily evaluated by encapsulating iodine to create a protecting partition (Table 9.4).

Diosady et al. (2002) developed DFS containing iodine (KI and KIO_3, 50 mg/kg) microcapsules made by spray-drying and FBC, mixed in a blender with either $C_4H_2FeO_4$ or $FeSO_4$ (1 g/kg Fe). The fortificants stability was then investigated in diverse formulations as a function of storage temperature and RH. The best result among the used dissimilar barriers to encapsulate iodine was for dextrin with a wall:core ratio of 1:200. The whole content of iodine in DFS fortified with $FeSO_4$ and KI lost during 30-day storage at 40°C and 60% RH, while a 93% loss for iodine with KIO_3 under the equal conditions was monitored. A percentage of 80.9 iodine in DFS containing KIO_3 and $C_4H_2FeO_4$ kept at ambient temperature and 79.7% after 7-month storage at 40°C and 60% RH. The maximum of iodine maintenance (98.4%–101.9%) was acquired in spray-dried microcapsules of $C_4H_2FeO_4$ and KI during 12-month storage at 40°C and 60% RH. These researchers

TABLE 9.4 Fe-Fortification of Iodized Salts using Mineral Microcapsules and their Effect on the Quality and Bioavailability

Salt Type	Mineral Type (Core)[a] Wall Materials (WM)	Encapsulation Method	Special Mention of Determined Aspects	Reference
Dual fortified salt	*Core*: $(C_4H_2FeO_4)$, KIO_3 *WM*: soy stearine, Hydroxypropyl methylcellulose (HPMC)	Spray-drying	Microencapsulation of $C_4H_2FeO_4$ had little effect on in-vitro Fe-bioavailability: more than 90% of Fe in the premixes was released during 2-h digestion in the simulated gastric fluid. In-vivo tests in rats have confirmed that the $C_4H_2FeO_4$ microencapsulated into a lipid is highly bioavailable, with a bioavailability of 95% relative to $FeSO_4$	Li et al. (2009)
Dual fortified salt	*Core*: $(C_4H_2FeO_4)$; KIO_3 *WM*: HPMC(10% w/w)	Extrusion agglomeration	Fe-bioavailability will not be reduced by this technique because the Fe^{3+} encapsulated in a readily digestible polymeric coating was stable	Li et al. (2010)
Dual fortified salt	*Core*: $(C_4H_2FeO_4)$; KIO_3 *WM*: CMC; maltodextrin (DE = 7)Arabic gum; HPMC	Spray-drying	Negative effects have been not reportedThe in-vitro bioavailability of the Fe on salt was acceptable. About 80% of the Fe dissolved in simulated stomach acid within 2 h	Romita et al. (2011)
Dual fortified salt	*Core*: $(C_4H_2FeO_4)$; KIO_3 *WM*: HPMC	Fluidized-bed agglomeration followed by lipid coating	A limited impact on the oxidation state of $C_4H_2FeO_4$ by the extrusion process led to a little Fe^{2+}–Fe^{3+} conversion in the extruded Fe particles, hence maintaining high Fe digestibility/bioavailability	Li et al. (2011)
Dual fortified salt	*Core*: $(C_4H_2FeO_4)$; KI, KIO_3 *WM*: Fully hydrogenated soy stearine	Spray-drying	Microcapsulation process can protect both I^- and the Fe^{3+} during distribution and retail in typical tropical conditions in Kenya's highlands and humid lowlands	Oshinowo et al. (2004)

(Continued)

TABLE 9.4 (Continued)

Salt Type	Mineral Type (Core)[a] Wall Materials (WM)	Encapsulation Method	Special Mention of Determined Aspects	Reference
Dual fortified salt	*Core*: ($FeSO_4$, $FeSO_4$); KIO_3 *WM*: Eudragit EPO, chitosan (CS)	Spray-drying	CS was suitable for producing Fe-premix for stable salt double fortified with iodine and Fe	Dueik and Diosady (2016)
Dual fortified salt	*Core*: ($C_4H_2FeO_4$); KI, KIO_3 *WM*: Modified starch (dextrin 7071), Gelatin, Hydrogenated mixture of mono- and di-glycerides	Spray-dryingFluidized bed drying	Spray-drying and was the most effective technique and encapsulating agent for the microencapsulation of iodine	Diosady et al. (2002)
Dual fortified salt	*Core*: ($FeSO_4 \cdot H_2O$, $C_4H_2FeO_4$, $Fe_4O_{21}P_6$, Electrolytic-Fe); KIO_3, KI *WM*: Stearic acid, Dextrin, Partially hydrogenated soybean/palm oil, Edible wax matrix, Cellulose and its derivatives	Spray-cooling	For most compounds, encapsulation did not protect against adverse sensory changes and iodine losses$C_4H_2FeO_4$-fortified salts were relatively more stable to iodine losses, but were related with unacceptable reddish brown color changes	Wegmüller et al. (2003)
Dual fortified salt	*Core*: ($C_4H_2FeO_4$); KIO_3 *WM*: HPMC	Extrusion agglomeration	All Fe premix formulations showed high particle density, good bioavailability and acceptable sensory attributes	Yadava et al. (2012)
Triple fortified salt	*Core*: (I (KIO_3), Fe ($Fe_4O_{21}P_6$), vitamin A ($C_{36}H_{60}O_2$)) *WM*: Hydrogenated palm oil containing 1% lecithin	Spray-cooling (spray chilling)	At 10 months, prevalence of vitamin A deficiency and Fe deficiency anemia was notably lower in the triple-fortified salt (TFS) group than in the iodized salt group	Zimmermann et al. (2004)
Triple fortified salt	*Core*: ($KIO_3 + Fe_4O_{21}P_6 + C_{36}H_{60}O_2$) *WM*: Hydrogenated palm fat	Spray-cooling	Fe, iodine, and vitamin A microcapsules fortified in salt showed no change in color and overall acceptability of TFS was good	Wegmüller et al. (2006)

[a]*Core material(s) has/have been placed in parenthesis.*

pointed out that although great progress had been made in chemical stability of iodine, encapsulation of this essential mineral was unfeasible because the produced microcapsules were very fine so that could easily separate from the much greater salt particles. Besides, a very extensive investment had already been made in salt iodization via spraying KIO_3 solution onto salt. Diosady et al. (2002) finally concluded that encapsulation of Fe-compounds in salt structure would be a more attractive achievement.

Diosady, Yusufali, Oshinowo, and Laleye (2006) also produced a DFS with un-encapsulated iodine and encapsulated Fe and then studied the storage stability of both fortificants under the environmental conditions in the retails placed in the highland (Nairobi) and costal (Mombasa) regions of Kenya. Results showed that the DFS with encapsulated $C_4H_2FeO_4$ and KI premix maintained 92% Fe and 90% iodine in Mombasa, whereas the mean corresponding values in Nairobi were 87 and 86%, respectively. Since encapsulated Fe in DFS avoids interactions between the used fortificants, the prepared DFS with $C_4H_2FeO_4$ premix was found resistant to color variation and can considerably help to enhance the bioavailability rate. Zimmermann, Zeder, Chaouki, Torresani, and Hurrell (2003) earlier demonstrated a decrease in undesirable sensory attributes and iodine loss along with maintaining significant bioavailability level could attain by placing a barrier around Fe via mineral encapsulation in DFS.

Wegmüller, Zimmermann, and Hurrell (2003) performed a great research work on the stability of indigenous salts (in Morocco and Côte d'Ivoire) dual fortified with iodine and 19 encapsulated- and unencapsulated-Fe compounds (e.g., fumarate, sulfate, pyrophosphate, and elemental Fe). Iodine content and color attributes of the dissimilar stored samples were analyzed during 6 months. Findings showed that the encapsulation process could not protect most compounds against iodine losses and unwanted organoleptic changes during the storage. Nonetheless, $Fe_4O_{21}P_6$ ($\sim$2.5 and 0.5 μm) particles were suitable to enrich salt. A case study on the storage stability of table DFS with Fe ($C_4H_2FeO_4$), and iodine (KI and KIO_3) in the mountainous and coastal areas of Kenya was conducted. In addition, high iodine (up to 90% or more) preserved after 3-month storage (Oshinowo, Diosady, Yusufali, & Laleye, 2004). They then prepared a DFS by combining encapsulated $C_4H_2FeO_4$ with either iodated salt or salt iodized with encapsulated-KI. A high retention in Fe^{2+} amount was obtained so that a slight amount of Fe^{2+} got oxidized to Fe^{3+} (<17%). The DFS having iodine and encapsulated $C_4H_2FeO_4$ was highly stable for 3-month storage because both iodine and Fe^{2+} were protected during distribution and retail in the studied environmental conditions.

Li and his colleagues during 2009–2011 conducted three distinct studies on in-vitro and in-vivo Fe-bioavailability and storage stability of iodine DFS fortified with free and encapsulated-$C_4H_2FeO_4$. In the initial investigation, they produced a premix containing $C_4H_2FeO_4$ for incorporation into salt by agglomerating $C_4H_2FeO_4$ into salt-size particles followed by encapsulation process. In-vitro bioavailability of the dissimilar formulas of Fe in DFS and encapsulated $C_4H_2FeO_4$ premixes prepared by various methods was subsequent scrutinized. For as much as the Fe release quantity in the premix formulations during 2-h digestion in the simulated gastric fluid was more than 90%; thus, the in-vitro Fe-bioavailability was slightly affected the materials and techniques applied in encapsulating $C_4H_2FeO_4$ (Li, Diosady, & Wesley, 2009). They found that use of TiO_2 as a masking agent in mixture with the coating solids can successfully hide reddish-brown color of $C_4H_2FeO_4$ and subsequently led to the improved overall sensory acceptability and iodine stability in the final product. Moreover, evaluation of in-vivo Fe-bioavailability in rats also revealed that the $C_4H_2FeO_4$ encapsulated into a lipid structure is extremely bioavailable, as bioavailability of this Fe salt compared to $FeSO_4$ determined about 95% (Li et al., 2009).

In the second investigation by Li et al. (2010), these researchers evaluated the iodine stability in iodized salt fortified with encapsulated Fe ($C_4H_2FeO_4$) developed by extrusion-based encapsulation method. The optimum formulation with extruded $C_4H_2FeO_4$ (10%) in order to achieve the lowest loss in iodine content (15%) composed of 30% (w/w) binder, 25% (w/w) TiO_2, 10% (w/w) water-soluble polymer of Methocel. In conflict, the uncoated Fe particles after their extruding showed a 50%–65% loss in iodine content of DFS. So, this practical method will deliver iodine and Fe in DFS through the normal salt distribution system. In the final research by Li et al. (2011), a possibility and optimization study on cold-forming extrusion process in a pilot scale was performed to agglomerate and encapsulate $C_4H_2FeO_4$ in DFS. Grain flours thus were interesting to generate extrudable doughs incorporating high amounts of $C_4H_2FeO_4$. Besides, an excellent in-vitro Fe-digestibility for all extruded Fe particles (300–700 μm) was found.

Yadava et al. (2012) using an extrusion agglomeration procedure prepared a Fe premix to fortify salt and after that optimized stages of the color-masking and polymer coating. Having mixed the Fe premix with iodized salt, the stability of iodine and Fe^{2+} during 90-day storage at 35°C and 60% RH was scrutinized. A high efficacy for water-soluble polymeric coatings to keep the micronutrients at the 10% encapsulation capacity was determined. Furthermore, all Fe premix formulations confirmed high bioavailability and desirable sensory attributes.

Romita et al. (2011) produced spray-dried encapsulated Fe ($C_4H_2FeO_4$) powders to incorporate into coarse iodized salt. $C_4H_2FeO_4$ was encapsulated in a 10% w/v suspension plus in the dissolved form (1.2% w/v) using several biopolymers. DFS was prepared by blending Fe capsules (1000 ppm) with iodized salt (100 ppm iodine), and the iodine stability in DFS was intermittently analyzed for 6 months when stored at room temperature (20% RH), and in severe environmental conditions (40°C and 60% RH). It was not surprising that iodine stability of the unencapsulated samples (60%) was lower than all the encapsulated samples (80%–90%) in DFS. As well, $C_4H_2FeO_4$ in the suspended and dissolved states oxidized 4% and 55%, respectively (Romita et al., 2011). Dueik and Diosady (2016) have recently designed a stable DFS containing iodine and spray-dried Fe microcapsules (~10 μm) using CS as an ideal covering agent. The developed microparticles could be properly attached to the surface of coarse salt when the initial moisture content of salt was 2.4%. Iodine content also retained 90% at 25°C and 70% at 45°C after 12-week storage. There was a high retention for Fe from encapsulating particles at pH 7; however, a high release level was determined at pH 1.

9.4.2.2 *Triple Fortified Salt*

Fortification programs in terms of the simultaneous addition of three essential micronutrients (Fe, iodine, and vitamin A) to a single foodstuff can be implemented with a low cost. Strong interactions among these components in metabolism can notably be effective in triple fortification of salt. Zimmermann et al. (2004) encapsulated three functional ingredients of KIO_3, micronized $Fe_4O_{21}P_6$, and retinyl palmitate into HPF to form a TFS containing 30 μg I, 2 mg Fe, and 60 μg vitamin A/g salt. A loss in iodine and vitamin A values (~12%–15%) from the color-stable TFS after 6-month storage was determined. Moreover, deficiency prevalence of vitamin A and Fe anemia were significantly lower in the group fed with TFS compared with the group fed with iodized salt at 10 months. Wegmüller et al. (2006) developed microcapsules comprising three ingredients of $Fe_4O_{21}P_6$, KI, and retinyl palmitate (vitamin A) in HPF and added them in salt to produce TFS. Fig. 9.13 shows images of the formed microcapsules using light and scanning electron (SEM) microscopy. Results showed that not only color changes in the TFS during 6-month storage were acceptable, but also iodine losses were around 20% similar to the iodized salt. Wegmüller et al. (2006) also reported a high stability for retinyl palmitate only with losses about 12% after 6-month storage. No significant differences in the overall sensory acceptability between TFS and iodized salt were also explored.

9.4.3 Use of Encapsulated Iron in Fortifying Cereals and Bakery Products

Jakel and Belshaw (1971) for the first time developed a preliminary process for $FeSO_4$ encapsulation into baking flour mixers. The produced fine-sized and white-color powders with free-flowing characters could highly keep during 6-month storage without any negative quality change. In later years, an investigation (Kongkachuichai, Kounhawej, Chavasit, & Charoensiri, 2007) on the production feasibility and qualities of physicochemical and sensory of instant noodles fortified with various forms of Fe fortificants ($FeSO_4$, ferric sodium ethylenediaminetetraacetate (NaFeEDTA), and encapsulated elemental Fe) was

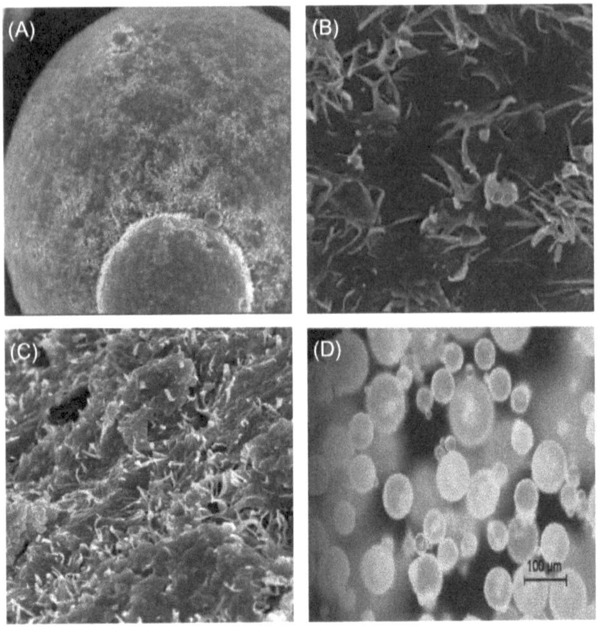

FIGURE 9.13 Surface (A, 700× magnified; B, 5000× magnified) and inner (C, 700× magnified) structure of sprayed microcapsules containing $Fe_4O_{21}P_6$, KI, and vitamin A analyzed by SEM and the microcapsule image obtained from LM (D, 100× magnification). *Reproduced from Wegmüller et al. (2006) with permission of Wiley and Sons.*

performed. Data showed that the noodles enriched with encapsulated-Fe (~5 mg Fe per serving of noodles) had a similar sensory score in overall acceptability with the unfortified and other enriched noodles. In addition, Kongkachuichai et al. (2007) realized that the oxidative (peroxide value) and physical (color) parameters were not affected by all the Fe-fortificants during 3-month storage at ambient temperature.

Biebinger et al. (2009) fortified wheat flours with the small-dense microcapsules of $FeSO_4$ and KIO_3 formed using wall material of HPF. The prepared microcapsules could potentially dominate adverse changes of sensory characteristics and Fe-deficiency of biscuits formulated with wheat flour. The mean iodine loss was ~25% during spray-cooling process; however, no assessable iodine losses in the baking process were produced. They reported that the efficiency of fortified wheat-based biscuits was compared to a non-fortified control by using a randomized feeding trial in young Kuwaiti women. There were no significant differences in taste and color attributes of the Fe microcapsule-fortified and unfortified biscuits. Fe-deficiency prevalence about of ~50% was reduced by consuming the fortified ones. Serum ferritin and urinary iodine levels also increased in group receiving the fortified samples. The Fe absorption from the encapsulated $FeSO_4$ was measured bout 11% (Biebinger et al., 2009), which was similar to the data obtained by Zimmermann et al. (2005), who investigated the effectiveness of wheat-based snacks fortified with un-encapsulated $FeSO_4$, electrolytic Fe, or hydrogen-reduced elemental Fe in averting Fe-deficiency occurrence in Thai women. They also reported that the absorption of administered Fe dose after 20 weeks was about 11%. The Fe bioavailability rate could not be affected by its encapsulation. Hurrell et al. (2010) also suggested that the encapsulated $FeSO_4$ or $C_4H_2FeO_4$ can be mixed to low-extraction ($\leq$0.8% ash) wheat flours at the same levels with the nonencapsulated compounds.

Kiskini, Kapsokefalou, Yanniotis, and Mandala (2010) have recently used both the form of Fe (soluble, insoluble, or encapsulated) and the bread (wheat or gluten-free) to scrutinize the effect of Fe ($Fe_4O_{21}P_6$, $FeSO_4$, NaFeEDTA, and reduced Fe) and gluten on critical quality properties of the final products. The Fe fortification significantly affected the physical properties of wheat bread, such as color, crust firmness, specific volume, cell number and uniformity, and also aroma sensory trait. Nonetheless, a considerable difference was observed between unfortified and Fe-fortified gluten-free breads in terms of color, crust firmness, cell number, odor, metallic taste, and stickiness characteristics. Furthermore, the encapsulation seemed to be inadequate in protecting Fe from being oxidized, as wheat and free-gluten breads fortified with encapsulated $FeSO_4$ meaningfully differed from unfortified ones as far as color, taste, and texture are concerned. Souto, Brasil, and Taddei (2008) earlier evaluated acceptability of bread fortified with encapsulated $FeSO_4$ by children of daycare centers of Sao Paulo, Brazil. Results showed that the children's acceptance of the Fe-fortified breads was considerably lower than that of unfortified samples. However, the fortified breads might be a practicable option in the inhibition of Fe-deficiency anemia in children.

Moretti et al. (2005) using a single-screw extruder developed simulated rice grains from a paste composed of rice flour, Fe compounds (0.5 and 1.0 g Fe/100 g), and 25% water during extrusion processes. Fe components at a ratio of 1:100 or 1:200 with natural rice grains were mixed to reach Fe concentration in the final product by 5 mg Fe/100 g. Incorporation of Fe [micronized dispersible $Fe_4O_{21}P_6$ ($\sim$0.5 μm), reagent-grade $Fe_4O_{21}P_6$, $Fe_4O_{21}P_6$ ($\sim$2.5 and $\sim$20 μm), electrolytic Fe, encapsulated $FeSO_4$ in liposome and HPF, and $FeSO_4$ (positive control)] into the rice grains was performed via cold extrusion process. The simulated rice seeds were nightly dried to give a $\sim$10% water content. The all Fe constituents except micronized $Fe_4O_{21}P_6$ (0.5–20 μm) led to a strong color dissimilarities in the uncooked and cooked rice grains. The minimum level of Fe loss ($\leq$2%) in washing stage for 30 min at 30°C earlier cooking was for the fortified rice seeds with $Fe_4O_{21}P_6$ and elemental Fe. Moretti et al. (2005) could finally produce Fe-fortified extruded rice seeds with excellent sensory characteristics and high bioavailability rate based on micronized $Fe_4O_{21}P_6$ in order to prevent/control Fe-deficiency in schoolchildren in India.

Li and his colleagues in two separated studies developed Ultra Rice having encapsulated $C_4H_2FeO_4$, micronized $Fe_4O_{21}P_6$, FeNaEDTA, thiamine, and dissimilar antioxidant blends. They proved that Fe-incorporation into Ultra Rice had no effect on the rates of Fe bioavailability and stability. In addition, the simulated rice seeds having the encapsulated $Fe_4O_{21}P_6$ had a satisfactory creamy-yellow color with more than 50% loss in thiamine quantity. However, use of unencapsulated $C_4H_2FeO_4$ led to an unacceptable reddish-brown color. The darker colored seeds with a higher in-vitro bioavailability were also produced by the both ingredients of FeNaEDTA and encapsulated $C_4H_2FeO_4$. Usage of $Fe_4O_{21}P_6$ particles or colloids could diminish thiamine losses and provided the physical and organoleptic characteristics during 32-week storage at 60% RH and 40°C. Ultra Rice formulations having FeNaEDTA had the least loss of thiamine, but oxidative rancidity was arisen. The most useful free-radical scavenging constituents in maintaining thiamine in both uncoated and encapsulated Fe during the storage, respectively, were BHA and BHT, and hydrophilic citric acid and sodium hexametaphophate (Li, Diosady, & Jankowski, 2008a,b).

Hotz et al. (2008) previously observed an 80% reduction in anemia and a 29% reduction in Fe deficiency prevalence in young Mexican women getting a daily portion of cooked Ultra Rice containing micronized $Fe_4O_{21}P_6$. An effective decrease in Fe deficiency and anemia in young Brazilian children (6–24-month old) by consuming Ultra Rice fortified with micronized $Fe_4O_{21}P_6$ (3.14 μm) was also reported by Beinner, Velasquez-Meléndez, Pessoa, and Greiner (2010).

9.4.4 Encapsulated Minerals in Fortifying Other Foods

Kim et al. (2006) developed encapsulated-Ca in liposome (L-Ca) with EPC and injected it into rabbit before slaughter to evaluate its effect on the meat ageing. L-Ca injection into rabbit could effectively reduce the meat ageing time without causing any contamination and/or physical shock. Choi, Decker, and McClements (2009) prepared a W/O/W emulsion to encapsulate Fe in the inner aqueous layer to inhibit oxidation rate. An insignificant generation of TBA reactive substances (TBARS) and a high Fe-EE (99.75%) were observed after preparing a water-in-corn oil emulsion with Tween 60 emulsifier. An emulsion based on fish oil was subsequently developed and mixed with the first emulsion droplets in order to examine the Fe impact on the stability of fish oil. Findings showed a rise in TBA values by interacting Fe with fish oil and by activating oxidation mechanisms.

Porrarud and Pranee (2010) by spray-drying method encapsulated natural green colorants of Zn–chlorophyll derivatives extracted from pandan leaf with three different coating materials (GA, Ms, and MD). The SEM micrographs illustrated spherical and smooth particles for powders encapsulated with wall material of Ms, while powders produced by encapsulating agents of MD and GA showed the surface shrinkage (Fig. 9.14). According to the results obtained from physicochemical and stability studies by Porrarud and Pranee (2010), the maximum greenness, total chlorophyll, and antioxidant levels were for Zn–chlorophyll microcapsules made of 30% Ms. Based on the first-order kinetic model, the resulted powder exhibited an extended predicted half-life (462 days) compared with the formed powders with GA (330 days) and MD (385 days).

Ferreira et al. (2011) investigated suitability of black beans fortified with Fe microparticles. These scientists formulated three samples of stewed black beans namely control sample without any encapsulated $FeSO_4$ (1), and bean samples with 5 (2) and 10 mg (3) encapsulated-Fe added to each portion spoon. Although there was a slight alteration between treatments (2) and (3), a better acceptability was found for the samples of control and treatment (1). Fantastic sufficiency of samples fortified by spray-dried $FeSO_4$ microencapsules could introduce them as an ideal candidate in order to prevent and/or control Fe-deficiency anemia.

Blanco-Rojo et al. (2011) assessed Fe status in menstruating women by consuming a Fe-fortified peach/apple juice. The core and wall components in producing spray-dried Fe microcapsules were $Fe_4O_{21}P_6$ and lecithin, respectively. The intake of both agents of carbohydrate and ascorbic acid and also

FIGURE 9.14 SEM images of encapsulated powders of spray-dried Zn–chlorophylls extracted form pandan leaf (wall material: (A) GA, (B) MD, and (C) Ms). *Reproduced from Porrarud and Pranee (2010).*

body mass index within natural ranges were increased with the consumption of peach–apple juice enriched with Fe. The fortified samples could improve Fe situation and may be applied to prevent/control Fe-deficiency anemia.

9.5 USE OF MINERAL NANOPARTICLES

In the recent decade, declining the inclusion amounts of minerals and enhancing their absorption in the digestive tract by decreasing their particle size in a nanoscale have been attracted by many scientists (Vijayakumar & Balakrishnan, 2014). The size of mineral NPs is usually less than 100 nm with a spherical morphology (Table 9.5). Nanominerals are extensively applied in differentiated subdivisions, such as agriculture, animal, food, and pharma industries. The mineral NPs synthesized using the different methods of physical, chemical, biological, and enzymatic have many advantages from industrial and biological standpoints. Table 9.5 exhibits technological, bioavailability, and toxicity features of some mineral NPs.

Se-NPs maybe are the most practical NPs used in the food industries so that they can potentially control growth and biofilm formation by foodborne pathogenic bacteria (Khiralla & El-Deeb, 2015). To prepare Se-NPs with a reduced toxicity risk, the diverse components including melatonin (Wang et al., 2005), PEG (Zheng et al., 2012), adenosine triphosphate (Zhang et al., 2013), various types of polysaccharides (Chen, Wong, Zheng, Bai, & Huang, 2008; Han et al., 2012; Zhang, Wang, & Zhang, 2010), and epigallocatechin-3-gallate (Wu et al., 2013) have been applied. As Se had numerous positive effects on the health, especially improvement of immune system and inhibition of cancer types because of its aptitude to prevent the growth of cancer cells via induction of S phase arrest (Luo, Wang, Bai, Chen, & Zheng, 2012), use of Se-NPs in drug and food formulations as dietary supplements can potentially enhance its functionality and bioaccessibility in the body (Wang, Taylor, Wang, Wan, & Zhang, 2012). Although most researchers reported that use of these NPs has rare toxicity in the body (Khiralla & El-Deeb, 2015; Wang et al., 2012; Zhang, Wang, & Xu, 2008), Zhang, Taylor, Wan, and Peng (2012) stated that the diet supplementations containing Se-NPs led to a significantly reduced bioavailability in Se-deficient mice.

Moreover, the fabricated NPs from inorganic nanometal oxides, such as magnesium oxide (MgO), zinc oxide (ZnO), and calcium oxide (CaO) have been studied as antimicrobial potential compounds (Akbar & Anal, 2014; Bajpai, Chand, & Chaurasia, 2012; Jin & He, 2011; Mirhosseini & Afzali, 2016; Premanathan, Karthikeyan, Jeyasubramanian, & Manivannan, 2011; Stoimenov, Klinger, Marchin, & Klabunde, 2002; Tang & Lv, 2014). Powerful antibacterial potential of the metal oxide-NPs is through disruption of the microbial cell membrane by the creating reactive oxygen species, for

TABLE 9.5 Technological Aspects, Bioavailability, and Toxicity of Some Mineral NPs

Mineral Type	Shape/Size	Technological Aspects	Bioavailability/Toxicity	References
Se	Spherical 10–50 nm	Controls growth and biofilm formation by six foodborne pathogens (MIC90 = 25 mg/mL)	No toxicity on *Artemia* larvae up to 100 mg/mL	Khiralla and El-Deeb (2015)
Se	Spherical 80 nm	The thermo-stability of Se-NPs is size-dependent: smaller Se-NPs being more resistant than larger Se-NPs to transformation into nanorods during heat treatment	Leads to significantly reduced bioavailability and phase II enzyme induction in Se-deficient mice	Zhang et al. (2012)
Se	Spherical 20–60 nm (~36 nm)	Se-NPs compared with Se compounds generally used in dietary supplements	Se-NPs have meaningfully lower toxicity, without compromising their ability to up-regulate seleno-enzymes at nutritional levels and induce phase II enzymes at supranutritional levels	Wang et al. (2012)
Se	Spherical 20–60 nm	A unique and indirect way for designing meat products with increased nutritive value and functionality is feeding of elemental Se-NPs obtained from probiotic bacteria fed to lambs resulted in enrichment of lamb meat with Se (1)	Se-NPs have similar efficiency with Se-methyl-seleno-cysteine (SeMSC), in increasing the activity of glutathione peroxidase (GPx), thioredoxin reductase (TrxR), and glutathione-*S*-transferase (GST) in mice. So, Se-NP is a potential chemo-preventive agent with reduced risk of Se toxicity (2)	[1]Ungvári et al. (2013) [2]Zhang et al. (2008)

(Continued)

TABLE 9.5 (Continued)

Mineral Type	Shape/Size	Technological Aspects	Bioavailability/Toxicity	References
Fe_2O_3 (iron oxide)	Not reported <50 nm	Use of quercetin along with Fe_2O_3 NPs in food applications	Fe_2O_3-NPs induced cellular damage and quercetin (50 μmol/L) plays a vital protective role in Fe_2O_3-NPs induced cytotoxicity and apoptotic death	Sarkar and Sil (2014)
MgO (magnesium oxide)	Hexagonal 10–50 nm	MgO-NPs in combination with nisin in milk damage the bacterial cell membrane, leading to a leakage of intracellular contents and finally their death. A protocol develops to decrease pasteurization temperatures and the required level of MgO-NP for pasteurizing milk and controlling pathogens (1)	Presenting a concerning report about the safety connected with MgO-NPs applications in consumer products due to their DNA damage, cell death, and oxidative damage effects (2)	Mirhosseini and Afzali (2016) Mahmoud et al. (2016)
ZnO (zinc oxide)	Spherical ≤100 nm (10–30 nm)	ZnO-NPs loaded active film of Ca alginate showed high antimicrobial ability against two foodborne pathogens (*S. typhimurium* and *S. aureus*) in ready-to-eat poultry meat, so that reduced the bacteria from log seven to zero within 10 days at 8 °C (1)	Significant changes in liver enzymes, oxidative stress, liver and renal tissue and sperm quality and quantity of adult male Wistar rats were found at concentration more than 50 mg/kg ZnO-NPs. But, the use of low doses requires further investigation (2)	Akbar and Anal (2014) Abbasalipourkabir et al. (2015)

$FePO_4$ (ferric phosphate)	Spherical 30.5/ 10.7 nm	Stable colloid of nanosized $FePO_4$ as a functional additive can be fortified into food and beverage formulations (1)	$FePO_4$-NPs have a solubility and relative bioavailability value. Reducing poorly soluble Fe compounds (e.g., $FePO_4$) to nanoscale might strongly enhance Fe-absorption and bioavailability (2,3)	Rossi, Velikov, and Philipse (2014) Rohner et al. (2007) Zimmermann et al. (2007)
$Ca_3(PO4)_2$ (calcium phosphate)	Spherical <100 nm	Greater potential of $Ca_3(PO4)_2$-NPs in poultry industry especially in feed management and in minimizing the mineral wastages	The bioavailability of $Ca_3(PO4)_2$-NPs in broiler chicken is 200% when compared to $CaHPO_4$	Vijayakumar and Balakrishnan (2014)
$CaCO_3$ (calcium carbonate) $Ca_3(C_6H_5O_7)_2$ (calcium citrate)[2]	Not reported ≤50 nm	Introducing $CaCO_3$-NP as an effective antimicrobial agent can be applied in industries related to food and agriculture because its MIC in broth was two times more than the MIC concentration in solid medium (1)	In-vivo studies indicate that administering $CaCO_3$- and $Ca_3(C_6H_5O_7)_2$-NPs are more bioavailable than micro ones and can enhance the serum Ca concentration and maintain the whole-body bone mineral density in mice (2)	Ataee et al. (2011) Huang et al. (2009)

instance, hydroxyl radicals, hydrogen peroxide, and superoxide on the oxide surface which can cause a leakage of intracellular components and lastly their cell death (Krishnamoorthy, Moon, Hyun, Cho, & Kim, 2012; Makhluf et al., 2005; Stoimenov et al., 2002).

Krishnamoorthy et al. (2012) reported that MgO-NPs are low-cost and nontoxic, and have high thermo-stability and biocompatibility. MgO as one of the six main Mg-agents are now diagnosed as safe by the USDA (21CFR184.1431); nonetheless, Mahmoud, Ezgi, Merve, and Özhan (2016) by investigating in-vitro toxicity of MgO-NPs on liver (HepG2), kidney (NRK-52E), intestine (Caco-2), and lung (A549) cell lines have lately found that these NPs can extremely lead to the oxidative damaging impacts, DNA destruction, and cell death. This concern should be thus provided a specific attention on their safety and application in consumer products. Apart from this fact, many researchers reported the promising news on use of MgO-NPs in helping therapy of various complications. A lower cytotoxic effect of MgO-NPs than NPs of ZnO and TiO_2 on human astrocytoma U87 cells was reported by Lai et al. (2008). In cancer therapy, Fe/MgO nanoshells as a magnetic resonance imaging agent were also applied by Boubeta et al. (2010). Fe/MgO-NPs in cancer therapy as magnetically mediated hyperthermia agents were used (Chalkidou et al., 2011). Also, Di, He, Sun, and Liu (2012) have recently reported the hopeful use of MgO-NPs in nanocryosurgery for tumor treatment. Corot, Robert, Idee, and Port (2006) also pointed out that NPs obtained from iron oxide (Fe_2O_3) are biocompatible and have less toxicity in natural systems compared to other metals. Weinstein et al. (2009) have recently used Fe_2O_3-NPs for different uses consisting of mineral complements, controlled release of therapeutic and nutraceutical biomaterials, and colorant agents particularly in colored cosmetic products. Sarkar and Sil (2014) discovered that 50 μmol/L quercetin can have an important protective role against cytotoxicity and apoptotic death brought by Fe_2O_3-NPs.

ZnO also as one of the five Zn compounds is introduced as generally recognized as safe by the USDA (21CFR182.8991). The incorporation possibility of antimicrobial ZnO-NPs into active packaging films in order to prevent/control the foodborne pathogens, especially in ready-to-eat meat formulations, was studied (Akbar & Anal, 2014). They found that these NPs not only have an interesting antibacterial potential but also can be considered as a Zn provenance for supplementing many foods. Despite CaO antimicrobial NPs, other Ca-based NPs, such as $Ca_3(PO4)_2$-, $CaCO_3$- and $Ca_3(C_6H_5O_7)_2$-NPs revealed promising functions in poultry, agriculture, and food industries because of their antimicrobial and functional properties. In addition, these Ca-based NPs with a great in-vivo bioavailability could effectively improve Ca level in serum and mineral density of whole-body bones (Ataee, Derakhshanpour, Mehrabi Tavana, & Eydi, 2011; Huang, Chen, Hsu, & Chang, 2009; Vijayakumar & Balakrishnan, 2014).

9.6 CONCLUSION AND FURTHER REMARKS

In the past two decades, numerous studies have focused on the production of fortified foods with bioactive compounds in order to provide the increasing demands for the consumption of healthy functional foods containing high amounts of vital minerals. Application of encapsulated mineral forms in the development of dual and TFSs and in particular dairy products led to a notable improvement in its bioavailability and release amounts in the body. In addition to higher absorption quantity of minerals, application of encapsulated minerals compared with their direct addition had other substantial advantages, such as strong protection or minimum loss of minerals against harsh environmental and processing conditions in the final products with enhanced quality properties. Although physical and mechanical techniques have been applied to encapsulate minerals, these methods were lower utilized to nanoencapsulation of these micronutrients compared with the chemical ones. In addition to the economic aspects, selecting an appropriate technology to micro- and nanoencapsulate minerals depends on the EE and physicochemical characteristics of produced capsules and also their effects on the bioavailability and functionality of fortified products. Some findings have shown that use of complexes between phenolic compounds (e.g., catechins) and minerals (e.g., Zn) to load into NPs can significantly increase the values of bioavailability and antioxidant activity. Since antimicrobial abilities of NPs and phenolic compounds have been proved, thus it can be recommended to form the stable complexes between minerals and other bioactive plant/bacterial metabolites (e.g., flavonoids and bacterial pigments) and subsequently micro/nanoencapsulate these synergistic complexes into the wall materials to enhance their antimicrobial potential. Although mineral microcapsules especially in formulation of some dairy and bakery products have been applied, use of mineral nanocapsules in enriching these strategic products has been rarely reported. Overall, a limited number of optimization studies have been conducted on the determination of optimal levels of operating and formulation parameters involved in various methods of micro- and nanoencapsulation of minerals. Since there has been a more attention on the encapsulation of Fe and Zn, it seems to be necessary to further scrutinize this process for other multifunctional minerals and incorporate them into food and drug matrices. To conclude, the toxicity and safety issues of new fabricated NPs should be also evaluated to take necessary approvals in fortifying the final products with nanoencapsulated minerals.

REFERENCES

Abbasalipourkabir, R., Moradi, H., Zarei, S., Asadi, S., Salehzadeh, A., Ghafourikhosroshahi, A., ... Ziamajidi, N. (2015). Toxicity of zinc oxide nanoparticles on adult male Wistar rats. *Food and Chemical Toxicology*, *84*, 154–160.

Abbasi, S., & Azari, S. (2011). Efficiency of novel iron microencapsulation techniques: Fortification of milk. *International Journal of Food Science and Technology, 46*, 1927–1933.

Akbar, A., & Anal, A. K. (2014). Zinc oxide nanoparticles loaded active packaging, a challenge study against *Salmonella typhimurium* and *Staphylococcus aureus* in ready-to-eat poultry meat. *Food Control, 38*, 88–95.

Akhtar, S., Anjum, F. M., Rehman, S. U., & Munir, A. S. (2010). Effect of storage and baking on mineral contents of fortified whole wheat flour. *Journal of Food Processing and Preservation, 34*, 335–349.

Al-Gawhari, F. J. (2016). Preparation of ferrous sulfate microcapsules as a sustained release dosage forms. *International Journal of Applied Pharmaceutics, 8*, 16–19.

Alvim, I. D., Souza, F. S., Koury, I. P., Jurt, T., & Dantas, F. B. H. (2013). Use of the spray chilling method to deliver hydrophobic components: physical characterization of microparticles. *Food Science and Technology (Campinas), 33*(Suppl. 1), 34–39.

Arce, A. (2016). *Effect of micro-encapsulated ferrous sulfate on Cheddar cheese composition, divalent cation balance and acceptability*. An M.Sc. Thesis in Food Science, Michigan State University, East Lansing, MI.

Ataee, R. A., Derakhshanpour, J., Mehrabi Tavana, A., & Eydi, A. (2011). Antibacterial effect of calcium carbonate nanoparticles on *Agrobacterium tumefaciens*. *Iranian Journal of Military Medicine, 13*, 65–70.

Azzam, M. A. (2009). Effect of fortification with iron–whey protein complex on quality of yoghurt. *Egyptian Journal of Dairy Science, 37*, 55–63.

Bagci, P. O., & Gunasekaran, S. (2016a). Iron-encapsulated cold-set whey protein isolate gel powder—Part 1: Optimisation of preparation conditions and in vitro evaluation. *International Journal of Dairy Technology, 69*, 1–10. Available from http://dx.doi.org/10.1111/1471-0307.12317.

Bagci, P. O., & Gunasekaran, S. (2016b). Iron-encapsulated cold-set whey protein isolate gel powder—Part 2: Effect of iron fortification on sensory and storage qualities of yoghurt. *International Journal of Dairy Technology, 69*, 1–8. Available from http://dx.doi.org/10.1111/1471-0307.12316.

Bajpai, K. S., Chand, N., & Chaurasia, V. (2012). Nano zinc oxide-loaded calcium alginate films with potential antibacterial properties. *Food and Bioprocess Technology, 5*, 1871–1881.

Beinner, M. A., Velasquez-Meléndez, G., Pessoa, M. C., & Greiner, T. (2010). Iron-fortified rice is as efficacious as supplemental iron drops in infants and young children. *Journal of Nutrition, 140*, 49–53.

Berginc, K. (2015). Pharmacokinetic interactions between drugs and dietary supplements: Carbohydrate, protein, vitamin and mineral supplements. In K. Berginc, & S. Kreft (Eds.), *Dietary supplements* (pp. 85–126). London: Elsevier.

Biebinger, R., Zimmermann, M. B., Al-Hooti, S. N., Al-Hamed, N., Al-Salem, E., Zafar, T., ... Hurrell, R. F. (2009). Efficacy of wheat-based biscuits fortified with microcapsules containing ferrous sulfate and potassium iodate or a new hydrogen-reduced elemental iron: A randomised, double-blind, controlled trial in Kuwaiti women. *British Journal of Nutrition, 102*, 1362–1369.

Blanco-Rojo, R., Perez-Granados, A. M., Toxqui, L., Gonzalez-Vizcayno, C., Delgado, M. A., & Vaquero, M. P. (2011). Efficacy of a microencapsulated iron pyrophosphate-fortified fruit juice: A randomised, double-blind, placebo-controlled study in Spanish iron-deficient women. *British Journal of Nutrition, 105*, 1652–1659.

Boccio, J. R., Zubillaga, M. B., Caro, R. A., Gotelli, C. A., Gotelli, M. J., & Weill, R. (1997). A new procedure to fortify fluid milk and dairy products with high-bioavailable ferrous sulfate. *Nutrition Reviews*, *55*, 240–246.

Bonnet, M., Cansell, M., Berkaoui, A., Ropers, M. H., Anton, M., & Leal-Calderon, F. (2009). Release rate profiles of magnesium from multiple W/O/W emulsions. *Food Hydrocolloids*, *23*, 92–101.

Boubeta, C. M., Balcells, L., Cristofol, R., Sanfeliu, C., Rodnguez, E., Weissleder, R., Piedrafita, S. L., et al. (2010). Self-assembled multifunctional Fe/MgO nanospheres for magnetic resonance imaging and hyperthermia. *Nanomedicine: Nanotechnology, Biology and Medicine*, *6*, 362.

Cavallini, D. C. U., & Rossi, E. A. (2009). Soy yogurt fortified with iron and calcium: Stability during the storage. *Alimentos e Nutrição Araraquara*, *20*, 7–13.

Chalkidou, A., Simeonidis, K., Angelakeris, M., Samaras, T., Boubeta, C. M., Balcells, L. l, ... Kalogirou, O. J. (2011). *In vitro* application of Fe/MgO nanoparticles as magnetically mediated hyperthermia agents for cancer treatment. *Journal of Magnetism and Magnetic Materials*, *323*, 775–780.

Chang, Y. H., Lee, S. Y., & Kwak, H. S. (2016). Physicochemical and sensory properties of milk fortified with iron microcapsules prepared with water-in-oil-in-water emulsion during storage. *International Journal of Dairy Technology*, *69*, 452–459.

Chen, T., Wong, Y. S., Zheng, W., Bai, Y., & Huang, L. (2008). Selenium nanoparticles fabricated in *Undaria pinnatifida* polysaccharide solutions induce mitochondria-mediated apoptosis in A375 human melanoma cells. *Colloids and Surfaces B: Biointerfaces*, *67*, 26–31.

Choi, S. J., Decker, E. A., & McClements, D. J. (2009). Impact of iron encapsulation within the interior aqueous phase of water-in-oil-in-water emulsions on lipid oxidation. *Food Chemistry*, *116*, 271–276.

Corot, C., Robert, P., Idee, J. M., & Port, M. (2006). Recent advances in iron oxide nanocrystal technology for medical imaging. *Advanced Drug Delivery Reviews*, *58*, 1471–1504.

de Kruif, C. G., Weinbreck, F., & de Vries, R. (2004). Complex coacervation of proteins and anionic polysaccharides. *Current Opinion in Colloid and Interface Science*, *9*, 340–349.

Desai, K. G. H., & Park, H. J. (2005). Recent developments in microencapsulation of food ingredients. *Drying Technology*, *23*, 1361–1394.

Devaraj, G. N., Parakh, S. R., Devraj, R., Apte, S. S., Ramesh Rao, B., & Rambhau, D. (2002). Release studies on niosomes containing fatty alcohols as bilayer stabilizers instead of cholesterol. *Journal of Colloid and Interface Science*, *251*, 360–365.

Di, D. R., He, Z. Z., Sun, Z. Q., & Liu, J. (2012). A new nano-cryosurgical modality for tumor treatment using biodegradable MgO nanoparticles. *Nanomedicine: Nanotechnology, Biology and Medicine*, *8*, 1233–1241.

Ding, B., Xia, S., Hayat, K., & Zhang, X. (2009). Preparation and pH stability of ferrous glycinate liposomes. *Journal of Agricultural and Food Chemistry*, *57*, 2938–2944.

Ding, B., Zhang, X., Hayat, K., Xia, S., Jia, C., Xie, M., & Liu, C. (2011b). Preparation, characterization and the stability of ferrous glycinate nanoliposomes. *Journal of Food Engineering*, *102*, 202–208.

Ding, B. M., Zhang, X. M., & Xia, S. Q. (2011a). Effectiveness of treatment of iron-deficiency anemia in rats with ferrous glycinate nanoliposomes. *Journal of Food Science and Biotechnology*, 1–11 , 2011–01

Diosady, L., Yusufali, R., Oshinowo, T., & Laleye, L. (2006). A study of storage and distribution of double fortified salts in Kenya. *Journal of Food Engineering*, *76*, 547–556.

Diosady, L. L., Alberti, J. O., & Venkatesh Mannar, M. G. (2002). Microencapsulation for iodine stability in salt fortified with ferrous fumarate and potassium iodide. *Food Research International*, *35*, 635–642.

Du, W. L., Niu, S. S., Xu, Y. L., Xu, Z. R., & Fan, C. L. (2009). Antibacterial activity of chitosan tripolyphosphate nanoparticles loaded with various metal ions. *Carbohydrate Polymers*, *75*, 385–389.

Dueik, V., & Diosady, L. L. (2016). Microencapsulation of iron in a reversed enteric coating using spray drying technology for double fortification of salt with iodine and iron. *Journal of Food Process Engineering*. Available from http://dx.doi.org/10.1111/jfpe.12376.

El-Sayed, M. M., Hassan, Z. M. R., Mervat, I. F., Awad, R. A., & Salama, H. H. (2015). Chitosan-whey protein complex (CS-WP) as delivery systems to improve bioavailability of iron. *International Journal of Applied and Pure Science and Agriculture*, *1*, 34–46.

Faridi Esfanjani, A., & Jafari, S. M. (2016). Biopolymer nano-particles and natural nano-carriers for nano-encapsulation of phenolic compounds. *Colloids and Surfaces B: Biointerfaces*, *146*, 532–543.

Ferreira, B. S., Cardoso, B. T., Pereira, H. V. R., Pierucci, A. P., Pedrosa, C., & Citelli, M. (2011). Acceptability of black beans (*Phaseolus vulgaris* L.) fortified with iron microparticles. *Revista Ceres*, *58*, 548–553.

Flaten, G. E., Bunjes, H., Luthman, K., & Brandl, M. (2006). Drug permeability across a phospholipid vesicle-based barrier 2. Characterization of barrier structure, storage stability and stability towards pH changes. *European Journal of Pharmaceutical Sciences*, *28*, 336–343.

Gharibzahedi, S. M. T., & Jafari, S. M. (2017). Nano-capsule formation by cyclodextrins. In S. M. Jafari (Ed.), *Nanoencapsulation technologies in the food and nutraceutical industries* (Chapter 7: pp. 187–262). Amsterdam: Elsevier.

Gharibzahedi, S. M. T., & Jafari, S. M. (2017a). The importance of minerals in human nutrition: Bioavailability, food fortification, processing effects and nanoencapsulation. *Trends in Food Science & Technology*, *62*, 119–132.

Gharibzahedi, S. M. T., & Mohammad Nabi, S. (2016). Characterizing the novel surfactant-stabilized nanoemulsions of stinging nettle essential oil: Thermal behaviour, storage stability, antimicrobial activity and bioaccessibility. *Journal of Molecular Liquids*, *224*, 1332–1340.

Gharibzahedi, S. M. T., Razavi, S. H., & Mousavi, S. M. (2015a). Optimal development of a new stable nutraceutical nanoemulsion based on the inclusion complex of 2-hydroxypropyl-β-cyclodextrin with canthaxanthin accumulated by *Dietzia natronolimnaea* HS-1 using ultrasound-assisted emulsification. *Journal of Dispersion Science & Technology*, *36*, 614–625.

Gharibzahedi, S. M. T., Rostami, H., & Yousefi, S. (2015b). Formulation design and physicochemical stability characterization of nanoemulsions of nettle (*Urtica dioica*) essential oil using a model-based methodology. *Journal of Food Processing and Preservation*, *39*, 2947–2958.

Ghosh, S. K. (2006). Functional coatings and microencapsulation: A general perspective. In S. K. Ghosh (Ed.), *Functional coating* (pp. 1–26). Weinheim: Wiley-VCH.

Gilliard Nkhata, S. (2013). *Iron fortification of yogurt and pasteurized milk*. An M.Sc. Thesis in Food Science, Michigan State University, East Lansing, MI.

Gouin, S. (2004). Microencapsulation: Industrial appraisal of existing technologies and trends. *Trends in Food Science & Technology*, *15*, 330–347.

Gülseren, İ., Fang, Y., & Corredig, M. (2012). Zinc incorporation capacity of whey protein nanoparticles prepared with desolvation with ethanol. *Food Chemistry*, *135*, 770–774.

Gupta, C., Chawla, P., & Arora, S. (2015a). Development and evaluation of iron microencapsules for milk fortification. *CyTA—Journal of Food*, *13*(1), 116–123.

Gupta, C., Chawla, P., Arora, S., Tomar, S. K., & Singh, A. K. (2015b). Iron microencapsulation with blend of gum arabic, maltodextrin and modified starch using modified solvent evaporation method—Milk fortification. *Food Hydrocolloids*, *43*, 622–628.

Gutiérrez, G., Matos, M., Barrero, P., Pando, D., Iglesias, O., & Pazos, C. (2016). Iron-entrapped niosomes and their potential application for yogurt fortification. *LWT—Food Science and Technology*, *74*, 550–556.

Han, J., Guo, X., Lei, Y., Dennis, B. S., Wu, S., & Wu, C. (2012). Synthesis and characterization of selenium–chondroitin sulfate nanoparticles. *Carbohydrate Polymers*, *90*, 122–126.

Hermida, L. G., Roig, A., Bregni, C., Sabés-Xamaní, M., & Barnadas-Rodríguez, R. (2011). Preparation and characterization of iron-containing liposomes: Their effect on soluble iron uptake by Caco-2 cells. *Journal of Liposome Research*, *21*, 203–212.

Hirotsuka, M., Taniguchi, H., Narita, H., & Kito, M. (1984). Calcium fortification of soy milk with calcium-lecithin liposome system. *Journal of Food Science*, *49*, 1111–1127.

Hosny, K. M., Banjar, Z. M., Hariri, A. H., & Hassan, A. H. (2015). Solid lipid nanoparticles loaded with iron to overcome barriers for treatment of iron deficiency anemia. *Drug Design, Development and Therapy*, *9*, 313–320.

Hotz, C., Porcayo, M., Onofre, G., García-Guerra, A., Elliott, T., Jankowski, S., & Greiner, T. (2008). Efficacy of iron-fortified Ultra Rice in improving the iron status of women in Mexico. *Food and Nutrition Bulletin*, *29*, 140–149.

Huang, S., Chen, J. C., Hsu, C. W., & Chang, W. H. (2009). Effects of nano calcium carbonate and nano calcium citrate on toxicity in ICR mice and on bone mineral density in an ovariectomized mice model. *Nanotechnology*, *20*, 375102.

Hughes, G. A. (2005). Nanostructure-mediated drug delivery. *Nanomedicine: Nanotechnology, Biology and Medicine*, *1*, 22–30.

Hurrell, R., Ranum, P., de Pee, S., Biebinger, R., Hulthen, L., Johnson, Q., & Lynch, S. (2010). Revised recommendations for iron fortification of wheat flour and an evaluation of the expected impact of current national wheat flour fortification programs. *Food and Nutrition Bulletin*, *31*, S7–S21.

Hurrell, R. F. (2002b). How to ensure adequate iron absorption from iron-fortified food. *Nutrition Reviews*, *60*, S7–S15.

Hurrell, R. F. (2002a). Fortification: Overcoming technical and practical barriers. *Journal of Nutrition*, *132*, 806S–812S.

IDFA (2015). *Cheese production. International Dairy Foods Association.* Retreived from http://www.idfa.org/news-views/media-kits/cheese/cheese-production. Accessed on September 12, 2016.

Jackson, L. S., & Lee, K. (1992a). The effect of dairy products on iron availability. *Critical Reviews in Food Science and Nutrition*, *31*, 259–270.

Jackson, L. S., & Lee, K. (1992b). Fortification of cheese with microencapsulated iron. *Cultured Dairy Products Journal*, *27*, 4–7.

Jafari, S. M., Fathi, M., & Mandala, I. (2015). *Emerging product formation. Food Waste Recovery: Processing Technologies and Industrial Techniques* (pp. 293–317). Elsevier Inc.

Jafari, S. M., He, Y., & Bhandari, B. (2007). Optimization of nano-emulsions production by microfluidization. *European Food Research and Technology*, *225*, 733–741.

Jakel, S. S., & Belshaw, F. (1971). Encapsulated ferrous sulphate protects baking mixers, flour from rancid flavors. *Food Proccessing*, *35*, 28–30.

Jalili, M. (2016). Chemical composition and sensory characteristics of Feta cheese fortified with iron and ascorbic acid. *Dairy Science & Technology*, *96*(4), 579–589.

Jayalalitha, V., Balasundaram, B., Palanidorai, A., & Naresh, K. C. (2012). Fortification of encapsulated iron in probiotic yoghurt. *International Journal of Agriculture: Research and Review*, *2*, 80–84.

Jiménez-Alvarado, R., Beristain, C. I., Medina-Torres, L., Román-Guerrero, A., & Vernon-Carter, E. J. (2009). Ferrous bisglycinate content and release in W1/O/W2 multiple emulsions stabilized by protein–polysaccharide complexes. *Food Hydrocolloids, 23*, 2425–2433.

Jin, T., & He, Y. P. (2011). Antibacterial activities of magnesium oxide (MgO) nanoparticles against foodborne pathogens. *Journal of Nanoparticles Research, 13*, 6877–6885.

Kapor, A. J., Nikolić, L. B., Nikolić, V. D., & Stanković, M. Z. (2012). The synthesis and characterization of iron(II) fumarate and its inclusion complexes with cyclodextrins. *Advanced Technologies, 1*, 7–15.

Katouzian, I., & Jafari, S. M. (2016). Nano-encapsulation as a promising approach for targeted delivery and controlled release of vitamins. *Trends in Food Science & Technology, 53*, 34–48.

Katuwavila, N. P., Perera, A. D., Dahanayake, D., Karunaratne, V., Amaratunga, G. A., & Karunaratne, D. N. (2016). Alginate nanoparticles protect ferrous from oxidation: Potential iron delivery system. *International Journal of Pharmaceutics, 513*, 404–409.

Khiralla, G. M., & El-Deeb, B. A. (2015). Antimicrobial and antibiofilm effects of selenium nanoparticles on some foodborne pathogens. *LWT—Food Science and Technology, 63*, 1001–1007.

Khosroyar, S., Akbarzade, A., Arjoman, M., Safekordi, A. A., & Mortazavi, S. A. (2012). Ferric–saccharate capsulation with alginate coating using the emulsification method. *African Journal of Microbiology Research, 6*, 2455–2461.

Khosroyar, S., Akbarzade, A., Mortazavi, M., Arjomand, M., Safekordi, A. A., & Mortazavi, S. A. (2013). The study of dispersion method in the formation of ferric saccharate microcapsules with alginate coating. *International Journal of Life science and Pharma Reviews, 2*, 325–334.

Khosroyar, S., & Arastehnodeh, A. (2007). Ferric saccharate microencapsulation with alginate coating using the coacervation method. *Afinidad, 80*, 121–127.

Kim, J. M., Kim, Y. J., Jeong, J., & Kim, C. J. (2006). Meat tenderizing effect of injecting encapsulated Ca^{2+} in liposome into rabbit before slaughter. *Bioscience, Biotechnology, and Biochemistry, 70*, 2381–2386.

Kim, S. J., Ahn, J., & Kwak, H. S. (2003). Microencapsulated iron for drink yoghurt fortification. *Asian Australian Journal of Animal Sciences, 16*, 581–587.

Kiskini, A., Kapsokefalou, M., Yanniotis, S., & Mandala, I. (2010). Effect of different iron compounds on wheat and gluten-free breads. *Journal of the Science of Food and Agriculture, 90*, 1136–1145.

Kongkachuichai, R., Kounhawej, A., Chavasit, V., & Charoensiri, R. (2007). Effects of various iron fortificants on sensory acceptability and shelf-life stability of instant noodles. *Food and Nutrition Bulletin, 28*, 165–172.

Kosaraju, S. L., Tran, C., & Lawrence, A. (2006). Liposomal delivery systems for encapsulation of ferrous sulfate: Preparation and characterization. *Journal of Liposome Research, 16*, 347–358.

Krishnamoorthy, K., Moon, J. Y., Hyun, H. B., Cho, S. K., & Kim, S. J. (2012). Mechanistic investigation on the toxicity of MgO nanoparticles toward cancer cells. *Journal of Materials Chemistry, 22*, 24610–24617.

Kwak, H. S., Ihm, M. R., & Ahn, J. (2001). Microencapsulation of β-galactosidase with fatty acid esters. *Journal of Dairy Science, 84*, 1576–1582.

Kwak, H. S., Ju, Y. S., Ahn, H. J., Ahn, J., & Lee, S. (2003a). Microencapsulated iron fortification and flavor development in cheddar cheese. *Asian-Australasian Journal of Animal Sciences, 16*, 1205–1211.

Kwak, H. S., Yang, K. M., & Ahn, J. (2003b). Microencapsulated iron for milk fortification. *Journal of Agricultural and Food Chemistry, 51*, 7770–7774.

Lai, J. C. K., Lai, M. B., Jandhyam, S., Dukhande, V. V., Bhushan, A., Daniels, C. K., & Leung, S. W. (2008). Exposure to titanium dioxide and other metallic oxide nanoparticles induces cytotoxicity on human neural cells and fibroblasts. *International Journal of Nanomedicine*, *3*, 533–545.

Leite, R. A., Lino, A. C. S., & Takahata, Y. (2003). Inclusion compounds between α-, β- and γ-cyclodextrins: Iron II lactate: A theoretical and experimental study using diffusion coefficients and molecular mechanics. *Journal of Molecular Structure*, *644*, 49–53.

Li, Y.O. (2009). Development of microencapsulation-based technologies for micronutrient fortification in staple foods for developing countries. Ph.D. Thesis in Chemical Engineering and Applied Chemistry, University of Toronto, Toronto, ON.

Li, Y. O., Diosady, L. L., & Jankowski, S. (2008a). Effect of iron compounds on the storage stability of multiple-fortified Ultra Rice®. *International Journal of Food Science and Technology*, *43*, 423–429.

Li, Y. O., Diosady, L. L., & Jankowski, S. (2008b). Stability of vitamin B1 in Ultra Rice® in the presence of encapsulated ferrous fumarate. *International Journal of Food Sciences and Nutrition*, *59*, 24–33.

Li, Y. O., Diosady, L. L., & Wesley, A. (2009). Iron *in vitro* bioavailability and iodine storage stability in double fortified salt (DFS). *Food and Nutrition Bulletin*, *30*, 327–335.

Li, Y. O., Diosady, L. L., & Wesley, A. S. (2010). Iodine stability in iodized salt dual fortified with microencapsulated ferrous fumarate made by an extrusion-based encapsulation process. *Journal of Food Engineering*, *99*, 232–238.

Li, Y. O., Yadava, D., Lo, K. L., Diosady, L. L., & Wesley, A. S. (2011). Feasibility and optimization study of using cold-forming extrusion process for agglomerating and microencapsulating ferrous fumarate for salt double fortification with iodine and iron. *Journal of Microencapsulation*, *28*, 639–649.

Lukaski, H. C. (2004). Vitamin and mineral status: Effects on physical performance. *Nutrition*, *20*, 632–644.

Luo, H., Wang, F., Bai, Y., Chen, T., & Zheng, W. (2012). Selenium nanoparticles inhibit the growth of HeLa and MDA-MB-231 cells through induction of S phase arrest. *Colloids and Surfaces B: Biointerfaces*, *94*, 304–308.

Luo, Y., Zhang, B., Cheng, W. H., & Wang, Q. (2010). Preparation, characterization and evaluation of selenite-loaded chitosan/TPP nanoparticles with or without zein coating. *Carbohydrate Polymer*, *82*, 942–951.

Lysionek, A. E., Zubillaga, M. B., Salgueiro, M. J., Pineiro, A., Caro, R. A., Weill, R., & Boccio, J. R. (2002). Bioavailability of microencapsulated ferrous sulfate in powdered milk produced from fortified fluid milk: A prophylactic study in rats. *Nutrition*, *18*, 279–281.

Madene, A., Jacquot, M., Scher, J., & Desobry, S. (2006). Flavour encapsulation and controlled release—A review. *International Journal of Food Science and Technology*, *41*, 1–21.

Mahmoud, A., Ezgi, Ö., Merve, A., & Özhan, G. (2016). *In vitro* toxicological assessment of magnesium oxide nanoparticle exposure in several mammalian cell types. *International Journal of Toxicology*, *35*, 429–437.

Majeed, H., Jamshaid Qazi, H., Safdar, W., & Fang, Z. (2013). Microencapsulation can be a novel tool in wheat flour with micronutrients fortification: Current trends and future applications—a review. *Czech Journal of Food Sciences*, *31*, 527–540.

Makhluf, S., Dror, R., Nitzan, Y., Abramovich, Y., Jelinek, R., & Gedanken, A. (2005). Microwave-assisted synthesis of nanocrystalline MgO and its use as a bacteriocide. *Advanced Functional Materials*, *15*, 1708–1715.

Marianecci, C., Di Marzio, L., Rinaldi, F., Celia, C., Paolino, D., Alhaique, F., ... Carafa, M. (2014). Niosomes from 80s to present: The state of the art. *Advances in Colloid and Interface Science*, *205*, 187–206.

Martin, A. H., & de Jong, G. A. H. (2012). Impact of protein pre-treatment conditions on the iron encapsulation efficiency of whey protein cold-set gel particles. *European Food Research and Technology*, *234*, 995–1003.

McClements, D. J., & Rao, J. (2011). Food-grade nanoemulsions: Formulation, fabrication, properties, performance, biological fate, and potential toxicity. *Critical Reviews in Food Science and Nutrition*, *51*, 285–330.

Mehansho, H., & Mannar, M. G. V. (1999). The mineral fortification in developing countries. In R. Hurrell (Ed.), *The mineral fortification of foods* (pp. 210–228). Leatherhead, England: Leatherhead International Ltd.

Min, K. A., Cho, J. H., Song, Y. K., & Kim, C. K. (2016). Iron casein succinylate–chitosan coacervate for the liquid oral delivery of iron with bioavailability and stability enhancement. *Archives of Pharmacal Research*, *39*, 94–102.

Mirhosseini, M., & Afzali, M. (2016). Investigation into the antibacterial behavior of suspensions of magnesium oxide nanoparticles in combination with nisin and heat against *Escherichia coli* and *Staphylococcus aureus* in milk. *Food Control*, *68*, 208–215.

Moran, L.L. (2013). Feasibility of spray chilling encapsulation for protection and stabilization of flavor compound-zinc ion complexes. *M.Sc. Thesis in in Food Science and Human Nutrition*, University of Illinois, Urbana-Champaign, IL, USA.

Moretti, D., Lee, T. C., Zimmermann, M. B., Nuessli, J., & Hurrell, R. F. (2005). Development and evaluation of iron-fortified extruded rice grains. *Journal of Food Science*, *70*, S330–S336.

Mozafari, M. R., Khosravi-Darani, K., Borazan, G. G., Cui, J., Pardakhty, A., & Yurdugul, S. (2008). Encapsulation of food ingredients using nanoliposome technology. *International Journal of Food Properties*, *11*, 833–844.

Navas-Carretero, S., Perez-Granados, A. M., Sarria, B., Schoppen, S., & Vaquero, M. P. (2007). Iron bioavailability from pate enriched with encapsulated ferric pyrophosphate or ferrous gluconate in rats. *Food Science and Technology International*, *13*, 159–163.

Naveen, S., & Kanum, F. (2014). Characterization and evaluation of iron nano-emulsion prepared by high speed homogenization. *International Journal of Biology, Pharmacy and Allied Sciences*, *3*, 45–55.

Oehlke, K., Adamiuk, M., Behsnilian, D., Gräf, V., Mayer-Miebach, E., Walz, E., & Greiner, R. (2014). Potential bioavailability enhancement of bioactive compounds using food-grade engineered nanomaterials: A review of the existing evidence. *Food & Function*, *5*, 1341–1359.

Okuro, P. K., Junior, F. E. d M., & Favaro-Trindade, C. S. (2013). Technological challenges for spray chilling encapsulation of functional food ingredients. *Food Technology and Biotechnology*, *51*, 171–182.

Oner, L., Kas, H. S., & Hincal, A. A. (1988). Studies on zinc sulphate microcapsules (1): Microencapsulation and *in vitro* dissolution kinetics. *Journal of Microencapsulation*, *5*, 219–223.

Oshinowo, T., Diosady, L. L., Yusufali, R., & Laleye, L. (2004). Stability of salt double-fortified with ferrous fumarate and potassium iodate or iodide under storage and distribution conditions in Kenya. *Food and Nutrition Bulletin*, *25*, 264–270.

Patil, J. S., Kamalapur, M. V., Marapur, S. C., & Kadam, D. V. (2010). Ionotropic gelation and polyelectrolyte complexation: The novel techniques to design hydrogel particulate sustained, modulated drug delivery system: A review. *Digest Journal of Nanomaterials and Biostructures*, *5*, 241–248.

Pham, T. T., Jaafar-Maalej, C., Charcosset, C., & Fessi, H. (2012). Liposome and niosome preparation using a membrane contactor for scale-up. *Colloids and Surfaces B: Biointerfaces*, *94*, 15–21.

Pistone, S., Qoragllu, D., Smistad, G., & Hiorth, M. (2015). Formulation and preparation of stable cross-linked alginate–zinc nanoparticles in the presence of a monovalent salt. *Soft Matter, 11*, 5765–5774.

Porrarud, S., & Pranee, A. (2010). Microencapsulation of Zn chlorophyll pigment from pandan leaf by spray drying and its characteristic. *International Food Research Journal, 17*, 1031–1042.

Premanathan, M., Karthikeyan, K., Jeyasubramanian, K., & Manivannan, G. (2011). Selective toxicity of zinc oxide nanoparticles toward Gram-positive bacteria and cancer cells by apoptosis through lipid peroxidation. *Nanomedicine Nanotechnology Biology and Medicine, 7*, 184–192.

Rice, W. H., & McMahon, D. J. (1998). Chemical, physical, and sensory characteristics of mozzarella cheese fortified using protein-chelated iron or ferric chloride. *Journal of Dairy Science, 81*, 318–326.

Rohner, F., Ernst, F. O., Arnold, M., Hilbe, M., Biebinger, R., Ehrensperger, F., ... Zimmermann, M. B. (2007). Synthesis, characterization, and bioavailability in rats of ferric phosphate nanoparticles. *Journal of Nutrition, 137*, 614–619.

Romita, D., Cheng, Y. L., & Diosady, L. L. (2011). Microencapsulation of ferrous fumarate for the production of salt double fortified with iron and iodine. *International Journal of Food Engineering*, 7 . Article 5. Available from http://dx.doi.org/10.2202/1556-3758.2122.

Rossi, L., Velikov, K. P., & Philipse, A. P. (2014). Colloidal iron(III) pyrophosphate particles. *Food Chemistry, 151*, 243–247.

Saeidy, S., Keramat, J., & Nasirpour, A. (2014a). Microencapsulation of calcium using water in oil in water double emulsion method. *Journal of Dispersion Science and Technology, 35*, 370–379.

Saeidy, S., Keramat, J., & Nasirpour, A. (2014b). Physicochemical properties of calcium-fortified soymilk with microencapsulated and chelated calcium salt. *European Food Research and Technology, 238*, 105–112.

Sarkar, A., & Sil, P. C. (2014). Iron oxide nanoparticles mediated cytotoxicity via PI3K/AKT pathway: Role of quercetin. *Food and Chemical Toxicology, 71*, 106–115.

Sharifi, A., Golestan, L., & Sharifzadeh Baei, M. (2013). Studying the enrichment of ice cream with alginate nanoparticles including Fe and Zn salts. *Journal of Nanoparticles, 2013*, 5, ID no. 754385.

Shewan, H. M., & Stokes, J. R. (2013). Review of techniques to manufacture micro-hydrogel particles for the food industry and their applications. *Journal of Food Engineering, 119*, 781–792.

Singh, S., Rathi, N., Angal, A., Parida, P., & Rautaray, D. (2016). Biofortification of food with minerals and vitamins encapsulated in silica. In S. Ranjan, N. Dasgupta, & E. Lichtfouse (Eds.), Nanoscience in food and agriculture 2, sustainable agriculture reviews (pp. 157–206). Basel, Switzerland: Springer International Publishing.

Soumya, R. S., Vineetha, V. P., Reshma, P. L., & Raghu, K. G. (2013). Preparation and characterization of selenium incorporated guar gum nanoparticle and its interaction with H9c2 cells. *PLoS ONE, 8*, 1–12.

Soumya, R. S., Vineetha, V. P., Salin Raj, P., & Raghu, K. G. (2014). Beneficial properties of selenium incorporated guar gum nanoparticles against ischemia/reperfusion in cardiomyoblasts (H9c2). *Metallomics, 6*, 2134–2147.

Souto, T. S., Brasil, A. L. D., & Taddei, J. A. D. C. (2008). Acceptability of bread fortified with microencapsulated iron by children of daycare centers in the south and east regions of Sao Paulo City, Brazil. *Revista de Nutrição, 21*, 647–657.

Stoimenov, P. K., Klinger, R. L., Marchin, G. L., & Klabunde, K. J. (2002). Metal oxide nanoparticles as bactericidal agents. *Langmuir, 18*, 6679–6686.

Subash, R., & Elango, A. (2015). Microencapsulated iron for fortification in yoghurt. *Food Science Research Journal, 6*, 258–262.

Subash, R., Elango, A., Pandiyan, C., Karthikeyan, N., & Kumaresan, G. (2015). Fortification of microencapsulated iron in yoghurt. *International Journal of Food and Nutritional Sciences, 4*, 201–205.

Tang, S. Y., & Sivakumar, M. (2013). A novel and facile liquid whistle hydrodynamic cavitation reactor to produce submicron multiple emulsions. *AIChE Journal, 59*, 155–167.

Tang, Z. X., & Lv, B. F. (2014). MgO Nanoparticles as antibacterial agent: Preparation and activity. *Brazilian Journal of Chemical Engineering, 31*, 591–601.

Taylor, T. M., & Davidson, P. M. (2005). Liposomal nanocapsules in food science and agriculture. *Critical Reviews in Food Science and Nutrition, 45*, 587–605.

Turchiuli, C., Fuchs, M., Bohin, M., Cuvelier, M. E., Ordonnaud, C., Peyrat-Maillard, M. N., & Dumoulin, E. (2005). Oil encapsulation by spray drying and fluidised bed agglomeration. *Innovative Food Science and Emerging Technologies, 6*, 29–35.

Uicich, R., Pizarro, F., Almeida, C., Diaz, M., Bocchio, J., Zubillaga, M., ... O'Donnell, A. (1999). Bioavailability of microencapsulated ferrous sulfate in fluid cow's milk. Studies in human beings. *Nutrition Research, 19*, 893–897.

Ungvári, E., Monori, I. N., Megyeri, A., Csiki, Z. N., Prokisch, J. Z., Sztrik, A., ... Benkő, I. (2013). Protective effects of meat from lambs on selenium nanoparticle supplemented diet in a mouse model of polycyclic aromatic hydrocarbon-induced immunotoxicity. *Food and Chemical Toxicology, 64*, 298–306.

Vijayakumar, M. P., & Balakrishnan, V. (2014). Evaluating the bioavailability of calcium phosphate nanoparticles as mineral supplement in broiler chicken. *Indian Journal of Science and Technology, 7*, 1475–1480.

Wagner, M. E., Spoth, K. A., Kourkoutis, L. F., & Rizvi, S. S. H. (2016). Stability of niosomes with encapsulated vitamin D3 and ferrous sulfate generated using a novel supercritical carbon dioxide method. *Journal of Liposome Research, 26*, 261–268.

Wang, D., Taylor, E. W., Wang, Y., Wan, X., & Zhang, J. (2012). Encapsulated nanoepigallocatechin-3-gallate and elemental selenium nanoparticles as paradigms for nanochemoprevention. *International Journal of Nanomedicine, 7*, 1711–1721.

Wang, H., Wei, W., Zhang, S. Y., Shen, Y. X., Yue, L., Wang, N. P., & Xu, S. Y. (2005). Melatonin–selenium nanoparticles inhibit oxidative stress and protect against hepatic injury induced by *Bacillus* Calmette-Guérin/lipopolysaccharide in mice. *Journal of Pineal Research, 39*, 156–163.

Watzke, H. J. (1998). Impact of processing on bioavailability examples of minerals in foods. *Trends in Food Science & Technology, 9*, 320–327.

Wegmüller, R., Zimmermann, M. B., Buhr, V. G., Windhab, E. J., & Hurrell, R. F. (2006). Development, stability, and sensory testing of microcapsules containing iron, iodine, and vitamin a for use in food fortification. *Journal of Food Science, 71*, 181–187.

Wegmüller, R., Zimmermann, M. B., & Hurrell, R. F. (2003). Dual fortification of salt with iodine and encapsulated iron compounds: Stability and acceptability testing in Morocco and Côte d'Ivoire. *Journal of Food Science, 68*, 2129–2135.

Weingartner, K. E., Nelson, Al, & Erdman, J. W., Jr. (1983). Effects of calcium addition on stability and sensory properties of soy beverage. *Journal of Food Science, 48*(256–257), 263.

Weinstein, J. S., Varallyay, C. G., Dosa, E., Gahramanov, S., Hamilton, B., Rooney, W. D., et al. (2009). Superparamagnetic iron oxide nanoparticles: Diagnostic magnetic resonance imaging and potential therapeutic applications in neurooncology and central nervous system inflammatory pathologies, a review. *Journal of Cerebral Blood Flow and Metabolism, 30*, 15–35.

Wu, S., Sun, K., Wang, X., Wang, D., Wan, X., & Zhang, J. (2013). Protonation of epigallocatechin-3-gallate (EGCG) results in massive aggregation and reduced oral bioavailability of EGCG-dispersed selenium nanoparticles. *Journal of Agricultural and Food Chemistry, 61*, 7268–7275.

Xia, S., & Xu, S. (2005). Ferrous sulfate liposomes: Preparation, stability and application in fluid milk. *Food Research International, 38*, 289–296.

Xu, Z., Liu, S., Wang, H., Gao, G., Yu, P., & Chang, Y. (2014). Encapsulation of iron in liposomes significantly improved the efficiency of iron supplementation in strenuously exercised rats. *Biological Trace Element Research, 162*, 181–188.

Yadava, D., Li, Y. O., Diosady, L. L., & Wesley, A. S. (2012). Optimisation of polymer coating process for microencapsulating ferrous fumarate for salt double fortification with iodine and iron. *Journal of Microencapsulation, 29*, 729–738.

Yuan, L., Geng, L., Ge, L., Yu, P., Duan, X., Chen, J., & Chang, Y. (2013). Effect of iron liposomes on anemia of inflammation. *International Journal of Pharmaceutics, 454*, 82–89.

Zahraei, M., Monshi, A., Shahbazi-Gahrouei, D., Amirnasr, M., Behdadfara, B., & Rostami, M. (2015). Synthesis and characterization of chitosan coated manganese zinc ferrite nanoparticles as MRI contrast agents. *Journal of Nanostructures, 5*, 77–86.

Zariwala, M. G., Elsaid, N., Jackson, T. L., Corral López, F., Farnaud, S., Somavarapu, S., & Renshaw, D. (2013). A novel approach to oral iron delivery using ferrous sulphate loaded solid lipid nanoparticles. *International Journal of Pharmaceutics, 456*, 400–407.

Zhang, B., Yang, T., Wang, Q., Zhang, G., Huo, J., Huang, J., & Wang, L. (2016). Fabrication of uniform alginate–agarose microcapsules loading $FeSO_4$ using water–oil–water–oil multiple emulsions system combined with premix membrane emulsification technique. *Colloids and Surfaces A: Physicochemical and Engineering Aspects, 498*, 128–138.

Zhang, D., & Mahoney, A. W. (1988). Effect of iron fortification on quality of cheddar cheese. *Journal of Dairy Science, 72*, 322–332.

Zhang, D., & Mahoney, A. W. (1989). Bioavailability of iron-milk-protein complexes and fortified cheddar cheese. *Journal of Dairy Science, 72*, 2845–2855.

Zhang, D., & Mahoney, A. W. (1990). Effect of iron fortification on quality of cheddar cheese. 2. Effects of aging and fluorescent light on pilot scale cheeses. *Journal of Dairy Science, 73*, 2252–2258.

Zhang, H., Jung, J., & Zhao, Y. (2016). Preparation, characterization and evaluation of antibacterial activity of catechins and catechins–Zn complex loaded β-chitosan nanoparticles of different particle sizes. *Carbohydrate Polymers, 137*, 82–91.

Zhang, H., & Zhao, Y. (2015). Preparation, characterization and evaluation of tea polyphenol-Zn complex loaded β-chitosan nanoparticles. *Food Hydrocolloids, 48*, 260–273.

Zhang, J., Taylor, E. W., Wan, X., & Peng, D. (2012). Impact of heat treatment on size, structure, and bioactivity of elemental selenium nanoparticles. *International Journal of Nanomedicine, 7*, 815–825.

Zhang, J., Wang, X., & Xu, T. (2008). Elemental selenium at nano size (nano-Se) as a potential chemopreventive agent with reduced risk of selenium toxicity: Comparison with se-methylselenocysteine in mice. *Toxicological Sciences, 101*, 22–31.

Zhang, S., Luo, Y., Zeng, H., Wang, Q., Tian, F., Song, J., & Cheng, W. H. (2011). Encapsulation of selenium in chitosan nanoparticles improves selenium availability and protects cells from selenium-induced DNA damage response. *Journal of Nutritional Biochemistry, 22*, 1137–1142.

Zhang, Y., Li, X., Huang, Z., Zheng, W., Fan, C., & Chen, T. (2013). Enhancement of cell permeabilization apoptosis-inducing activity of selenium nanoparticles by ATP surface decoration. *Nanomedicine*, *9*, 74–84.

Zhang, Y., Wang, J., & Zhang, L. (2010). Creation of highly stable selenium nanoparticles capped with hyperbranched polysaccharide in water. *Langmuir*, *26*, 17617–17623.

Zheng, S., Li, X., Zhang, Y., Xie, Q., Wong, Y. S., Zheng, W., & Chen, T. (2012). PEG-nanolized ultrasmall selenium nanoparticles overcome drug resistance in hepatocellular carcinoma HepG2 cells through induction of mitochondria dysfunction. *International Journal of Nanomedicine*, *7*, 3939–3949.

Zimmermann, M., Rohner, F., Ernst, F., Biebinger, R., Ehrensperger, F., Pratsinis, S., & Hurrell, R. (2007). Synthesis, characterization and bioavailability of ferric phosphate nanoparticles. *The FASEB Journal*, *21*, A1113–A1114.

Zimmermann, M. B., Wegmueller, R., Zeder, C., Chaouki, N., Biebinger, R., Hurrell, R. F., & Windhab, E. (2004). Triple fortification of salt with microcapsules of iodine, iron, and vitamin A. *The American Journal of Clinical Nutrition*, *80*, 1283–1290.

Zimmermann, M. B., Winichagoon, P., Gowachirapant, S., Hess, S. Y., Harrington, M., Chavasit, V., ... Hurrell, R. F. (2005). Comparison of the efficacy of wheat-based snacks fortified with ferrous sulfate, electrolytic iron, or hydrogen-reduced elemental iron: Randomized, double-blind, controlled trial in Thai women. *The American Journal of Clinical Nutrition*, *82*, 1276–1282.

Zimmermann, M. B., Zeder, C., Chaouki, N., Torresani, T., & Hurrell, R. F. (2003). Dual fortification of salt with iodine and microencapsulated iron: A randomized, double blind, controlled trial in Moroccan schoolchildren. *The American Journal of Clinical Nutrition*, *77*, 425–432.

Chapter 10

Release, Characterization, and Safety of Nanoencapsulated Food Ingredients

Seid Mahdi Jafari[1], Afshin Faridi Esfanjani[1], Iman Katouzian[1,2] and Elham Assadpour[1]
[1]*Gorgan University of Agricultural Sciences and Natural Resources, Gorgan, Iran,* [2]*Nano-encapsulation in the Food, Nutraceutical, and Pharmaceutical Industries Group (NFNPIG), Universal Scientific Education and Research Network (USERN), Tehran, Iran*

10.1 RELEASE OF NANOENCAPSULATED FOOD INGREDIENTS

The controlled and targeted release of food ingredients plays a key role in their bioavailability (McClements, 2014). The bioactive food ingredients are released into the body following consumption; nevertheless, the untargeted delivery system releases these substances in an undesirable manner. The nanoencapsulation systems have successfully been applied to overcome this barrier. Indeed, the nanoencapsulation techniques are the best delivery systems used for encasing bioactive ingredients that can provide an efficient absorption of the ingredients into the right place. After reaching the targeted tissue, the active ingredients must enter the circulatory system. In regard to this task, bioactive ingredients are first passed through the mucus (a slimy agent used for the aim of lubrication, protection, etc. by the mucous membranes). Delivery systems provide a controlled release of bioactive ingredients according to the equilibrium between the kinetics of their adhesion and the rate of mucus erosion. The amount of the mucous adhered to the epithelial surface is measured by the associated equilibrium between the rate of secretion regulated by the epithelial tissue together with the mechanical shear iatrogenic through the method of digestion (Lai, Wang, & Hanes, 2009; Shaikh, Singh, Garland, Woolfson, & Donnelly, 2011). Therefore, it is necessary to understand the accurate adhesion mechanism of nanocarriers so as to boost the standard of controlled-release formulations and processes. Associate improved adhesion is often obtained by applying functional molecules [e.g., chitosan (CS), lectin] to

Nanoencapsulation of Food Bioactive Ingredients. DOI: http://dx.doi.org/10.1016/B978-0-12-809740-3.00010-6

the formulation of delivery systems. Food and/or drug delivery are accomplished once the bioactive molecules reach the target cells.

10.1.1 Controlled-Release Mechanisms

Concerning the importance of release in delivery systems, it is necessary to have the basic information about the mechanisms of release. There are a variety of definitions considering the term "release mechanism," for instance, it has been used to describe the ways of transportation or release of drugs (Kranz, Ubrich, Maincent, & Bodmeier, 2000; Sansdrap & Moës, 1997). Also, the release mechanism can be used to portray a process or event, which determines the release rate. A broad range of mechanisms have been developed to achieve both temporal and sustained controlled release of delivery systems. In general, as it is shown in Fig. 10.1, the release procedure of active ingredients from delivery systems is classified into partitioning, dissolution, osmosis, diffusion, swelling, and erosion (Raval, Parikh, & Engineer, 2010).

Dissolution involves the transfer of a nutrient from its solid phase into the surrounding medium, such as water phase, polymer network, or tissue. Most active ingredients need to be dissolved in the surrounding medium and freely diffuse within that media before they can act on their target receptors.

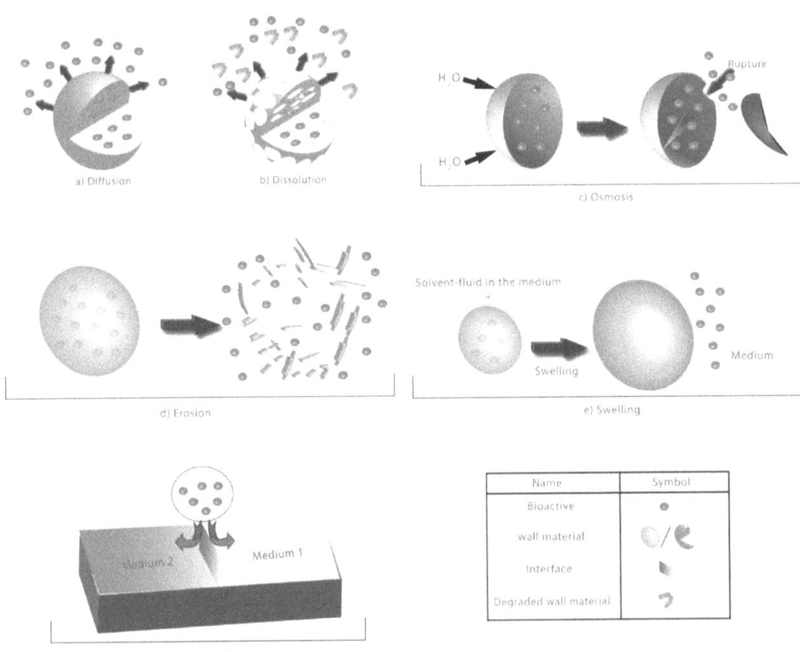

FIGURE 10.1 Different mechanisms for the release of bioactive compounds.

Incorporation of active ingredients in nanoencapsulation systems provides a retarded dissolution and finally a programmable release of active ingredients is achieved. For example, a slower dissolution of active ingredients can be obtained by applying a coating polymer or matrix (Uhrich, Cannizzaro, Langer, & Shakesheff, 1999).

Diffusion describes the way in which the ingredients are released. The molecule and the medium are two main factors determining the diffusion process. As an example, large molecules generally diffuse more slowly than small ones and thus diffusion should be decelerated in viscous liquids. The release of active ingredients based on diffusion may also be controlled by loading them in appropriate nanovehicles. For instance, the diffusion of entrapped active ingredients is inhibited in a matrix of the polymeric particles as an encapsulation system; therefore, active ingredients must be transported from tortuous pathways (Fredenberg, Wahlgren, Reslow, & Axelsson, 2011; Uhrich et al., 1999). In other words, diffusion is dependent on the porous structure of the delivery systems.

Osmotic pumping may be a development that occurs once osmotic pressure is caused by water absorption, which initiates the transport of the nutraceutical. Here, the type of the transport is known as convection but not diffusion. Moreover, in polymer particles, osmotic pressure is caused by water absorption that induces rupture of the polymer (Jonnalagadda & Robinson, 2000).

Erosion is another release mechanism (e.g., release of active ingredients without their transport). It results in identical profiles of active release and erosion of the vehicle, suggesting that the active ingredients are homogeneously distributed throughout the delivery system. To point out, in the polymeric particle delivery systems, active ingredients, which are very close to the surface, are released without transport due to polymer erosion (Fredenberg et al., 2011).

Swelling is another release mechanism resulting in the uptake of water by delivery systems, with an increase in volume. The process of swelling is similar to osmosis since water enters the delivery systems relatively rapidly. Swelling of a polymer may be influenced by nutrients or other ingredients acting as osmolytes, which draw water into the polymer. This process can be induced via the selection of suitable polymers plus the modification of external parameters, such as temperature and pH (Huang & Brazel, 2001; Korsmeyer, Gurny, Doelker, Buri, & Peppas, 1983).

The partition coefficients of active ingredients have the main effect on the driving forces for release. Actually, when active ingredients encounter an interface between two phases, the ratio of active ingredient solubility in these phases determines the partition coefficient. The ingredients with high partition coefficient have low aqueous solubility and the large ingredients with low partition coefficient reveal low membrane permeability due to their poor lipophilicity (Shang, Zhang, Du, & Venkatraman, 2008).

10.1.2 Analytical Methods for Measuring Release

The entrapment of active ingredients in nanoencapsulation systems during storage and their following release in targeted zones are the main aims of the delivery design. Therefore, it is necessary to apply analytical methods that measure the concentration of active ingredients within nanoencapsulation systems and to provide information about their release from the nanocarrier into the surrounding medium. These methods can commonly be classified into destructive and nondestructive methods.

The main *destructive methods* used for measuring the concentration of active ingredient in the nanoencapsulation systems are dehydration, centrifugation, gravitational separation, filtration, and dialysis (Nielsen, 1998). The dialysis is one of the common destructive methods, which can be applied to measure the release of active ingredients. For this purpose, the nanoencapsulated ingredient is placed within a dialysis bag that has pores smaller than the nanoencapsulated bioactive but larger than the bioactive ingredient. Thereupon, in order to obtain the release profile of active ingredients during storage, the concentration of active ingredients released from the nanoencapsulation systems and into the liquid surrounding the dialysis bag is measured over time (McClements, 2014).

On the other hand, the *nondestructive methods* are ideal methods for measuring the relative concentration of an active ingredient inside and outside of the nanoencapsulated particles. In this case, nuclear magnetic resonance (NMR) methods can measure the release of active ingredients from nanoencapsulation systems in different physical environments within multiphase materials (McClements, 2014). Fluorescence spectroscopy can also provide the fundamental information about the microenvironment identification of active ingredients within nanoencapsulation systems. The fluorescence image provides information about the behavior of nanocapsules in the gastrointestinal tract (GIT), and their uptake could develop a fundamental understanding of the distribution profile of nanocapsules within in vivo systems. Detection of fluorescence emission from fluorophores in the whole-body of living small animals is performed by using the sensitive camera for providing fluorescence images. Long emissions at the near-infrared region are generally preferred to overcome the photon attenuation in living tissues (Rao, Dragulescu-Andrasi, & Yao, 2007). The concentration of specific ionic ingredients (such as Na^+ and Ca^{2+}) that are incorporated in nanoencapsulation systems can be measured via ion-selective electrodes (Kotanen, Wilson, Wilson, Ishihara, & Guiseppi-Elie, 2012).

10.2 BIOLOGICAL FATE OF NANOENCAPSULATED FOOD INGREDIENTS

Nanoencapsulation of food ingredients has been applied to a variety of purposes including targeted release in a specific zone of GIT, increasing oral bioavailability, protection against environmental and process stresses as well as masking undesirable flavors. Given these points, released ingredients from encapsulation systems should be absorbed by the human body (McClements, 2014). It is important to consider that the purpose of absorption is the optimal absorption and not the maximum rate of absorption. Overall, an optimal amount of active ingredients is sufficient for maximum usage of them. Engineering nanoencapsulation systems with the ability to provide controlled and targeted release via optimal absorption are necessary to improve the bioavailability of bioactives (Omenn et al., 1996).

The availability of bioactive ingredients (e.g., phenolic ingredients and antioxidants) in food formulations is related to the term of *bioavailability*. For the most part, bioavailability incorporates a range of parts, clutch absorption, distribution, metabolism, and excretion (ADME) including biochemical and physiological effects. There are many routes for the administration of bioactive ingredients. In pharmacology, drugs are commonly administered by injection, usually intravenous, subcutaneously, intramuscularly, and intraperitoneal. However, oral administration is the best route in terms of patient compliance (Markovsky et al., 2012; Szakács, Varadi, Özvegy-Laczka, & Sarkadi, 2008). On the other hand, free food bioactive ingredients and formulated food products containing bioactive ingredients are orally administered. Nevertheless, the entry of bioactive ingredients into systemic circulation is restricted by several barriers, such as acidic conditions of the stomach, brush border membrane, and proteases in the gut lumen, metabolism by liver enzymes (the "first-pass effect") and tightly bound intestinal epithelial cells (enterocytes). Recently, these restrictions have been overcome by applying a variety of nanoencapsulation systems such as nanoemulsions, nanoliposomes, and polymeric NPs. The nanoencapsulated ingredients can pass the intestinal epithelium either between the enterocytes (paracellular route) or through the cells via the transcytosis mechanism. Nonloaded ingredients can also pass the epithelium by diffusing through the cells (transcellular route), as shown in Fig. 10.2. For instance, in the paracellular route, molecules transport from around or between cells. Besides, transportation of macromolecules and polar ingredients is difficult by intercellular tight junction (Madhav, Shakya, Shakya, & Singh, 2009; Markovsky et al., 2012).

Physicochemical properties of the nanoencapsulation systems such as size, shape, surface charge, and decorations, along with their mechanical properties, have a major role in the fate of nanoencapsulated ingredients in blood stream including biodistribution, targeting, vascular dynamics, uptake, clearance, release kinetics, and degradation. The free bioactive ingredients as low-molecular-weight substances have an untargeted release by diffusing through the normal vasculature to any tissue and entering the cells. The untargeted release of nutraceuticals can be resolved by exerting them in nanoencapsulation systems, such as biodegradable polymeric NPs. The gaps in angiogenic vessels are the route of nanoencapsulated bioactive ingredients as macromolecules into tissues. Therefore, nanoencapsulation systems can

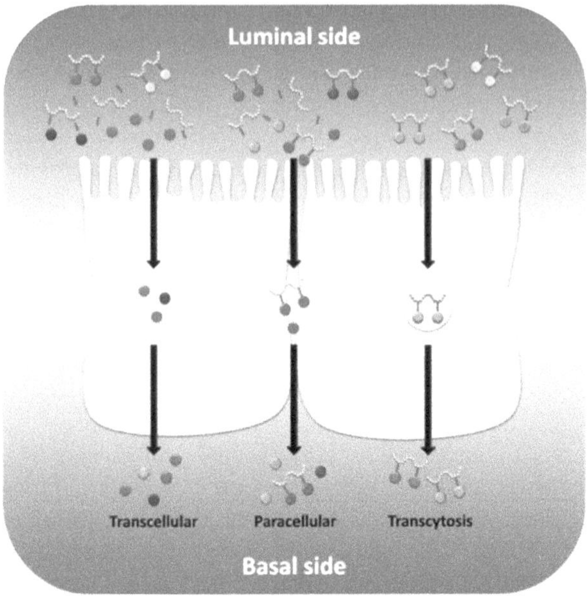

FIGURE 10.2 Transepithelial delivery. Polymer–drug conjugates can cross the epithelium through either transcytosis (through the cells) or via paracellular (between the cells) route. Free drugs can also diffuse through the cells (transcellular route). *Reprinted with permission from Markovsky et al. (2012).*

provide selectivity for target tissues by altered biodistribution compared with the unencapsulated form of them (Matsumura & Maeda, 1986).

The particle size is one of the important characterization tests of nanoencapsulated ingredients, which has a major effect on its distribution. In general, prolonged circulation, accumulation in the target tissue, and enhanced diffusion in the tissue can be obtained by applying carriers within the particle size range of 20–100 nm (Alexis, Pridgen, Molnar, & Farokhzad, 2008). The shape of nanoencapsulated ingredients also has a major influence on its distribution and performance. For example, linear polymeric carriers have flexible random coil structure and can pass through gaps smaller than their hydrodynamic diameter and penetrate into tissues better than the more structured spherical carriers (Markovsky et al., 2012). On the other hand, the charge of nanoencapsulation systems is important in dictating cellular fate. For example, in regard to a slight negative charge of the cell membrane and main role of electrostatic attractions in cell uptake, the biodistribution and cellular uptake of cationic carriers (e.g., positive charge of NPs) are faster than neutral or negative charge carriers (Thorek & Tsourkas, 2008).

Predicting the metabolic fate of nanoencapsulation systems is extremely difficult due to the complex nature of the processes. For instance, the nanoencapsulation systems based on polymers can undergo cleavage of bonds in the polymeric backbone, either hydrolytically or enzymatically, biodegradation with the degradation products, such as amino acids and saccharides, absorbed in the biochemical pathways of the body (Mehvar, 2000; Nair & Laurencin, 2007).

The act of engineering nanoencapsulation systems for encasing bioactive ingredients will not only deliver the bioactive ingredients to its target site and release procedure, but it will also excrete from the body. The high-molecular weight of nanoencapsulated ingredients inhibits their elimination from the body. Generally, the large molecules remain in the tissue after cellular death or undergo exocytosis. Later, they are eliminated through returning into the blood via the lymphatic circulation and glomerular filtration in the kidney (Seymour, Duncan, Strohalm, & Kopeček, 1987).

10.3 MUCOADHESIVE DELIVERY SYSTEMS

The bioactive ingredients that are consumed via the oral route for reaching the target tissue must adhere to mucus and cross the mucosal layer. As a matter of fact, the mucus layer is the first barrier for the nanoencapsulated ingredients. This layer covers the majority of the epithelium; therefore, it retards the transport of bioactive ingredients by preventing the direct adhesion to the epithelial cells. Hence, the balance between the kinetics of nanoencapsulation systems adhesion and the rate of erosion of the mucus are key parameters determining their delivery to the targeted cells. In addition, the delivery systems for bioactive ingredients must enter the bloodstream to trigger the targeted release. For this purpose, the nanovehicles must adhere to the mucus layer and then diffuse across it (Allen & Flemström, 2005).

10.3.1 Characterization of the Mucus Layer

Mucus is a complex hydrogel. The main constituents of the mucus are proteins, lipids, carbohydrates, antibodies, salts, bacteria, and cellular debris (Cone, 2009).

Viscoelasticity is the key characterization factor for the mucus layer; accordingly, two types of mucus include loosely adherent mucus layer and firmly adherent mucus layer, differing in terms of viscoelasticity, are secreted. A more firm mucus layer can be observed underneath this loose mucus. The difference of loose and firm mucus is underlined due to their rheological properties. Applying a low shear force can remove the loose mucus; therefore, loose mucus acts as a lubricant agent and facilitates the movement of materials in the gut. On the other hand, the firm layer is more resistant to applied shear and can form a firm protective barrier and a stable unstirred layer, which is resistant to shear (Allen & Flemström, 2005; Laine, Takeuchi, & Tarnawski, 2008). The role of mucus layer in different parts of the gut has an essential role in the thickness of loose and firm mucus. For example, a thick layer of loose mucus in the small intestine is associated with the lubricant activity of mucus (Ensign, Cone, & Hanes, 2012).

pH also varies throughout the gut. In the stomach, mucus pH varies from 1 to 2 on the luminal surface to nearly neutral pH at the epithelial surface. The pH is rapidly yet only partially neutralized in the duodenum and steadily increases through the rest of the gastrointestinal (GI) tract, reaching a pH of 7–8 in the colon and rectum (Ensign et al., 2012).

Hydrophobicity is the main property of mucus layer that results in the protection of the epithelium. This feature emanates from the presence of lipids, glycerol glucolipids, and ceramides in mucus (Allen & Flemström, 2005; Lichtenberger, 1995).

10.3.2 Stages of Mucoadhesion

The macromolecules, such as nanoencapsulation systems that adhere onto the mucus, are commonly named mucoadhesive. The adhesion mechanism of mucoadhesive onto the surface of mucus is complex and not well understood yet. In general, the adhesion mucoadhesive to the mucus is performed in two stages including the contact stage and the consolidation stage (Fig. 10.3A). The first stage starts by contact between mucoadhesive and mucus layer with spreading and swelling of the formulation. The consolidation step is based on the activation of the mucoadhesive materials by the presence of moisture. Indeed, moisture plasticizes the system, permitting the mucoadhesive molecules to interrupt free and to converge by weak van der Waals and hydrogen bonds (Morales & McConville, 2011; Smart, 2005).

Typically, the consolidation step is explained by two theories: the diffusion theory and the dehydration theory. Diffusion theory is based on the interaction of mucoadhesive materials and the glycoproteins of the mucus mutually by means of the interpenetration of their chains and the building of secondary bonds. On the contrary, dehydration theory is based on an osmotic balance between mucoadhesive materials and mucus (Fig. 10.3B). When a material capable of being readily verifiable in an aqueous environment is brought into contact with the mucus can cause its dehydration due to the difference in osmotic pressure. Later, the mixture of mucoadhesive materials and mucus is formed with the transport of water into the formulation until the osmotic balance is reached between the two layers (Morales & McConville, 2011; Smart, 2005).

10.3.3 Mucoadhesion Theories

The theory of mucoadhesion is very complex and the chemical and physical basis of mucoadhesion are not yet well understood. However, generally, there are six theories that can be applied to explain this complex process.

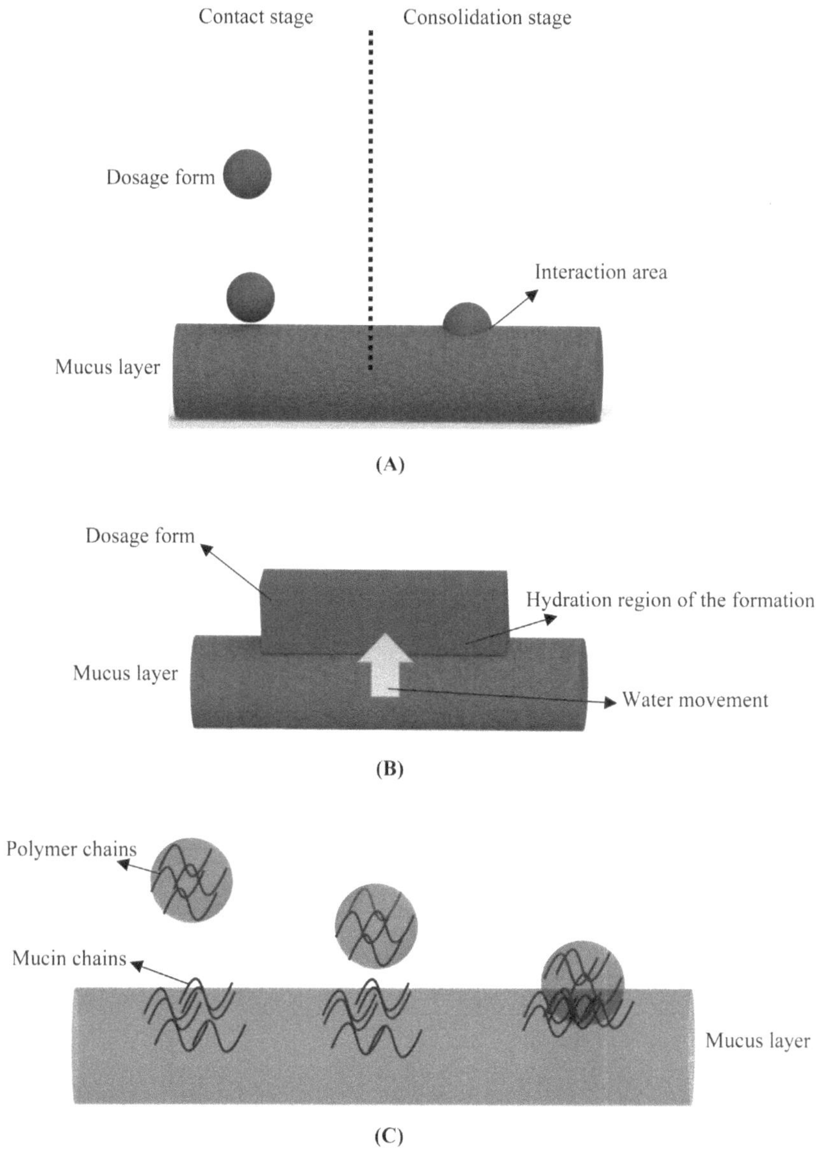

FIGURE 10.3 (A) Two steps of the mucoadhesion process; (B) dehydration theory of mucoadhesion; and (C) secondary interactions resulting from interdiffusion of polymer chains of mucoadhesive material and of mucus.

They include electronic, adsorption, wetting, diffusion, fracture, and mechanical theories (Boddupalli, Mohammed, Nath, & Banji, 2010; Smart, 2005).

Electronic theory: In this theory, it is assumed that the mucoadhesive materials and the mucus have opposing electrical charges. Therefore, a building of a double electronic layer at the interface is achieved through the transfer of electrons across the interface. These attractive forces applied to the mucoadhesive material maintain on the surface of the mucus layer upon intimate contact (Smart, 2005).

Adsorption theory: This theory is based on the secondary interactions between a mucoadhesive material and the molecules of the mucus. These interactions include van der Waals and hydrogen bonds as well as electrostatic attraction or hydrophobic interactions. Different interactions may end up in associate intense global adhesion (Smart, 2005).

Wetting theory: This theory is applied to the liquid mucoadhesive materials and is based on the ability of affinity of liquid mucoadhesive material to spread spontaneously onto a surface of mucus as a requirement for the event of adhesion. This liquid affinity for a surface can be found by using the measurement techniques, such as the contact angle. Hence, for providing adequate spreadability, the contact angle should be equal or close to zero. The spreading coefficient (S_{AB}) can be calculated using the difference between the surface energies γ_B and γ_A of the mucus and liquid mucoadhesive material and the interfacial energy γ_{AB}, using the following equation (Smart, 2005):

$$S_{AB} = \gamma_B - \gamma_A - \gamma_{AB} \tag{10.1}$$

S_{AB} should be positive for the liquid mucoadhesive material to spread spontaneously over the mucus and consequently, an intimate contact and a strong adhesive bond are favored.

The adhesion work (W_A), the energy required to separate the liquid mucoadhesive material and mucus, can be given by (Smart, 2005):

$$W_{AB} = \gamma_B + \gamma_A - \gamma_{AB} \tag{10.2}$$

Diffusion theory: This theory is based on interdiffusion of polymer chains into the mucus layer and by the diffusion of strands of the mucin molecules into the bioadhesive polymer (Fig. 10.3C). This process relies on the concentration gradients plus the available molecular chain lengths. It is mainly enhanced by the presence of similar functional groups between bioadhesive polymer and mucin molecules (Smart, 2005).

Mechanical theory: This theory is based on the interaction of liquid mucoadhesive material into irregularities on a rough surface of mucus. The roughness of the surface provides an increased contact area (Boddupalli et al., 2010; Smart, 2005).

Fracture theory: It analyzes the forces required to separate two surfaces after adhesion is established. In this theory, one should assume that the detachment happens at the interface. However, failure usually happens at the weakest part that is often a cohesive failure at intervals one among the adhering surfaces. This theory is taken into account to be acceptable for the calculation of fracture strength of adhesive bonds involving rigid mucoadhesive materials (Boddupalli et al., 2010; Smart, 2005).

10.4 CHARACTERIZATION OF NANOENCAPSULATED FOOD INGREDIENTS

There are many different nanoencapsulation systems, such as nanoemulsions, nanoliposomes, solid lipid nanocarriers (SLNs), nanostructured lipid carriers (NLCs), polymeric-based NPs, micelles, etc. that can be used for nanoencapsulation of food ingredients (Esfanjani & Jafari, 2016; Katouzian & Jafari, 2016). Identification characterization of these systems can provide useful information about their storage and application of them in food formulations. The morphology, size, surface properties, and physicochemical properties are the important characterization parameters of nanoencapsulated food ingredients that can mainly affect the stability along with the functional and physicochemical properties of them. Also, information about the characterization of nanoencapsulated ingredients is necessary so as to fortify the food formulations. Certainly, improving the knowledge in the area of nanoencapsulation characterization results in the better design of a delivery system without an adverse change in the quality of fortified foods (e.g., appearance). Consequently, the best methods should be applied to determine the characterization of nanoencapsulation systems. In this section, we provide applicable information about common and typical instruments and analytical methods for the analysis of nanoencapsulated ingredients plus their characterization.

10.4.1 Morphology

Morphology is one of the main characterization features of nanoencapsulation systems in a way that most of their properties such as functional properties, physicochemical state, and stability are influenced by their morphology. Necessary information about structure (e.g., surface properties including smoothness, wrinkles, voids, etc.) and shape (e.g., spherical, nanoboxes, nanoclusters, nanoreefs, nanotubes, etc.) of nanocapsules can be obtained by preparing images and interpreting them by microscopic studies. The optical microscope is the most common type of the microscope; however, it cannot provide a high-resolution image from particles smaller than 500 nm (McClements, 2014). Therefore, it is important to employ modern and accurate instruments like electron microscopy, atomic force microscopy (AFM), and confocal laser scanning microscopy (CLSM) for studying the morphology of nanoencapsulated ingredients (Gupta & Wells, 2004; McClements, 2014).

10.4.1.1 Electron Microscopy

Electron microscopy is based on a beam of electrons and provides as much as a 1000-fold increase in resolving power. The high energy of electron beams is provided from the cathode and accelerated into the anode through the electron field, focused by electrostatic and electromagnetic lenses, and then transmitted through a specimen that is in part transparent to electrons and in part scatters them out of the beam. When it emerges from the specimen, the electron beam carries information about the structure of the specimen which is magnified by the objective lens of the microscope. The magnified electron image is provided by projecting onto a fluorescent viewing screen coated with a phosphor or scintillator material like zinc sulfide (Egerton, 2005). Electron microscopy comprises transmission electron microscopy (TEM) and scanning electron microscopy (SEM). Their schematic representations are illustrated in Fig. 10.4A and B (Egerton, 2005; Egli, 2010).

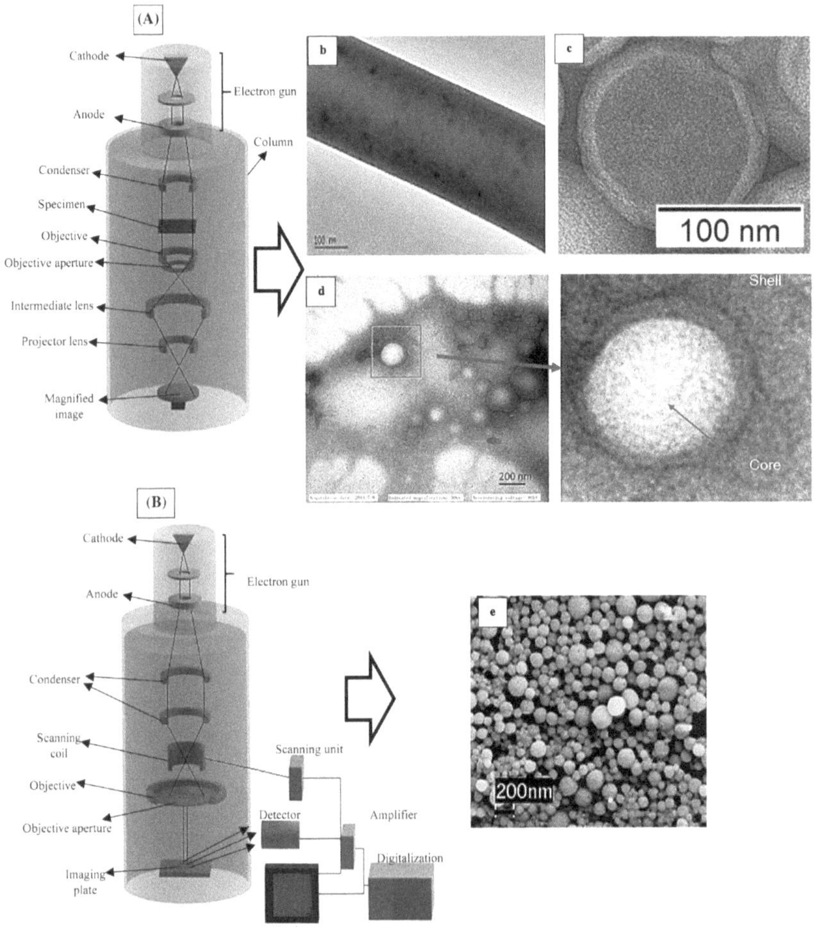

FIGURE 10.4 Schematic diagram of transmission electron microscope (TEM) (A). TEM image of gallic acid-loaded zein electrospun fibers (a) reprinted with permission from Neo et al. (2013). Copyright © 2016, ACS, L-carnosine-encapsulated nanoliposomes (b) reprinted with permission from Maherani et al. (2012). Copyright © 2016, ACS, triple-layered nanocapsules contain curcumin (d) reprinted with permission from Abbas et al. (2014). Copyright © 2016, ACS, Schematic diagram of scanning electron microscope (SEM) (B) SEM images of zein nanoparticles (e) reprinted with permission from Zhong and Jin (2009). Copyright © 2016, ACS.

The high-resolution image (at a magnification of $\times 10^3 - \times 10^6$ and a resolution smaller than 1 nm) of nanoencapsulation systems in biological and living systems can be obtained by interacting particles and electrons via TEM. As an example, Neo et al. (2013) employed TEM for studying the microstructure of zein nanofibers containing gallic acid. The TEM-provided image of nanofibers showed that the sub-micron fibers present a beadless morphology, which displayed smooth surfaces without visible separation of particles from the fiber matrix (Fig. 10.4b). The average diameters of the gallic acid-loaded fibers varied from 327 to 387 nm. The provided TEM image also showed the deposition and distribution of gallic acid particles within the fiber mats under high resolution (Fig. 10.4b). Gallic acid-loaded zein electrospun fibers exhibited dark patches throughout the fibers.

In a study conducted by Maherani, Arab-Tehrany, Kheirolomoom, Cleymand, and Linder, 2012), TEM was utilized for providing information about nanoliposomes loaded with natural dipeptide antioxidants. The provided image by TEM confirmed that the ultimate nanoliposomes were in the form of large unilamellar vesicles (Fig. 10.4b). Also, the bilayer structure of nanoliposomes was clearly visible in these micrographs (defined as closed continuous bilayer structures).

TEM can provide basic information about the position of walls in nanoencapsulation systems, especially in multilayer systems. For instance, TEM images of prepared nanocapsules were obtained by observing the thickness of the shell and the morphology of core material, as shown in Fig. 10.4c. These polymeric nanocapsules were prepared by sequential deposition of food-grade polyelectrolytes through the self-assembling process onto the oil [medium-chain triglycerides (MCT)] droplets enriched with curcumin (CU) (lipophilic bioactive compound) (Abbas, Bashari, Akhtar, Li, & Zhang, 2014).

On the other hand, the SEM provides an image (with a magnification of $\times 10 - \times 500{,}000$ at a resolution of less than 1–20 nm) about the surface of a dried form of nanoencapsulation systems, such as nanoencapsulated powders and fibers (Lametschwandtner, Lametschwandtner, & Weiger, 1990). For example, the prepared SEM image from zein NPs as food-grade nanoscalar delivery systems showed that most particles were spherical and had diameters smaller than 200 nm (Fig. 10.4e). The magnitudes measured from SEM images were similar to those from dynamic light scattering (DLS) but manifested a monotonic decrease with an increase in homogenization rate.

10.4.1.2 Confocal Laser Scanning Microscopy

High resolution and efficient depth image of nanoencapsulation systems can be obtained by using CLSM as scanning probe microscopies. The principle of CLSM is based on detection of focused light generated by the interaction of the illuminating beam with the small sample illumination, by point illumination in which the pinhole spatially filters allow only focused light to reach the detector (Fig. 10.7A). The addition of a fluorescent dye into nanoencapsulation systems without fluorescence is needed to study their morphology by CLSM (Aguilera & Stanley, 1999).

Determination of the wall materials deposition on nanoencapsulation systems (e.g., oil droplet in oil in water nanoemulsions and nanocapsules) is the main application of CLSM. Abbas et al. (2014) prepared images of nanoencapsulated CU in CS-coated nanoemulsions and carboxymethyl cellulose (CMC)-coated nanocapsules using CLSM in order to provide information about the deposition of wall materials on the core materials by marking the CS molecules and oily core with Nile Red and fluorescein isothiocyanate (FITC) dyes, respectively. After excitation caused by a laser beam, Nile Red dye gave a red color to the lipids and oils while FITC induced a green color to proteins and CS due to the presence of the amino groups. The oily droplet containing CU with Nile Red and CS layer (green) marked with FITC are shown in Fig. 10.5a and b. Fig. 10.5c shows oily droplet containing CU which is completely coated with CS and has a spherical shape. Abbas et al. (2014) analyzed the deposition of CS−CMC polymers on the starch-stabilized oil droplets. Fig. 10.5f is an overlay image of Fig. 10.5d (nanocapsule marked with Nile Red fluorescence dye [red]) and Fig. 10.5e (fluorescence from FITC [green]). It illustrates yellowish-green polymeric shell as spherical and stable nanocapsules consisting of oil core and starch−CS−CMC-based polymeric shell.

10.4.1.3 Atomic Force Microscopy

The visualization in three dimensions can be achieved by AFM. Resolution in the vertical axis, or *Z*, is limited by the vibration environment of the instrument, whereas resolution in the horizontal axis, or *X*−*Y*, is limited by the diameter of tip utilized for scanning. AFC provides the main information about structures of nanoencapsulation systems at the atomic and molecular level by scanning the surface of samples with a sharp tip, without any contact. As explained schematically in Fig. 10.5B, mounted tip on a cantilevered arm is closely held against the surface of the sample. The ventricular movement of a tip from the surface due to the repulsive forces is detected by the optical system that senses the spacing between the probe and the sample and provides a correction signal. By producing an amplified signal of the minute deflections of the cantilever, it is possible to obtain an image of its structure (Binnig, Quate, & Gerber, 1986; Jalili & Laxminarayana, 2004).

The main advantages of AFM include providing an image from nanomaterials in both lateral and vertical positions without any sample preparation, the possibility of operation under air, or fluid in addition to both lateral and vertical resolution (Jalili & Laxminarayana, 2004; Liang, Mao, & Ng, 2004). Torres-Giner, Gimenez, and Lagarón (2008) applied AFM to study the surface topography of zein nanofibers prepared via electrospinning method. AFM image revealed that the morphology of the fibers is not homogeneous

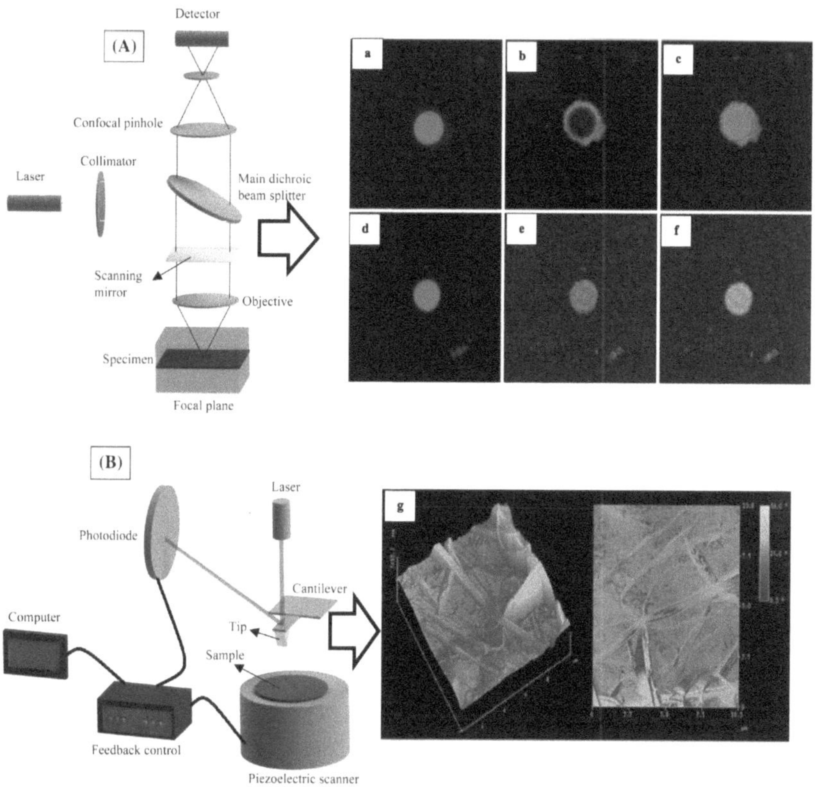

FIGURE 10.5 Schematic diagram of confocal laser scanning microscope (CLSM) (A). CLSM images of chitosan-coated nanoemulsion (NEII) (a–c) and nanocapsules (d–f) reprinted with permission from Abbas et al. (2014). Copyright © 2016, ACS. *z*-Stack confocal images of NEII droplets (a) oily core (red, pictures a, c and d) marked with Nile Red, (b) chitosan layer (green, pictures b, c and e) marked with FITC. (c) Superposition of image (a) and (b), showing oily droplet inside chitosan layer. *z*-Stack confocal images of nanocapsules (d) oily core (red, pictures a, c and d) marked with Nile Red, (e) chitosan–CMC shell (green, pictures b, c and e) marked with FITC. (f) Superposition of image (d) and (e) (yellowish in color), showing oily droplet encapsulated inside the polymeric shell. Sematic diagram of atomic force microscope (AFM) (B). AFM height (left) and phase (right) images of zein fibers electrospun under the so-called standard conditions, reprinted with permission from Torres-Giner et al. (2008). Copyright © 2016, ACS.

or wholly tubular (Fig. 10.5g). In addition, the ribbons with two tubes at the sides were observed in AFM image.

10.4.2 Size

Most of the characteristics of nanoencapsulation systems comprising stability, appearance, rheology, functional properties, and release are influenced by their size and size distribution. Furthermore, information about the size of nanoencapsulation systems is a determining factor in the fortification of food formulations. For instance, an optimized nanoencapsulation system (e.g., nanoemulsions) below 100 nm needs to be designed in order to enrich the transparent beverages. The information about the size of nanoencapsulation systems can be offered as particle size distribution, polydispersity index, or the central tendency and width of the distribution. Besides, there are some main parameters that can be expressed as the Sauter, or volume mean diameter, span, and Z-average size (Hassellöv, Readman, Ranville, & Tiede, 2008; Jafari, He, & Bhandari, 2007; Mohammadi, Jafari, Assadpour, & Esfanjani, 2016; Mohammadi, Jafari, Esfanjani, & Akhavan, 2016):

- Surface area moment (Sauter) mean diameter, $D_{32} = \left(\left(\Sigma n_i \times d_i^3\right) / \left(\Sigma n_i \times d_i^2\right)\right)$
- Volume or mass moment (Brouckere) mean diameter, $D_{43} = \left(\left(\Sigma n_i \times d_i^4\right) / \left(\Sigma n_i \times d_i^3\right)\right)$
- Distribution width of droplet sizes (Span), smaller span shows a narrower distribution,
- Span = ([$d(v,90)$ − $d(v,10)$]/d(v,50)). In this formula, $d(v,10)$, $d(v,50)$, and $d(v,90)$ are diameters at 10%, 50%, and 90% cumulative volume, respectively. In other words, [$d(v,90)$ − $d(v,10)$] is the range of the data and $d(v,50)$ is the median diameter.
- *Z*-average size, an intensity weighted average attributed to certain methods, $D_n = \left(\left(\Sigma n_i \times d_i^6\right) / \Sigma n_i d_i^5\right)$.

In general, the volume mean diameter represents the existence of big particles, while surface mean diameter (D_{32}) is associated with the smaller particles. When there is a big difference between D_{32} and D_{43}, it means that the size distribution is bimodal with two peaks and when they are close together, it shows that we have only one peak (Jafari, He, & Bhandari, 2006). Accordingly, a bimodal and one pick size distribution on nanoemulsions produced by rotor–stator homogenization and sonication can be observed in Fig. 10.6 (Jafari et al., 2007).

This section provides some common and fundamental analytical instruments that can be used for providing mentioned information about the size and size distribution of nanoencapsulation systems.

10.4.2.1 Static Light Scattering

Static light scattering is widely used for studying shape, size, size distribution, and aggregate structure size distribution of nanoencapsulation systems. It is based on the light scattering pattern. As shown in Fig. 10.7a, when a ray of light passes through a dilute spread particulate sample, large particles scatter lightweight at little angles relative to it and little particles scatter light at massive angles. The particle size is then calculated by analyzing the angular scattering intensity data through a software which is based on Mie theory of light scattering (Brown, 1996).

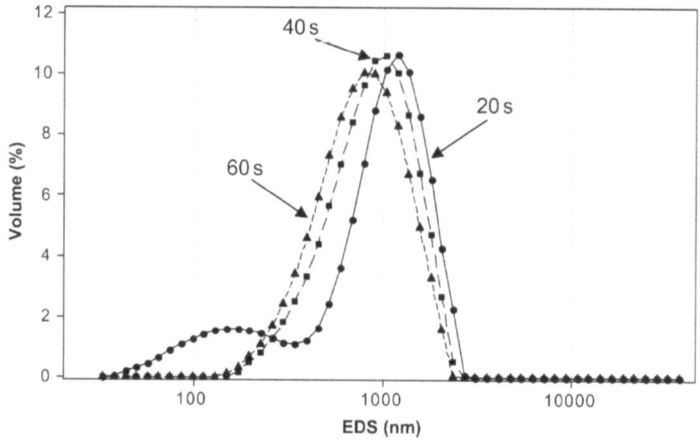

FIGURE 10.6 Effect of preemulsion preparation method and sonication time on emulsion droplet size and size distribution. Coarse emulsions prepared by IKA mixer. *Reprinted with permission from Jafari et al. (2007). Copyright © 2016, ACS.*

10.4.2.2 Dynamic Light Scattering

DLS is typically used for measuring the size and size distribution of particles, emulsions, or molecules typically in the submicron region, which has been dispersed or dissolved in a liquid. Also, particle concentration range and microrheology of small particles can be obtained by DLS. The basic principle relies on changing the wavelength of the incoming light: The sample in Brownian motion is exposed to a laser beam and the fluctuations of the scattered light are detected at a known scattering angle θ by a fast photon detector (Fig. 10.7b) (Pecora, 2000).

10.4.2.3 Gravitational Settling and Centrifugation

Gravitational settling and centrifugation is the basis of many particle size techniques including light scattering, electrical conductivity, ultrasonic velocity, NMR, or X-rays. These techniques are based on the velocity at which particles move in a gravitational or centrifugal field: Larger particles move more rapidly than smaller ones, and therefore measurements of particle velocity can be used to determine the particle size using a suitable mathematical model. For example, sedimentation or creaming velocity owing to gravity can be applied for measuring the size of large particles, whereas that of smaller particles can be determined by measuring their velocity in a strong centrifugal field (Mcclements, 2007, 2014).

10.4.2.4 Laser-Induced Breakdown Detection

Laser-induced breakdown detection (LIBD) is one of the most sensitive instruments that can be used for measuring the size of colloidal NPs with diameters ranging from approximately 5 to 1000 nm. In particular, LIBD is

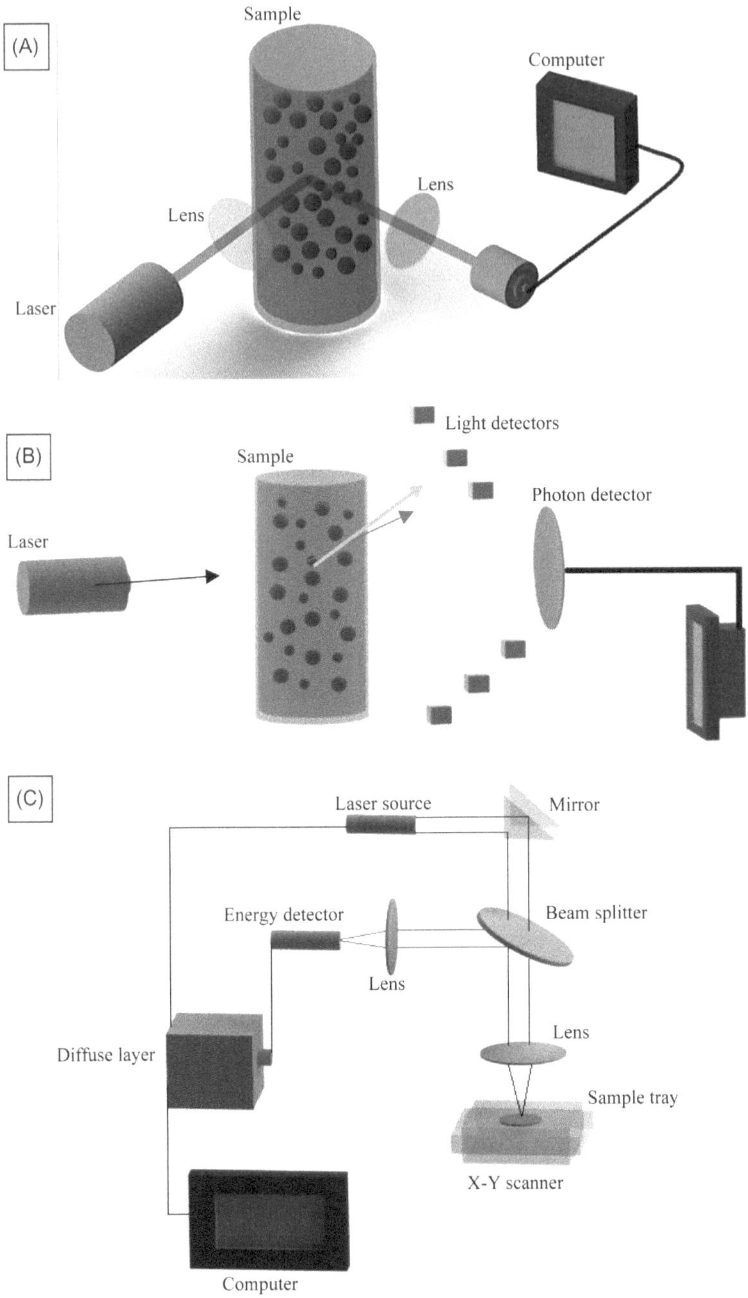

FIGURE 10.7 Schematic diagram of analytical instruments for measuring size of nanocapsules; dynamic light scattering (A), static light scattering (B), laser-induced breakdown detection (C).

being developed for detecting extremely small particles less than 100 nm in diameter, which is not easily detectable by using the currently available commercial devices that adopt the measurement of a light scattering intensity. LIBD is based on the plasma generation (or breakdown) on single particles by a focused, pulsed laser beam that can be used for detection of colloids as a sensitive method. Plasma formation is occurred by passing nanoencapsulation systems through the focal volume of the optical cell. A microscope connected to a triggered charge-coupled-device (CCD) camera or a piezo-electric crystal attached to the cuvette can be used for detection of plasma light emission (Fig. 10.7c). The detection data by CCD or piezo-electric crystal are presented as breakdown probability (BP). Since BP for a given laser energy depends both on particle concentration and size, it is necessary to elucidate both (Bundschuh, Knopp, & Kim, 2001).

10.4.3 Electrical Characteristics

The electrical features of nanoencapsulation systems emanate from employing a variety of ionic materials, such as free fatty acids, phospholipids, proteins, and ionic polysaccharides in their formulation. In other words, type, location, and concentration of any ionized species present at their surfaces as well as the ionic composition and physical properties of the surrounding liquid have a key role on electrical characteristics of nanoencapsulation systems. Different properties of nanoencapsulation systems such as stability, release, and physicochemical properties are influenced by their electrical properties. Moreover, electrical characterization can provide information about the ability of nanoencapsulation systems to adhere to solid surfaces, biological surfaces after ingestion, and aggregating of particles as a result of electrostatic repulsion (Hunter, 2001; McClements, 2015).

Commonly, surface charge properties of nanoencapsulation systems are described by zeta-potential due to better representation and easy measurement of the electrical characteristics. It is defined as the distance away from the particle surface below which any counterions remain strongly attached to the particle when it moves in an electrical field (Hunter, 2001; McClements, 2004; Zhang, Zhang, & McClements, 2016).

Typically, the electrical properties of a particle are characterized by measuring the z-potential versus pH profile. Zeta-potential has values that typically range from +100 to −100 mV and colloidal particles with zeta-potentials of greater than +30 mV or less than −30 mV are considered strongly cationic and anionic, consecutively (Clogston & Patri, 2011). Generally, a high zeta-potential (either positive or negative) maintains a stable system. A higher electric charge on the surface of the droplets of emulsion will prevent aggregation of the droplets in buffer solution because of the strong repellent forces among particles (Honary & Zahir, 2013). Furthermore, zeta-potential can be measured by applying different analytical techniques like electrophoresis, laser Doppler velocimetry, and zeta-sizer instrument (Malvern, UK) (Clogston & Patri, 2011; Hunter, 2001; McClements, 2004).

10.4.4 Surface Composition

Chemical characteristics of the surface layer of nanoencapsulation systems can be obtained by X-ray photoelectron spectroscopy (XPS) as it provides an electron image from samples by scanning of a focused X-ray beam. As shown in Fig. 10.8A, the electrons are re-emitted from the surface of the sample exposed to X-ray. The kinetic energy and number of the emitted electrons provide a composition and electronic state of the samples (Fig. 10.8A) (Andrade, 1985).

The common application of XPS in nanoencapsulation systems is the determination of free compounds on the surface of the produced capsules. Jafari, Assadpoor, Bhandari and He (2008) identified the unloaded fish oils by analyzing the surface composition of the nanoencapsulated powders using the XPS. Regarding the XPS study, the relative atomic concentration (%) of carbon, oxygen, and nitrogen in the surface layer of the powder sample was explored. As a consequence, the percent composition (relative coverage) of the different components, such as fish oil in the surface layer of powder prepared from whey protein concentrate (WPC) and maltodextrin, was calculated by using the relation in a matrix formula as

$$CE = \alpha CS + \beta CP + \gamma CO$$

$$OE = \alpha OS + \beta OP + \gamma OO$$

$$NE = \alpha NS + \beta NP + \gamma NO$$

where C, O, and N are the relative atomic concentrations of carbon, oxygen, and nitrogen in encapsulated powder (C_E, O_E, and N_E), maltodextrin (C_S, O_S, and N_S), protein, i.e., WPC (C_P, O_P, and NP), and fish oil (C_O, O_O, NO),

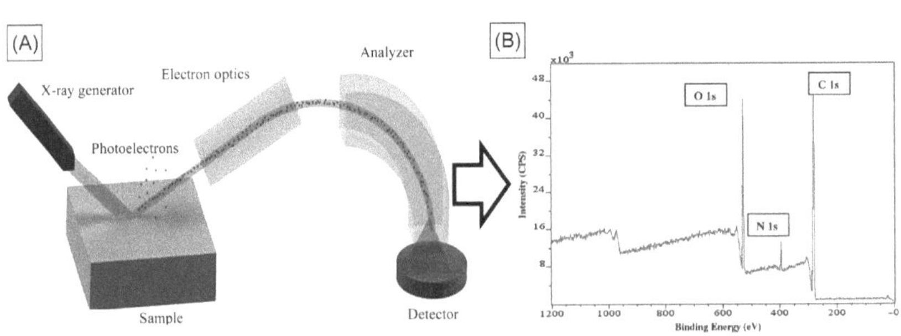

FIGURE 10.8 Schematic diagram of X-ray photoelectron spectroscopy (XPS) (A). Example of survey scan of XPS spectra obtained for fish oil (20 wt%) encapsulated powder containing WPC and maltodextrin in the proportion of 1:3 (B) reprinted with permission from Jafari et al. (2008).

which are values obtained from the areas of the C 1s, O 1s, and N 1 s at the XPS peaks (Fig. 10.8B).

10.4.5 Physicochemical Characterization

Physicochemical properties of nanoencapsulation systems are important considering the stability and functional features of these systems. Food-grade nanoencapsulation systems are composed of a variety of materials including proteins, carbohydrates, and lipids. The physical properties of these materials can be changed as they undergo process stresses (e.g., temperature). For example, some of the physical changes of these materials comprising conformational changes or sol–gel transitions of proteins and carbohydrates, phase transitions of surfactants and phospholipids, glass–rubbery transitions of solid particles (McClements, 2014). Therefore, it is important to have relative information in regard to the main and common analytical methods for determining physical properties of nanoencapsulation systems.

10.4.5.1 X-ray Diffraction

Identification of a crystalline material, unit cell dimensions, and loading of food ingredients in nanoencapsulation systems can be obtained by using X-ray diffraction (XRD) (Fathi, Varshosaz, Mohebbi, & Shahidi, 2013; McClements & McClements, 2016; Neo et al., 2013). In a study conducted by Hosseini, Zandi, Rezaei, and Farahmandghavi (2013), XRD was used for providing a crystallographic structure of CS powder, CS NPs, and the oregano essential oil (OEO)-loaded CS NPs. As shown in Fig. 10.9Aa, a high pick of CS indicated a high degree of crystallinity and there is no peak in the diffractograms of CS NPs, reflecting the destruction of the native CS packing structure (Fig. 10.9Ab). The presence of 1-weak peak in a diffraction spectrum of OEO-loaded CS NPs has confirmed the loading of OEO within CS NPs (Fig. 10.9Ac). In addition, they came to a conclusion that the broad peak of CS NPs may be caused by the cross-linking reaction between CS and tripolyphosphate (TPP), which may destroy the crystalline structure of CS.

10.4.5.2 Differential Scanning Calorimetry

Differential scanning calorimetry (DSC) as a thermoanalytical technique is based on measurements of the enthalpy changes (heat released or absorbed) of samples that are exposed to temperature variations like the heating or cooling process. This technique can provide fundamental information about the glass transition temperature (T_g), heat capacity (C_p), melting point (T_m), and recrystallization times of nanoencapsulation systems (Fang, Li, Yang, Liu, & Wu, 2009; Jenning, Thünemann, & Gohla, 2000). For example, DSC was applied to study thermo-analytical properties of the prepared SLNs and NLCs by using monocaprate as the solid lipid, MCT as the liquid oil, and

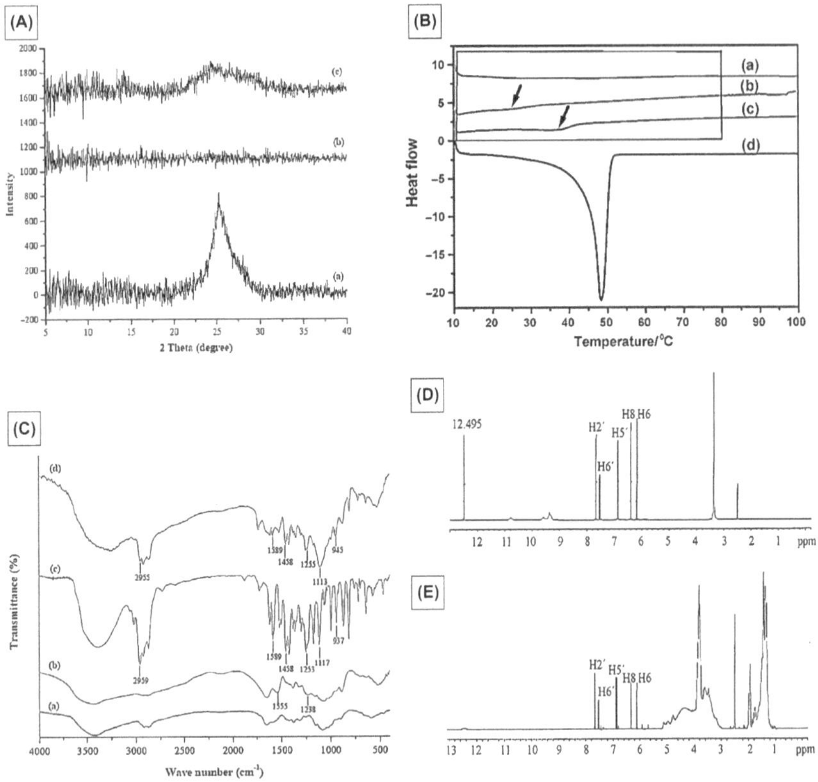

FIGURE 10.9 (A) XRD patterns of (a) chitosan powder, chitosan (b) nanoparticles, and (c) oregano essential oil (OEO)-loaded chitosan nanoparticles, reprinted with permission from Hosseini et al. (2013). (B) Differential scanning calorimetry curves (a) NLC-30 (30 wt% MCT), (b) NLC-10 (10 wt% MCT), (c) SLN, (d) monocaprate (MC), reprinted with permission from Lin et al. (2007) (C). FTIR spectra of (a) chitosan powder, (b) chitosan nanoparticles, (c) OEO and (d) OEO-loaded chitosan nanoparticles with chitosan to OEO, reprinted with permission from Hosseini et al. (2013). H NMR spectra of (D) quercetin and the (E) quercetin-loaded nanoparticles, reprinted with permission from Wu et al. (2008).

polyethylene glycol (PEG) sorbitan monooleate (Tween 80) as the emulsifier (Lin et al., 2007). The results of this study showed a depressing in melting point by turning of bulk matrix material into SLN and NLC (Fig. 10.9B). The melting of amorphous solids requires much less energy than crystalline substances. Therefore, they found that the NLCs had a less ordered crystal or amorphous lipid matrix compared to SLNs.

10.4.5.3 Fourier Transform Infrared Spectroscopy

The principle of Fourier transform infrared spectroscopy (FTIR) is adsorption of light in the infrared region of the electromagnetic spectrum by the most molecules of the sample. The resulting spectrum represents the molecular absorption and transmission, creating a molecular fingerprint of the sample with absorption peaks which correspond to the frequencies of vibrations between the bonds of the atoms making up the material (Esfanjani, Jafari, Assadpoor, & Mohammadi, 2015; Sarmento, Ferreira, Veiga, & Ribeiro, 2006). The identification of different materials, determination of the amount of material present in the sample by measuring the size of the peak, determination of number of components in a mixture, and specifying the quality or consistency of a sample are the main advantages of FTIR. Therefore, FTIR can be exerted as a fast and sensitive analytical method to acquire the main information about the characterization of nanoencapsulation systems (Silva, Cerqueira, & Vicente, 2012).

Hosseini et al. (2013) implemented FTIR for analyzing the OEO-loaded–CS NPs by a two-step method, i.e., oil-in-water emulsion and ionic gelation of CS with sodium TPP. They showed that the electrostatic interaction between NH^{3+} groups of CS and phosphoric groups of TPP was formed within the NPs, shifting the peak of amide I (NH_2 bending) from 1647 to 1651 cm^{-1} and appearing new peaks at 1238 (C–O–C stretch) and 1555 cm^{-1} (amide II) compared to spectra of CS powder (Fig. 10.9Cb). The results showed that there was no interaction between OEO and CS NPs considering the similarity spectra of pure OLE and OEO-loaded CS NPs (Fig. 10.9Cc,d). Besides, the increasing intensity of the CH stretching peak at 2867–2955 cm^{-1} in spectra of OEO-loaded CS NPs in comparison to CS NPs indicated an increase in the content of ester groups from OEO molecules; therefore, their results confirmed the encapsulation of OEO in CS NPs.

10.4.5.4 Nuclear Magnetic Resonance

NMR analysis is based on measuring the magnetic properties of certain atomic nuclei. The irradiation of the pulse of radiofrequency electromagnetic onto nanoencapsulation systems causes the alternation of resonance frequency by hydrogen nuclei and further provides information about the electronic structure of the sample. Therefore, the physical state of nanoencapsulation systems can be determined by changes in resonance. NMR data reflect simultaneously the chemical nature and the molecular mobility of individual system components frequency (Hoch & Stern, 1996).

Wu et al. (2008) incorporated quercetin (QU) in NPs prepared by using the nanoprecipitation technique with Eudragit E (EE) and polyvinyl alcohol (PVA) as carriers. They analyzed the hydrogen bonding between QU and PVA/EE in QU-loaded-nanoparticle (QUEN). The spectrum of QUEN showed that there was not intramolecular hydrogen bonding and also aromatic protons (H6 and H8) of QU were obviously shifted to upfield in comparison to QU spectrum (Fig. 10.9D). These results demonstrated that the A-ring of QU formed intermolecular hydrogen bonding with the PVA–EE carriers.

10.4.5.5 Optical Characterization

Information about the optical characterization of nanoencapsulated ingredients is important for their application, especially in the fortification of food formulations. The optical characterization of nanoencapsulation systems depends on different factors, such as particle sizes of suspensions, concentrations, or refractive index contrasts. Measuring the selective absorption of visible light in either transmission or reflectance modes provides main information about optical properties of nanoencapsulation systems: the lower the transmittance, the higher the turbidity, and the greater the reflectance, the lighter the material (McClements, 2014).

The optical properties of nanoencapsulation systems can be obtained by using a variety of analytical methods, such as UV–visible spectrophotometers and image processing techniques. These methods provide information about the optical properties of nanoencapsulation systems via $L^*a^*b^*$ system. In this system, the a^* and b^* values are color coordinates, where $+a^*$ is the red direction, $-a^*$ is the green direction, $+b^*$ is the yellow direction, and $-b^*$ is the blue direction. The opacity of a colloidal dispersion can, therefore, be characterized by the lightness (L^*), while the overall color intensity can be characterized by the chroma: $C = (a^{*2} + b^{*2})^{1/2}$ (Hosseini et al., 2013; McClements, 2002a, 2002b).

10.5 SAFETY AND TOXICITY OF FOOD NPs EMPLOYED IN FOOD PRODUCTS

The physiological pathways of NPs entry into the body include skin permeation, inhalation, and oral consumption. As a matter of fact, the characteristics of the NPs as well as their extent directly affect the ADME profile in human and other living systems (Katouzian & Jafari, 2016). As NPs enter the GIT via the oral route, they remain isolated from blood by epithelial cells. NPs may translocate to the other organs through the connective tissues located beneath the epithelial layer. Moreover, the absorption of the NPs seems to be different in different parts of the GIT, as it is high in the small intestine and relatively low in the buccal cavity (Roblegg et al., 2012). Deleterious effects have been observed as a result of NPs accumulation in the intestinal cells, such as inflammatory bowel disease (Tetley, 2007), besides other studies revealed that some metal NPs can be aggregated in rats' (Jani, McCarthy, & Florence, 1994) and fish (Zhang et al., 2007) intestine and transferred to other organs causing inflammation responses. The aftermath of inflammation and lacerations could be the high rate of NPs' uptake within the intestine. Moreover, food NPs present in the intraepithelial zone can compete in absorbance with normal food due to their high surface area and extra reactivity (Fröhlich & Roblegg, 2012).

The safety of produced nanocapsules plus their physicochemical and functional properties are of utmost importance in the production process and their health outcomes. Monitoring of the safety of nanocapsules is also essential during their fabrication. Thus, the identification and implementation of safety analysis techniques by precise and correct instruments are necessary for the development of nanoencapsulation technologies for the food and nutraceutical industries.

Concerning the possible toxicity and risks of digested NPs, it should be considered their physicochemical alterations triggered by pH, ionic strength, carbohydrate, lipid, and the protein content of the body fluids (Fröhlich & Roblegg, 2012). When the NPs enter the blood stream, they react with blood cells as well as proteins of plasma; afterward, they enter the cells and can modify DNA of the cells, therefore changing their performance. In spite of the low rate of NPs absorption in the body, they can be accumulated in organs in long term and cause cytotoxicity and immunological responses. Currently, not many studies have been performed regarding the toxicity and absorption profile of food-relevant NPs. Therefore, clinical studies and in vitro tests should be carried out to fill the gap existing in safety of these edible engineered NPs (Bellmann et al., 2015). In this chapter, mostly the safety and toxicity potential of food-relevant and edible nanostructures are discussed and chiefly the recent investigations are sought. Moreover, the characterization methods and instruments are explored in detail.

10.5.1 Genotoxicity and Epigenetic Regulations Modified by NPs

According to the tiny size of NPs, they are likely to enter the nucleus of cells and to interact with the chromosomes resulting in abnormal protein production and the anonymous aftermath. The mechanisms of how NPs cause genotoxicity are not yet clarified in depth. Meanwhile, a nomenclature has been suggested for the classes of genotoxicity. Primary genotoxicity happens when NPs enter the cells and impart their genotoxicity effect, while secondary genotoxicity is actually caused by the chemical production of oxidant agents due to the activation of body cells like phagocytes following inflammation or ulcerations (Gonzalez & Kirsch-Volders, 2016). The two types of genotoxicity can be either indirect or direct. Regarding the first case, the cellular objective is the DNA, whereas the second case is not related to that. Studies have been carried out to obtain information about DNA hypermethylation and the following expression of genes, especially for the DNA methyltransferases (DNMTs) (Dai, Tang, Li, Zhao, & Xiao, 2015; Pirela et al., 2016; Xiao, Wang, & Han, 2016). However, epigenetics has suggested that hypomethylation can induce both overexpression of genes plus their instability. Therefore, it is necessary to connect the epigenetic and genetic fields to have a better understanding of the chemical reactions of NPs inducing toxicity (Gonzalez & Kirsch-Volders, 2016).

Epigenetics is a field of genetic refers to the study of cellular and physiological phenotypic characteristic differences emanating from the environment, which turns the genes on and off and thus altering the gene expression (Corujo, Mas, Malinverni, Di Croce, & Buschbeck, 2016; Moore, 2015). Currently, the epigenetics procedures are not investigated in depth and the role of these processes in health and appearance of diseases should be analyzed accurately (Jaenisch & Bird, 2003). Here, some of the influences of NPs on genetic structures are explored. Also, Fig. 10.10 elucidates the basic factors regulating the epigenetic phenomena.

10.5.1.1 DNA Methylation

This process takes place after the replication of DNA as it bonds with a methyl (CH_3) group. In most mammals, this reaction occurs at the 5′ zone of cytosines within the sequence of CpG (dinucleotide). However, the CpG groups present in the promoter domain plus the initial exon of most genes remain intact. CpG sites are related to the whole house-keeping genes along with some genes set for tissues (Larsen, Gundersen, Lopez, & Prydz, 1992). Methylation of DNA is responsible for imprinting, formation of the embryo, inactivation of X chromosome, and supporting the genome integrity (Smith & Meissner, 2013). Hypermethylation of CpG islands is also directly related to chromosome imbalance plus oncogenesis. Whenever tumor suppressors are inactivated by epigenetic factors, uncontrolled cell proliferation may cause malignancy. It has been reported that methylation in other parts of the gene rather than the promoter causes transcription elongation as well as modifying splicing (Shenker & Flanagan, 2012). DNMTs enzyme undertakes the methylation process within the genome in which the methyl group is transferred from *S*-adenosyl-L-methionine to the CpG dinucleotide's cytosine. A group of DNMTs is associated with the methylation of DNA together with its sustenance. Currently, a compound (5-hydroxymethyl-2′-deoxycytidine) has been characterized, which acts as an epigenetic regulator in the neuronal performance. As an example, the process of de novo DNA methylation (methylation induced by de novo methyltransferase enzymes) that turns the genes on and off is illustrated in Fig. 10.11.

Although the nonstandard DNA methylation occurs rarely, cases have been observed causing complex diseases. There also some challenges like the relationship and reactions among different epigenetic mechanisms that must be addressed (Portela & Esteller, 2010).

10.5.1.2 Histone Alteration

Chromatin is a structure in cell nucleus formed by histone proteins together with the associated DNA. Nucleosome is known as the building block of

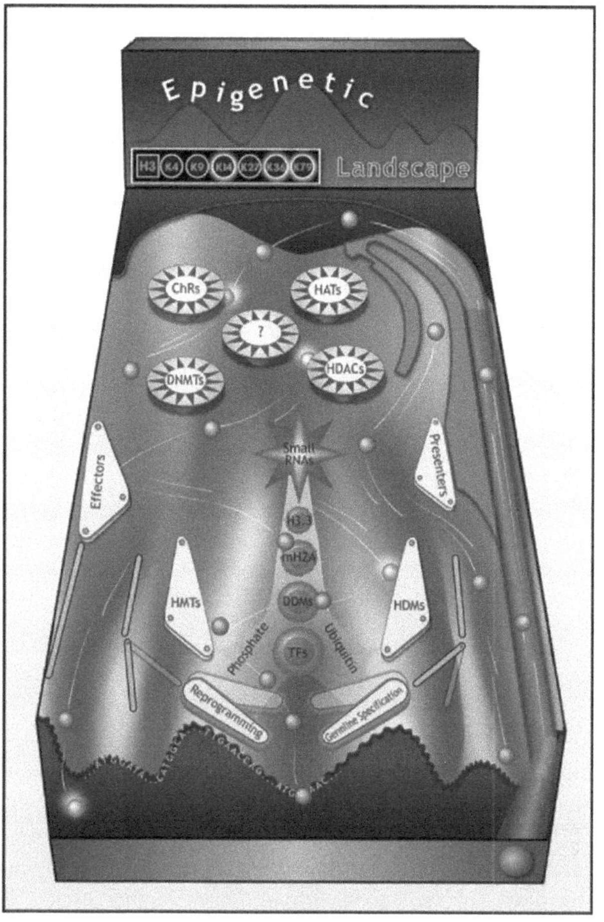

FIGURE 10.10 In this illustration, the pinballs are considered as cells, which are directed through different pathways by various known factors affecting the epigenetic process. Molecular events do not obey a specific rule and the sequences remain unknown. Effector proteins are able to specify the particular histone alterations. However, presenters do impart substrate specificity concerning the histone-alternating enzymes. H3.3 and macroH2A are represented as typical histone variants imparting in transcriptional activation or suppression, consecutively. Other histone and nonhistone proteins are not indicated in this illustration for simplicity. The genetic component of the epigenetic regulation are ChR, chromatin remodelers; DNMTs, DNA methyltransferases; HATs, histone acetyltransferases; HDACs, histone deacetylases; HMTs, HDMs, histone methyltransferases; DDMs, DNA demethylases. *Reprinted with permission from Goldberg, Allis, and Bernstein (2007).*

chromatin structure, which includes a 147-bp DNA, coiled around a set of histone core proteins. In this conformation, two H3–H4 histone dimers are covered by two pairs of H2A–H2B proteins. Hetero and euchromatin structures are developed by the spacing present in nucleosomes. Following the transcription process, histones may be chemically modified to alter the

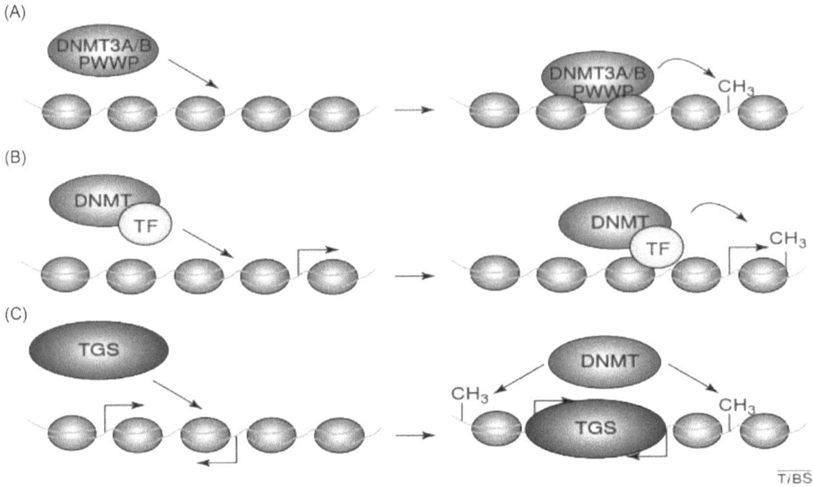

FIGURE 10.11 De novo DNA methylation. (A) The PWWP zone of the DNMT3a and DNMT3b enzymes is needed to target DNA methyltransferase activity to the zones of pericentromeric heterochromatin. The PWWP zone works as a DNA or chromatin-targeting role. (B) Transcription factors (TF) are able to interact with DNA methyltransferase enzymes (DNMT) and to employ methyltransferase activity as a part of the molecular slicing in order to turn off gene expression. (C) De novo DNA methylation might be affected by transcriptional gene slicing (TGS) routes replying to RNAi signals. The molecular pathways that silence the genes are yet unclear. *Reprinted with permission from Klose and Bird (2006).*

chromatin morphology along with its performance. These modifications include methylation, phosphorylation, ubiquitination, etc. (Suganuma & Workman, 2008). The theory of histone code signifies the role of histone modifications in controlling transcription. Methylation and acetylation of histone lysine residues are typical modifications occurring at the posttranslational phase. High rate of acetylation yields an open chromatin configuration plus the onset of transcription, while methylation can either suppress or activate the transcription phase (Rice & Allis, 2001). Several enzymes modify the activities of histone including histone deacetylases, acetyltransferases, methyltransferases (HMTs) plus demethylases (HDMs). Accordingly, the atypical modifications may result in improper gene transcription (Thompson, Guppy, Sawchuk, Davie, & McManus, 2013).

10.5.1.3 miRNAs as Gene Expression Modifiers

Protein expression can be regulated by a process known as RNA interference. microRNA or simply miRNA is one of the well-known molecules responsible for the regulation of gene expression as they react with mRNA and alter their performance, which can lead to the change in mRNA's stability or its disintegration, also in some cases gene transcription is elevated

(Mathers, Strathdee, & Relton, 2009). Moreover, miRNAs are able to affect the chromatin structure through the regulation of histone modifiers. RNA interference is employed in eukaryotic cells to inactivate transposons (jumping genes) following transcription so as to shield cells against DNA damage. Currently, over 1050 miRNAs are identified within human genomes affecting the transcription factors as well as receptors and messengers (Griffiths-Jones, Saini, van Dongen, & Enright, 2008). The atypical expression of miRNAs causes lethal diseases and disorders, such as cancer and Alzheimer (Adams, Kasinski, & Slack, 2014; Lau et al., 2013; Yoshino et al., 2013).

10.5.1.4 In Vivo and In Vitro Assays

It is proved that environmental parameters alter the epigenetic patterns and may cause diseases; however, few researches have been done in regard to the epigenetic reactions of NPs (Shyamasundar, Ng, Lanry Yung, Dheen, & Bay, 2015; Stoccoro, Karlsson, Coppedè, & Migliore, 2013). At first, investigations centering on the modification of gene expression via NPs were analyzed without the consideration of epigenetic factors. Several NPs were tested for their effects on the DNA methylation and gene expression. According to some studies, the impacts of NPs on cell genome include up and downregulation caused by DNA damage, cell cycle advancement, protein-folding genes, DNA restoration, emergence of inflammatory factors plus other responses (Alexis et al., 2008; Dawson & Kouzarides, 2012; Jesus et al., 2015; Maisanaba, Jordá-Beneyto, Cameán, & Jos, 2016; Zheng, Wang, Yang, & Wu, 2014). Also, enhanced oxidative stress, cell membrane impairment, lipid peroxidation are hazardous reactions triggered by NPs in living cells (Asadpour, Sadeghnia, Ghorbani, Sedaghat, & Boroushaki, 2016; Li et al., 2011). The results of recent in vitro and in vivo studies are summarized in Table 10.1.

Maronpot et al. (2016) employed cyclodextrin glucanotransferase (CGTase) as the encapsulant for alpha-glycosyl isoquercitrin (hydrophilic form of QU) and tested it for genotoxicity using B6C3F1 female and male rats. The micronucleus assay and comet tests were carried out to detect the possible genotoxicity in rats. Regarding the in vivo tests, no DNA damage was observed in the rats in their organs (intestine, gut and liver) and therefore they concluded that CGTase is a safe nanocarrier for being implemented in numerous practices.

Madureira et al. (2016) assessed the genotoxicity of SLNPs (prepared by witepsol and carnauba waxes) loaded with rosmarinic acid (RA) using in vivo assays. Lymphocytes were tested for their DNA alteration plus viability and apoptosis. Accordingly, Wistar rats were gavaged with free and entrapped RA, following this process samples were isolated from their blood, urine, and several tissues. In effect, doses lower than 1.5 mg/mL presented no toxicity; however, concentrations higher than that revealed genotoxicity revealed by necrosis of lymphocytes.

TABLE 10.1 The Recent In Vitro and In Vivo Tests Used to Investigate the Possible Genotoxicity Effects of Organic and Food-Grade Nanostructures

In Vitro Tests			
Nanostructure	**Treated Cells**	**Result of Exposure**	**Reference**
Polyamidoamine (PAMAM) dendrimers	PLHC-1 cells	ROS production increased with a linear trend	Naha and Byrne (2013)
Nanoemulsion	*Allium cepa* cells	DNA damage observed due to the presence of phytol	Islam et al., (2016)
Cellulose nanocrystals	BEAS 2B cells	CNC had no genotoxic or immunotoxic activities unlike its macro scale counterpart	Catalán et al. (2015)
In Vivo Tests			
Nanostructure	**Animal Model**	**Result of Exposure**	**Reference**
Cyclodextrin glucanotransferase	B6C3F1 female and male rats	The nanovehicle was considered safe via conducting comet assays in different organs	Maronpot et al. (2016)
Solid lipid nanoparticles (SLNs)	Wistar rats	Levels of rosmarinic acid higher than 1.5 mg/mL incited necrosis of lymphocytes	Madureira et al. (2016)
Chitosan (CS) coated with phosphatidylserine	Balb/c mice	No chromosome or DNA damage was observed as a result of exposure	Zheng et al. (2015)

Also, Zheng, Chen, Jin, Ye, and Liu (2015) prepared a food-grade nanodelivery system composed of chitosan (CS) coated with phosphatidylserine as the shell material and CU as the core material. The nanovehicle was tested for possible genotoxicity (micronucleus and chromosome aberration assay) on five test groups of healthy Balb/c mice and cytotoxicity on the embryonic cells of human kidney (HEK 293). All in all, no chromosome or DNA modification was observed due to exposure of the engineered food-grade nanocarrier.

Naha and Byrne (2013) analyzed the genotoxicity of nanoscale polyamidoamine (PAMAM) dendrimers via in vitro tests. The nanosized organic dendrimer was applied to PLHC-1 cells (a fish hepatoma cell line) and results suggested that reactive oxygen species (ROS) production increased with a linear trend and the genotoxicity could be associated with the total surface amino groups and formation of PAMAM.

In another study, Islam et al. (2016) prepared a nanoemulsion loaded with phytol and evaluated its potential genotoxicity via the comet assay on *Allium cepa* cells. They observed that DNA damage was induced by phytol.

Moreover, Catalán et al. (2015) explored the genotoxic and cytotoxic effects of cellulose nanocrystals (CNC) with the length and width of 135 and 7.3 nm, respectively, together with commercial microcrystalline cellulose (MCC) with the size of 50 μm. In vitro tests were performed using BEAS 2B human bronchial epithelial cells. As a result, CNC was detected to have no genotoxic or immunotoxic activities but the microscale MCC triggered inflammatory responses in the treated cells. The authors suggested that in vivo assays seem necessary to examine the secondary genotoxic effects of CNC, thus obtaining accurate results of the possible genotoxicity that CNC can incite.

Scholars speculate that the typical genetic variants are not sufficient to identify the causes of genetic disorders and epigenetic factors have inevitable impacts of the loci of genetic disorders (Farh et al., 2015; Zannas, Provençal, & Binder, 2015). During embryogenesis, cell differentiation is mediated by the epigenetic factors and later protein expression is controlled by these factors. Epigenetic structures are dissimilar in different cells and can be modified via environmental exposure. These genetic modifications are transferred to the next generation via cell proliferation. The idea of progressive modifications of epigenetic markings has been suggested by Tang et al. (2008). According to this scientist, the epigenetic marks, which are formed in the time of embryogenesis, undergo a series of alterations; thus, susceptibility to health risks and illnesses may be modified. Consequently, the relationship of epigenetic factors with environmental and nutritional factors may bring about benefits, such as retardation of prevention of complicated diseases (Shvedova et al., 2014).

Several growth and neurological diseases are emanated from aberrant epigenetic regulations, such as Albright's hereditary osteodystrophy (Feinberg, 2007), Prader Willi and Beckwith–Wiedemann disorders (Horsthemke & Wagstaff, 2008; Maher & Reik, 2000). Other examples of diseases comprise the improper methylation of repeated/gene-related sequences like Martin–Bell and ICF syndromes (Weng, An, Shin, Song, & Ming, 2013) as well as mutations caused by aberrant acetylation and phosphorylation in histones, which leads to Rubinstein–Taybi and Coffin–Lowry syndromes consecutively (Lopez-Atalaya, Valor, & Barco, 2014; Millichap, 2013).

Although there is a great deal of information in the link between environmental exposure and epigenetic modifications, the data to interpret the mechanism of induced diseases are missing. Nevertheless, cancer was the first disease identified in connection with epigenetics (Feinberg & Vogelstein,

1983). Teschendorff et al. (2012) noted that the variability of DNA methylation has a stronger impact on cancer progression rather than solely the modifications made in DNA. This operation introduces a principal disruption for keeping the epigenetic integrity in cancer cells. In fact, this process brings about useful information for tumor diagnosis and treatment.

Today, the biggest obstacle in the way of determining the genotoxicity of NPs (especially food-grade NPS) is related to the NPs' physicochemical properties along with their characterization in the testing environment. Also, the discrepancy between dose metrics and experiment systems has a great impact on conducted studies (Singh et al., 2009). In order to facilitate the genotoxicity assessment of NPs, the following practices are recommended.

1. The physicochemical properties of exerted food-grade NPs should be determined earlier or during the genotoxicity experiments. Some of the crucial physicochemical features, which affect the genotoxicity, are shape, size, chemical and crystal structure, specific surface area, charge, degree of solubility plus agglomeration status. Moreover, the examinations should involve the stability of food-relevant NPs in a test environment.
2. Standard assays should be advanced to cover the identification of NPs because most genotoxicity tests deal with the common chemical agents and microscale elements. Mutation tests including *Hprt* or MLA assay can be employed to determine the mutagenicity of NPs as Ames assay seems inappropriate in regard to the nanoscale materials. Furthermore, cytotoxicity aids to determine accurate genotoxicity results, particularly in high concentrations that both cytotoxicity and genotoxicity are observed.
3. Proper controls need to be incorporated within NPs genotoxicity assays. NPs' ADME profile is not similar to their macroscale counterparts and therefore have different bioavailability and half-life (Lockman, Koziara, Mumper, & Allen, 2008). Positive bulk materials may expose different bioavailability. For example, they can reach a particular organ, while the NP might be insoluble and cannot enter the target cell. To sum up, nanoscale positive controls can yield accurate and useful information for identifying the possible risks and hazards of NPs.
4. Currently, few studies have been carried out to explore the chronic and subchronic toxicity profile of NPs; thus, the long-term effects of NPs should be considered in the planned experiments. For instance, Catalán et al. (2016) analyzed the genotoxicity of nanofibrillated cellulose (NFC) with the diameter of 10–25 nm on the lungs and bone marrow of mice administered by inhalation. Altogether, they suggested that NFC caused acute inflammation together with DNA damage in the lungs but systemic genotoxicity was not observed in the bone marrow of mice.
5. If effective approaches, such as CometChip technology, are employed, the identification of damaged DNA and isolation of a lot of NPs will be

simplified. Also, other markers play important roles in the genotoxicity of NPs. Accordingly, it has been reported that various molecular pathways, epigenetic modifications plus autophagocytosis have considerable impacts on the genotoxicity of NPS (Boland, Hussain, & Baeza-Squiban, 2014; Roy et al., 2014).

10.6 POSSIBLE HEALTH RISKS AND CARCINOGENESIS INDUCED BY NPs IN SOME ORGANS AND BODY SYSTEMS

10.6.1 Neurotoxicity

NPs are capable of reaching the CNS (central neural system) and causing toxicity due to their large surface area and small size. Moreover, they enter the CNS via specific routes in their translocation. Nevertheless, the biokinetics of all NPs are not the same; in fact, they differ according to their chemical structure, size, solubility, shape, aggregation rate, etc. (Landsiedel et al., 2012). Accordingly, these factors determine the absorbance, translocation from the entry point to the target cells and the potential toxicity or injury that these tiny particles may induce (Oberdörster, Oberdörster, & Oberdörster, 2005a). The interactions among NPs and body fluids plus proteins are of utmost importance as they determine their distribution throughout the body.

Inhalation is considered as one the main entry routes of NPs present in the atmosphere due to mass production in industrial sectors. Upon inhalation, NPs are distributed in different zones of the respiratory system; however, their translocation to other parts of the body depends on their size. If the size of inhaled NPs is 1 nm, they are accumulated nonuniformly in the respiratory system (10% in the tracheobronchial region, 90% in the nasopharyngeal region and none in the alveolar zone), whereas for particles in the range of 5 nm, the accumulation is uniform in all the three areas. Concerning particles in the size of 20 nm, nearly half of the deposition exists in the alveolar region (Oberdörster et al., 2005a). When the NPs are accumulated, they are ready to be translocated to other parts of the body, especially the CNS through various mechanisms, unlike their macroscale counterparts. In general, the clearance of foreign molecules is carried out by the alveolar macrophages; nevertheless, NPs seem to avoid this defense system and reach other parts of the biological system (Gill et al., 2007; Yang, Peters, & Williams, 2008). Several investigations have been conducted to examine the clearance of NPs via the phagocycotic system in rats. The results revealed that after 24 hours about 20% of NPs within the size range of 15–20 or 80 nm could be eliminated from the lungs by the macrophages (Geiser, 2010; Oberdürster, 2000; Qie et al., 2016).

Digestive system is another path, which leads to the circulatory system. In the respiratory system, NPs are trapped in the sticky mucus and pushed up to the throat by the mucociliary escalator. The slurry can be swallowed and enter the GI tract. Also, NPs can be ingested straightly from food, nevertheless studies revealed that uptake of nanomaterials in the GI tract did not expose considerable absorption and translocation to other organs. For instance, fullerenes of C_{60} were orally consumed by rats, and it was observed that almost 98% were excreted, present in feces during a 48-hour period (Yamago et al., 1995). In a study conducted by Jani, Halbert, Langridge, and Florence (1990), the uptake of polystyrene NPs by the GI cells was estimated to be 6.6% and 5.8% for NPs with the size of 50 nm and 100 nm, respectively. Skin permeation is also another route that macro/NPs can spread to the dermis. Flexing of intact skin and damaged skin allows micro/NPs to enter dermis and influence the biological system. As the substances get in the dermis, they are distributed via the lymphatic system. Clares et al. (2014) used three different nanocarriers (nanoemulsions, liposomes, and SLNPs) for enhancing the skin delivery of retinyl palmitate and following the process they will get in the circulatory system. As a result, liposomes skin retention were determined to be higher than the other vehicles.

Absorption of NPs by the sensory nerve endings in contact with the air, such as olfactory nerves in the nose and sensory nerve endings in the skin and their transmission to CNS and ganglionic bodies is a less common but interesting pathway of NPs entry to the neural system. Several metal NPs are reported to enter the CNS via these routes; however, no food-relevant nanostructures have been reported to enter the CNS regarding this pathway (Karmakar, Zhang, & Zhang, 2014).

Furthermore, Picone et al. (2016) synthesized a nanogel matrix comprising carboxyl functionalized poly(*N*-vinyl pyrrodine) delivering insulin to the brain in order to protect the neurons from Alzheimer's disease. The prepared nanogels passed the blood–brain barrier (BBB) and were considered as promising carriers to heal neurodegenerative disorders. (Yuan, Hu, and Gao (2015) suggested that exposure of rat models to polysorbate 80-modified CS NPs (TmCS-NPs) for 7 days caused significant body weight loss along with neural apoptosis and necrosis. Besides, the expression of glial fibrillary acidic protein declined in the cerebellum of mice being injected with TmCS-NPs. In another study, all in all, most of the studies explored above represent the olfactory nerve as a route for NPs to enter the CNS, whenever humans are exposed to the NPs in the environment or industrial sites. After all, this information is not new since it was shown that viruses, especially, the polio virus ($\sim$30 nm) can enter the CNS via olfactory bulb and nerve (Oberdörster et al., 2005a).

Compared to the other organs, the brain is equipped with an exclusive barrier to shield the brain against external agents known as the BBB. The BBB is responsible for the regulation of the concentration plus the composition of fluids around the brain (Obermeier, Daneman, & Ransohoff, 2013). The BBB is placed in the endothelial cells of the microvessels of cerebrum in brain coming in contact via tight junctions. The permeability of BBB can be compared to the plasma membrane as they both do not possess vesicular transport. Therefore, hydrophobic compounds can get through the BBB simply (Bujak, Struck-Lewicka, Kaliszan, Kaliszan, & Markuszewski, 2015). According to the minuscule size of NPs, currently, they are employed in medicinal and pharmaceutical applications as they can pass through the BBB. Indeed, they are exerted as robust drug delivery candidates (Wohlfart, Gelperina, & Kreuter, 2012).

10.6.1.1 Neurotoxicity Mechanisms Stimulated by NPs

Reduction in the size of particles renders them more reactive as their surface area increases exponentially. The size reduction can also have toxicological effects on the biological system. For example, the electron donor and acceptor domains interaction with molecular oxygen might be altered due to high reactivity of NPs. Electron uptake results in the generation of the superoxides (O_2^-), which further creates extra ROS (Fu, Xia, Hwang, Ray, & Yu, 2014).

It has been proved that NPs can induce inflammatory responses and cause toxicity in lung cells compared to their macroscale counterparts that possess the same chemical structure and concentration (Oberdörster, Oberdörster, & Oberdörster, 2005b; Wittmaack, 2007). In a study run by Aragao-Santiago et al. (2016), the toxicity of poly(D,L-lactide-*co*-glycolide) (PLGA) with a layer of CS, PVA, and poloxamer 188 (PF68) were tested via in vivo assays in the lungs of mice. Also, the pulmonary toxicity of rutile and anatase types of TiO_2 were tested as inorganic nanomaterials and compared with the biodegradable nanoformulations. In effect, PLGA NPs did not cause considerable changes in the inflammatory markers. However, anatase type of TiO_2 was present in lung cells even after 48 hours and stimulated the secretion of proinflammatory cytokines.

(Nel, Xia, Mädler, & Li, 2006) suggested that toxic effects of nanomaterials commonly include ROS formation, oxidative stress, mitochondrial distress, protein degradation, DNA damage, inflammation, neoantigen formation plus modification of normal cell cycle. The most typical procedure related to NP toxicity is ROS formation and oxidative stress illustrated in Fig. 10.12. Usually, ROS are generated in mitochondria during respiratory chain at low dose and is degenerated by antioxidants like glutathione (GSH) and enzymes with antioxidant properties. Yet, as the ROS production rate increases during exposure to NPs, the natural antioxidant agents cannot overcome ROS (Nel, 2005). Currently, the accurate mechanism of ROS production is not fully recognized. The suggested mechanisms presented by Oberdörster et al. (2005a) is as follows: (1) excitation of fullerenes under visible light forces the intersystem to generate free electrons, (2) metabolism of NPs cause redox reactions in cell, and (3) inflammation stimulation in vivo resulting in the oxyradical release (Oberdörster et al., 2005a).

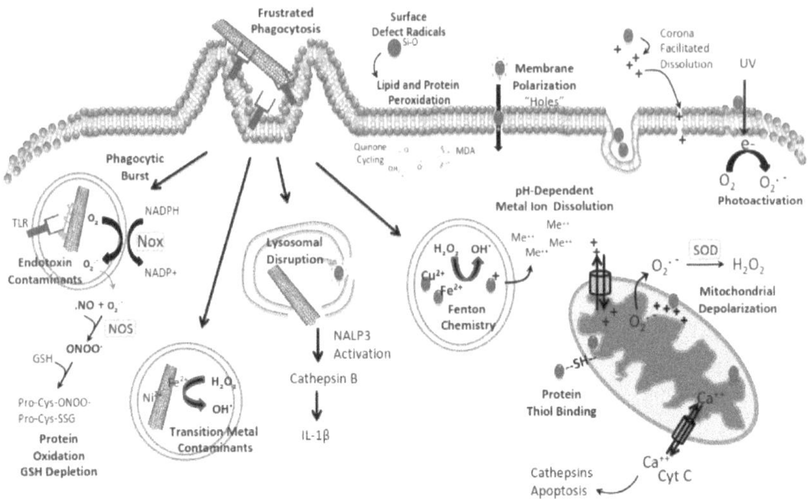

FIGURE 10.12 Representation of cellular oxidative stress generated by nanomaterials. Here, the mechanisms leading to the generation of ROS by the nanomaterials are exhibited in bold. *Reprinted with permission from Kodali and Thrall (2015).*

As stated before, mitochondria are considered as the redox-active organelles and therefore ROS production can be altered by NPs aggregation in this organelle dominating antioxidant defenses (Wilson, Natarajan, Hayward, Khalimonchuk, & Kidambi, 2015). Also, Li et al. (2003) noted that NPs enter the mitochondrial transition pore triggers the programmed cell death through proapoptotic elements. Moreover, in vitro studies have approved the role of NPs in proinflammatory and oxidative stress in sample cell lines (Monteiller et al., 2007; Xia et al., 2006). Inflammation and oxidative stress are induced by numerous mechanisms: (1) the high surface area of NPs impose oxidative stress in cell and enhance the calcium concentration along with the activation of particular genes, (2) transition metals exonerated from particles lead to oxidative stress and following calcium increase plus gene activation, (3) genes are activated by the transition metals released from the NPs, and (4) the dispersion of NPs within the mitochondria leads to oxidative stress (Oberdörster et al., 2005a).

Today, it is necessary to understand the interactions between NPs and biological systems and to examine their possible toxicological effects. Maynard et al. (2006) has suggested five future research trends in the field of nanotoxicology, which are described below.

1. Manufacturing devices in order to evaluate the exposure rate of nanomaterials in different media.
2. Promote novel methods to determine the toxicity of nanomaterials.
3. Establish predictive models via computer software and programming to determine the effect of nanomaterials on the environment as well as public health.

4. Develop effective systems for assessing the impact of nanomaterials on human health during a lifetime period.
5. Prepare strategic plans to establish researches centering on adverse impacts of nanomaterials on biological systems.

In comparison to the whole nanotoxicity research fields, neurotoxicity investigations are limited, especially in the food industry and food-grade nanovehicles. Most of the studied NPs include metallic NPs, such as silver, copper, aluminum, gold, etc. (Schrand et al., 2010). These NPs can pass through the BBB and even damage the BBB obstacle integrity (Sharma & Sharma, 2007). As these NPs reach the brain, they are likely to cause ROS generation, oxidative damage, and even neurodegeneration. Finally, the careful design of novel methods and in vivo toxicity assays are needed to study the neurotoxicity characterization better and to predict its risk in biological media.

10.6.2 Hepatotoxicity

When human come in contact with nanomaterials via skin or are exposed to these tiny particles by inhalation or ingestion, NPs can be transferred to the other organs in the body through the circulatory system (Kermanizadeh, Balharry, Wallin, Loft, & Møller, 2015). Hence, NPs can target different organs in the living system.

Liver is one of the most essential organs in which many crucial body functions are performed. There were studies being conducted to examine the potential toxicity of some NPs on the liver (Baky, Fadda, Al-Rasheed, & Ahmed, 2013; Garcia et al., 2014; Volkovova et al., 2015). In a study performed by (Saleh, 2015), they employed nanochitosan (NC) together with quercitine (Q) to suppress the hepatotoxic activity of CCl_4 in female albino rats. When animal models were treated with CCl_4 a dramatic increase was observed in the levels of liver enzymes including aspartate aminotransferase (AST), alanine aminotransferase (ALT), alkaline phosphatase (ALP), alpha-fetoprotein (AFP), carcinoembryonic antigen (CEA), and the level of lipid peroxidation. Also, CCl_4-induced genotoxic effects on liver cells. In conclusion, Q and NC presented protective behavior against CCl_4 compound, particularly the pathological and DNA fragmentation of rat's liver. It should be noted that higher doses of NC plus Q incited a greater protective behavior.

Later, Gopee et al. (2007) analyzed the uptake of NPs in mouse. In this assay, 9-week-old mice were injected with quantum dots coated with PEG at the level of 48 pmol/animal. After 24 hours, deposit of quantum dots in the kidney, liver lymph nodes, and spleen were detected. Also, studies conducted on the pig skin showed that quantum dots can pass through the skin barrier, while in human it is mostly accumulated in the follicles of hair and does not cross the stratum corneum (the outermost layer of epidermis) (Ryman-Rasmussen, Riviere, & Monteiro-Riviere, 2006; Zhang, William, Colvin, &

Monteiro-Riviere, 2008). Xue et al. (2012) examined the antiinflammatory and hepatotoxicity effects of *Tripterygium wilfordii* Hook f-loaded (TWHF) solid lipid NPs (SLNs) on 64 healthy male rats. The orally administered TWHF-SLN and TWHF group (120, 60, 30 mg/kg) alone during 24 days were tested for their liver injury concerning the serum ALT, AST, ALP, etc. using an autoanalyzer. Furthermore, the levels of superoxide dismutase, GSH peroxidase malondialdehyde, and xanthine oxidase were assessed by commercial kits. In essence, the results revealed no significant damage to the liver when TWHF was loaded with SLNs and that the SLN delivery system is able to enhance the antiinflammatory impact of TWHF, whereas the unloaded TWHF may pose hepatotoxicity effects.

In another study, Selim et al. (2007) injected rabbit models with superparamagnetic magnetite NPs entrapped by lactobionic acid and reported their accumulation in liver cells.

As mentioned above, liver is an organ in which many nanomaterials can deposit and cause further damage and toxicity. Meanwhile, some food-grade nanostructures have shown the potential for the delivery of deleterious drugs without any harmful effects on the liver. Thus, they are currently being employed as practical nanocarriers in the pharmaceutical industries.

10.6.2.1 In Vitro Tests

The in vitro data is not very broad and therefore more studies should be carried out in this field in near future. Some of the in vitro studies in regard to the hepatotoxicity potential of food-relevant NPs are as follows.

Abdelwahab et al. (2013) fabricated a nanodelivery system for thymoquinone (TQ) as the bioactive compounds of *Nigella sativa* coated by NLCs. The NLCs comprised palm oil, olive oil, and phosphatidylcholine as the solid phase and sorbitol plus Tween 80 for the liquid phase. Hepatotoxic tests were performed on rats and WRL-68 liver cell lines of human and no acute hepatotoxic impacts were detected. Also, the removal half-life of TQ was about 4.4933 ± 0.015 hours suggesting that TQ is a proper candidate for extravascular and oral administration.

Yu and Huang (2013) evaluated the toxic effects of three types of nanoemulsions prepared from Tween 20, whey protein isolate and modified starch on HepG2 cells. Results showed that nanoemulsions comprising modified starch and whey protein isolate affected the proliferation and viability of the HepG2 cells, while Tween 20 did not exhibit a significant change in the cell's cycle.

Ochi, Amoabediny, Rezayat, Akbarzadeh, and Ebrahimi (2016) employed a pegylated nanoliposomes carrier to entrap two herbal drugs (silibinin and glycyrrhizic acid) and tested the nanodelivery system on HePG2 cell line. Results depicted that the therapeutic impacts of nanoencapsulated herbal components were enhanced compared to the free compounds.

10.6.2.2 In Vivo Tests

The in vivo tests being undertaken in recent years implies that some NPs cause hepatotoxicity in rodents, while other food-grade systems loaded with bioactives may prevent this process. As an example, Lam et al. (2015) implemented bovine serum albumin, glucose instead of glutaraldehyde as a green nanocapsule formulation to deliver berberine to the liver cells of mice at doses of 1 and 2 μg/g to prevent liver fibrosis.

Also, Coradini et al. (2015) nanoencapsulated resveratrol and CU in lipid–core nanocapsules and applied them to treat arthritis in model rats. Interfacial deposition of preformed polymer was used to obtain the nanocapsules. Model rats were injected with the prepared formulation in the right paw. Altogether, when the bioactive compounds were nanoencapsulated, they did not modify the hepatic biomarkers, such as ALP, AST, and ALT, and thus this formulation is an efficient means to treat inflammation disorders like arthritis imposing no hepatotoxic effects.

In another investigation, Ramos et al. (2015) investigated the possible cyto/hepatotoxicity of D-limonene in Wistar rats as animal models. This flavor is a volatile oil derived from citrus and is implemented in the nanoencapsulation food systems as a flavor. A reduction in the ALT level was observed in the animals treated with 75 mg/kg of D-limonene. Also, morphological lesions like kupffer cell hyperplasia, incipient fibrosis, and microvesicular necrosis along with steatosis was noticed, which could be directly related to the usage of D-limonene.

According to the in vivo studies on the rodent models mentioned above, some of the NPs are likely to deposit in the liver and to cause tissue damage. Nevertheless, the safety information and detrimental effects of NPs are limited and therefore more research should be conducted in this area. Moreover, the hepatotoxic potential of organic engineered nanomaterials (ENMs) employed in different sectors especially the food industry should be accurately studied mainly via the designed in vivo and in vitro tests as they provide us useful information. Last but not least, a series of cost-effective and high-throughput procedures should be identified for the novel products which are released in the food sector to assure their safety of consumption.

10.7 GLOBAL NANOFOOD REGULATIONS

Peters et al. (2014) provided a literature in regard to the nanofood legislations. A definition of nanomaterials was prepared by the European Commission (EU) in 2011 (European Commission, 2011). Also, the ESFA scientific committee released a guidance material for the safety of applying nanoscience in the food industry. Thereafter, a draft standard was developed for the novel food by European Commission in 2013 (Coles & Frewer, 2013). All of these scripts and documents determine the plans of safety aspects to be applied in the food nanotechnology branch. Likewise, strategies are considered for hazard analysis of the nanomaterials employed in the chemical industry safety by Organization for Economic Co-operation and Development. A comprehensive review in the field of regulations and legislations for nanofood products is included in safety chapter of the book entitled "Nanoencapsulation in the food and nutraceutical industries."

10.8 FUTURE TRENDS

Possible toxicity and health risks related to the ENMs in food systems should be characterized initially before the production processes. Regarding the nanofood products, it needs to be determined whether the product is safe with the currently available data or further tests should be designed in order to assess the safety of these functional products. According to the latest data on ENMs and their long-term impact on body, some safe applications of ENMs include:

1. nanomaterials, which are generally considered safe for human consumption rendering no or low toxicity like ZnO, Ag, nanomicelles with soap components, etc.;
2. ENMs that are different from known toxic materials in shape and biopersistence, such as asbestos;
3. encapsulated ENMs that render low exposures due to their delivery system; and
4. simulated or modeled products, which are assessed through realistic dose tests for human consumption.

As a whole, fabricating nanofood products must include a safety evaluation process in which phase studies regarding exposure, toxicity should be conducted to acquire reliable information that guarantees consumer's safety.

Furthermore, together with the development of safety evaluation techniques, characterization methods also need to be advanced so as to determine the essential physicochemical parameters (size, shape, surface area, zeta-potential, etc.) responsible for the toxicity of nanofood materials. Also, in vitro and in vivo assays are of great importance to understand the possible toxicity of nanomaterials in cells and living organisms with the priority of in vitro tests.

Recently, in vivo studies have been performed by the FDA and the national toxicology program to characterize the impact of nanomaterials on living systems. These nanomaterials include nanosilver plus nanogold and their ADME profile is being analyzed. Moreover, the FDA is conducting in vitro tests to evaluate the genotoxicity of these miniature materials and know about their possible mutagenicity or carcinogenicity involving TiO_2

and Ag. In regard to these two nanomaterials, Ag has many applications because of its biocidal properties.

Private sectors along with governmental agencies are studying the safety of nanofood products. As an illustration, the Institute for food Technologists and Grocery Manufacturer's Association are cooperating with the Nanotechnology Characterization Laboratory to explore the applications of nanomaterials in food as additives and packaging materials as well as testing their possible toxicity. Here, some of the initiatives active in the field of food nanotechnology are summarized in Table 10.2.

Albeit these investigations will yield useful information on nanofood products, there is still a long way to go within this novel field of science. Moreover, the effect of nanomaterials in the digestive system and their ADME profile in the GI tract needs to be carefully studied as the literature in this field is still rare and more questions need to be answered in this area. Finally, the regulatory issues legislated for nanofood products is still in its infancy and must be completed by the cooperation of all the food authorities worldwide.

TABLE 10.2 List of Some Institutes, Research Centers, and Public Sectors Active in the Field of Food Nanotechnology

No	Name	URL
1	Iran Nanotechnology Initiative Council	http://theiranproject.com/blog/tag/iran-nanotechnology-initiative-council-inic/
2	National Nutrition and Food Technology Research Institute	http://www.nnftri.ac.ir/eng/
3	Food Nanotechnology Research Group	http://www.chester.ac.uk/node/33739
4	Hydrocolloids Research Centre	https://www.chester.ac.uk/isfi/centres-groups/hrc
5	Functional Foods Research Centre	https://www.chester.ac.uk/ifsi/centres-groups/ffrc
6	Restricted Diets Research Centre	http://www.chester.ac.uk/ifsi/centres-groups/rdrc
7	Centre for Science Communication	http://www.chester.ac.uk/scicomm
8	NoWFOOD	http://www.nowfood.co.uk/

(Continued)

TABLE 10.2 (Continued)

No	Name	URL
9	US Food and Drug Administration	http://www.fda.gov/ScienceResearch/SpecialTopics/Nanotechnology/default.htm
10	National Nanotechnology Initiative	http://www.nano.gov/
11	Institute of Food Research, UK	http://www.ifr.ac.uk/SPM/nanotechnology.html
12	Food and Agriculture Organization of the United Nations	http://www.fao.org/food/food-safety-quality/a-z-index/nano/en/
13	United States Department of Agriculture	http://www.csrees.usda.gov/nanotechnology.cfm
	National Institute of Food and Agriculture	
14	Nanowerk	http://www.nanowerk.com/nanotechnology-in-food.php
15	Korean Food and Drug Administration	http://www.kfda.go.kr
16	Nanostructural Analysis Network Organisation, Major National	http://www.nano.org.au
17	National Food Research Institute	http://www.nfri.affrc.go.jp/english/ourroles/index.html
18	National Institute for Nanotechnology	http://nint-innt.nrc-cnrc.gc.ca/home/index_e.html
19	National Institute for the Physics of Matter	http://www.infm.it/
20	German Federal Institute for Risk Assessment	http://www.bfr.bund.de/cd/template/index_en
21	French Research Network in Micro and NanoTechnologies	http://www.rmnt.org/EN/index.html
22	European Nanotechnology Gateway	http://www.nanoforum.org
23	Center for Nanotechnology	http://www.ipt.arc.nasa.gov/index.html
24	Project on Emerging Nanotechnologies	http://www.nanotechproject.org
25	National Science Foundation	http://www.nsf.gov/crssprgm/nano

10.9 CONCLUSION

Similar to other novel technologies, nanotechnology also needs regulatory issues and guidance for the safe use of its products especially in the food and beverage sector. Considering the rapid growth of food nanotechnology more data should be gathered in order to predict what sort of products and related principles FDA and other well-known agencies are considering in the time coming. Nanofood products regulatory issues and challenges are more magnified due to their nanosize and less information in regard to these tiny particles. Currently, there is no comprehensive data referring to the effects of basic physicochemical factors that affects nanotoxicity like shape, surface area, porosity, etc. As a result, a planned scheme is required in order to obtain fundamental data for the toxicity of nanomaterials plus the connection between the nanomaterial's parameters, which influence the possible nanotoxicity. Accordingly, instruments should be developed to identify the properties of NPs and to characterize their fate in the body concerning their ADME profile until they are excreted from the body or they may deposit in different organs as they are transferred to different parts of the body via the circulatory system. Last but not least, following dissolution and transfer process of food-NPs in organs as well as body fluids, in vitro and in vivo assays are considered as the key experiments to determine the toxicity of nanomaterials employed in food products.

REFERENCES

Abbas, S., Bashari, M., Akhtar, W., Li, W. W., & Zhang, X. (2014). Process optimization of ultrasound-assisted curcumin nanoemulsions stabilized by OSA-modified starch. *Ultrasonics Sonochemistry*, *21*, 1265–1274.

Abdelwahab, S. I., Sheikh, B. Y., Taha, M. M. E., How, C. W., Abdullah, R., Yagoub, U., ... Eid, E. (2013). Thymoquinone-loaded nanostructured lipid carriers: Preparation, gastroprotection, in vitro toxicity, and pharmacokinetic properties after extravascular administration. *International Journal of Nanomedicine*, *8*, 2163–2172.

Adams, B. D., Kasinski, A. L., & Slack, F. J. (2014). Aberrant regulation and function of microRNAs in cancer. *Current Biology*, *24*, R762–R776.

Aguilera, J. M., & Stanley, D. W. (1999). *Microstructural principles of food processing and engineering* (2nd ed). Gaithersburg, MD: Aspen Publishers, Inc.

Alexis, F., Pridgen, E., Molnar, L. K., & Farokhzad, O. C. (2008). Factors affecting the clearance and biodistribution of polymeric nanoparticles. *Molecular Pharmaceutics*, *5*, 505–515.

Allen, A., & Flemström, G. (2005). Gastroduodenal mucus bicarbonate barrier: Protection against acid and pepsin. *American Journal of Physiology—Cell Physiology*, *288*, C1–C19.

Andrade, J. D. (1985). X-ray photoelectron spectroscopy (XPS). In J. D. Andrade (Ed.), *Surface and interfacial aspects of biomedical polymers. Vol. 1. Surface chemistry and physics* (pp. 105–195). New York, NY: Plenum Press.

Aragao-Santiago, L., Hillaireau, H., Grabowski, N., Mura, S., Nascimento, T. L., Dufort, S., ... Fattal, E. (2016). Compared in vivo toxicity in mice of lung delivered biodegradable and non-biodegradable nanoparticles. *Nanotoxicology*, *10*, 292–302.

Asadpour, E., Sadeghnia, H. R., Ghorbani, A., Sedaghat, M., & Boroushaki, M. T. (2016). Oxidative stress-mediated cytotoxicity of zirconia nanoparticles on PC12 and N2a cells. *Journal of Nanoparticle Research*, *18*, 1–13.

Baky, N. A., Fadda, L., Al-Rasheed, N. M., & Ahmed, R. (2013). Role of quercetin and L-arginine in alleviating zinc oxide nanoparticle hepatotoxicity in rats. *Chiang Mai Journal of Science*, *40*, 577–592.

Bellmann, S., Carlander, D., Fasano, A., Momcilovic, D., Scimeca, J. A., Waldman, W. J., ... Pereira, D. I. (2015). Mammalian gastrointestinal tract parameters modulating the integrity, surface properties, and absorption of food-relevant nanomaterials. *Wiley Interdisciplinary Reviews: Nanomedicine and Nanobiotechnology*, *7*, 609–622.

Binnig, G., Quate, C. F., & Gerber, C. (1986). Atomic force microscope. *Physical Review Letters*, *56*, 930.

Boddupalli, B. M., Mohammed, Z. N., Nath, R. A., & Banji, D. (2010). Mucoadhesive drug delivery system: An overview. *Journal of Advanced Pharmaceutical Technology & Research*, *1*, 381.

Boland, S., Hussain, S., & Baeza-Squiban, A. (2014). Carbon black and titanium dioxide nanoparticles induce distinct molecular mechanisms of toxicity. *Wiley Interdisciplinary Reviews: Nanomedicine and Nanobiotechnology*, *6*, 641–652.

Brown, W. (1996). *Light scattering: Principles and development*. New York, NY: Oxford University Press.

Bujak, R., Struck-Lewicka, W., Kaliszan, M., Kaliszan, R., & Markuszewski, M. J. (2015). Blood–brain barrier permeability mechanisms in view of quantitative structure–activity relationships (QSAR). *Journal of Pharmaceutical and Biomedical Analysis*, *108*, 29–37.

Bundschuh, T., Knopp, R., & Kim, J. (2001). Laser-induced breakdown detection (LIBD) of aquatic colloids with different laser systems. *Colloids and Surfaces A: Physicochemical and Engineering Aspects*, *177*, 47–55.

Catalán, J., Ilves, M., Järventaus, H., Hannukainen, K. S., Kontturi, E., Vanhala, E., ... Norppa, H. (2015). Genotoxic and immunotoxic effects of cellulose nanocrystals in vitro. *Environmental and Molecular Mutagenesis*, *56*, 171–182.

Catalán, J., Rydman, E., Aimonen, K., Hannukainen, K.-S., Suhonen, S., Vanhala, E., ... Sneck, A. (2016). Genotoxic and inflammatory effects of nanofibrillated cellulose in murine lungs. *Mutagenesis*, *32*(1), 23–31.

Clares, B., Calpena, A. C., Parra, A., Abrego, G., Alvarado, H., Fangueiro, J. F., & Souto, E. B. (2014). Nanoemulsions (NEs), liposomes (LPs) and solid lipid nanoparticles (SLNs) for retinyl palmitate: Effect on skin permeation. *International Journal of Pharmaceutics*, *473*, 591–598.

Clogston, J. D., & Patri, A. K. (2011). Zeta potential measurement. *Methods in Molecular Biology*, *697*, 63e70.

Coles, D., & Frewer, L. (2013). Nanotechnology applied to European food production—A review of ethical and regulatory issues. *Trends in Food Science & Technology*, *34*, 32–43.

European Commission (2011). Commission recommendation of 18 October 2011 on the definition of nanomaterial (2011/696/EU). *Official Journal of the European Communities: Legis*, 38–40.

Cone, R. A. (2009). Barrier properties of mucus. *Advanced Drug Delivery Reviews*, *61*, 75–85.

Coradini, K., Friedrich, R. B., Fonseca, F. N., Vencato, M. S., Andrade, D. F., Oliveira, C. M., ... Pohlmann, A. R. (2015). A novel approach to arthritis treatment based on resveratrol and curcumin co-encapsulated in lipid–core nanocapsules: In vivo studies. *European Journal of Pharmaceutical Sciences*, *78*, 163–170.

Corujo, D., Mas, G., Malinverni, R., Di Croce, L., & Buschbeck, M. (2016). Barcelona Conference on Epigenetics and Cancer 2015: Coding and non-coding functions of the genome. *Epigenetics*, *11*, 95–100.

Dai, C., Tang, S., Li, D., Zhao, K., & Xiao, X. (2015). Curcumin attenuates quinocetone-induced oxidative stress and genotoxicity in human hepatocyte L02 cells. *Toxicology Mechanisms and Methods*, *25*, 340–346.

Dawson, M. A., & Kouzarides, T. (2012). Cancer epigenetics: From mechanism to therapy. *Cell*, *150*, 12–27.

Egerton, R. F. (2005). *Physical principles of electron microscopy*. New York, NY: Springer.

Egli, M. (2010). Diffraction techniques in structural biology. *Current Protocols in Nucleic Acid Chemistry*, Available from http://dx.doi.org/10.1002/0471142700.nc0713s41.

Ensign, L. M., Cone, R., & Hanes, J. (2012). Oral drug delivery with polymeric nanoparticles: The gastrointestinal mucus barriers. *Advanced Drug Delivery Reviews*, *64*, 557–570.

Esfanjani, A. F., & Jafari, S. M. (2016). Biopolymer nano-particles and natural nano-carriers for nano-encapsulation of phenolic compounds. *Colloids and Surfaces B: Biointerfaces*, *146*, 532–543.

Esfanjani, A. F., Jafari, S. M., Assadpoor, E., & Mohammadi, A. (2015). Nano-encapsulation of saffron extract through double-layered multiple emulsions of pectin and whey protein concentrate. *Journal of Food Engineering*, *165*, 149–155.

Fang, G., Li, H., Yang, F., Liu, X., & Wu, S. (2009). Preparation and characterization of nano-encapsulated *n*-tetradecane as phase change material for thermal energy storage. *Chemical Engineering Journal*, *153*, 217–221.

Farh, K. K.-H., Marson, A., Zhu, J., Kleinewietfeld, M., Housley, W. J., Beik, S., ... Shishkin, A. A. (2015). Genetic and epigenetic fine mapping of causal autoimmune disease variants. *Nature*, *518*, 337–343.

Fathi, M., Varshosaz, J., Mohebbi, M., & Shahidi, F. (2013). Hesperetin-loaded solid lipid nanoparticles and nanostructure lipid carriers for food fortification: Preparation, characterization, and modeling. *Food and Bioprocess Technology*, *6*, 1464–1475.

Feinberg, A. P. (2007). Phenotypic plasticity and the epigenetics of human disease. *Nature*, *447*, 433–440.

Feinberg, A. P., & Vogelstein, B. (1983). Hypomethylation distinguishes genes of some human cancers from their normal counterparts. *Nature*, *301*, 89–92.

Fredenberg, S., Wahlgren, M., Reslow, M., & Axelsson, A. (2011). The mechanisms of drug release in poly(lactic-*co*-glycolic acid)-based drug delivery systems—A review. *International Journal of Pharmaceutics*, *415*, 34–52.

Fröhlich, E., & Roblegg, E. (2012). Models for oral uptake of nanoparticles in consumer products. *Toxicology*, *291*, 10–17.

Fu, P. P., Xia, Q., Hwang, H.-M., Ray, P. C., & Yu, H. (2014). Mechanisms of nanotoxicity: Generation of reactive oxygen species. *Journal of Food and Drug Analysis*, *22*, 64–75.

Garcia, S. C., Guterres, S. S., Bubols, G. B., Bulcao, R. P., Charao, M. F., & Pohlmann, A. R. (2014). *Polymeric nanoparticles: In vivo toxicological evaluation, cardiotoxicity, and hepatotoxicity. Nanotoxicology* (pp. 299–324). New York: Springer.

Geiser, M. (2010). Update on macrophage clearance of inhaled micro-and nanoparticles. *Journal of Aerosol Medicine and Pulmonary Drug Delivery*, *23*, 207–217.

Gill, S., Löbenberg, R., Ku, T., Azarmi, S., Roa, W., & Prenner, E. J. (2007). Nanoparticles: Characteristics, mechanisms of action, and toxicity in pulmonary drug delivery—A review. *Journal of Biomedical Nanotechnology*, *3*, 107–119.

Goldberg, A. D., Allis, C. D., & Bernstein, E. (2007). Epigenetics: A landscape takes shape. *Cell*, *128*(4), 635–638.

Gonzalez, L., & Kirsch-Volders, M. (2016). Biomonitoring of genotoxic effects for human exposure to nanomaterials: The challenge ahead. *Mutation Research/Reviews in Mutation Research*, *768*, 14–26.

Gopee, N. V., Roberts, D. W., Webb, P., Cozart, C. R., Siitonen, P. H., Warbritton, A. R., ... Howard, P. C. (2007). Migration of intradermally injected quantum dots to sentinel organs in mice. *Toxicological Sciences*, *98*, 249–257.

Griffiths-Jones, S., Saini, H. K., van Dongen, S., & Enright, A. J. (2008). miRBase: Tools for microRNA genomics. *Nucleic Acids Research*, *36*, D154–D158.

Gupta, A. K., & Wells, S. (2004). Surface-modified superparamagnetic nanoparticles for drug delivery: Preparation, characterization, and cytotoxicity studies. *IEEE Transactions on Nanobioscience*, *3*, 66–73.

Hassellöv, M., Readman, J. W., Ranville, J. F., & Tiede, K. (2008). Nanoparticle analysis and characterization methodologies in environmental risk assessment of engineered nanoparticles. *Ecotoxicology*, *17*, 344–361.

Hoch, J. C., & Stern, A. S. (1996). *NMR data processing*. New York: Wiley-Liss.

Honary, S., & Zahir, F. (2013). Effect of zeta potential on the properties of nano-drug delivery systems—A review (Part 1). *Tropical Journal of Pharmaceutical Research*, *12*, 255–264.

Horsthemke, B., & Wagstaff, J. (2008). Mechanisms of imprinting of the Prader–Willi/Angelman region. *American Journal of Medical Genetics, Part A*, *146*, 2041–2052.

Hosseini, S. F., Zandi, M., Rezaei, M., & Farahmandghavi, F. (2013). Two-step method for encapsulation of oregano essential oil in chitosan nanoparticles: Preparation, characterization and in vitro release study. *Carbohydrate Polymers*, *95*, 50–56.

Huang, X., & Brazel, C. S. (2001). On the importance and mechanisms of burst release in matrix-controlled drug delivery systems. *Journal of Controlled Release*, *73*, 121–136.

Hunter, R. J. (2001). *Foundations of colloid science* (2nd ed). Oxford, UK and New York, NY: Oxford University Press.

Islam, M. T., Streck, L., Silva, S. W. C., Paz, M. F. C. J., Júnior, A. L. G., Vinícius, M., & de Moura Dantas, S. M. M. (2016). Evaluation of toxic, cytotoxic, genotoxic and mutagenic effects of phytol and its nanoemulsion: Data-in-brief. *World Journal of Pharmacy and Pharmaceutical Sciences*, *5*(9), 410–418.

Jaenisch, R., & Bird, A. (2003). Epigenetic regulation of gene expression: How the genome integrates intrinsic and environmental signals. *Nature Genetics*, *33*, 245–254.

Jafari, S., He, Y., & Bhandari, B. (2006). Nano-emulsion production by sonication and microfluidization—A comparison. *International Journal of Food Properties*, *9*, 475–485.

Jafari, S. M., Assadpoor, E., Bhandari, B., & He, Y. (2008). Nano-particle encapsulation of fish oil by spray drying. *Food Research International*, *41*, 172–183.

Jafari, S. M., He, Y., & Bhandari, B. (2007). Production of sub-micron emulsions by ultrasound and microfluidization techniques. *Journal of Food Engineering*, *82*, 478–488.

Jalili, N., & Laxminarayana, K. (2004). A review of atomic force microscopy imaging systems: Application to molecular metrology and biological sciences. *Mechatronics*, *14*, 907–945.

Jani, P., Halbert, G. W., Langridge, J., & Florence, A. T. (1990). Nanoparticle uptake by the rat gastrointestinal mucosa: Quantitation and particle size dependency. *Journal of Pharmacy and Pharmacology*, *42*, 821–826.

Jani, P. U., McCarthy, D. E., & Florence, A. T. (1994). Titanium dioxide (rutile) particle uptake from the rat GI tract and translocation to systemic organs after oral administration. *International Journal of Pharmaceutics*, *105*, 157–168.

Jenning, V., Thünemann, A. F., & Gohla, S. H. (2000). Characterisation of a novel solid lipid nanoparticle carrier system based on binary mixtures of liquid and solid lipids. *International Journal of Pharmaceutics*, *199*, 167–177.

Jesus, D. R., Barbosa, L. N., Prando, T. B. L., Martins, L. F., Gasparotto, F., Guedes, K. M. R., ... Gasparotto Junior, A. (2015). Ninety-day oral toxicity assessment of an alternative biopolymer for controlled release drug delivery systems obtained from cassava starch acetate. *Evidence-Based Complementary and Alternative Medicine*, *2015*, 390416.

Jonnalagadda, S., & Robinson, D. H. (2000). A bioresorbable, polylactide reservoir for diffusional and osmotically controlled drug delivery. *AAPS PharmSciTech*, *1*, 26–34.

Karmakar, A., Zhang, Q., & Zhang, Y. (2014). Neurotoxicity of nanoscale materials. *Journal of Food and Drug Analysis*, *22*, 147–160.

Katouzian, I., & Jafari, S. M. (2016). Nano-encapsulation as a promising approach for targeted delivery and controlled release of vitamins. *Trends in Food Science & Technology*, *53*, 34–48.

Kermanizadeh, A., Balharry, D., Wallin, H., Loft, S., & Møller, P. (2015). Nanomaterial translocation—The biokinetics, tissue accumulation, toxicity and fate of materials in secondary organs—A review. *Critical Reviews in Toxicology*, *45*, 837–872.

Klose, R. J., & Bird, A. P. (2006). Genomic DNA methylation: The mark and its mediators. *Trends in Biochemical Sciences*, *31*(2), 89–97.

Kodali, V., & Thrall, B. D. (2015). Oxidative stress and nanomaterial-cellular interactions. In S. M. Roberts, J. P. Kehrer, & L.-O. Klotz (Eds.), *Studies on experimental toxicology and pharmacology* (pp. 347–367). Cham: Springer International Publishing.

Korsmeyer, R. W., Gurny, R., Doelker, E., Buri, P., & Peppas, N. A. (1983). Mechanisms of solute release from porous hydrophilic polymers. *International Journal of Pharmaceutics*, *15*, 25–35.

Kotanen, C. N., Wilson, A. N., Wilson, A. M., Ishihara, K., & Guiseppi-Elie, A. (2012). Biomimetic hydrogels gate transport of calcium ions across cell culture inserts. *Biomedical Microdevices*, *14*, 549–558.

Kranz, H., Ubrich, N., Maincent, P., & Bodmeier, R. (2000). Physicomechanical properties of biodegradable poly(D,L-lactide) and poly(D,L-lactide-*co*-glycolide) films in the dry and wet states. *Journal of Pharmaceutical Sciences*, *89*, 1558–1566.

Lai, S. K., Wang, Y.-Y., & Hanes, J. (2009). Mucus-penetrating nanoparticles for drug and gene delivery to mucosal tissues. *Advanced Drug Delivery Reviews*, *61*, 158–171.

Laine, L., Takeuchi, K., & Tarnawski, A. (2008). Gastric mucosal defense and cytoprotection: Bench to bedside. *Gastroenterology*, *135*, 41–60.

Lam, P.-L., Kok, S.-L., Gambari, R., Kok, T.-W., Leung, H.-Y., Choi, K.-L., ... Lam, K. (2015). Evaluation of berberine/bovine serum albumin nanoparticles for liver fibrosis therapy. *Green Chemistry*, *17*, 1640–1646.

Lametschwandtner, A., Lametschwandtner, U., & Weiger, T. (1990). Scanning electron microscopy of vascular corrosion casts—Technique and applications: Updated review. *Scanning Microscopy*, *4*, 889–940, discussion 941.

Landsiedel, R., Fabian, E., Ma-Hock, L., Wohlleben, W., Wiench, K., Oesch, F., & van Ravenzwaay, B. (2012). Toxico-/biokinetics of nanomaterials. *Archives of Toxicology*, *86*, 1021–1060.

Larsen, F., Gundersen, G., Lopez, R., & Prydz, H. (1992). CpG islands as gene markers in the human genome. *Genomics*, *13*, 1095–1107.

Lau, P., Bossers, K., Salta, E., Frigerio, C. S., Barbash, S., Rothman, R., ... Papadopoulou, A. S. (2013). Alteration of the microRNA network during the progression of Alzheimer's disease. *EMBO Molecular Medicine*, *5*, 1613–1634.

Li, N., Sioutas, C., Cho, A., Schmitz, D., Misra, C., Sempf, J., ... Nel, A. (2003). Ultrafine particulate pollutants induce oxidative stress and mitochondrial damage. *Environmental Health Perspectives*, *111*, 455.

Li, S., Wang, H., Qi, Y., Tu, J., Bai, Y., Tian, T., ... Lu, Z. (2011). Assessment of nanomaterial cytotoxicity with SOLiD sequencing-based microRNA expression profiling. *Biomaterials*, *32*, 9021–9030.

Liang, X., Mao, G., & Ng, K. S. (2004). Probing small unilamellar EggPC vesicles on mica surface by atomic force microscopy. *Colloids and Surfaces B: Biointerfaces*, *34*, 41–51.

Lichtenberger, L. (1995). The hydrophobic barrier properties of gastrointestinal mucus. *Annual Review of Physiology*, *57*, 565–583.

Lin, X., Li, X., Zheng, L., Yu, L., Zhang, Q., & Liu, W. (2007). Preparation and characterization of monocaprate nanostructured lipid carriers. *Colloids and Surfaces A: Physicochemical and Engineering Aspects*, *311*, 106–111.

Lockman, P. R., Koziara, J. M., Mumper, R. J., & Allen, D. D. (2008). Nanoparticle surface charges alter blood–brain barrier integrity and permeability. *Journal of Drug Targeting*, *12* (9–10), 635–641.

Lopez-Atalaya, J. P., Valor, L. M., & Barco, A. (2014). Epigenetic factors in intellectual disability: The Rubinstein-Taybi syndrome as a paradigm of neurodevelopmental disorder with epigenetic origin. *Progress in Molecular Biology and Translational Science*, *128*, 139–176.

Madhav, N. S., Shakya, A. K., Shakya, P., & Singh, K. (2009). Orotransmucosal drug delivery systems: A review. *Journal of Controlled Release*, *140*, 2–11.

Madureira, A. R., Nunes, S., Campos, D. A., Fernandes, J. C., Marques, C., Zuzarte, M., ... Sarmento, B. (2016). Safety profile of solid lipid nanoparticles loaded with rosmarinic acid for oral use: In vitro and animal approaches. *International Journal of Nanomedicine*, *11*, 3621.

Maher, E. R., & Reik, W. (2000). Beckwith–Wiedemann syndrome: Imprinting in clusters revisited. *The Journal of Clinical Investigation*, *105*, 247–252.

Maherani, B., Arab-Tehrany, E., Kheirolomoom, A., Cleymand, F., & Linder, M. (2012). Influence of lipid composition on physicochemical properties of nanoliposomes encapsulating natural dipeptide antioxidant L-carnosine. *Food Chemistry*, *134*, 632–640.

Maisanaba, S., Jordá-Beneyto, M., Cameán, A. M., & Jos, Á. (2016). Effects of two organomodified clays intended to food contact materials on the genomic instability and gene expression of hepatoma cells. *Food and Chemical Toxicology*, *88*, 57–64.

Markovsky, E., Baabur-Cohen, H., Eldar-Boock, A., Omer, L., Tiram, G., Ferber, S., ... Satchi-Fainaro, R. (2012). Administration, distribution, metabolism and elimination of polymer therapeutics. *Journal of Controlled Release*, *161*, 446–460.

Maronpot, R. R., Hobbs, C. A., Davis, J., Swartz, C., Boyle, M., Koyanagi, M., ... Hayashi, S.-m (2016). Genetic and rat toxicity studies of cyclodextrin glucanotransferase. *Toxicology Reports*, *3*, 381–392.

Mathers, J. C., Strathdee, G., & Relton, C. L. (2009). Induction of epigenetic alterations by dietary and other environmental factors. *Advances in Genetics*, *71*, 3–39.

Matsumura, Y., & Maeda, H. (1986). A new concept for macromolecular therapeutics in cancer chemotherapy: Mechanism of tumoritropic accumulation of proteins and the antitumor agent smancs. *Cancer Research*, *46*, 6387–6392.

Maynard, A. D., Aitken, R. J., Butz, T., Colvin, V., Donaldson, K., Oberdörster, G., ... Stone, V. (2006). Safe handling of nanotechnology. *Nature*, *444*, 267–269.

McClements, D. J. (2004). *Food emulsions: Principles, practice and techniques* (2nd ed). Boca Raton, FL: CRC Press.

McClements, D. J. (2002a). Colloidal basis of emulsion color. *Current Opinion in Colloid & Interface Science, 7*, 451 455.

McClements, D. J. (2002b). Theoretical prediction of emulsion color. *Advances in Colloid and Interface Science, 97*, 63–89.

Mcclements, D. J. (2007). Critical review of techniques and methodologies for characterization of emulsion stability. *Critical Reviews in Food Science and Nutrition, 47*, 611–649.

McClements, D. J. (2014). *Nanoparticle- and microparticle-based delivery systems: Encapsulation, protection and release of active compounds*. Boca Raton, FL: CRC Press.

McClements, D. J. (2015). *Food emulsions: Principles, practices, and techniques* (3rd ed). Boca Raton, FL: CRC Press.

McClements, J., & McClements, D. J. (2016). Standardization of nanoparticle characterization: Methods for testing properties, stability, and functionality of edible nanoparticles. *Critical Reviews in Food Science and Nutrition, 56*, 1334–1362.

Mehvar, R. (2000). Dextrans for targeted and sustained delivery of therapeutic and imaging agents. *Journal of Controlled Release, 69*, 1–25.

Millichap, J. G. (2013). *Neurological syndromes—A compendium for clinicians*, New York.

Mohammadi, A., Jafari, S. M., Assadpour, E., & Esfanjani, A. F. (2016). Nano-encapsulation of olive leaf phenolic compounds through WPC–pectin complexes and evaluating their release rate. *International Journal of Biological Macromolecules, 82*, 816–822.

Mohammadi, A., Jafari, S. M., Esfanjani, A. F., & Akhavan, S. (2016). Application of nano-encapsulated olive leaf extract in controlling the oxidative stability of soybean oil. *Food Chemistry, 190*, 513–519.

Monteiller, C., Tran, L., MacNee, W., Faux, S., Jones, A., Miller, B., & Donaldson, K. (2007). The pro-inflammatory effects of low-toxicity low-solubility particles, nanoparticles and fine particles, on epithelial cells in vitro: The role of surface area. *Occupational and Environmental Medicine, 64*, 609–615.

Moore, D. S. (2015). *The developing genome: An introduction to behavioral epigenetics*. Oxford: Oxford University Press.

Morales, J. O., & McConville, J. T. (2011). Manufacture and characterization of mucoadhesive buccal films. *European Journal of Pharmaceutics and Biopharmaceutics, 77*, 187–199.

Naha, P. C., & Byrne, H. J. (2013). Generation of intracellular reactive oxygen species and genotoxicity effect to exposure of nanosized polyamidoamine (PAMAM) dendrimers in PLHC-1 cells in vitro. *Aquatic Toxicology, 132*, 61–72.

Nair, L. S., & Laurencin, C. T. (2007). Biodegradable polymers as biomaterials. *Progress in Polymer Science, 32*, 762–798.

Nel, A. (2005). Air pollution-related illness: Effects of particles. *Science, 308*, 804–806.

Nel, A., Xia, T., Mädler, L., & Li, N. (2006). Toxic potential of materials at the nanolevel. *Science, 311*, 622–627.

Neo, Y. P., Ray, S., Jin, J., Gizdavic-Nikolaidis, M., Nieuwoudt, M. K., Liu, D., & Quek, S. Y. (2013). Encapsulation of food grade antioxidant in natural biopolymer by electrospinning technique: A physicochemical study based on zein–gallic acid system. *Food Chemistry, 136*, 1013–1021.

Nielsen, S. S. (1998). *Food analysis* (2nd ed). Gaithersburg, MD: Aspen Publication.

Oberdörster, G., Oberdörster, E., & Oberdörster, J. (2005a). Nanotoxicology: An emerging discipline evolving from studies of ultrafine particles. *Environmental Health Perspectives, 113*(7), 823–839.

Oberdörster, G., Oberdörster, E., & Oberdörster, J. (2005b). Nanotoxicology: An emerging discipline evolving from studies of ultrafine particles. *Environmental Health Perspectives, 113*, 823–839.

Oberdürster, G. (2000). Toxicology of ultrafine particles: In vivo studies. *Philosophical Transactions of the Royal Society of London A: Mathematical, Physical and Engineering Sciences*, *358*, 2719–2740.

Obermeier, B., Daneman, R., & Ransohoff, R. M. (2013). Development, maintenance and disruption of the blood–brain barrier. *Nature Medicine*, *19*, 1584–1596.

Ochi, M. M., Amoabediny, G., Rezayat, S. M., Akbarzadeh, A., & Ebrahimi, B. (2016). In vitro co-delivery evaluation of novel pegylated nano-liposomal herbal drugs of silibinin and glycyrrhizic acid (nano-phytosome) to hepatocellular carcinoma cells. *Cell Journal (Yakhteh)*, *18*, 135.

Omenn, G. S., Goodman, G. E., Thornquist, M. D., Balmes, J., Cullen, M. R., Glass, A., ... Williams, J. H., Jr (1996). Effects of a combination of beta carotene and vitamin A on lung cancer and cardiovascular disease. *New England Journal of Medicine*, *334*, 1150–1155.

Pecora, R. (2000). Dynamic light scattering measurement of nanometer particles in liquids. *Journal of Nanoparticle Research*, *2*, 123–131.

Peters, R., Brandhoff, P., Weigel, S., Marvin, H., Bouwmeester, H., Aschberger, K., ... Botelho, F. (2014). *Inventory of Nanotechnology Applications in the Agricultural, Feed and Food Sector*. CFT/EFSA/FEED/2012/01.

Picone, P., Ditta, L. A., Sabatino, M. A., Militello, V., San Biagio, P. L., Di Giacinto, M. L., ... Giacomazza, D. (2016). Ionizing radiation-engineered nanogels as insulin nanocarriers for the development of a new strategy for the treatment of Alzheimer's disease. *Biomaterials*, *80*, 179–194.

Pirela, S. V., Miousse, I. R., Lu, X., Castranova, V., Thomas, T., Qian, Y., ... Demokritou, P. (2016). Effects of laser printer-emitted engineered nanoparticles on cytotoxicity, chemokine expression, reactive oxygen species, DNA methylation, and DNA damage: A comprehensive in vitro analysis in human small airway epithelial cells, macrophages, and lymphoblasts. *Environmental Health Perspectives*, *124*, 210.

Portela, A., & Esteller, M. (2010). Epigenetic modifications and human disease. *Nature Biotechnology*, *28*, 1057–1068.

Qie, Y., Yuan, H., von Roemeling, C. A., Chen, Y., Liu, X., Shih, K. D., ... Jiang, W. (2016). Surface modification of nanoparticles enables selective evasion of phagocytic clearance by distinct macrophage phenotypes. *Scientific Reports*, *6*, 26269.

Ramos, C. A. F., Sá, R. d C. d S., Alves, M. F., Benedito, R. B., de Sousa, D. P., Margareth de Fátima, F., ... de Almeida, R. N. (2015). Histopathological and biochemical assessment of D-limonene-induced liver injury in rats. *Toxicology Reports*, *2*, 482–488.

Rao, J., Dragulescu-Andrasi, A., & Yao, H. (2007). Fluorescence imaging in vivo: Recent advances. *Current opinion in Biotechnology*, *18*, 17–25.

Raval, A., Parikh, J., & Engineer, C. (2010). Mechanism of controlled release kinetics from medical devices. *Brazilian Journal of Chemical Engineering*, *27*, 211–225.

Rice, J. C., & Allis, C. D. (2001). Histone methylation versus histone acetylation: New insights into epigenetic regulation. *Current Opinion in Cell Biology*, *13*, 263–273.

Roblegg, E., Fröhlich, E., Meindl, C., Teubl, B., Zaversky, M., & Zimmer, A. (2012). Evaluation of a physiological in vitro system to study the transport of nanoparticles through the buccal mucosa. *Nanotoxicology*, *6*, 399–413.

Roy, R., Singh, S. K., Chauhan, L., Das, M., Tripathi, A., & Dwivedi, P. D. (2014). Zinc oxide nanoparticles induce apoptosis by enhancement of autophagy via PI3K/TOR inhibition. *Toxicology Letters*, *227*, 29–40.

Ryman-Rasmussen, J. P., Riviere, J. E., & Monteiro-Riviere, N. A. (2006). Penetration of intact skin by quantum dots with diverse physicochemical properties. *Toxicological Sciences*, *91*, 159–165.

Saleh, R.H.A. (2015). Possible protective effect of nano-chitosan and quercetin against oxidative stress in experimentally induced liver toxicity in rats. *CU Theses*.

Sansdrap, P., & Moës, A.-J. (1997). In vitro evaluation of the hydrolytic degradation of dispersed and aggregated poly(D,L-lactide-*co*-glycolide) microspheres. *Journal of Controlled Release*, *43*, 47–58.

Sarmento, B., Ferreira, D., Veiga, F., & Ribeiro, A. (2006). Characterization of insulin-loaded alginate nanoparticles produced by ionotropic pre-gelation through DSC and FTIR studies. *Carbohydrate Polymers*, *66*, 1–7.

Schrand, A. M., Rahman, M. F., Hussain, S. M., Schlager, J. J., Smith, D. A., & Syed, A. F. (2010). Metal-based nanoparticles and their toxicity assessment. *Wiley Interdisciplinary Reviews: Nanomedicine and Nanobiotechnology*, *2*, 544–568.

Selim, K. K., Ha, Y.-S., Kim, S.-J., Chang, Y., Kim, T.-J., Lee, G. H., & Kang, I.-K. (2007). Surface modification of magnetite nanoparticles using lactobionic acid and their interaction with hepatocytes. *Biomaterials*, *28*, 710–716.

Seymour, L., Duncan, R., Strohalm, J., & Kopeček, J. (1987). Effect of molecular weight (Mw) of *N*-(2-hydroxypropyl) methacrylamide copolymers on body distribution and rate of excretion after subcutaneous, intraperitoneal, and intravenous administration to rats. *Journal of Biomedical Materials Research*, *21*, 1341–1358.

Shaikh, R., Singh, T. R. R., Garland, M. J., Woolfson, A. D., & Donnelly, R. F. (2011). Mucoadhesive drug delivery systems. *Journal of Pharmacy and Bioallied Sciences*, *3*, 89.

Shang, L., Zhang, S., Du, H., & Venkatraman, S. S. (2008). A novel approach for the control of drug release rate through hydrogel membrane: I. Effect of drug immobilization on drug release rate by copolymerization method. *European Journal of Pharmaceutics and Biopharmaceutics*, *68*, 715–723.

Sharma, H. S., & Sharma, A. (2007). Nanoparticles aggravate heat stress induced cognitive deficits, blood–brain barrier disruption, edema formation and brain pathology. *Progress in Brain Research*, *162*, 245–273.

Shenker, N., & Flanagan, J. (2012). Intragenic DNA methylation: Implications of this epigenetic mechanism for cancer research. *British Journal of Cancer*, *106*, 248–253.

Shvedova, A. A., Yanamala, N., Kisin, E. R., Tkach, A. V., Murray, A. R., Hubbs, A., … Kagan, V. E. (2014). Long-term effects of carbon containing engineered nanomaterials and asbestos in the lung: One year postexposure comparisons. *American Journal of Physiology—Lung Cellular and Molecular Physiology*, *306*, L170–L182.

Shyamasundar, S., Ng, C. T., Lanry Yung, L. Y., Dheen, S. T., & Bay, B. H. (2015). Epigenetic mechanisms in nanomaterial-induced toxicity. *Epigenomics*, *7*, 395–411.

Silva, H. D., Cerqueira, M. Â., & Vicente, A. A. (2012). Nanoemulsions for food applications: Development and characterization. *Food and Bioprocess Technology*, *5*, 854–867.

Singh, N., Manshian, B., Jenkins, G. J., Griffiths, S. M., Williams, P. M., Maffeis, T. G., … Doak, S. H. (2009). NanoGenotoxicology: The DNA damaging potential of engineered nanomaterials. *Biomaterials*, *30*, 3891–3914.

Smart, J. D. (2005). The basics and underlying mechanisms of mucoadhesion. *Advanced Drug Delivery Reviews*, *57*, 1556–1568.

Smith, Z. D., & Meissner, A. (2013). DNA methylation: Roles in mammalian development. *Nature Reviews Genetics*, *14*, 204–220.

Stoccoro, A., Karlsson, H. L., Coppedè, F., & Migliore, L. (2013). Epigenetic effects of nano-sized materials. *Toxicology*, *313*, 3–14.

Suganuma, T., & Workman, J. L. (2008). Crosstalk among histone modifications. *Cell*, *135*, 604–607.

Szakács, G., Varadi, A., Özvegy-Laczka, C., & Sarkadi, B. (2008). The role of ABC transporters in drug absorption, distribution, metabolism, excretion and toxicity (ADME–Tox). *Drug Discovery Today*, *13*, 379–393.

Tang, W.-Y., Newbold, R., Mardilovich, K., Jefferson, W., Cheng, R. Y., Medvedovic, M., & Ho, S.-M. (2008). Persistent hypomethylation in the promoter of nucleosomal binding protein 1 (Nsbp 1) correlates with overexpression of Nsbp 1 in mouse uteri neonatally exposed to diethylstilbestrol or genistein. *Endocrinology*, *149*, 5922–5931.

Teschendorff, A. E., Jones, A., Fiegl, H., Sargent, A., Zhuang, J. J., Kitchener, H. C., ... Widschwendter, M. (2012). Epigenetic variability in cells of normal cytology is associated with the risk of future morphological transformation. *Genome Medicine*, *4*, 24.

Tetley, T. (2007). Health effects of nanomaterials. *Biochemical Society Transactions*, *35*, 527–531.

Thompson, L. L., Guppy, B. J., Sawchuk, L., Davie, J. R., & McManus, K. J. (2013). Regulation of chromatin structure via histone post-translational modification and the link to carcinogenesis. *Cancer and Metastasis Reviews*, *32*, 363–376.

Thorek, D. L., & Tsourkas, A. (2008). Size, charge and concentration dependent uptake of iron oxide particles by non-phagocytic cells. *Biomaterials*, *29*, 3583–3590.

Torres-Giner, S., Gimenez, E., & Lagarón, J. M. (2008). Characterization of morphology and thermal properties of zein prolamine nanostructures obtained by electrospinning. *Food Hydrocolloids*, *22*, 601–614.

Uhrich, K. E., Cannizzaro, S. M., Langer, R. S., & Shakesheff, K. M. (1999). Polymeric systems for controlled drug release. *Chemical Reviews*, *99*, 3181–3198.

Volkovova, K., Handy, R. D., Staruchova, M., Tulinska, J., Kebis, A., Pribojova, J., ... Dusinska, M. (2015). Health effects of selected nanoparticles in vivo: Liver function and hepatotoxicity following intravenous injection of titanium dioxide and Na-oleate-coated iron oxide nanoparticles in rodents. *Nanotoxicology*, *9*, 95–105.

Weng, Y.-L., An, R., Shin, J., Song, H., & Ming, G.-l (2013). DNA modifications and neurological disorders. *Neurotherapeutics*, *10*, 556–567.

Wilson, C. L., Natarajan, V., Hayward, S. L., Khalimonchuk, O., & Kidambi, S. (2015). Mitochondrial dysfunction and loss of glutamate uptake in primary astrocytes exposed to titanium dioxide nanoparticles. *Nanoscale*, *7*, 18477–18488.

Wittmaack, K. (2007). In search of the most relevant parameter for quantifying lung inflammatory response to nanoparticle exposure: Particle number, surface area. *Environmental Health Perspectives*, 187–194.

Wohlfart, S., Gelperina, S., & Kreuter, J. (2012). Transport of drugs across the blood–brain barrier by nanoparticles. *Journal of Controlled Release*, *161*, 264–273.

Wu, T.-H., Yen, F.-L., Lin, L.-T., Tsai, T.-R., Lin, C.-C., & Cham, T.-M. (2008). Preparation, physicochemical characterization, and antioxidant effects of quercetin nanoparticles. *International Journal of Pharmaceutics*, *346*, 160–168.

Xia, T., Kovochich, M., Brant, J., Hotze, M., Sempf, J., Oberley, T., ... Nel, A. E. (2006). Comparison of the abilities of ambient and manufactured nanoparticles to induce cellular toxicity according to an oxidative stress paradigm. *Nano Letters*, *6*, 1794–1807.

Xiao, D., Wang, H., & Han, D. (2016). Single and combined genotoxicity effects of six pollutants on THP-1 cells. *Food and Chemical Toxicology*, *95*, 96–102.

Xue, M., Jiang, Z.-Z., Wu, T., Li, J., Zhang, L., Zhao, Y., ... Yang, S.-y (2012). Anti-inflammatory effects and hepatotoxicity of Tripterygium-loaded solid lipid nanoparticles on adjuvant-induced arthritis in rats. *Phytomedicine*, *19*, 998–1006.

Yamago, S., Tokuyama, H., Nakamura, E., Kikuchi, K., Kananishi, S., Sueki, K., … Ambe, F. (1995). In vivo biological behavior of a water-miscible fullerene: 14C labeling, absorption, distribution, excretion and acute toxicity. *Chemistry & Biology*, *2*, 385–389.

Yang, W., Peters, J. I., & Williams, R. O. (2008). Inhaled nanoparticles—A current review. *International Journal of Pharmaceutics*, *356*, 239–247.

Yoshino, H., Seki, N., Itesako, T., Chiyomaru, T., Nakagawa, M., & Enokida, H. (2013). Aberrant expression of microRNAs in bladder cancer. *Nature Reviews Urology*, *10*, 396–404.

Yu, H., & Huang, Q. (2013). Investigation of the cytotoxicity of food-grade nanoemulsions in Caco-2 cell monolayers and HepG2 cells. *Food Chemistry*, *141*, 29–33.

Yuan, Z.-Y., Hu, Y.-L., & Gao, J.-Q. (2015). Brain localization and neurotoxicity evaluation of polysorbate 80-modified chitosan nanoparticles in rats. *PLoS One*, *10*, e0134722.

Zannas, A. S., Provençal, N., & Binder, E. B. (2015). Epigenetics of posttraumatic stress disorder: Current evidence, challenges, and future directions. *Biological Psychiatry*, *78*, 327–335.

Zhang, L. W., William, W. Y., Colvin, V. L., & Monteiro-Riviere, N. A. (2008). Biological interactions of quantum dot nanoparticles in skin and in human epidermal keratinocytes. *Toxicology and Applied Pharmacology*, *228*, 200–211.

Zhang, X., Sun, H., Zhang, Z., Niu, Q., Chen, Y., & Crittenden, J. C. (2007). Enhanced bioaccumulation of cadmium in carp in the presence of titanium dioxide nanoparticles. *Chemosphere*, *67*, 160–166.

Zhang, Z., Zhang, R., & McClements, D. J. (2016). Encapsulation of β-carotene in alginate-based hydrogel beads: Impact on physicochemical stability and bioaccessibility. *Food Hydrocolloids*, *61*, 1–10.

Zheng, N.-G., Wang, J.-L., Yang, S.-L., & Wu, J.-L. (2014). Aberrant epigenetic alteration in Eca9706 cells modulated by nanoliposomal quercetin combined with butyrate mediated via epigenetic-NF-κB signaling. *Asian Pacific Journal of Cancer Prevention*, *15*, 4539–4543.

Zheng, Y., Chen, Y., Jin, L.-W., Ye, H.-Y., & Liu, G. (2015). Cytotoxicity and genotoxicity in human embryonic kidney cells exposed to surface modify chitosan nanoparticles loaded with curcumin. *AAPS PharmSciTech*, *17*(6), 1347–1352.

Zhong, Q., & Jin, M. (2009). Zein nanoparticles produced by liquid–liquid dispersion. *Food Hydrocolloids*, *23*(8), 2380–2387.

Index

Note: Page numbers followed by "*f*" and "*t*" refer to figures and tables, respectively.

A

Absorption, 434
  to epithelium cells, 134
  Fe absorption, 380–381
  in gastrointestinal tract, 44
  intestinal absorption of phytosterols, 300–301
  of NPs, 434
  of nutrients, 5
  of visible light, 424
Absorption, distribution, metabolism, and excretion (ADME), 405, 424, 441
Acceptable daily intake (ADI), 208
Acetic acid, 185*t*, 186
Acetylation, 426–428
2-Acetylpyridine (2-APri), 342
Acid hydrolysis, 35
Acrylic acid, 42
Active ingredients, 402–404
S-Adenosyl-L-methionine, 426
Adequate intake (AI), 147–148
ADHD. *See* Attention deficit hyperactivity disorder (ADHD)
Adhesion
  mechanism, 401–402
  mucoadhesive, 408
  work, 410
ADI. *See* Acceptable daily intake (ADI)
ADME. *See* Absorption, distribution, metabolism, and excretion (ADME)
Adsorption theory, 410
AFM. *See* Atomic force microscopy (AFM)
AI. *See* Adequate intake (AI)
ALA. *See* α-Linolenic acid (ALA)
Alanine aminotransferase (ALT), 437, 439
Alginate-based nanogels, 41, 312
Aliphatic aldehydes, 263
Alkali reagents, 40–41, 311–312
Alkaline pH, 45, 263, 304–305
Alkaline phosphatase (ALP), 437
Allyl isothiocyanate, 205
α cyclodextrins, 171, 202
Alpha-glycosyl isoquercitrin, 429
α-helix, 284–285
α-lactalbumin (α-La), 16–17, 43, 312–313
  nanotubes, 313
α-linolenic acid (ALA), 103–104
αs_1-casein (αs_1-CN), 304
ALT. *See* Alanine aminotransferase (ALT)
Ames assay, 432
AMF. *See* Anhydrous milk fat (AMF)
Amidination, 309
Amine-based cross-linking, 42
Amino acid(s), 304
  chains, 300
  composition, 300
  flavors derived from, 264, 265*t*
(1–4)-2-Amino-2-deoxy-β-D-glucan. *See* Chitosan
Ammonium ferric citrate ($(NH_4)_5Fe(C_6H_4O_7)_2$), 345
Amorphous solids, 421–422
Amperometric biosensor, 309
Amphipathic lipids, 9
Amplified signal, 414
Amylose nanocarriers, 23–24
Angiogenic vessels, 405–406
Anhydrous milk fat (AMF), 244
Animal sources
  chitosan, 187–190
  lactoferrin, 190–191
  lysozyme, 190
Anthocyanins, 227*t*, 229–230
Antimicrobial(s), 183. *See also* Food antimicrobial agents
  agents, 197–198
  encapsulated antimicrobials, 210–211
  hydrophobic antimicrobials, 193
  resistance, 303

Antioxidants, 63, 64*f*, 65–66, 298
 classification and sources, 67*t*
 nanoencapsulation process of, 69–73
  biopolymer-based technologies for, 85–91
  equipment-based technologies for, 91–93
  lipid-based formulations for, 73–85
2-APri. *See* 2-Acetylpyridine (2-APri)
Aqueous food systems, 208–209
Aroma compounds. *See* Epoxy compounds
Ascorbic acid, 148
Aspartate aminotransferase (AST), 437
Astaxanthin (ASTA), 242–243
 astaxanthin-generated fine globular particles, 249
 astaxanthin-loaded NLCs, 245
 fabricated nanoliposomes, 240–241
Atomic force microscopy (AFM), 205–206, 411, 414–415
Attention deficit hyperactivity disorder (ADHD), 106
Avidin, 39

B

Bacillus cereus, 201
Bacillus circulans, 21
Bacillus licheniformis, 42–43
Bacillus macerans, 21
Bakery products, 72
 encapsulated iron in, 379–382
 flavors derived from, 267–268
Basil oil, 191
BBB. *See* Blood–brain barrier (BBB)
BCD. *See* β-Cyclodextrin (BCD)
Bead-milling process, 45–46
Beet pectin (BP), 91
Benzoic acid, 185*t*, 186–187
Berberis vulgaris, 233–234
β cyclodextrins, 171, 202
Beta vulgaris.. *See* Red beet (*Beta vulgaris*)
β-carotene, 148, 170–171
 β-carotene–loaded NLCs, 244
β-casein (B-CN), 246–247, 304–305
 β-casein-lysozyme nanoparticles, 305
β-cyclodextrin polymer (CDP), 286
β-cyclodextrin (BCD), 90, 125–126, 285–286, 285*f*
β-lactoglobulin (BLG), 86–89, 169, 311, 313
Betacyanins, 230
Betalains, 230
Betaxanthins, 230
Beverages, 72–73
Bioactive bearing foods, 298
Bioactive compounds, 5, 65, 145, 167, 232, 298, 308, 312, 439
 chitosan, 312
 molecular degradation, 299, 301–302
 protein nanogels, 311–312
Bioactive food ingredients, 231, 401–402
Bioactive ingredients, 405, 407
Bioactive lipids, 300–301
Bioactive peptides, 300, 314–315
Bioadhesion, 306
Bioavailability, 63, 65, 405
 Fe-and Ca-encapsulation for fortifying dairy products, 366*t*
 Fe-bioavailability, 362–363, 377–378, 380–381
 of food colorants, 231–232
 of lutein-LMWC nanostructures, 248
 mineral NPs, 385*t*
 of nanoemulsified oils, 116–117
 of vitamin D_3, 165
Biocatalytic efficiency, 308–309
Biokinetics, 433
Biological molecules
 biopolymer based technologies for nanoencapsulation, 311–314
 equipment based technologies for nanoencapsulation, 307–311
 lipid formulation technologies for nanoencapsulation, 302–304
 natural nanocarriers for nanoencapsulation, 304–307
 safety of nanoencapsulation systems loaded with, 316
Biomacromolecules, 339
Biopolymer-based nanoencapsulation
 complexation of biopolymers, 36–39
  molecular forces between biopolymers, 36–38
  polysaccharide–polysaccharide complexes, 39
  protein–polysaccharide interactions, 38–39
  protein–protein complexes, 39
 of fish oil and essential fatty acids, 129–130
 of food antimicrobial agents, 200–202
  carbohydrate-based biopolymers, 201–202
  protein-based biopolymers, 200–201
 nanogels, 40–42

nanotubes, 42–43
single biopolymer NPs, 33–35
techniques, 33–43
technologies for nanoencapsulation of biological molecules
alginate-based nanogels, 312
chitosan-based nanogels, 312
nanotubes, 312–313
protein nanogels, 311–312
starch nanoparticles, 313–314
Biopolymer(s), 1, 91–92, 208–209
biopolymeric nanocarriers, 284–287
nanoparticles, 168–169, 361–362
food colorants nanoencapsulation by, 247–248
technologies, 85–91
Biosynthetic pathways, 297
BLG. *See* β-Lactoglobulin (BLG)
Blood–brain barrier (BBB), 434–435, 437
Body fluids plus proteins, 433
Boron (B), 333
Bottom-up method, 2–3, 194
Bovine serum albumin (BSA), 16–17, 166–167, 313, 315, 439
BP. *See* Beet pectin (BP); Breakdown probability (BP)
Brain, 434–435
development and function, 105–106
Brassica oleracea L. *See* Red cabbage (*Brassica oleracea* L.)
Breakdown probability (BP), 417–419
Breast milk, 106
Brownian motion, 115–116, 417
BSA. *See* Bovine serum albumin (BSA)

C

CA. *See* Cellulose acetate (CA)
Calcium (Ca), 16–18, 333, 436
binding with ions, 19
Ca-fortification, 370
Calcium alginate, 202
Calcium caseinate, 18
Calcium chloride ($CaCl_2$), 335
Calcium lactate ($C_6H_{10}CaO_6$), 344
Calcium oxide (CaO), 384–388
Cancer, 63, 106, 431–432
therapy, 388
Canola oil, 169
Canthaxanthin, 170–171
CaO. *See* Calcium oxide (CaO)
Capsaicin, 281
Carbohydrate(s), 64–65, 313–314
carbohydrate-based biopolymers, 201–202
carbohydrate-based products, 208–209
nanoemulsions stabilized by, 78
Carbon chain ($-CH_2-$), 103
Carbon nanotubes, 2
Carboxy methyl cellulose sodium salt (Na-CMC), 247–248
Carboxylic acid (–COOH), 103
Carboxymethyl cellulose-coated nanocapsules (CMC-coated nanocapsules), 414
Carcinogenesis/carcinogenicity, 440–441
hepatotoxicity, 435–437
neurotoxicity, 433–437
Carotenoids, 66, 170–171, 226–229, 227*t*, 231–232
Carvacrol, 203–204
Casein micelles (CMs), 16–17, 89, 246–247, 304–305, 313
Casein protein (CSP), 287, 304
Casein(s), 16–20, 200–201
nanocapsules, 304–305
nanodelivery vehicles, ingredients for preparing, 18–19
nanoencapsulation by, 125
nanoparticles, 89, 168–169, 246–247
phosphoproteins, 304
preparation methods, 19–20
Caseinates, 310
Catalase (CAT), 66
CCD. *See* Charge-coupled-device (CCD)
CD-IC. *See* Cyclodextrin inclusion (CD-IC)
CDP. *See* β-cyclodextrin polymer (CDP)
CDs. *See* Cyclodextrins (CDs)
CEI technique. *See* Conventional ethanol injection technique (CEI technique)
Cell differentiation, 431
Cell genome, 429
Cell viability analysis, 362
Cellulose acetate (CA), 172
Cellulose nanocrystals (CNC), 431
Central neural system (CNS), 433
Centrifugal extrusion, 277
CGTase. *See* Cyclodextrin glucanotransferase (CGTase)
Charge-coupled-device (CCD), 417–419
Cheddar cheese, 372–374
Cheese, 372–374
flavors derived from, 266–267
Chelators, 190
Chemical characterization methods, 133

Chemical methods. *See also* Physicomechanical methods
coacervation, 349–350
emulsification, 351–353
fatty acid esters, 347–349
liposome entrapment, 343–346
modified solvent evaporation, 350
niosome entrapment, 347
salt-induced cold gelation, 353
synthesis of WPI–Fe gel particles, 354*f*
Chemical polymer(s), 1
nanogels, 41
Chemical procedures for flavors encapsulation, 270–272
co-crystallization, 271–272
coacervation, 270–271
Chitosan (CS), 41, 89–90, 169, 187–190, 201–202, 302, 340–341, 401–402, 430
Chitosomes, 170–171
Chloride (Cl), 333
Chlorophylls, 226, 227*t*
Cholesterol, 10–11, 154, 198–199
Chromatin, 426–428
Chromium (Cr), 333
Chronic diseases, 297
Citrus peel pectin, 166–167
CLA. *See* Conjugated linoleic acid (CLA)
CLSM. *See* Confocal laser scanning microscopy (CLSM)
CMC-coated nanocapsules. *See* Carboxymethyl cellulose-coated nanocapsules (CMC-coated nanocapsules)
CMs. *See* Casein micelles (CMs)
CNC. *See* Cellulose nanocrystals (CNC)
CNS. *See* Central neural system (CNS)
Co-crystallization, 271–272
Coacervation, 154–155, 270–271, 271*f*, 349–350, 364–365
Coated nanocrystals. *See* Encapsulated nanocrystals
Coaxial process, 277
Cob (III) alamins, 153–154
Cobalt (Co), 333
Coffin–Lowry syndrome, 431
Colloidal calcium phosphate, 16–17
Colloidal delivery systems, 145
Color of food materials, 223
Colorants, 223
Comet tests, 429
CometChip technology, 432–433
Complex biopolymer nanoparticles, 91
Complex coacervation, 37–38, 271, 271*f*, 349–350
Complex-forming side-chains, 307
Confectionery sector, 72–73
Confocal laser scanning microscopy (CLSM), 411, 413–414
Conjugated linoleic acid (CLA), 24
Convection, 403
Conventional ethanol injection technique (CEI technique), 347
Copper (Cu), 333
Coronary heart disease, 105
CoS. *See* Cosurfactant (CoS)
Cosolubility, 37
Cosurfactant (CoS), 80
Cow milk, 304
CpG islands, 426
Cross-linking, 129
disulfide-based, 42
ethyl hexanoate[4]-loaded nanoparticles preparation, 287
gelation techniques based on, 353
method, 314
Crystallinity, 245, 313–314, 421
CS. *See* Chitosan (CS)
CSP. *See* Casein protein (CSP)
CU. *See* Curcumin (CU)
Cur-CD. *See* Curcumin-loaded γ-hydroxypropyl CD (Cur-CD)
Cur-M. *See* Curcumin-loaded micelles (Cur-M)
Cur-M loaded in hydrogels (Cur-M-H), 248–249
Cur-NLCs. *See* Curcumin-loaded NLCs (Cur-NLCs)
Cur-SLNs. *See* SLNs loaded with curcumin (Cur-SLNs)
Curcumin (CU), 413
Curcumin-loaded micelles (Cur-M), 248–249
Curcumin-loaded NLCs (Cur-NLCs), 244
Curcumin-loaded γ-hydroxypropyl CD (Cur-CD), 246
Cyanidin (cyd), 229–230
Cyclic oligosaccharides, 306
Cyclodextrin glucanotransferase (CGTase), 21, 306, 429
Cyclodextrin inclusion (CD-IC), 283
Cyclodextrins (CDs), 16, 20–23, 90, 171, 202, 245, 283, 306, 359
α-CD, 307

ingredients for preparing CD nanocarriers, 21–22
molecular inclusion complexes with, 246
nanocapsules, 305–306
nanoencapsulation by, 125–126
nanoparticles, 90
preparation methods, 22–23
cyd. *See* Cyanidin (cyd)
Cytotoxicity, 432

D

D-Glucose units, 313–314
D-α-Tocopherol polyethylene glycol 1000 succinate (TPGS), 247–248
Dairy products, fortification of
cheese, 372–374
Fe-and Ca-encapsulation for fortifying dairy products, 366*t*
milk, 365–370
yogurt, 370–372
Dairy sector, 72
DC-cholesterol. *See* [*N*-(*N*′,*N*′-Dimethylaminoethane)-carbamoyl] cholesterol (DC-cholesterol)
DCP. *See* Dicetyl phosphate (DCP)
Debaryomyces hansenii, 266–267
Degradation, 301
Dehydration theory, 408
Deleterious effects, 424
Delphinidin (dpd), 229–230
Dendrimeters, 46–47
Desolvation method, 33–35
Destructive methods, 404
Detergent removal method, 12
DFS. *See* Double fortified salt (DFS)
DHA. *See* Docosahexaenoic acid (DHA)
DHPM. *See* Dynamic high-pressure microfluidization (DHPM)
Dialysis, 404
in removal of surfactants, 12
sac approach, 165–166
Dicetyl phosphate (DCP), 198–199
Diet
daily diet, 64–66, 299
dietary sources of minerals, 333
dietary supplements, 384
fatty acids in, 103–104
vitamins, 146
Differential scanning calorimetry (DSC), 133, 243, 284, 421–422
Diffusion, 403
Diffusion theory, 408, 410
Digestive system, 433–434
Diketones, 263
Dilution methods, 207–208
[*N*-(*N*′,*N*′-Dimethylaminoethane)-carbamoyl] cholesterol (DC-cholesterol), 358
Dimethylformamide (DMF), 270
1,2-Dioleoyl-3-sn-phosphatidylcholine, 122
1,2-Dipalmitoyl-3-sn-phosphatidylcholine, 122
Dipalmitoylphosphatidylcholine (DPPC), 198–199
Disk diffusion method, 207–208
Dissolution, 402–403
Disulfide-based cross-linking, 42
DLS. *See* Dynamic light scattering (DLS)
DMF. *See* Dimethylformamide (DMF)
DNA, 301
damage, 431–433
methylation, 431–432
molecules, 309–310
DNA methyltransferases (DNMTs), 425–426
DNMTs. *See* DNA methyltransferases (DNMTs)
Docosahexaenoic acid (DHA), 103–104
1-Dodecanol, 347
Double capillarity extrusion instruments
centrifugal extrusion, 277
dual capillary, 277
recycling centrifugal extrusion, 277–278
Double emulsions, 165
Double fortified salt (DFS), 339–340
Double nanoemulsions, 6
dpd. *See* Delphinidin (dpd)
DPPC. *See* Dipalmitoylphosphatidylcholine (DPPC)
Draft standard, 439–440
Drawing-out precipitation, 34
Droplet
drying of, 31–32
generation, 31
Drug-delivery system, 1–2, 29
Dry thin film method (DTFM), 240–241
DSC. *See* Differential scanning calorimetry (DSC)
DTFM. *See* Dry thin film method (DTFM)
Dual capillary process, 277
Duplex emulsion. *See* Double nanoemulsions
Dynamic dialysis method, 315
Dynamic high-pressure microfluidization (DHPM), 82
Dynamic light scattering (DLS), 278–281, 413, 417

E

EAR. *See* Estimated average requirement (EAR)
EE. *See* Encapsulation efficiency (EE); Eudragit E (EE)
EFSA. *See* European Food Safety Authority (EFSA)
EGCG. *See* Epigallocatechin-3-gallate (EGCG)
Egg phosphatidyl choline (EPC), 345–346, 346*f*
Eicosapentaenoic acid (EPA), 103–104
EIM. *See* Ethanol injection method (EIM)
EIP method. *See* Emulsion inversion point method (EIP method)
Elastin-like polypeptides, 308
Electrical characteristics, nanoencapsulated food ingredients, 419–420
Electrical conductivity, 26
Electrohydrodynamic atomization. *See* Electrospraying
Electrohydrodynamic encapsulation methods, 308
Electrohydrodynamic processes, 202–203, 307–308
  enzyme immobilization based on, 308–309
  nucleic acid encapsulation based on, 309–310
  protein encapsulation based on, 310–311
Electromagnetic spectrum, 422–423
Electron, 435
Electron beams, 411–412
Electron field, 411–412
Electron microscopy, 411–413
Electronic theory, 409–410
Electrospinning, 24–27, 72, 91–92, 307–308, 310–311
  nanoencapsulation by, 126–128, 127*f*
  for nanoencapsulation of flavors, 283–284
  preparation of electrospun fibers, 25–27
  technique, 172, 202–203, 247
Electrosprayed nanoparticles, 171–173, 311
Electrospraying, 24, 27–29, 72, 91–92, 307–308, 311
  nanoencapsulation by, 128
  for nanoencapsulation of flavors, 283–284
  preparation of electrosprayed nanocarriers, 28–29
  technique, 202–203, 247
Electrospun, 310–311
  amaranth protein, 311
  fibers, 308
    ambient conditions, 27
    polymer solution properties, 26
    preparation, 25–27
    processing conditions, 26–27
  nanofibers, 311
Electrostatic force, 25
Electrostatic interaction, 17–18, 37–39, 129–130, 423
Electrostatic stabilizers, 14–15
Embryogenesis, 431
Emulsification, 351–353
Emulsifiers, 7
Emulsion inversion point method (EIP method), 79, 117–118
Emulsion(s), 302
  system, 74
  techniques, 153–154
  titration, 282
Encapsulated antimicrobials, 210–211
Encapsulated colorants application in food formulations, 250–251, 251*t*
Encapsulated iodine, salt fortification with
  dual fortified salt, 374–379
  Fe-fortification of iodized salts, 375*t*
  triple fortified salt, 379
Encapsulated iron
  in fortifying cereals and bakery products, 379–382
  salt fortification with
    dual fortified salt, 374–379
    triple fortified salt, 379
Encapsulated minerals, 365–384. *See also* Minerals
  encapsulated iron in fortifying cereals and bakery products, 379–382
  fortification of dairy products with, 365–374
  in fortifying other foods, 382–384
  salt fortification with encapsulated iron and iodine, 374–379
Encapsulated nanocrystals, 44–45
Encapsulated powder, 420–421
Encapsulated protein hydrolysate, 315
Encapsulated-Ca in liposome (L-Ca), 382
Encapsulation, 47, 261. *See also* Nanoencapsulation
  of flavors
    chemical procedures for, 270–272
    mechanical procedures for, 272–278
  by nanoemulsions, 113–120
  by nanoliposomes, 120–122
  of natural colors, 232–234, 235*t*

by NLCs, 123–125
by SLNs, 122–123
systems, 299–300
techniques, 232
chemical methods, 343–353
of minerals, 334–353
physicomechanical methods, 335–343
to preparing encapsulated minerals, 336*t*
technology, 65, 107
Encapsulation efficiency (EE), 206–207, 233, 281, 335
Engineered nanomaterials (ENMs), 439–440
Engineering nanoencapsulation systems, 404–405, 407
ENMs. *See* Engineered nanomaterials (ENMs)
Enrichment, 148
Enthalpy, 421–422
Enzymatic antioxidants, 66
Enzymatic hydrolysis process, 262
Enzyme(s), 299
immobilization, 308–309
stability of, 299
EOs. *See* Essential oils (EOs)
EPA. *See* Eicosapentaenoic acid (EPA)
EPC. *See* Egg phosphatidyl choline (EPC)
Epigallocatechin-3-gallate (EGCG), 78, 86–89
Epigenetic(s), 425–426
epigenetic regulations
DNA methylation, 426, 428*f*
histone alteration, 426–428
miRNAs as gene expression modifiers, 428–429
modified by NPs, 425–433, 427*f*
in vivo and in vitro assays, 429–433, 430*t*
factors, 426, 431
Epoxy compounds, 267
Equilibrium solution. *See* Macromolecule-depleted phase
Equipment based technologies
for nanoencapsulation, 307–311
enzyme immobilization, 308–309
nucleic acid encapsulation, 309–310
protein encapsulation based, 310–311
for nanoencapsulation of phenolics and antioxidants, 91–93
electro-spinning and electro-spraying, 91–92
nano-spray drying, 93
Equipment based-nanoencapsulation
of fish oil and essential fatty acids, 126–129
electrospinning, 126–128, 127*f*
electrospraying, 128
nanospray dryer, 128–129
of food antimicrobials, 202–203
Erosion, 403
Escherichia coli, 200–201
Essential fatty acids, 103–113
biopolymer-based nanoencapsulation, 129–130
equipment based-nanoencapsulation, 126–129
health benefits of, 104–106
brain development and function, 105–106
cancer, 106
coronary heart disease, 105
infant health, 106
mental health, 106
RA, 105
lipid formulation nanoencapsulation of, 113–125
nanoencapsulation
of fish oil and, 107–113
technologies, 131
natural nanocarriers for nanoencapsulation of, 125–126
source of, 104, 104*t*
stability of fish oil and, 106–107
Essential oils (EOs), 81–82, 85–86, 191, 193
Estimated average requirement (EAR), 147–148
Ethanol injection method (EIM), 240–241
3-Ethoxy-4-hydroxybenzaldehyde. *See* Ethyl-vanillin (EV)
Ethyl-vanillin (EV), 284
Euchromatin structure, 426–428
Eudragit E (EE), 423
European Food Safety Authority (EFSA), 69
EV. *See* Ethyl-vanillin (EV)
Evolutionary stresses, 297
Extrusion, 276, 277*f*, 342–343

F

Fabricating nanofood products, 440
Fat-soluble pigments, 226–229
Fatty acid esters (FAEs), 335, 347–349
Fe encapsulation, 348*f*
morphology of $FeSO_4$ microcapsules, 349*f*

Fatty acids, 103
FBC. *See* Fluidized bed coating (FBC)
FDA. *See* US Food and Drug Administration (FDA)
Fe-encapsulated cold-set WPI (WPI-Fe), 372
Ferric chloride ($FeCl_3$), 348–349
Ferric pyrophosphate ($Fe_4O_{21}P_6$), 342
Ferric sodium ethylenediaminetetraacetate (NaFeEDTA), 379–380
Ferripolyphosphate–whey protein (FIP–WP), 373
Ferrous bisglycinate ($C_4H_8FeN_2O_4$), 351
Ferrous fumarate ($C_4H_2FeO_4$), 339, 343*f*
Ferrous sulfate (FeSO4), 339
Fe–WP. *See* Whey protein-chelated Fe (Fe–WP)
Fiber morphology, 26
Fibrinogen, 310–311
Fick's law, 205
Fickian diffusion, 204–205, 361
FIP–WP. *See* Ferripolyphosphate–whey protein (FIP–WP)
First-pass effect, 405
Fish oil, 72, 104–105, 420–421
  biopolymer-based nanoencapsulation, 129–130
  equipment based-nanoencapsulation, 126–129
  fish oil-loaded nanocapsules, safety and regulatory issues of, 134
  nanoencapsulation, 107–113
    lipid formulation, 113–125
    natural nanocarriers, 125–126
    technologies for, 131
  stability, 106–107
Fish sarcoplasmic protein fibers (FSP fibers), 311
FITC dye. *See* Fluorescein isothiocyanate dye (FITC dye)
Flavonoid dyes, 229–230
Flavors
  from amino acids, 264
  classification of food bioflavors, 262–263
  encapsulation, 261
  in food products, 266–268
  from lipids, 264–266
  mechanical procedures for encapsulation
    double capillarity extrusion instruments, 277–278
    extrusion, 276
    freeze drying, 274–275
    spray chilling/cooling, 275–276
    spray drying and spray coating, 272–274, 272*f*
  microencapsulation of food flavors, 268–278
  nanoencapsulation technologies for food flavors, 278–287
  from saccharides, 263, 264*t*
  simple of microspheres and microcapsules, 262*f*
Flow rate, 26–27
Fluidized bed coating (FBC), 154, 335, 339–340
Fluorescein isothiocyanate dye (FITC dye), 414
Fluorescence
  emission, 404
  image, 404
  spectroscopy, 404
Fluoride (F), 333
Fluorophores, 404
Folate cyanocobalamin. *See* Vitamin B_{12}
"Folded" protein, 310
Food antimicrobial agents, 183
  analysis and characterization of antimicrobial-loaded nanocapsules, 205–208
  controlled release of antimicrobial agents, 204–205
  nanoencapsulated food antimicrobial agents
    applications, 208–211
    safety, 208
  nanoencapsulation, 184, 194–204
  natural antimicrobial agents, 187–193
  problems and opportunities of natural antimicrobials, 193–194
  traditional chemical preservative agents, 184–187
Food bioactive components, 2
Food bioflavors
  classification, 262–263
  primary flavors, 262
  secondary flavors, 263
Food colorants, 223
  classification, 224–231, 225*t*
    betalains, 230
    carotenoids, 226–229, 227*t*
    chlorophylls, 226, 227*t*
    flavonoid dyes, 229–230
    heme pigments, 226, 227*t*
    miscellaneous colorants, 230–231
  encapsulated colorants application in food formulations, 250–251, 251*t*

encapsulation of natural colors, 232–234, 235*t*
nanoencapsulation, 234–250, 237*t*
by biopolymer nanoparticles, 247–248
lipid-based nanocarriers, 239–245
nature-inspired nanocarriers, 245–247
other methods, 248–250
by special equipment, 247
safety aspects of nanoencapsulated food colorants, 250
stability and bioavailability, 231–232
Food flavors
microencapsulation of, 268–278
nanoencapsulation technologies for, 278–287
Food formulations, encapsulated colorants application in, 250–251, 251*t*
Food fortification, 145, 333–334
Food industry, 89–90
encapsulated minerals, 365–384
encapsulated iron in fortifying cereals and bakery products, 379–382
fortification of dairy products, 365–374
in fortifying other foods, 382–384
salt fortification with encapsulated iron and iodine, 374–379
Food ingredients, 16, 24, 29
Food packaging, 149
Food processing conditions, vitamin retention under, 149–150, 150*t*
Food products, 72, 229–230, 261
flavors in, 266–268
bakery products, flavors derived from, 267–268
cheeses, flavors derived from, 266–267
fruits and vegetables, flavors derived from, 268
red meat, flavors derived from, 267
vitamin fortification in, 148
Food safety, 184
Food Standards Australia and New Zealand (FSANZ), 315–316
Food-grade
nanoencapsulation systems, 421
nanostructures, 438
nanovehicles, 155
Food-processing methods, 307
Foodborne disease, 210–211
Fortification of dairy products
cheese, 372–374
Fe-and Ca-encapsulation for fortifying dairy products, 366*t*
milk, 365–370
yogurt, 370–372
Fortifying cereals and bakery products, encapsulated iron in, 379–382
Fourier transform infrared spectroscopy (FTIR), 130, 422–423
Fracture theory, 410
Free bioactive ingredients, 405–406
Free radicals, 65–66
Freeze-drying, 274–275, 275*f*, 335–338
Fresh produces and vegetables, 210–211
Fruits, flavors derived from, 268
FSANZ. *See* Food Standards Australia and New Zealand (FSANZ)
FSP fibers. *See* Fish sarcoplasmic protein fibers (FSP fibers)
FTIR. *See* Fourier transform infrared spectroscopy (FTIR)

G

Gallic acid-loaded zein electrospun fibers, 413
γ cyclodextrins, 171, 202
γ-PGA. *See* Poly γ-glutamic acid (γ-PGA)
Ganglionic bodies, 434
Gas chromatography olfactory device (GC-O device), 266
Gastrointestinal tract (GI tract), 145, 302, 404–405, 408, 441
gastrointestinal content, 44
gastrointestinal processing, 315–316
GI cells, 433–434
GC-O device. *See* Gas chromatography olfactory device (GC-O device)
Gelation techniques, 353
Gene expression, 63, 426
miRNAs as modifiers, 428–429
via NPs, 429
Gene therapy, 301
Gene-targeted therapy, 1–2
Generally recognized as safe (GRAS), 186–187, 250
Genetic disorders, 431
Genotoxicity, 425–433, 427*f*, 440–441
DNA methylation, 426, 428*f*
histone alteration, 426–428
miRNAs as gene expression modifiers, 428–429
in vivo and in vitro assays, 429–433
Geraniol, 284
GI tract. *See* Gastrointestinal tract (GI tract)
GIA. *See* Global Industry Analysts (GIA)

GIT. *See* Gastrointestinal tract (GI tract)
Glass–rubbery transitions, 421
Glial fibrillary acidic protein, 434–435
Global Industry Analysts (GIA), 70–73
Global nanofood regulations, 439–440
Globular proteins, 310
Glutaraldehyde, 439
Glutathione (GSH), 65–66, 435
  reductase, 66
Glutathione peroxidase (GSHpx), 66
Glycyrrhizic acid, 438
Gold, 2, 437
Gossypol, 230–231
Gram-negative bacteria, 191
Gram-positive bacteria, 191–193
GRAS. *See* Generally recognized as safe (GRAS)
Gravitational settling and centrifugation, 417
"Green" method, 201
GSH. *See* Glutathione (GSH)
GSHpx. *See* Glutathione peroxidase (GSHpx)
Gums, 202

H

Halloysite ($Al_2Si_2O_5(OH)_4 \cdot 2H_2O$), 43
HC. *See* Hi-Cap 100 (HC)
HDL. *See* High-density lipoproteins (HDL)
HDMs. *See* Histone demethylases (HDMs)
Heme pigments, 226, 227*t*
Hemoglobin, 226
Hepatotoxicity, 435–437
  in vitro tests, 438
  in vivo tests, 439
Hetero structure, 426–428
Hexafluoro-2-propanol (HFP), 310–311
Hi-Cap 100 (HC), 78
High energy methods, 74–79
High-density lipoproteins (HDL), 105
High-pressure homogenization (HPH), 12–13, 15–16, 46, 242, 247, 358–359
Histone demethylases (HDMs), 426–428, 427*f*
Histone methyltransferases (HMTs), 426–428, 427*f*
Homogenization, 360–361
2-HPβCD. *See* 2-Hydroxypropyl-β-cyclodextrin (2-HPβCD)
HPF. *See* Hydrogenated palm fat (HPF)
HPH. *See* High-pressure homogenization (HPH)
HPMC. *See* Hydroxypropyl methylcellulose (HPMC)
Hprt, 432
HPγCD. *See* 2-Hydroxypropyl-γ-cyclodextrin (HPγCD)
HSPC. *See* Hydrogenated phosphatidylcholine (HSPC)
Human growth hormones, 315
Human nervous system, 105–106
HY molecules. *See* Hypericin molecules (HY molecules)
Hyaluronic acid nanogels, 312
Hydrogels, 40
Hydrogenated palm fat (HPF), 342
Hydrogenated phosphatidylcholine (HSPC), 358
Hydrolyze α-lactalbumin, 313
Hydrophilic
  solutes, 11–12
  vitamins, 146
Hydrophobicity, 305, 313–314, 408
  hydrophobic antimicrobials, 193, 200
  hydrophobic compounds, 239, 434–435
  hydrophobic interactions, 19
  hydrophobic proteins, 303–304
  hydrophobic/hydrophilic interactions, 303
Hydroxyl groups (–OH), 263
5-Hydroxymethyl-2′-deoxycytidine, 426
Hydroxypropyl methylcellulose (HPMC), 339–340
2-Hydroxypropyl-β-cyclodextrin (2-HPβCD), 286
2-Hydroxypropyl-γ-cyclodextrin (HPγCD), 246
Hypericin molecules (HY molecules), 286
Hypermethylation, 426
Hypertension, 105
Hypomethylation, 425–426

I

ICs. *See* Inclusion complexes (ICs)
ICS. *See* Iron casein succinylate (ICS)
IgG. *See* Immunoglobulin (IgG)
Image processing techniques, 424
Immobilization based on electrohydrodynamic processes, enzyme, 308–309
Immunoglobulin (IgG), 315
In vitro
  antimicrobial activity evaluation, 207–208
  assays, 429–433
  tests, 438
In vivo
  antimicrobial activity evaluation, 207–208

assays, 429–433
studies, 440–441
tests, 439
Inclusion complexes (ICs), 90
Industrial utilization, 299
Infant health, 106
Inflammation, 436
Inflammatory bowel disease, 424
Infrared spectra (IR spectra), 309
Inhalation, 433
Inorganic nanomaterials, 2
Inorganic nanometal oxides, 384–388
Insulin, 434
Intermolecular hydrogen bonding, 304
Intestinal phase, 315–316
Intra-molecular hydrogen bonding, 304
Intraepithelial zone, 424
Iodine (I), 333
Ionotropic gelation, 362–364
IR spectra. *See* Infrared spectra (IR spectra)
Iron (Fe), 333
Fe-deficiency anemia, 358
fortification, 371–372
microcapsules using emulsification method, 352*f*
Iron casein succinylate (ICS), 364–365
Iron oxide (Fe_2O_3), 388
2-Isobutyl-3 methoxypyrazine, 261
Isoelectric point (pI), 37–38
Isopropyl alcohol, 276

K

Ketones, 263
KI. *See* Potassium iodide (KI)
Kluyveromyces lactis, 266–267
Korsmeyer–Peppas model, 205

L

LA. *See* Linoleic acid (LA)
Lactic acid, 185*t*, 186
Lactic acid bacteria (LAB), 266
Lactobacillus acidophilus, 371
Lactoferrin (LF), 39, 169, 190–191, 302
Large microencapsulated $FeSO_4$ (LMFS), 373–374
Large unilamellar vesicles (LUVs), 9
L-ascorbic acid, 167
Laser diffraction technique (LD technique), 243
Laser energy, 417–419
Laser light scattering technique, 205–206
Laser-induced breakdown detection (LIBD), 417–419
L-Ca. *See* Encapsulated-Ca in liposome (L-Ca)
LCT. *See* Long-chain triglycerides (LCT)
LD technique. *See* Laser diffraction technique (LD technique)
LDL. *See* Low-density lipoproteins (LDL)
LE. *See* Loading efficiency (LE)
Leafy vegetables, 226, 232
Lecithin, 154
food-grade W/O microemulsions based on, 81
liposome system, 365–369
soy lecithin, 10, 120
Leuprolide, 315
LF. *See* Lactoferrin (LF)
LIBD. *See* Laser-induced breakdown detection (LIBD)
D-Limonene, 439
Linoleic acid (LA), 130
Lipid formulation(s), 77*f*
nanoemulsions, 194–198
nanoencapsulation of fish oil and essential fatty acids
encapsulation by NLCs, 123–125
encapsulation by SLNs, 122–123
nanoemulsions, encapsulation by, 113–120
nanoliposomes, encapsulation by, 120–122
for nanoencapsulation of phenolics and antioxidants, 73–85, 75*t*
in nanoemulsions, 74–82
in nanoliposomes, 82–83
NLCs, 84–85
in solid lipid nanoparticles, 83–84
nanoliposomes, 198–199
SLN, 199
techniques, 197*f*
technologies for nanoencapsulation
nanoemulsions, 302
nanoliposomes, 303
nanostructured lipid carriers, 303–304
Lipid nanocapsules (LNCs), 244–245
Lipid-based nanocarriers for food color nanoencapsulation, 239–245
nanoemulsions, 241–242
nanoliposomes, 239–241
NLCs, 243–245
SLNs, 242–243

Lipid-based nanoencapsulation, 204–205
  nanoemulsions, 5–8
  nanolipid carriers, 12–16
  nanoliposomes, 8–12
  techniques, 5–16
Lipid(s), 1, 73, 198–199
  autoxidation, 267
  flavors derived from, 264–266
  matrices, 315
  oxidation, 106–107
Lipo-soluble vitamins, 303
Lipolysis, 266–267
Lipophilic bioactive agents, 5–6
Lipoplex nanoparticles, 309–310
Lipopolysaccharide (LPS), 190
Lipoprotein cholesterol, 300
Liposome(s), 9, 82, 120, 198–199
  bacteriocins, 303
  carriers, 154
  encapsulation, 303, 344–345
  entrapment, 343–346
    liposomal microcapsules formation, 345*f*
    phase-contrast microscopy images of EPC–liposomes, 346*f*
  enzymes, 303
  microcapsules formation, 345*f*
  skin retention, 433–434
Liquid lipid nanoparticles (LLNs), 242–243
Liquid lipids, 73–74
Liquid oil, 14
Liquid–liquid dispersion, 6
Listeria monocytogenes, 200
Liver, 437
LLNs. *See* Liquid lipid nanoparticles (LLNs)
LMFS. *See* Large microencapsulated $FeSO_4$ (LMFS)
LMWC nanoparticles. *See* Low molecular weight chitosan nanoparticles (LMWC nanoparticles)
LNCs. *See* Lipid nanocapsules (LNCs)
Loading capacity, antimicrobial-loaded nanocapsules, 206–207
Loading efficiency (LE), 91
Long-chain triglycerides (LCT), 241–242
Low energy methods, 74
Low molecular weight chitosan nanoparticles (LMWC nanoparticles), 248
Low-density lipoproteins (LDL), 105, 300–301
Low-energy methods, 79–82, 80*f*
Low–energy emulsification technique, 165–166
LPS. *See* Lipopolysaccharide (LPS)
Lutein, 170–171, 226–229, 240–241
  lutein-LMWC nanoparticles, 248
  natural antioxidant, 84–85
  NLCs for nanoencapsulation of, 243
LUVs. *See* Large unilamellar vesicles (LUVs)
Lycopene, 170–171
Lymphocytes, 429
Lyophilisation. *See* Freeze-drying
Lysozyme, 39–41, 190, 311–312

M

Macrominerals, 333
Macromolecular micelles. *See* Nanogels
Macromolecule-depleted phase, 37–38
Macromolecules, 408
Magnesium (Mg), 333
Magnesium oxide (MgO), 384–388
Magnified electron image, 411–412
Maltodextrin (MD), 247, 273, 350, 420–421
Malvidin (mvd), 229–230
Manganese (Mn), 333
Mangiferin, 230–231
Marine organism, 297
MBC. *See* Minimum bactericidal concentration (MBC)
MCC. *See* Microcrystalline cellulose (MCC)
MCT. *See* Medium chain triglycerides (MCT)
MD. *See* Maltodextrin (MD)
Mechanical theory, 410
Medium chain triglycerides (MCT), 73–74, 118, 165, 241, 351, 413
Mental health, essential fatty acids use in, 106
Mentha piperita, 204
Mesquite gum (MG), 351
Methylation, 426–428
MF. *See* Microfluidization (MF)
MG. *See* Mesquite gum (MG)
MgO. *See* Magnesium oxide (MgO)
MIC. *See* Minimum inhibition concentration (MIC)
Micelles, 304–305
Microbial sources, 192–193
  natamycin, 193
  nisin, 192–193
Microcapsules, 261–262, 262*f*, 342
Microcrystalline cellulose (MCC), 431
Microemulsion method, 15
  HPH, 15–16
Microemulsions, 5–6, 79–80
Microencapsulation, 334. *See also* Nanoencapsulation

of food flavors, 268–278, 269*t*, 270*f*
chemical procedures for encapsulation of flavors, 270–272
mechanical procedures for encapsulation of flavors, 272–278
techniques, 145
of vitamins, 150–155, 151*f*, 152*t*
coacervation, 154–155
emulsion techniques, 153–154
fluidized bed coating, 154
liposome carriers, 154
spray chilling and cooling, 153
spray drying technique, 151–153
Microfluidics devices, 24
Microfluidization (MF), 354–358
Microfluidizer, 12
Microminerals, 333
Micronucleus assay, 429
Microscope, 417–419
Microspheres, 261–262, 262*f*
Mie theory, 416
Milk, 365–370
milk proteins, nanoencapsulation of flavors by, 287
serum proteins, 314–315
Minerals, 333. *See also* Encapsulated minerals
encapsulation techniques, 334–353
food fortification, 333–334
micro/nanoencapsulation, 334
nanoencapsulation, 354–365
nanoparticles, 384–388, 385*t*
production processes/technologies of the mineral ions/salts-nanocapsules, 355*t*
Miniemulsions. *See* Nanoemulsions
Minimum bactericidal concentration (MBC), 207–208
Minimum inhibition concentration (MIC), 187–190
miRNAs as gene expression modifiers, 428–429
Miscellaneous colorants, 230–231
Mitochondria, 435–436
MLA assay, 432
MLVs. *See* Multilamellar large vesicles (MLVs)
Modified Fickian model, 204–205
Modified solvent evaporation, 350
Molecular degradation, 299, 301–302
Molecular forces between biopolymers, 36–38
complex coacervation, 37–38
cosolubility, 37
Molecular inclusion complexes with cyclodextrins, 246
Molecular nanoencapsulation, 20, 359
Molecular weight, 26
Molybdenum (Mo), 333
Monosaccharides, 263
Montmorillonite nanoclay, 203–204
Morphology in nanoencapsulation systems, 411–415
AFM, 414–415
CLSM, 413–414, 415*f*
electron microscopy, 411–413, 412*f*
Mucoadhesion theories, 408–410
Mucoadhesive delivery systems, 407–410
characterization of mucus layer, 407–408
mucoadhesion theories, 408–410
stages of mucoadhesion, 408, 409*f*
Mucus agent, 401–402
Mucus layer characterization, 407–408
Multilamellar large vesicles (MLVs), 170–171
Multiphase colloidal dispersions, 302
Multiple emulsion. *See* Double nanoemulsions
Mutagenicity, 440–441
Mutation tests, 432
mvd. *See* Malvidin (mvd)
Myoglobin, 226

N

Na-CMC. *See* Carboxy methyl cellulose sodium salt (Na-CMC)
Nano-organogels, 166–167
Nanocapsules, 85, 305–306, 425
controlled release
of antimicrobial agents, 204–205
of fatty acids from, 131–132
instrumental analysis and characterization of nanocapsules containing fish oil, 132–134
chemical characterization methods, 133
other characterization methods, 134
physical characterization methods, 132–133
safety and regulatory issues of fish oil-loaded nanocapsules, 134
Nanocarriers, 3–5, 4*t*, 401–402, 404, 433–434
amylose, 23–24
ingredients for preparing CD, 21–22
preparation of electrosprayed, 28–29
Nanochitosan (NC), 437

Nanocomponents, 316
Nanocomposites of food antimicrobial agents, 203–204
Nanocrystals, 43–46, 245, 247
encapsulated (coated) nanocrystals, 44–45
nanocapsules, 305–306
preparation methods, 45–46
bead-milling process, 45–46
combination technologies, 46
high-pressure homogenization, 46
Nanoemulsification of food flavors, 278–283
Nanoemulsions (NEs), 2, 5–8, 72–73, 155–166, 194–198, 241–242, 302–304, 358–359, 416, 431, 438
encapsulation by, 113–120, 117*f*
ingredients for preparing nanoemulsions, 6–7
emulsifiers, 7
oil, 6–7
stabilizers, 7
nanoencapsulation of phenolics and antioxidants, 74–82
high-energy methods, 74–79
low-energy methods, 79–82
preparation methods, 7–8
Nanoencapsulated food antimicrobial agents
applications of, 208–211
safety of, 208
Nanoencapsulated food ingredients
biological fate, 404–407, 406*f*
characterization, 411–424
electrical characteristics, 419–420
morphology, 411–415
physicochemical characterization, 421–424
size, 415–419, 417*f*, 418*f*
surface composition, 420–421
future trends, 440–442, 441*t*
global nanofood regulations, 439–440
health risks and carcinogenesis induced by NPS, 433–439
mucoadhesive delivery systems, 407–410
release, 401–404
analytical methods for measuring release, 404
controlled-release mechanisms, 402–403, 402*f*
safety and toxicity of food NPS employed in food products, 424–433
Nanoencapsulation, 65, 69, 116*f*, 184, 334, 404. *See also* Encapsulation; Microencapsulation
bioactive compounds, 298
biopolymer based technologies, 85–91, 129–130, 311–314
biosynthetic pathways, 297
charge of, 406
electrical features, 419
enzymes, 299
equipment based technologies, 91–93, 126–129, 307–311
of fish oil and essential fatty acids, 107–113, 108*t*, 113*f*, 114*t*
controlled release of fatty acids from nanocapsules, 131–132
instrumental analysis and characterization of nanocapsules, 132–134
safety and regulatory issues of fish oil-loaded nanocapsules, 134
technologies, 131
of flavors
by biopolymeric nanocarriers, 284–287
by milk proteins, 287
of food antimicrobial agents, 194–204, 195*t*
applications of nanoencapsulated food antimicrobial agents, 208–211
biopolymer-based nanoencapsulation of food antimicrobial agents, 200–202
equipment-based nanoencapsulation of food antimicrobials, 202–203
lipid-formulation techniques, 194–199, 197*f*
nanocomposites of food antimicrobial agents, 203–204
high-resolution image, 413
kinetics, 407
lipid formulation technologies, 302–304
lipid-based formulations for, 73–85
metabolic fate, 406–407
of minerals, 354–365
biopolymer nanoparticles, 361–362
coacervation, 364–365
cyclodextrin inclusion, 359–360
ionotropic gelation, 362–364
nanoemulsification, 358–359
nanoliposomes, 354–358
SLNs, 360–361
molecular degradation, 299
nanoencapsulated phenolics, 70–73
of natural food colorants, 234–250, 237*t*
by biopolymer nanoparticles, 247–248
lipid-based nanocarriers, 239–245
nature-inspired nanocarriers, 245–247
safety aspects, 250

by special equipment, 247
natural nanocarriers, 125–126, 304–307
nucleic acids, 301–302
peptides, 300
physicochemical properties, 405–406
phytosterols, 300–301
process of phenolic compounds and antioxidants, 69–73, 71*t*
release of peptides from nanoscale delivery systems, 314–316
safety of nanoencapsulation systems, 316
techniques, 2–5, 3*t*, 43–47, 145, 149–150, 204, 248–250
bioactive ingredients, 1
biopolymer-based nanoencapsulation techniques, 33–43
dendrimeters, 46–47
electrospraying and electrospinning, 283–284
for food flavors, 278–287, 279*t*
lipid-based nanoencapsulation techniques, 5–16
nanocrystals, 43–46
nanoemulsification, 278–283
nanomedicine, 1–2
nature-inspired nanoencapsulation techniques, 16–24
special-equipment-based nanoencapsulation techniques, 24–33
of vitamins, 155–173
Nanoencapsulation systems
physicochemical characterization, 421–424
DSC, 421–422
FTIR, 422–423
NMR, 423
optical characterization, 424
XRD, 421, 422*f*
Nanofibers (NF), 171–173, 283, 308–309
Nanofibrillated cellulose (NFC), 432
Nanofibrils, 313
Nanofood products, 441
Nanogels, 40–42, 204
alginate-based, 312
chitosan-based, 312
ingredients to preparing nanogels, 40–41
chemical polymer nanogels, 41
polysaccharide nanogels, 41
protein nanogels, 40–41
matrix, 434
preparation methods, 41–42
protein, 311–312
Nanohydrogels, 166–167
Nanolipid carriers, 12–16
ingredients for preparing SLNs and NLCs, 13–15
preparation methods, 15–16
microemulsion method, 15
Nanoliposomes, 8–12, 121, 169–171, 198–199, 239–241, 302, 413
encapsulation by, 120–122
ingredients for preparing nanoliposomes
cholesterol, 10–11
phenolic compounds, 11
phospholipids, 10
solvent, 11
nanoencapsulation of phenolics and antioxidants in, 82–83
preparation methods, 11–12
detergent removal method, 12
reverse phase method, 12
solvent injection method, 12
thin film hydration, 11–12
Nanomaterials, 439–440
Nanomedicine, 1–2
NanoMorph technology, 45
Nanoparticles (NPs), 2, 69, 85, 354
employed in food products, safety and toxicity of food, 424–433
genotoxicity and epigenetic regulations, 425–433, 427*f*, 428*f*
fabrication of TP–Zn complex and TP loaded CS-NPs, 364*f*
health risks and carcinogenesis induced by NPS, 433–439
hepatotoxicity, 435–437
neurotoxicity, 433–437
particle size distribution of CS-NPs loaded Cu^{2+}, Zn^{2+}, Mn^{2+}, and Fe^{2+}, 363*f*
Nanophytosomes, 9–10
Nanoprecipitation, 35, 249
Nanoscale delivery systems, release of peptides from, 314–316
Nanospheres, 33, 85
Nanospray dryer, 24, 91
nanoencapsulation by, 128–129
Nanospray drying, 29–33, 72, 93, 203, 247
optimizing nanospray-drying process, 32–33
preparation of nanospray-dried particles, 31–32
droplet generation, 31
drying of droplets, 31–32

Nanospray drying (*Continued*)
particle collection, 32
Nanostructured lipid carriers (NLCs), 2, 13, 72, 123, 167–168, 239, 243–245, 303–304, 411, 421–422, 438
encapsulation by, 123–125
ingredients for preparing SLNs and, 13–15
liquid oil, 14
solid lipid, 14
stabilizer, 14–15
nanoencapsulation of phenolics and antioxidants in, 84–85
Nanosuspensions, 2
Nanotechnology, 69, 316
strategies, 204, 208
Nanotechnology Characterization Laboratory, 441
Nanotoxicology, 436–437
Nanotubes, 42–43, 312–313
Nanovehicle, 430
Natamycin, 193
Natural antimicrobial agents, 184, 187–193
animal sources, 187–191
antimicrobial spectrum of selected, 188*t*
microbial sources, 192–193
plant sources, 191–192
problems and opportunities of, 193–194
Natural food colorants. *See* Food colorants
Natural food colors, 224–225
Natural nanocarriers for nanoencapsulation
casein nanocapsules, 304–305
caseins, nanoencapsulation by, 125
CDs nanocapsules, 305–306
cyclodextrins, nanoencapsulation by, 125–126
nanocrystal nanocapsules, 305–306
Nature-inspired nanocarriers for food color nanoencapsulation, 245–247
caseins nanoparticles, 246–247
molecular inclusion complexes with cyclodextrins, 246
nanocrystals, 247
Nature-inspired nanoencapsulation techniques, 16–24
amylose nanocarriers, 23–24
caseins, 16–20
cyclodextrins, 20–23
NC. *See* Nanochitosan (NC)
Nephelium lappaceum.. *See* Rambutan seed (*Nephelium lappaceum*)
NEs. *See* Nanoemulsions (NEs)
Neural diseases, 146–147
Neurological diseases, 431
Neurotoxicity, 433–437
mechanisms stimulated by NPs, 436*f*
NF. *See* Nanofibers (NF)
NFC. *See* Nanofibrillated cellulose (NFC)
Nicotinic acid. *See* Vitamin B_3
Nile Red dye, 414
Niosome entrapment, 347
Nisin, 192–193, 303
Nisin plus EDTA, 303
NLCs. *See* Nanostructured lipid carriers (NLCs)
NMR. *See* Nuclear magnetic resonance (NMR)
No observed adverse effect level (NOEL), 208
Non-Fickian diffusion model, 204–205
Noncovalent bonding, 39
Nondestructive methods, 404
Nonenzymatic antioxidants, 66
Nonleafy vegetables, 226
Nonloaded ingredients, 405
Nonprotein components, 299
Nonresponsive nanogels, 40
Nonthermal process, 308–309
Nontoxic ingredients, 306
Nonviral gene-delivery vectors, 309–310
NPs. *See* Nanoparticles (NPs)
Nuclear magnetic resonance (NMR), 404, 423
Nucleic acids, 301–302
encapsulation based on electrohydrodynamic processes, 309–310
Nucleosome, 426–428
Nutraceutical delivery system, 1–2

O

Octenyl succinic anhydride (OSA), 201
2-Octyl-dodecanoyl-6-*O*-ascorbid acid, 167
OEO. *See* Oregano essential oil (OEO)
OEO-loaded–CS NPs, 423
Oil, 6–7
sector, 73
Oil droplet (O), 6
Oil-in-water nanoemulsions (O/W nanoemulsions), 5–6, 72–77, 241–242, 281
Oil-in-water-in-oil emulsion (O/W/O emulsion), 6, 79
Oil-soluble colorants, 224
Oleic acid, 103–104

Olfactory nerves, 434
Oligodeoxynucleotide, 309–310
Omega-3 fatty acids, 73, 283
Opacity, 424
Optical characterization, 424
Optical microscope, 411
Optimized nanoencapsulation system, 415–416
Oral administration, 405
Orange essential oils, 81–82
Oregano essential oil (OEO), 421
Organic acids, 184
Organic polymers, 168
OSA. *See* Octenyl succinic anhydride (OSA)
Osmolytes, 403
Osmotic pressure, 403
Osmotic pumping, 403
Ostwald ripening process, 165–166
Ovalbumin, 40–41, 311–312
O/W nanoemulsions. *See* Oil-in-water nanoemulsions (O/W nanoemulsions)
O/W/O emulsion. *See* Oil-in-water-in-oil emulsion (O/W/O emulsion)
Oxidation, 106
  lipid, 106–107
Oxidative stress, 436

P

PAMAM. *See* Polyamidoamine (PAMAM)
Pantothenic acid. *See* Vitamin B_5
Parathyroid hormone (PTH), 166
Particle collection, 32
PBS. *See* Phosphate buffer solution (PBS)
PCL. *See* Poly(ε-caprolactone) (PCL)
PCS. *See* Photon correlation spectroscopy (PCS)
PDI. *See* Polydispersivity index (PDI)
Pediocin AcH, 303
PEG. *See* Polyethylene glycol (PEG)
PEI. *See* Polyethyleneimine (PEI)
Pelargonidin (plg), 229–230
Penaeus japonicas.. *See* Shrimps (*Penaeus japonicas*)
PEO. *See* Polyethylene oxide (PEO)
Peonidin (pnd), 229–230
PEO–PLGA particles, 315
Pepsin, 299
Peptides, 300, 303–304
  conjugation, 312
  release of peptides from nanoscale delivery systems, 314–316
Petunidin (ptd), 229–230
PG. *See* Propylene glycol (PG); Purity Gum 2000 (PG)
PGMS. *See* Propylene glycol monostearate (PGMS)
PGPR. *See* Polyglycerol polyricinoleate (PGPR)
PGU. *See* Purity Gum Ultra (PGU)
pH, 408
Pharmacology, 405
Phase inversion, 117–118, 197–198, 244–245
Phase inversion composition (PIC), 79
Phase inversion temperature (PIT), 79, 244, 282
Phenolic compounds, 11, 63, 64*f*, 66–68, 191–192
  biopolymer-based technologies for nanoencapsulation, 85–91
  classification and sources, 67*t*
  equipment-based technologies for nanoencapsulation, 91–93
  lipid-based formulations for nanoencapsulation, 73–85
  nanoencapsulation process, 69–73
Phosphate buffer solution (PBS), 169
Phosphatidylcholine, 10
Phosphatidylethanolamine, 122
Phospholipids, 10, 154
Phospholipon 100 H, 198–199
Phospholipon 90 H, 198–199
Phosphorus (P), 333
Phosphorylation, 431
Photon correlation spectroscopy (PCS), 243
Photosynthetic land organism, 297
Physical characterization methods, 132–133
Physicomechanical methods. *See also* Chemical methods
  extrusion, 342–343
  fluidized bed coating, 339–340
  freeze-drying method, 335–338
  spray-chilling/cooling, 341–342, 341*f*
  spray-drying method, 340–341
Phytochrome, 231
Phytosterols, 300–301
pI. *See* Isoelectric point (pI)
PIC. *See* Phase inversion composition (PIC)
Piezo-electric crystal, 417–419
Pigments, 226, 229–231
PIT. *See* Phase inversion temperature (PIT)
PLA. *See* Polylactides (PLA)
Plant extracts, 191–192

Plant sources
essential oils, 191
plant extracts, 191–192
Plant stanols, 300
Plasma formation, 417–419
plg. *See* Pelargonidin (plg)
PLGA. *See* Poly (lactide-co-glycolide) (PLGA)
PLHC-1 cells, 430–431
PLS. *See* Pro-liposome (PLS)
PMS. *See* Premenstrual syndrome (PMS)
pnd. *See* Peonidin (pnd)
PNIPAM. *See* Poly-*N*-isopropylacrylamide (PNIPAM)
Poly (ethyleneimine), 312
Poly (lactide-co-glycolide) (PLGA), 247–249, 435
Poly (vinylpyrrolidone), 312
Poly γ-glutamic acid (γ-PGA), 90
Poly-*N*-isopropylacrylamide (PNIPAM), 41, 312
Poly(ε-caprolactone) (PCL), 131
Polyacrylonitrile nanofibers, 172
Polyamidoamine (PAMAM), 430–431
Polydispersivity index (PDI), 242–243
Polyelectrolytes, 362
Polyethylene glycol (PEG), 167, 249, 342–343, 421–422
Polyethylene oxide (PEO), 41, 311–312
Polyethyleneimine (PEI), 41
Polyglycerol monostearate (PGMS). *See* Propylene glycol monostearate (PGMS)
Polyglycerol polyricinoleate (PGPR), 351
Polyl(actic-*co*-glycolic) acid copolymers, 44–45
Polylactides (PLA), 44–45, 270
Polymer, 439
concentration, 26
materials, 305–306
solution, 307–308
properties, 26
Polymerization, 40
Polypeptide nanogels, 311–312
Polyphenols, 65–66
Polysaccharides, 7–8, 201, 261
nanogels, 41
NPs, 35
starch NPs, 35
polysaccharide–polysaccharide complexes, 39
Polysorbate 80-modified CS NPs (TmCS-NPs), 434
Polyunsaturated fatty acids (PUFAs), 103–104, 298
esterified phytosterols, 300–301
Polyvinyl alcohol (PVA), 41, 83, 247–248, 283, 309, 312, 339–340, 423
Polyvinylpyrrolidone (PVP), 41, 247–248
Potassium (K), 333
Potassium citrate ($C_6H_5K_3O_7$), 370
Potassium iodate (KIO_3), 342
Potassium iodide (KI), 339
Power intensity, 304–305
Pragmatic encapsulation technology, 233
Precoil, 168
Premenstrual syndrome (PMS), 146–147
Premix membrane emulsification, 351–352
Primary flavors, 262
Pro-liposome (PLS), 354–358
Propylene glycol (PG), 165–166
Propylene glycol monostearate (PGMS), 243, 344–345
Protease enzymes, 313
Protein(s), 8, 261, 310, 420–421
encapsulation based on electrohydrodynamic processes, 310–311
expression, 428–429
fibers, 310
nanoemulsions stabilized by, 77–78
nanogels, 40–41, 311–312
NPs, 33–35
protein-based biopolymers, 200–201
protein–polysaccharide interactions, 38–39
sequences in, 300
Proteolysis, 266–267
Provitamins, 146
ptd. *See* Petunidin (ptd)
PTH. *See* Parathyroid hormone (PTH)
PUFAs. *See* Polyunsaturated fatty acids (PUFAs)
Pulmonary toxicity, 435
Purity Gum 2000 (PG), 78
Purity Gum Ultra (PGU), 78
PVA. *See* Polyvinyl alcohol (PVA)
PVP. *See* Polyvinylpyrrolidone (PVP)
Pyridoxine. *See* Vitamin B_6

Q

Quercetin (QT), 244–245, 247, 423

Quercetin-loaded cationic NLC (QT-CNLC), 244–245
Quercetin-loaded nanoemulsions (QT-NE), 242
Quercitine (Q), 437
Quinone, 230–231
QU-loaded-nanoparticle (QUEN), 423

R

RA. *See* Rheumatoid arthritis (RA); Rosmarinic acid (RA)
Radical scavenging activity studies, 234
Radiofrequency electromagnetic onto nanoencapsulation systems, 423
Rambutan seed (*Nephelium lappaceum*), 167–168
Rancidity, 107
Rapid expansion of supercritical solutions (RESS), 249
RDA. *See* Recommended dietary allowance (RDA)
Reactive oxygen species (ROS), 430–431, 435
Recommended dietary allowance (RDA), 147–148
Recycling centrifugal extrusion, 277–278
Red beet (*Beta vulgaris*), 230
Red cabbage (*Brassica oleracea* L.), 243
Red meat, flavors derived from, 267
Redbook of Colorants, 250
Reflectance modes, 424
Relative humidities (RHs), 342–343
"Release mechanism", 402
Repulsive forces, 414
Resonance frequency, 423
Respiratory system, 433–434
Response-surface methodology (RSM), 351
RESS. *See* Rapid expansion of supercritical solutions (RESS)
Resveratrol (RSV), 64–65
Reverse dialysis sac approaches, 165–166
Reverse phase method, 12
Reverse-phase evaporation (RPE), 344
Rheumatoid arthritis (RA), 105
RHs. *See* Relative humidities (RHs)
Riboflavin. *See* Vitamin B_2
RNA interference, 428–429
ROS. *See* Reactive oxygen species (ROS)
Rosmarinic acid (RA), 429
RPE. *See* Reverse-phase evaporation (RPE)
RSM. *See* Response-surface methodology (RSM)
RSV. *See* Resveratrol (RSV)
Rubinstein–Taybi syndromc, 431

S

SA. *See* Stearic acid (SA); Stearylamine (SA)
Saccharides, flavors derived from, 263, 264*t*
Safety, 223
  aspects of nanoencapsulated food colorants, 250
  food, 184
  of nanoencapsulated food antimicrobial agents, 208
  of nanoencapsulation systems, 316
  and regulatory issues of fish oil-loaded nanocapsules, 134
  and toxicity of food NPS employed in food products, 424–433
Salt fortification with encapsulated iron and iodine
  dual fortified salt, 374–379
  triple fortified salt, 379
Salt-induced cold gelation, 353
Saturated fatty acids, 103
SC. *See* Sodium caseinate (SC)
Scanning electron microscopy (SEM), 133, 205–206, 309, 379, 411–413
SCF. *See* Supercritical fluid (SCF)
SDA. *See* Sodium diacetate (SDA)
SDS. *See* Sodium dodecyl sulfate (SDS)
Se-NPs, 384
Secondary flavors, 263
Sedimentation, 417
SEDS approach, 246
Segregative phase separation, 37
Selenium (Se), 333
Self-assembly, 166–167
  and reassembly of caseins, 19–20
Self-emulsification method, 117–118, 197–198
SEM. *See* Scanning electron microscopy (SEM)
Sensory nerve, 434
Shear force, 407
Shrimps (*Penaeus japonicas*), 154
Silibinin, 438
Silk, 310–311
  fibroin nanofibers, 172–173
Silver, 2
Single biopolymer NPs, 33–35
  polysaccharide NPs, 35
  protein NPs, 33–35

Single-screw extruder, 381
Single-walled carbon nanotubes (SWCNTs), 250
SLNPs, 429, 433–434
SLNs. *See* Solid–lipid nanoparticles (SLNs)
SLNs loaded with curcumin (Cur-SLNs), 242
Small micro-encapsulated $FeSO_4$ (SMFS), 373–374
Small molecule surfactants, nanoemulsions stabilized by, 78–79
Small unilamellar vesicles (SUVs), 9, 170–171, 198–199
SMFS. *See* Small micro-encapsulated $FeSO_4$ (SMFS)
SMP. *See* Sucrose monopalmitate (SMP)
SNPs. *See* Starch nanoparticles (SNPs)
SOD. *See* Superoxide dismutase (SOD)
Sodium acetate, 186
Sodium caseinate (SC), 18, 77, 108*t*, 129–130, 248, 304–305
Sodium diacetate (SDA), 186
Sodium dodecyl sulfate (SDS), 119–120, 247–248
Sodium selenite (Na_2SeO_3), 362
Sodium tripolyphosphate, 169
Sol–gel transitions, 421
Solid food matrix, 209–210
Solid lipid nanocarriers. *See* Solid–lipid nanoparticles (SLNs)
Solid lipids, 14, 73–74
Solid phase microextraction gas chromotography (SPME-GC), 281
Solidification process, 14
Solid–lipid nanoparticles (SLNs), 2, 12–13, 72, 122–123, 167–168, 199, 239, 242–243, 303–304, 360–361, 411, 421–422, 437–438
  $FeSO_4$-loaded SLNs fabrication, 360*f*
  ingredients for preparing SLNs and NLCs, 13–15
    liquid oil, 14
    solid lipid, 14
    stabilizer, 14–15
  nanoencapsulation of phenolics and antioxidants in, 83–84
Solubility, 306
Solvent, 11
  diffusion method, 244
  injection method, 12
Sonication, 40
SOR. *See* Surfactant-to-oil (SOR)
Sorbic acid, 185*t*, 187
SOW systems. *See* Surfactant, oil, and water systems (SOW systems)
Soy lecithin, 10
Soy protein, 169
Soy protein isolate (SPI), 248, 287
Soybean phosphatidylcholine (SPC), 84
SPC. *See* Soybean phosphatidylcholine (SPC)
Special-equipment-based nanoencapsulation techniques, 24–33
  electrospinning, 25–27
  electrospraying, 27–29
  nanospray drying, 29–33
SPI. *See* Soy protein isolate (SPI)
Spinneret
  orifice diameter, 27
  tip to collector distance, 27
SPME-GC. *See* Solid phase microextraction gas chromotography (SPME-GC)
Spontaneous emulsification, 117–118, 165–166, 197–198
Spray
  chilling and cooling method, 153, 275–276, 341–342, 341*f*
  coating, 272–274, 272*f*, 274*f*
  dried powders, 224
  drying, 128–129, 151–153, 240–241, 246–247, 272–274, 272*f*, 274*f*, 340–341
  spray-cooling process, 380–381
Spreading coefficient (S_{AB}), 410
Stability, 271–272
  of food colorants, 231–232
Stabilizers, 7, 14–15
Stabilizing agent, 73
Standard assays, 432
Standardization, 148
Staphylococcus aureus, 201
Starch, 201, 284–285
Starch nanoparticles (SNPs), 35, 313–314
Static light scattering, 416, 418*f*
Stearic acid (SA), 242–243, 360–361
Stearylamine (SA), 198–199
Streptomyces natalensis, 193
Submicron emulsions. *See* Nanoemulsions
Sucrose monopalmitate (SMP), 282
Sulfur (S), 333
Supercritical fluid (SCF), 121
  approach, 246
Superoxide dismutase (SOD), 66
Surface composition, 420–421, 420*f*
Surface mean diameter, 416

Surfactant, oil, and water systems (SOW systems), 282
Surfactant-to-oil (SOR), 282
Surfactants, 1, 79–80
Suspension coating, 154
SUVs. *See* Small unilamellar vesicles (SUVs)
SWCNTs. *See* Single-walled carbon nanotubes (SWCNTs)
Swelling, 403
Synthetic colorants, 223, 225

T

"Taylor cone", 25, 126–127, 127*f*, 171–172, 308
TBA. *See* Thiobarbituric acid (TBA)
TBARS. *See* Thiobarbituric acid reactive substances (TBARS)
TDS. *See* Thiamine dilauryl sulfate (TDS)
Tea polyphenol (TP), 363–364
TEM. *See* Transmission electron microscopy (TEM)
TFE. *See* Trifluoroethanol (TFE)
TFS. *See* Triple fortified salt (TFS)
TGA. *See* Thermogravimetric analysis (TGA)
Thermo-analytical properties, 421–422
Thermodynamic compatibility. *See* Complex coacervation
Thermodynamic incompatibility, 37
Thermogravimetric analysis (TGA), 133
Thermophilic organisms, 299
Thiamin. *See* Vitamin B_1
Thiamine dilauryl sulfate (TDS), 156
Thin film hydration, 11–12
Thiobarbituric acid (TBA), 350
Thiobarbituric acid reactive substances (TBARS), 118–119, 382
Thyme oils, 191
  emulsion, 210–211
Thymoquinone (TQ), 438
Time-consuming process, 335
Titanium dioxide (TiO_2), 339–340
TmCS-NPs. *See* Polysorbate 80-modified CS NPs (TmCS-NPs)
"Top-down" method, 2–3, 194
TP. *See* Tea polyphenol (TP)
TPGS. *See* D-α-Tocopherol polyethylene glycol 1000 succinate (TPGS)
TPP. *See* Tripolyphosphate (TPP)
TQ. *See* Thymoquinone (TQ)
Trace minerals, 333
Traditional chemical preservative agents, 184–187
  action of, 185*f*
  structure and properties of, 185*t*
Transcription process, 426–428
Transglutaminase, 305
Transition metals, 436
Transmission, 424
Transmission electron microscopy (TEM), 133, 205–206, 411–413
Trifluoroethanol (TFE), 310–311
Triglycerides, 105
Triple fortified salt (TFS), 342, 379
Tripolyphosphate (TPP), 90, 362, 421
Tripterygium wilfordii Hook f-loaded group (TWHF group), 437–438
  TWHF-SLN group, 437–438
Trypsin, 299
Tumor suppressor, 426
Tween 80, 247
TWHF group. *See Tripterygium wilfordii* Hook f-loaded group (TWHF group)

U

Ultra Rice formulations, 382
Ultrasound probe, 12
Unilamellar vesicles (ULVs), 9, 121
Unsaturated fatty acids, 103
US Food and Drug Administration (FDA), 186–187, 223, 310, 315–316, 440–441
UV–Vis spectra, 309, 424

V

V-type amylase, 284–285
VBBP. *See* Vitamin B_6-bearing pullulan (VBBP)
VC. *See* Vitamin C (VC)
VDN. *See* Vitamin D-nanoemulsions (VDN)
Vegetables, 210–211
  flavors derived from, 268
  oils, 79–80
Very low-density lipoproteins (VLDL), 105
Viscoelasticity, 407–408
Vitamin B_1, 146, 149, 170
Vitamin B_2, 146, 149, 231
Vitamin B_3, 146
Vitamin B_5, 146
Vitamin B_6-bearing pullulan (VBBP), 167
Vitamin B_6, 146
Vitamin B_{12}, 146, 231
Vitamin C (VC), 65–66, 146, 149, 153, 166–167

Vitamin D, 146
Vitamin D-nanoemulsions (VDN), 166
Vitamin E, 66, 149
vitamin E-bearing nanoemulsions, 165–166
Vitamin K1 (VK1), 166
Vitamin(s)
biopolymer nanoparticles, 168–169
cyclodextrins, 171
deficiency outcomes, 147
dosage of vitamin consumption, 147–148
fortification in food products, 148
microencapsulation, 150–155, 151*f*, 152*t*
coacervation, 154–155
emulsion techniques, 153–154
fluidized bed coating, 154
liposome carriers, 154
spray chilling and cooling, 153
spray drying technique, 151–153
nanoemulsions, 155–166
nanoencapsulation, 155–173
for water-insoluble vitamins, 161*t*
for water-soluble vitamins, 157*t*
nanofibers and electrosprayed nanoparticles, 171–173
nanohydrogels and nano-organogels, 166–167
nanoliposomes, 169–171
nanostructured delivery system types, 156*f*
nutritional properties and classification, 146–150
SLNs and NLCs, 167–168
vitamin retention under food processing conditions, 149–150, 150*t*
VK1. *See* Vitamin K1 (VK1)
VLDL. *See* Very low-density lipoproteins (VLDL)
Voltage, 26
Volume mean diameter, 416

W

"Wall material", 149–150, 232–233
Water absorption, 403
Water droplets (W_1), 6
Water-in-oil nanoemulsions (W/O nanoemulsions), 5–6, 74, 241
Water-in-oil-in-water emulsions (W/O/W emulsions), 6, 79, 302
Water-soluble compound, 154–155
Water-soluble vitamins, 153, 157*t*, 303
Weibull model, 165
Wetting theory, 410
Whey protein, 77–78
Whey protein concentrate (WPC), 77–78, 172, 241, 351, 420–421
Whey protein isolate (WPI), 77–78, 120, 287, 311
solution, 351
WPI-coated nanoliposomes, 240–241
WPI/PEO composite nanofibers, 311
Whey protein-chelated Fe (Fe–WP), 371
Whipping instability, 91–92
Wistar rats, 429
W/O nanoemulsions. *See* Water-in-oil nanoemulsions (W/O nanoemulsions)
W/O/W emulsions. *See* Water-in-oil-in-water emulsions (W/O/W emulsions)
WPC. *See* Whey protein concentrate (WPC)
WPI. *See* Whey protein isolate (WPI)
WPI-Fe. *See* Fe-encapsulated cold set WPI (WPI-Fe)

X

X-ray diffraction (XRD), 421
X-ray photoelectron spectroscopy (XPS), 420–421, 420*f*
Xanthones, 230–231

Y

Yogurt, 370–372

Z

Zein, 39, 201
nanofibers, 414–415
nanoparticles, 85–86
Zeta-potential, 419–420
Zinc (Zn), 333
ions, 167
Zinc chloride ($ZnCl_2$), 342
Zinc oxide (ZnO), 384–388

Kolibri 360 GmbH
Ferdinand-Jühlke-Straße 7
99095 Erfurt, Deutschland
produktsicherheit@kolibri360.de